D1162323

Nuclear and Radiochemistry

Nuclear and

Radiochemistry

Second Edition

Gerhart Friedlander

Senior Chemist, Brookhaven National Laboratory

Joseph W. Kennedy

Late Professor of Chemistry, Washington University, St. Louis

Julian Malcolm Miller

Professor of Chemistry, Columbia University

John Wiley & Sons, Inc., New York · London · Sydney

Library of Congress Catalog Card Number: 64-20066
Printed in the United States of America

Preface

This book had its genesis in 1949 when *Introduction to Radiochemistry* appeared as "a textbook for an introductory course in the broad field of radiochemistry." In 1955, when a considerably revised and enlarged version was published, the title was changed to *Nuclear and Radiochemistry* to reflect more accurately, albeit somewhat ungrammatically, the growing division between nuclear chemists, who are concerned with the properties and reactions of atomic nuclei, and radiochemists, who are concerned with the application of the properties of radioactive nuclei to the solution of chemical problems. Despite this growing division, or perhaps because of it, we continue to feel that "nuclear chemistry and radiochemistry interact strongly with each other, and indeed are so interdependent that their discussion together is almost necessary in an introductory textbook."

Thus the general purpose of the book and the spirit in which we have written have remained the same. Yet this second edition of *Nuclear and Radiochemistry* differs from the first at least as much as that first edition differed from *Introduction to Radiochemistry*. The book is still intended as a text for an advanced undergraduate or first-year graduate course; but a significant aspect of its modernization is found in the greater depth of treatment accorded to many topics, particularly theoretical ones. This change, we believe, is in keeping with the improved preparation of chemistry students and also reflects the increasing sophistication in the research of nuclear chemists. Again the book has increased in size by roughly one third, and, with all the new material that we felt compelled to add, the growth would have been even greater had we not also deleted much that seemed no longer essential. The suggestions and comments of many colleagues have aided us greatly in deciding on the choice of subject matter to be included.

We have sorely missed the wisdom, knowledge, and judgment of Joseph W. Kennedy in the preparation of this new edition. His untimely death in 1957 deprived our profession of one of its leaders and us of a beloved friend and colleague. In spite of the many changes this book bears the unmistakable imprint of his authorship, and we hope that our revisions would have met with his approval.

The changes are too numerous to list in detail. Apart from the many new and rewritten sections which reflect our attempt to keep pace with the evolution of subject matter, there has been a rather important reorganization of material. We have enlarged the phenomenological discussion of atomic nuclei in chapter 2 to include some of the elementary material on radioactive decay modes and on nuclear reactions. The detailed treatment of these subjects, at a somewhat more advanced level and with considerably more emphasis on the theoretical framework than was given in the first edition, has now been moved to chapters 8 and 10, with a new chapter 9 on Nuclear Models in between. Following chapter 2 are now the chapters on Equations of Radioactive Growth and Decay, Interactions of Radiations with Matter, Radiation Detection Devices, and Statistical Considerations in Radioactivity Measurements. Most instructors who use this book will, in our opinion, wish to cover the subject matter of the first six chapters, although the relative emphasis may, of course, vary widely. A course designed primarily as an introduction to radiochemistry would, in addition, certainly include chapter 7 (Tracers in Chemical Application). On the other hand, this chapter might well be omitted if emphasis were to be placed on nuclear chemistry whereas the subject matter of chapters 8 to 10 would form the backbone of such a course. Chapters 11 and 12 will also be most useful to those primarily interested in nuclear matters. The last three chapters may be considered as optional material for either type of course. The new chapter 13 (Nuclear Processes as Chemical Probes) has been added in the hope that it will tempt chemists into exploring further how those nuclear processes which are affected by chemical environment can in fact be exploited in chemical studies.

As before, each chapter closes with a set of exercises and a list of references. As we already said in the preface to our 1949 book, the exercises "are intended as an integral part of the course, and only with them does the text contain the variety of specific examples which we consider necessary for an effective presentation." The references are listed by initial of first author and a serial number (B2, M5, etc.), and most of them are referred to at specific places in the text. However, we have by no means attempted to give complete documentation of the material covered and have, for the most part, given references to comprehensive review articles and books rather than to original research papers. General references which call attention to supplementary reading on the subject matter of an entire chapter or a major section are marked with an asterisk.

Although we have emphasized the use of this book as a text in a formal course, we have also sought to preserve and enhance its usefulness as a ready reference source for research workers.

It is a pleasure to acknowledge the help and advice we have received from many colleagues. We are especially grateful to Drs. J. P. Blewett, R. Davis, Jr., R. W. Dodson, B. M. Foreman, J. R. Grover, J. Hudis, J. Hummel, H. Kouts, T. Norris, A. M. Poskanzer, L. P. Remsberg, N. Sutin, and J. Weneser, each of whom read part of the manuscript, called our attention to errors and inaccuracies, and made valuable suggestions. Needless to say, any mistakes that have remained are entirely our responsibility. We shall be grateful to readers who will call errors to our attention. Brookhaven National Laboratory and Columbia University deserve our thanks for facilitating the time-consuming task of this revision.

Blue Point, N.Y.
Leonia, N.J.
May 1964

GERHART FRIEDLANDER
JULIAN MALCOLM MILLER

Contents

Radioactivity

A. DISCOVERY OF RADIOACTIVITY

Becquerel's Discovery. The more or less accidental series of events which led to the discovery of radioactivity depended on two especially significant factors: (1) the mysterious X rays discovered about one year earlier by W. C. Roentgen produced fluorescence (the term phosphorescence was preferred at that time) in the glass walls of X-ray tubes and in some other materials; and (2) Henri Becquerel had inherited an interest in phosphorescence from both his father and grandfather. The father, Edmund Becquerel (1820–1891), had actually studied phosphorescence of uranium salts, and about 1880 Henri Becquerel prepared potassium uranyl sulfate, $K_2UO_2(SO_4)_2 \cdot 2H_2O$, and noted its pronounced phosphorescence excited by ultraviolet light. Thus in 1895 and 1896, when several scientists were seeking the connection between X rays and phosphorescence and were looking for penetrating radiation from phosphorescent substances, it was natural for Becquerel to experiment along this line with the potassium uranyl sulfate.

It was on February 24, 1896, that Henri Becquerel reported his first results: after exposure to bright sunlight, crystals of the uranyl double sulfate emitted a radiation which blackened a photographic plate after penetrating black paper, glass, and other substances. During the next few months he continued the experiments, obtaining more and more puzzling results. The effect was as strong with weak light as with bright sunlight; it was found in complete darkness and even for crystals prepared and always kept in the dark. The penetrating radiation was emitted by other uranyl and also uranous salts, by solutions of uranium salts, and even by what was believed to be metallic uranium, and in each case with an intensity proportional to the uranium content. Proceeding by analogy with a known property of X rays, Becquerel observed that the penetrating rays from uranium would discharge an electroscope. All these results were obtained in the early part of 1896 (B1). Although Becquerel and others continued investigations for several years, the knowledge gained in this phase of the new science was summarized in 1898, when Pierre and Marie Sklodowska Curie concluded that the uranium rays

Table 1-1   Some uranium and thorium minerals

Name	Composition	Uranium Content	Thorium Content	Color and Form
Uraninite (pitchblende)	Uranium oxide, UO_2 to U_3O_8, with rare-earth and other oxides	60–80%	0–10%	Grayish, greenish, or brownish black; cubic, amorphous
Thorianite	Thorium and uranium oxides, $(Th, U)O_2$, with UO_3 and rare-earth oxides	4–40%	30–82%	Gray, brownish or greenish gray, black; cubic
Carnotite	Potassium uranyl vanadate, $K(UO_2)VO_4 \cdot nH_2O$	$\cong 45\%$		Yellow; hexagonal, rhombic
Monazite	Phosphates of the cerium earths and thorium, $CePO_4 + Th_3(PO_4)_4$		Up to 16%	Red, brown, yellowish brown; monoclinic
Pilbarite	Thorium lead uranate and silicate	$\cong 25\%$	$\cong 25\%$	Yellow
Autunite	Calcium uranyl phosphate, $Ca(UO_2)_2(PO_4)_2 \cdot 8H_2O$	$\cong 50\%$		Greenish yellow; rhombic
Thorite or organite	Thorium ortho-silicate, $ThSiO_4$		Up to 70%	Brown or black (thorite), orange-yellow (orangite); tetragonal

were an atomic phenomenon characteristic of the element, and not related to its chemical or physical state, and introduced the name "radioactivity" for the phenomenon.

The Curies.   Much new information appeared during the year 1898, mostly through the work of the Curies.   Examination of other elements led to the discovery, independently by Mme. Curie and G. C. Schmidt, that compounds of thorium emitted rays similar to those from uranium.   A very important observation was that some natural uranium ores were even more radioactive than pure uranium, and more active than a chemically similar "ore" prepared synthetically.   The chemical decomposition and fractionation of such ores constituted the first exercise in radiochemistry and led immediately to the discovery of polonium—as a new substance observed only through its intense radioactivity—and of radium, a highly radioactive substance recognized as a new element and soon identified spectroscopically.   The Curies and their co-workers had found radium in the barium fraction separated chemically from pitchblende (a dark almost black ore containing about 75 per cent U_3O_8), and they learned that it could be concentrated from the barium by repeated fractional crystallization of the chlorides, the radium salt remaining preferentially in the mother liquor.   By 1902 Mme. Curie reported the isolation of 100 mg of radium chloride spectroscopically free from barium and gave 225 as the approximate atomic weight of the element.   (The work had started with about two tons of pitchblende, and the radium isolated represented about a 25 per cent yield.)   Still later Mme. Curie redetermined the atomic weight to be 226.5 (the latest tabulated value is 226.05) and also prepared radium metal by electrolysis of the fused salt.

Becquerel in his experiments had shown that uranium, in the dark and not supplied with energy in any known way, continued for years to emit rays in undiminished intensity.   E. Rutherford had made some rough estimates of the energy associated with the radioactive rays; the source of this energy was quite unknown.   With concentrated radium samples the Curies made measurements of the resulting heating effect, which they found to be about 100 cal per hour per gram of radium.   The evidence for so large a store of energy not only caused a controversy among the scientists of that time but also helped to create a great popular interest in radium and radioactivity.   (An interesting article in the *St. Louis Post-Dispatch* of October 4, 1903, speculated on this inconceivable new power, its use in war, and as an instrument for destruction of the world.)

Early Characterization of the Rays.   The effect of radioactive radiations in discharging an electroscope was soon understood in terms of the ionization of the air molecules, as J. J. Thomson and others were developing a knowledge of this subject in their studies of X rays.   The use of the amount of ionization in air as a measure of the intensity of radiations was developed into a more precise technique than the photographic blackening, and this technique was employed in the Curie laboratory, where ionization currents were measured with an electrometer.   In 1899 Rutherford began a study of the properties of the rays themselves, using a similar instrument.   Measurements of the absorption of the rays in metal foils showed that there were two components. One component was absorbed in the first few thousandths of a centimeter of aluminum and was named α radiation; the other was absorbed considerably in roughly 100 times this thickness of aluminum and was named β radiation. For the β rays Rutherford found that the ionization effect was reduced to the fraction $e^{-\mu d}$ of its original value when d centimeters of absorber were interposed; the absorption coefficient μ was about 15 cm^{-1} for aluminum and increased with atomic weight for other metal foils.

Rutherford at that time believed that the absorption of the α radiation also followed an exponential law and listed for its absorption coefficient in aluminum the value $\mu = 1600$ cm^{-1}.   About a year later Mme. Curie found that

μ was not constant for α rays but increased as the rays proceeded through the absorber. This was a surprising fact, since one would have expected that any inhomogeneity of the radiation would result in early absorption of the less penetrating components with a corresponding decrease in absorption coefficient with distance. In 1904 the concept of a definite range for the α particles (they were recognized as particles by that time) was proposed and demonstrated by W. H. Bragg. He found that several radioactive substances emitted α rays with different characteristic ranges.

The recognition of the character of the α and β rays as streams of high-speed particles came largely as a result of magnetic and electrostatic deflection experiments. In this way the β rays were seen to be electrons moving with almost the velocity of light. At first the α rays were thought to be undeviated by these fields. More refined experiments did show deflections, from which the ratio of charge to mass was calculated to be about half that of the hydrogen ion, with the charge positive, and the velocity was calculated to be about one tenth that of light. The suggestion that the α particle was a helium ion was immediately made, and it was confirmed after much more study. The presence of helium in uranium and thorium ores had already been noticed and was seen to be significant in this connection. A striking demonstration was later made, in which α rays were allowed to pass through a very thin glass wall into an evacuated glass vessel; within a few days sufficient helium gas appeared in the vessel to be detected spectroscopically.

Before the completion of these studies of the α and β rays, an even more penetrating radiation, not deviated by a magnetic field, was found in the rays from radioactive preparations. The recognition of this γ radiation as electromagnetic waves, like X rays in character if not in energy, came rather soon. For a long time no distinction was made between the nuclear γ rays and some extranuclear X rays which often accompany radioactive transformations.[1]

Rutherford and Soddy Transformation Hypothesis. In the course of radioactivity measurements on thorium salts, Rutherford observed that the electrometer readings were sometimes quite erratic. During 1899 it was determined that the cause of this effect was the diffusion through the ionization chamber of a radioactive substance emanating from the thorium compound. Similar effects were obtained with radium compounds. Subsequent studies, principally by Rutherford and F. Soddy, showed that these emanations were inert gases of high molecular weight, subject to condensation at about $-150°C$. Another radioactive substance, actinium, had been separated from pitchblende in 1899, and it too was found to give off an active emanation.

The presence of the radioactive emanations from thorium, radium, and actinium preparations was a very fortunate circumstance for advancement of

---

[1] In the nomenclature of this book concerning radioactive decay processes the term γ rays includes only nuclear electromagnetic radiation; accompanying X rays are designated as such, even though this is not an entirely uniform practice in the literature.

knowledge of the real nature of radioactivity. Essentially the inert-gas character of these substances made radiochemical separations not only an easy process but also one which forced itself on the attentions of these early investigators. Two very significant consequences of the early study of the emanations were (1) the realization that the activity of radioactive substances did not continue forever but diminished in intensity with a time scale characteristic of the substance; and (2) the knowledge that the radioactive processes were accompanied by a change in chemical properties of the active atoms. The application of chemical separation procedures, especially by W. Crookes and by Rutherford and Soddy, in 1900 and the succeeding years revealed the existence of other activities with characteristic decay rates and radiations, notably uranium X, which is separated from uranium by precipitation with excess ammonium carbonate (the uranyl carbonate redissolves in excess carbonate through formation of a complex ion), and thorium X, which remains in solution when thorium is precipitated as the hydroxide with ammonium hydroxide. In each case it was found that the activity of the X body decayed appreciably in a matter of days and that a new supply of the X body appeared in the parent substance in a similar time. It was also shown that both uranium and thorium, when effectively purified of the X bodies and other products, emitted only α rays and that uranium X and thorium X emitted β rays.

By the spring of 1903 Rutherford and Soddy had reached an excellent understanding of the nature of radioactivity and published their conclusions that the radioactive elements were undergoing spontaneous transformation from one chemical atom into another, that the radioactive radiations were an accompaniment of these changes, and that the radioactive process was a subatomic change within the atom. However, it should be remembered that the idea of the atomic nucleus did not emerge until eight years later and that in 1904 Bragg was attempting to understand the α particle as a flying cluster of thousands of more or less independent electrons.

Statistical Aspect of Radioactivity. In 1905 E. von Schweidler used the foregoing conclusions as to the nature of radioactivity and formulated a new description of the process in terms of disintegration probabilities. His fundamental assumption was that the probability p of a particular atom of a radioactive element disintegrating in a time interval Δt is independent of the past history and the present circumstances of the atom; it depends only on the length of the time interval Δt and for sufficiently short intervals is just proportional to Δt; thus $p = \lambda \Delta t$, where λ is the proportionality constant characteristic of the particular species of radioactive atoms. The probability of the given atom not disintegrating during the short interval Δt is $1 - p = 1 - \lambda \Delta t$. If the atom has survived this interval, then its probability of not disintegrating in the next like interval is again $1 - \lambda \Delta t$. By the law for compounding such probabilities the probability that the given atom will survive the first interval and also the second is given by $(1 - \lambda \Delta t)^2$; for n such intervals this survival probability is $(1 - \lambda \Delta t)^n$. Setting $n \Delta t = t$, the total time, we obtain for the survival probability $(1 - \lambda t/n)^n$. Now the probability that the atom will remain unchanged after time t is just the value of this quantity when Δt is made indefinitely small; that is, it is the limit of $(1 - \lambda t/n)^n$ as n approaches infinity. Recalling that $e^x = \lim\limits_{n \to \infty} (1 + x/n)^n$, we obtain $e^{-\lambda t}$ for the limiting value. If we consider not one atom, but a large initial number N_0 of the radioactive atoms, then we may take the fraction remaining unchanged after time t to be $N/N_0 = e^{-\lambda t}$, where N is the number of unchanged atoms at time t. This exponential law of decay is just that which Rutherford had already found experimentally for the simple isolated radioactivities.

A more detailed discussion of the statistical nature of radioactivity is presented in Chapter 6.

B. RADIOACTIVE DECAY AND GROWTH

In the preceding section we mentioned that the decay of a radioactive substance followed the exponential law $N = N_0 e^{-\lambda t}$, where N is the (large) number of unchanged atoms at time t, N_0 is the number present when $t = 0$, and λ is a constant characteristic of the particular radioactive species. This will be recognized as the rate law for any unimolecular reaction, and, of course, it should be expected in view of the nature of the radioactive process. It may be derived if the decay rate, $-dN/dt$, is set proportional to the number of atoms present: $-dN/dt = \lambda N$. (This is to say that we expect twice as many disintegrations per unit time in a sample containing twice as many atoms, etc.) On integration the result is $\ln N = -\lambda t + a$, and the constant of integration a is evaluated from the limit $N = N_0$ when $t = 0$: $a = \ln N_0$. Combining these terms, we have: $\ln (N/N_0) = -\lambda t$, or $N/N_0 = e^{-\lambda t}$.

The constant λ is known as the decay constant for the radioactive species. As may be seen from the differential equation, it is the fraction of the number of atoms transformed per unit time, provided the time unit is chosen short enough so that N remains essentially constant in that interval. In any case, λ has the dimensions of a reciprocal time and is most often expressed in reciprocal seconds. It is to be noticed that for most radioactive substances no attempt to alter λ through variation of ordinary experimental conditions, such as temperature, chemical change, pressure, and gravitational, magnetic, or electric fields, has ever given a detectable effect.[2]

The characteristic rate of a radioactive decay may very conveniently be given in terms of the half-life $t_{1/2}$, which is the time required for an initial (large) number of atoms to be reduced to half that number by transformations.

---

[2] The exceptional cases in which slight changes of λ have been achieved are considered in chapter 13, p. 456.

Thus, at the time $t = t_{1/2}$, $N = N_0/2$ and

$$\ln \tfrac{1}{2} = -\lambda t_{1/2}, \quad \text{or} \quad t_{1/2} = \frac{\ln 2}{\lambda} = \frac{0.69315}{\lambda}.$$

In practical work with radioactive materials the number of atoms N is not directly evaluated, and even the rate of change dN/dt is usually not measured absolutely. The usual procedure is to determine, through its electric, photographic, or other effect, a quantity proportional to λN; we may term this quantity the activity $\mathbf{A}$, with $\mathbf{A} = c\lambda N = c(-dN/dt)$. The coefficient c, which we may term the detection coefficient, will depend on the nature of the detection instrument, the efficiency for the recording of the particular radiation in that particular instrument, and the geometrical arrangement of sample and detector. Care must be taken to keep these factors constant throughout a series of measurements. We may now write the decay law as it is commonly observed, $\mathbf{A} = \mathbf{A}_0 e^{-\lambda t}$.

The usual procedure for treating measured values of $\mathbf{A}$ at successive times is to plot log $\mathbf{A}$ versus t; for this purpose semilog paper (with a suitable number of decades) is most convenient. Now λ could be found from the slope of the resulting straight line corresponding to the simple decay law; however, in this procedure there is a possibility for the confusion of units or of different logarithm bases. It is more convenient to read from the plot on semilog paper the time required for the activity to fall from any value to half that value; this is the half-life $t_{1/2}$.

In this discussion we have considered only the radioactivity corresponding to the transformation of a single atomic species; however, the daughter substance resulting from the transformation may itself be radioactive, with its own characteristic radiation and half-life as well as its own chemical identity. Indeed, among the naturally occurring radioactive substances this is the more common situation, and in chapter 3 we shall treat quite complicated radioactive growths and decays. For the moment, consider the decay of the substance uranium I, or U_I. This species of uranium is an α-particle emitter, with $t_{1/2} = 4.51 \times 10^9$ years. The immediate product of its transformation is the radioactive substance uranium X_1, or UX_1, a β emitter with half-life 24.1 days (cf. figure 1-1). As already mentioned, the parent uranium may be separated from the daughter atoms by precipitation of the daughter with excess ammonium carbonate. The daughter precipitate will show a characteristic activity, which will decay with the rate indicated; that is, it will be half gone in 24.1 days, three fourths gone in 48.2 days, seven eighths gone in 72.3 days, etc. The parent fraction will, of course, continue its α activity as before but will for the moment be free of the β radiations associated with the daughter. However, in time, new daughter atoms will be formed, and the daughter activity in the parent fraction will return to its initial value, with a time scale corresponding to the rate of decay of the isolated daughter fraction.

In an undisturbed sample containing N_1 atoms of U_I, a steady state is established in which the rate of formation of the daughter UX_1 atoms (number N_2) is just equal to their rate of decay. This means that $-dN_1/dt = \lambda_2 N_2$ in this situation because the rate of formation of the daughter atoms is just the rate of decay of the parent atoms. Using the earlier relation, we have then $\lambda_1 N_1 = \lambda_2 N_2$, with λ_1 and λ_2 the respective disintegration constants. This is sometimes more convenient in terms of the two half-lives: $N_1/(t_{1/2})_1 = N_2/(t_{1/2})_2$. This state of affairs is known as secular equilibrium. No account is taken of the decrease of N_1 with time, since the fraction of U_I atoms transformed even throughout the life of the experimenter is completely negligible. In general, this situation obtains whenever a short-lived daughter results from the decay of a very long-lived parent. The same relation, $\lambda_1 N_1 = \lambda_2 N_2 = \lambda_3 N_3$, etc., may be used when several short-lived products arise from successive decays, beginning with a long-lived parent, provided again that the material has been undisturbed (that is, no daughter substances removed or allowed to escape) for a long enough time for secular equilibrium to be established.

C. NATURALLY OCCURRING RADIOACTIVE SUBSTANCES

Uranium, Thorium, and Actinium Series. All elements found in natural sources with atomic number greater than 83 (bismuth) are radioactive. They belong to chains of successive decays, and all the species in one such chain constitute a radioactive family or series. Three of these families include all the natural activities in this region of the periodic chart. One has U_I (mass 238 on the atomic weight scale) as the parent substance, and after 14 transformations (8 of them by α-particle emission and 6 by β-particle emission) reaches a stable end product, radium G (lead with mass 206); this is known as the uranium series (which includes radium and its decay products). Since the atomic mass is changed by four units in α decay and by only a small fraction of one unit in β decay, the various masses found in members of the family differ by multiples of 4, and a general formula for the approximate masses is $4n + 2$, where n is an integer. Therefore the uranium series is known also as the $4n + 2$ series. Figure 1-1 shows the members and transformations of the uranium series.

Thorium (mass 232) is the parent substance of the $4n$, or thorium series, with lead of mass 208 as the stable end product. This series is shown in figure 1-2. The $4n + 3$, or actinium, series has actino-uranium, AcU (uranium of mass 235), as the parent and lead of mass 207 as the stable end product. This series is shown in figure 1-3.

The fairly close similarity of the three series to one another and in their relations to the periodic chart is interesting and helpful in remembering the decay modes of and nomenclature for the active bodies. Actually, these his-

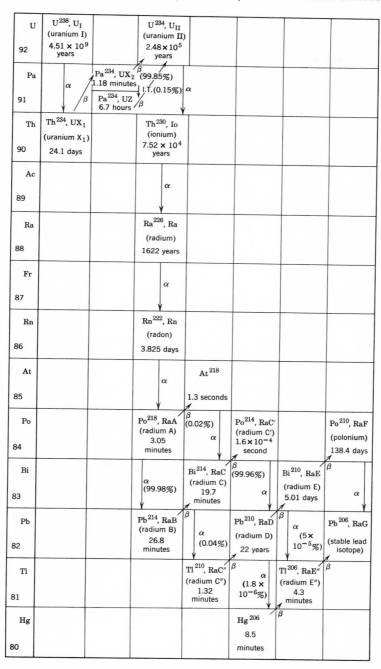

Fig. 1-1   The uranium series.

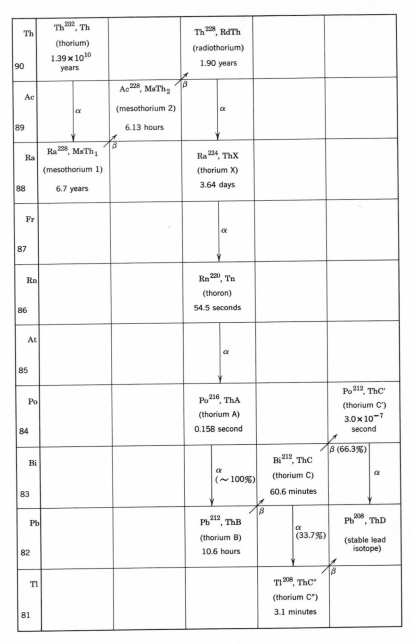

Fig. 1-2   The thorium series.

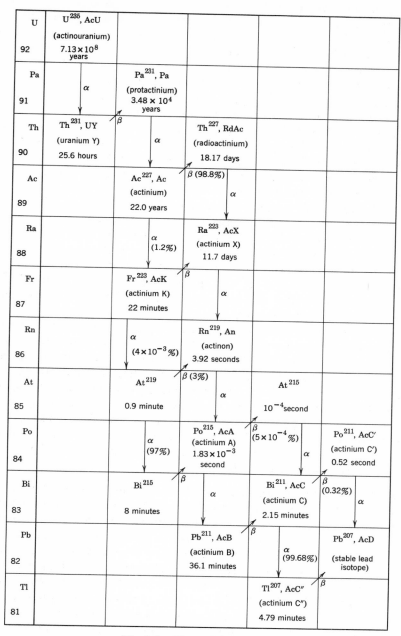

Fig. 1-3   The actinium series.

torical names are gradually becoming obsolete, and the designations of chemical element and atomic mass are becoming standard; already we are more familiar with U^{238}, U^{235}, and U^{234} than with U_I, AcU, and U_{II}. (This trend is favored by the fact that names like UX_1 and RaD do not immediately suggest that these substances are chemically like thorium and lead, respectively; also, in some of the early literature the nomenclature is different from current usage, which leads to some confusion. On the other hand, many of the historical names like RaA, RaB indicate immediately positions in the decay chain.)

The existence of branching decays in each of the three series should be noticed. As more sensitive means for the detection of low-intensity branches have become available, more branching decays have been discovered. For example, the occurrence of astatine of mass 219 in a 5×10^{-3} per cent branch of the actinium series was recognized only in 1953. With further refinements in technique, additional branchings will undoubtedly be found.

One important result of the unraveling of the radioactive decay series was the conclusion reached as early as 1910, notably by Soddy, that different radioactive species of different atomic weights could nevertheless have identical chemical properties. This is the origin of the concept of isotopes which we have already used implicitly in writing such symbols as U^{235} and U^{238} for uranium of atomic weights 235 and 238. Further discussion of isotopes is deferred until chapter 2, section B.

In each of the three families there is an isotope of element number 86, known as radon (sometimes called emanation). These rare gas radioactivities, radon, thoron, and actinon, are the emanations which we mentioned earlier and which were so important for the early understanding of radioactivity. It is because of the gaseous character of these substances that their descendants, the A, B, C products, etc., of the three families, can be so readily isolated from their longer-lived precursors. These descendants of the emanations are referred to as *active deposits*. The active deposit from any of the three radioactive series may be collected by exposure of any object, or more efficiently of a negatively charged electrode, to the emanation.

Other Naturally Occurring Radioactivities. Since the discovery of radioactivity nearly every known element has at one time or another been examined for evidences of naturally occurring radioactivity. In 1906 N. R. Campbell and A. Wood discovered weak β radioactivity in both potassium and rubidium, and for about 25 years they remained the only known radioactive elements outside the three decay series. In 1932 G. Hevesy and M. Pahl reported a radioactivity in samarium, and more recently several other naturally occurring radioactivities have been found. The presently known natural radioactivities other than those of the uranium, thorium, and actinium series are listed in table 1-2 with some of their properties. In some of these elements the particular isotope responsible for the radioactivity occurs in very small abundance, and in other cases the half-lives are extremely long. Either of these factors

Table 1-2 Additional naturally occurring radioactive substances

Active Substance	Type of Disinte- gration	Half-Life (Years)	Per Cent Isotopic Abundance	Stable Disintegration Products
K^{40}	β, EC*	1.27×10^9	0.012	Ca^{40}, Ar^{40}
V^{50}	β, EC*	6×10^{15}	0.24	Cr^{50}, Ti^{50}
Rb^{87}	β	5.7×10^{10}	27.8	Sr^{87}
In^{115}	β	5×10^{14}	95.7	Sn^{115}
Te^{123}	EC*	1.2×10^{13}	0.87	Sb^{123}
La^{138}	EC,* β	1.1×10^{11}	0.089	Ba^{138}, Ce^{138}
Ce^{142}	α	$\cong 5 \times 10^{15}$	11.07	Ba^{138}
Nd^{144}	α	2.4×10^{15}	23.85	Ce^{140}
Sm^{147}	α	1.1×10^{11}	14.97	Nd^{143}
Gd^{152}	α	1.1×10^{14}	0.20	Sm^{148}
Lu^{176}	β	3×10^{10}	2.59	Hf^{176}
Hf^{174}	α	2×10^{15}	0.18	Yb^{170}
Re^{187}	β	6×10^{10}	62.9	Os^{187}
Pt^{190}	α	7×10^{11}	0.013	Os^{186}

* The symbol EC stands for electron capture. This type of decay is briefly described in the following section and more fully treated in chapter 2.

makes such activities difficult to detect. Without question, additional natural radioactivities will be discovered.

In attempts to extend the search for new radioactivities to very low intensity levels difficulty arises from the general background of radiations present in every laboratory. In part, this general background is due to the presence of traces of uranium, thorium, potassium, etc., and in large part to the cosmic radiation of unknown origin, which is discussed in chapter 15. The cosmic rays reach every portion of the earth's surface; their intensity is greater at high altitudes but persists measurably even in deep caves and mines. The magnitude of the background effect is indicated in the discussion of radiation-detection instruments in chapter 5, section A4. In recent years there have been occasional temporary increases in background radiation due to scattered residues from large-scale atomic and thermonuclear explosions.

D. ARTIFICIALLY PRODUCED RADIOACTIVE SUBSTANCES

Historical Development. The naturally occurring radioactive substances were the only ones available for study until 1934. In January of that year I. Curie (daughter of Mme. Curie) and F. Joliot announced that boron and aluminum could be made radioactive by bombardment with the α rays from polonium (J1).   This very important discovery of artificially produced radioactivity came in the course of their experiments on the production of positrons by bombardment of these elements with α particles.   The positron had been discovered only two years earlier by C. D. Anderson as a component of the cosmic radiation; it is a particle much like the electron but positively charged. A number of laboratories quickly found that positrons could be produced in light elements by α-ray bombardment.   The Curie-Joliot discovery was that the boron and aluminum targets continued to emit positrons after removal of the α source and that the induced radioactivity in each case decayed with a characteristic half-life (reported as 14 minutes for B, 3.25 minutes for Al).

Much earlier, in 1919, Rutherford had produced nuclear transmutations by α-particle bombardment (R1), and the new phenomenon of induced radioactivity was therefore quickly understood in terms of the production of new unstable nuclei.   The unstable nucleus N^{13} is produced from boron (the stable nitrogen nuclei are N^{14} and N^{15}); from aluminum the product is P^{30} (the only stable phosphorus is P^{31}).   These are examples of but one of the many types of nuclear reactions now known to produce radioactive products and discussed in chapter 2, section F, and in chapter 10.

At the time that artificial radioactivity was discovered several laboratories had developed and put into operation devices for the acceleration of hydrogen ions and helium ions to energies at which nuclear transmutations were produced.   In addition, the discovery of the neutron in 1932 and the isolation of deuterium in 1933 made available two additional bombarding particles which turned out to be especially useful for the production of induced activities.   In the 30 years following Curie's and Joliot's discovery there occurred an almost unbelievably rapid growth of the new field.   The number of known artificially produced radioactive species reached 200 in 1937, about 450 in 1944, about 650 in 1949, about 1000 in 1954, and exceeded 1300 in 1963.   At least one radioactive isotope is now known for every element in the periodic table, and some elements have as many as 20 or more.   The measured half-lives range from milliseconds[3] to many millions of years.   Many artificially produced radioactivities have found important applications in such diverse fields as chemistry, physics, biology, medicine, and engineering.   A table of presently known radioactive species may be found in appendix E, and more extensive tabulations of their properties are available (S1, N1, G1).

Types of Radioactive Decay.   Although the first artificial radioactive substances decayed by positron emission, this is not the only or even the most common type of decay.   Alpha-particle emitters are found also, but only

---

[3] Modern electronic techniques have made possible the measurement of even much shorter half-lives, down to about 10^{-11} second.   Many γ emitters with such short half-lives are known but will not be considered as separate radioactive species for the present purpose (cf. chapter 2, section B, and chapter 8, section D).

among the heavier elements.   Ordinary β decay, as in the natural radioactive series, is commonly found throughout the range of the periodic table.   In this type of decay negative electrons are emitted and the atomic number is increased by one unit.   Positron emission results in a decrease by one unit in atomic number.   There is also another type of decay in which the atomic number is decreased by one unit, in this case by spontaneous incorporation into the nucleus of one of the atomic electrons (most often one from the K shell of the atom).   In these three processes the atomic mass remains essentially constant

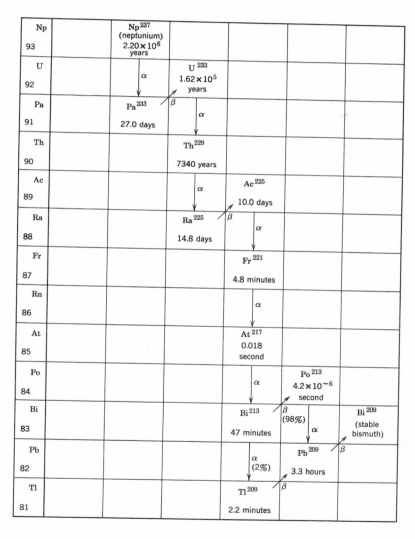

Fig. 1-4   The $4n + 1$ series.

while the atomic number changes; all three are generally classed as β-decay processes and may be distinguished by the names negatron emission, positron emission, and electron capture (or sometimes K capture).

Synthetic Elements and the $4n + 1$ Series. Not only has it been possible by transmutation techniques to produce radioactive isotopes of every known element, but also a number of elements not found in nature have been synthesized. In each case the new element has first been recognized in unweighably small amounts detectable only by its radioactivity; however, macroquantities of many of these new elements have now been prepared. The best known of the synthetic elements is probably plutonium, an element which within less than five years of its discovery was available in sufficient quantities to serve as an ingredient in atomic bombs. Up to 1963 eleven new elements beyond uranium in the periodic table had been produced by artificial transmutations (S3), as had the elements technetium (atomic number 43) and promethium (number 61) which are not known to occur naturally on the earth. The subject of the new elements is treated more fully in chapter 7.

In past years there has been speculation why elements beyond uranium are not found in nature. The most plausible explanation is that there is no transuranium species of sufficiently long half-life to have survived since the original formation of the elements. Similarly, there has been much comment about the nonexistence of a $4n + 1$ radioactive series in nature. Among the artificially produced radioactivities in the heavy-element region, a well-developed $4n + 1$ series has been prepared and investigated (S2). The part of this series that begins with its longest-lived member, neptunium of mass 237, is displayed in figure 1-4. This decay sequence has a general resemblance to the three natural radioactive families and has as its stable end product bismuth of mass 209.

REFERENCES

B1   H. Becquerel, *Compt. rend.* **122,** 422, 501, 559, 689, 762, 1086 (1896).

G1   General Electric Company, Chart of the Nuclides, sixth edition, 2nd printing, Schenectady, December 1962.

J1   F. Joliot and I. Curie, "Artificial Production of a New Kind of Radio-Element," *Nature* **133,** 201 (1934).

*J2   G. E. M. Jauncey, "The Early Years of Radioactivity," *Am. J. Phys.* **14,** 226 (1946).

*M1   S. Meyer and E. Schweidler, *Radioaktivität,* B. G. Teubner, Berlin, 1927.

N1   National Academy of Sciences—National Research Council, *Nuclear Data Sheets,* Washington, 1958–1963.

R1   E. Rutherford, "Collision of α Particles with Light Atoms, IV. An Anomalous Effect in Nitrogen," *Phil. Mag.* **37,** 581 (1919).

*R2   E. Rutherford, J. Chadwick, and C. D. Ellis, *Radiations from Radioactive Substances,* Cambridge University Press, 1930.

S1   D. Strominger, J. M. Hollander, and G. T. Seaborg, "Table of Isotopes," *Revs. Modern Phys.* **30,** 535 (1958).

S2   G. T. Seaborg, "The Neptunium ($4n + 1$) Radioactive Family," *Chem. Eng. News* **26,** 1902 (1948).

S3   G. T. Seaborg, *The Transuranium Elements,* Yale University Press, New Haven, 1958.

EXERCISES

1. One hundred milligrams of Ra would represent what percentage yield from exactly 2 tons of a pitchblende ore containing 75 per cent U_3O_8?   *Answer:* 26 per cent.

2. Calculate the rate of energy liberation (in calories per hour) for 1.00 g of pure radium free of its decay products.   What can you say about the actual heating effect of an old radium preparation?   *Answer to first part:* 25 cal/hr.

3. A certain active substance (which has no radioactive parent) has a half-life of 8.0 days.   What fraction of the initial amount will be left after (a) 16 days, (b) 32 days, (c) 4 days, (d) 83 days?   *Answer:* (a) 0.25.

4. How long would a sample of radium have to be observed before the decay amounted to 1 per cent?   (Neglect effects of radium A, B, C, etc., on the detector.)

5. Find the number of disintegrations of uranium I atoms occurring per minute in 1 mg of ordinary uranium, from the half-life of U_I, $t_{\frac{1}{2}} = 4.51 \times 10^9$ years.

6. How many β disintegrations occur per second in 1.00 g of pitchblende containing 70 per cent uranium?   You may assume that there has been no loss of radon from the ore.   *Answer:* 53,000/sec.

7. If 1 g of radium is separated from its decay products and then placed in a sealed vessel, how much helium will accumulate in the vessel in 60 days?   Express the answer in cubic centimeters at STP.   *Answer:* 0.028 cm^3.

8. What is the natural radioactivity in disintegrations per minute per milligram of ordinary potassium chloride (KCl)?   *Answer:* 1.0 d min^{-1} mg^{-1}.

2

Atomic Nuclei

A. ATOMIC STRUCTURE

Early Views. At the time the phenomenon of radioactivity was discovered the chemical elements were regarded as unalterable; they were thought to retain their identities throughout all chemical and physical processes. This view became untenable when it was recognized that radioactive disintegration involved the transformation of one element into another. As a result of J. J. Thomson's discovery of the electron in 1897, it had already become clear that atoms, until then regarded as the indivisible building blocks of matter, must have some structure. From experiments on the scattering of X rays and electrons by matter, Thomson and others concluded that the number of electrons per atom was approximately equal to the atomic weight (actually more nearly equal to half the atomic weight, as established by C. G. Barkla in 1911). This conclusion, together with Thomson's determination of the electron mass as approximately one two-thousandth of the mass of a hydrogen atom, led to the assumption that most of the mass of an atom must reside in its positively charged parts.

The problem that remained to be solved was: how are the positive and negative charges distributed inside the atom? Thomson in 1910 proposed a model of atomic structure in which positive electricity is uniformly distributed throughout the volume of the atom (atomic dimensions were known to be of the order of 10^{-8} cm) and the electrons occupy certain stable positions in this heavier, positively charged jelly.

Alpha-Particle Scattering. Thomson's model was soon found quite inadequate to account for experiments carried out by Rutherford and his co-workers on the scattering of α particles by thin metal foils. If a collimated beam of α particles is allowed to strike a thin film of matter, some of the particles are deflected from their original direction in passing through the film. This scattering is clearly caused by the electrostatic forces between the positively charged α particle and the positive and negative charges in the atoms of the scattering material. It is readily seen that the passage of α particles through

18

Thomson atoms would result in rather small deflections because the charges in the scattering atoms are so diffuse. In 1908 experiments by H. Geiger on scattering in extremely thin foils had indeed shown that single encounters between α particles and atoms resulted predominantly in very small scattering angles (of the order of 1°).

Shortly afterward, however, experiments of Geiger and E. Marsden proved that deflections through much larger angles (in fact, ranging well above 90°) occurred much more frequently than could be accounted for by single scattering encounters with the weak electric fields of Thomson atoms. Multiple scattering in successive collisions was first invoked to explain these results; but Rutherford showed by simple probability arguments that such multiple encounters were orders of magnitude too rare to account for the observed frequency of large-angle deflections. This realization led him to reject the Thomson atom and to propose a different atomic model which rapidly became universally accepted.

The Nuclear Model of the Atom. In his classic paper of 1911 (R1) Rutherford postulated that the observed large-angle scattering was due to single scattering processes and that these large-angle scatterings could be produced only by an intense electric field; consequently the positive charge and most of the mass of the atom had to be concentrated in a very small region, later known as the nucleus. A number of electrons sufficient to balance the positive charge was thought to be distributed over a sphere of atomic dimensions. Rutherford then proceeded to show that for large-angle scatterings of α particles by this sort of atom the effect of the electrons was negligible compared to that of the central charge. Considering this central charge (Ze) of the atom and the charge ($Z_\alpha e = 2e$) of the α particle as point charges, Rutherford then merely assumed the force between them at any distance d to be given by Coulomb's law: $F = Ze \cdot Z_\alpha e/d^2$. On this basis, and with the additional simplifying assumption that the nucleus is sufficiently heavy to be considered at rest during the encounter, Rutherford showed that the path of an α particle in the field of a nucleus is a hyperbola with the nucleus at the external focus. From the conditions of conservation of momentum and energy and from the geometric properties of the hyperbola he then derived[1] his celebrated scattering formula which relates the number $n(\phi)$ of α particles falling on a unit area at a distance r from the scattering point to the scattering angle ϕ (the angle between the directions of incident and scattered particle):

$$n(\phi) = n_0 \frac{Nt}{16r^2} \left(\frac{Ze \cdot Z_\alpha e}{\frac{1}{2}M_\alpha v_\alpha^2} \right)^2 \frac{1}{\sin^4 (\phi/2)}, \tag{2-1}$$

where n_0 is the number of incident α particles, t is the thickness of scatterer, N is the number of nuclei per unit volume of scatterer, and M_α and v_α are the mass and initial velocity of the α particle.

[1] For detailed derivations, see, for example R2 (p. 191) and K1 (p. 53).

The specific predictions of the Rutherford formula, particularly the inverse proportionality of the number of scattered particles detected per unit area to the square of the α-particle energy and to the fourth power of the sine of half the scattering angle, were quickly subjected to experimental test, principally by Geiger and Marsden. The agreement was found to be excellent for heavy-element scatterers and, provided the theory was suitably modified for the case of nuclei not at rest during impact, for light-element scatterers also.

Nuclear Charge and Atomic Number. The experimental verification of the scattering formula led to a general acceptance of Rutherford's picture of the atom as consisting of a small positively charged nucleus containing nearly the entire mass of the atom and surrounded by a distribution of negatively charged electrons. In addition, the scattering law made it possible to study the magnitude of the nuclear charge in the atoms of a given element because the scattering intensity depends on the square of the nuclear charge. It was by the method of α-particle scattering that nuclear charges were first determined, and this work led to the suggestion that the atomic number Z of an element, until then merely a number indicating its position in the periodic table, was identical with the nuclear charge (expressed in units of the electronic charge e). This suggestion was subsequently confirmed by H. G. Moseley's work on the X-ray spectra of the elements (M1). Moseley showed that the frequencies of the K X-ray emission lines increase regularly from element to element when the elements are arranged in the order of their appearance in the periodic system. The relation between frequency and atomic weight showed irregular variations, but when each element was assigned an "atomic number" Z, according to its position in the periodic table, Moseley noticed that the square root of the K X-ray frequency was proportional to $Z - 1$. He identified the atomic number with the number of unit charges on the nucleus. This number (which is also the number of extranuclear electrons in the neutral atom) was thus shown to be closely related to the chemical properties of an element.

Following the acceptance of Rutherford's nuclear model of the atom, the further understanding of atomic structure developed rapidly through the study of X-ray and optical spectra and culminated in Niels Bohr's theory of 1913 and its later wave-mechanical modifications. Further discussion of the extra-nuclear features of atomic structure is, however, outside the scope of this book.[2]

B. COMPOSITION OF NUCLEI

Nuclear Size and Density. The α-particle scattering experiments of Rutherford and his school not only confirmed the nuclear model and led to the

---

[2] The interested reader is referred to standard works on atomic structure, e.g., M. Born, *Atomic Physics*, Blackie, London, 1951; F. K. Richtmyer, E. H. Kennard, and T. Lauritsen, *Introduction to Modern Physics*, 5th ed., McGraw Hill, New York, 1955; H. Semat, *Introduction to Atomic and Nuclear Physics*, Rinehart, New York, 1954.

determination of approximate values of nuclear charges but also gave information on the sizes of nuclei. We can readily derive the distance of closest approach for an α particle (mass M_α and charge $2e$) of velocity v to a nucleus of charge Ze. At a distance d between nucleus and α particle, where the velocity of the α particle, as a result of the Coulomb field, has been slowed to v', conservation of energy demands that

$$\tfrac{1}{2}M_\alpha v^2 = \tfrac{1}{2}M_\alpha v'^2 + \frac{2Ze^2}{d}.$$

The minimum distance d_0 is reached in a head-on collision at the point where the α particle reverses its direction and where therefore $v' = 0$. Thus

$$d_0 = \frac{4Ze^2}{M_\alpha v^2} = 1.4 \times 10^5 \frac{Z}{v^2} \text{ cm}.$$

Velocities of α particles from natural radioactive sources range from about 1.3×10^9 to about 1.9×10^9 cm per second, so that d_0 becomes $(4 - 8) \times 10^{-14}Z$ cm, which corresponds to about $(0.5 - 1) \times 10^{-12}$ cm for Al, $(1 - 2) \times 10^{-12}$ cm for Cu, and $(3 - 6) \times 10^{-12}$ cm for Au.

From the fact that no deviations from the predictions of the Rutherford scattering formula were observed with natural α particles incident on heavy elements (Cu, Ag, Au), it was concluded that the Coulomb force law around these nuclei holds for these small distances and that therefore the nuclei could be considered as having radii no larger than these values. With aluminum (and other light elements), deviations from Rutherford scattering did occur, and the distances at which the repulsive forces appeared to become weaker than predicted from Coulomb's law ($\sim 7 \times 10^{-13}$ cm for Al) were thought to represent nuclear radii.

Since the nucleus contains almost the entire mass of the atom but has dimensions of the order of 10^{-12} cm (whereas atomic dimensions as determined, for example, from gas kinetics are about 10^{-8} cm), it follows that nuclei are very much denser than ordinary matter; the density of nuclear matter is in the neighborhood of 10^{14} g per cm^3 or 10^8 tons per cm^3.

Once it was established that almost the entire mass of an atom resides in its nucleus and that the atoms of each element have nuclei of characteristic charge, it became evident that radioactive transformations are, in fact, nuclear processes. The newly discovered nuclei thus could not be regarded as indivisible entities, but they had to have some structure of their own. The forces holding nuclei together had to be strong and of short range; their exact nature was (and is) not well understood.

Isotopes and Integral Atomic Weights. The existence of isotopes, as we saw briefly in chapter 1, became evident when different radioactive bodies in the naturally occurring decay series, for example, RaB, AcB, and ThB, were found to exhibit identical chemical properties (in the case mentioned, the properties of lead). This discovery led to a search for the existence of isotopes in nonradioactive elements. In early experiments with ion deflections in magnetic and electric fields J. J. Thomson showed in 1913 that neon consisted of two isotopes with atomic weights about 20 and 22 (now a third neon isotope of atomic weight 21 is known). Subsequently, it was found, principally as a result of F. W. Aston's pioneering work with his mass spectrograph, that the great majority of the elements consisted of mixtures of isotopes and that the atomic weights of the individual isotopes were almost exactly integers. This "whole number rule" of Aston's naturally led to a revival, in modern terms, of the hypothesis proposed a hundred years earlier by Prout, namely that all elements are built up of hydrogen.

The nucleus of the common hydrogen atom, called a proton, is the simplest known nucleus. Its positive charge is equal in magnitude to the negative charge on an electron, 4.8030×10^{-10} electrostatic units (esu). The mass of a proton is approximately equal to that of a hydrogen atom and therefore nearly equal to 1 on the atomic weight scale.

Proton-Electron Hypothesis. With both the masses and charges of nuclei known to be multiples (or very nearly exact multiples) of the mass and charge of a proton, it was natural to suppose that all nuclei were built up of protons. However, the mass number (integer nearest the atomic weight) of every element or isotope beyond hydrogen is larger than its atomic number; in fact it is generally at least twice as large. Therefore it was necessary to assume that a nucleus of mass number A and atomic number Z contains, in addition to the A protons accounting for its mass, $A - Z$ electrons to make the net positive charge Z. A nucleus of nitrogen ($A = 14$, $Z = 7$) would, on this hypothesis, consist of 14 protons and 7 electrons, a nucleus of U^{238} ($Z = 92$) of 238 protons and 146 electrons.

The idea that electrons exist in nuclei arose rather naturally from the phenomenon of β emission observed in radioactive decay. The emission of α particles was explained in terms of either the existence in nuclei or the formation at the moment of decay of tightly bound aggregates of 4 protons and 2 electrons.

It was nevertheless realized that the existence of free electrons in the nucleus led to certain difficulties, some of which stem from considerations of angular-momentum conservation and statistics and are mentioned in the following sections, where these nuclear properties are discussed. Another difficulty has to do with the requirement that, to be thought of as contained in a nucleus, an electron must have a de Broglie wavelength λ not much larger than nuclear dimensions; when we put $\lambda = 10^{-12}$ cm and calculate the corresponding momentum $mv = h/\lambda$, we obtain a value of 6.6×10^{-15} erg-sec per cm for the momentum or $\sim 2 \times 10^{-4}$ erg for the corresponding kinetic energy (cf. appendix B for the relativistic relations between energy and momentum). This is more than an order of magnitude larger than the observed kinetic energies of β particles, a result that seemed incompatible with the existence of free electrons in nuclei and led to the conclusion that electrons in nuclei must be tightly bound to much heavier aggregates. As already mentioned, α-particle clusters and other combinations of protons and electrons were therefore proposed. Already in 1920 Rutherford had suggested the existence in nuclei of the "neutron," a close combination of a proton and an electron.

Discovery of the Neutron. Many fruitless attempts were made to find evidence for the neutron postulated by Rutherford. Success finally came in 1932 to J. Chadwick (C1) in the course of his investigations of a very penetrating radiation previously observed by several other experimenters when they bombarded beryllium and boron with α particles. When this radiation was found to be capable of ejecting energetic protons from hydrogen-containing substances such as paraffin, the previously held view that one was dealing with high-energy γ rays became untenable unless one was willing to relinquish the laws of conservation of energy and momentum. Chadwick showed that, on the other hand, all the evidence was compatible with the assumption that the radiation consisted of neutrons, i.e., neutral particles of zero charge and of approximately the mass of protons. From the known masses and observed energies of the reactants and products in the reaction $B^{11} + He^4 \rightarrow N^{14} + n$ (n = neutron), Chadwick estimated the mass of the neutron as being slightly less than the combined masses of proton and electron. He therefore initially considered a neutron as consisting of "a proton and an electron in close combination," very much in the way Rutherford had postulated. Later more precise measurements showed the neutron mass to be actually about 0.08 per cent larger than the mass of a hydrogen atom.

Being electrically neutral, neutrons do not cause any primary ionization in passing through matter and are therefore not so readily detected as charged particles. Furthermore, they are not stable in the free state but undergo β decay, disintegrating into protons and electrons with a half-life of about 13 minutes. It was probably for these reasons that neutrons escaped discovery for so long, although, as Chadwick recognized already in his first paper on the neutron, they are undoubtedly among the constituents of all nuclei (other than those of hydrogen).

Proton-Neutron Hypothesis. Because of the difficulties we mentioned earlier, the proton-electron hypothesis was quickly discarded after the discovery of the neutron and replaced by the now accepted proton-neutron hypothesis of nuclear composition. According to this picture, the number of protons in a nucleus equals its atomic number Z, and the total number of neutrons and protons (collectively called nucleons) equals its mass number A. Therefore the neutron number N equals $A - Z$. Thus the nucleus of N^{14} is thought to contain seven protons and seven neutrons.

The atomic numbers of the known elements range from 1 for hydrogen to 103 for the most recently discovered transuranium element lawrencium. Nuclei with neutron numbers 0 to 156 are known. The known mass numbers range from 1 to 257.   The difference $N - Z$ (or $A - 2Z$) between the number of neutrons and protons in a nucleus is referred to as its neutron excess or isotopic number.

The symbol used to denote a nuclear species is the chemical symbol of the element with the atomic number as a left subscript and the mass number as a superscript (in the United States usually written to the right, elsewhere to the left); for example, $_2\text{He}^4$, $_{27}\text{Co}^{59}$, $_{92}\text{U}^{235}$.   The atomic number is often omitted because it is uniquely determined by the chemical symbol.

Isotopes and Nuclides.   As we have already mentioned, atomic species of the same atomic number, that is, belonging to the same element but having different mass numbers, are called *isotopes*.   In the nuclei of the different isotopes of a given element the number of protons characteristic of that element is combined with different numbers of neutrons.   For example, a $_{17}\text{Cl}^{35}$ nucleus contains 17 protons and 18 neutrons, whereas a $_{17}\text{Cl}^{37}$ nucleus contains 17 protons and 20 neutrons.   Deuterium, a rare isotope of hydrogen, has a nucleus containing one proton and one neutron.

As a result of mass-spectrographic investigations we now know that the elements with atomic numbers between 1 and 83 have on the average more than three stable isotopes each.   Some elements such as beryllium, phosphorus, arsenic, and bismuth, each have a single stable nuclear species, whereas tin, for example, has as many as 10 stable isotopes.

The stable isotopes of a given element generally occur together in constant proportions.   This accounts for the fact that atomic weight determinations on samples of a given element from widely different sources generally agree within experimental errors.   However, there are some notable exceptions to this rule of constant isotopic composition.   One is the variation in the abundances of lead isotopes, especially in ores containing uranium and thorium.   Depending on the age and composition of such ores, the end products of the three radioactive families, Pb^{206}, Pb^{207}, and Pb^{208}, and the nonradiogenic Pb^{204} may occur in different proportions.   Similarly, the isotope Sr^{87} has been found to have an abnormally high abundance in rocks that contain rubidium; this is explained by the fact that Rb^{87} is a naturally occurring β emitter and decays to Sr^{87}.

Helium from gas wells probably has its origin in radioactive processes (α disintegrations) and contains a much smaller proportion of the rare isotope He^3 than does atmospheric helium.   Water from various sources shows slight variations in the H^1/H^2 ratio.   This is in some cases due to the fact that heavy water has a slightly lower vapor pressure than ordinary water and is therefore concentrated by evaporation.   The enrichment of H^2 in the water of the Dead Sea and in certain vegetables is ascribed to this cause.   The waters which show abnormally high H^2 concentrations usually also have slightly higher $\text{O}^{18}/\text{O}^{16}$ ratios than normal.   Another cause for small variations in isotopic composition is that chemical equilibria are slightly dependent on the molecular weights of the reactants, and this may lead to isotopic enrichments in the course of reactions occurring in nature. For example, the slight enrichment of C^{13} in limestones relative to some other sources of carbon comes about because the equilibrium in the reaction between CO_2 and water to form carbonic acid lies somewhat further toward the side of carbonic acid for $C^{13}O_2$ than for $C^{12}O_2$. The effects of isotopic substitution on equilibria and rates of chemical reactions are discussed in chapter 7, section C.

The word *isotope* has been used also in a broader sense to signify any particular nuclear species characterized by its A and Z values. In this meaning it should be and now generally is replaced by the word *nuclide* suggested by T. P. Kohman and defined as "a species of atom characterized by the constitution of its nucleus, in particular by the number of protons and neutrons in its nucleus."

Isobars, Isotones, and Isomers. Atomic species having the same mass number but different atomic numbers are called *isobars*. A few examples of isobars are $_{32}Ge^{76}$ and $_{34}Se^{76}$; $_{52}Te^{130}$, $_{54}Xe^{130}$, and $_{56}Ba^{130}$; $_{80}Hg^{204}$ and $_{82}Pb^{204}$.

Atomic species having the same number of neutrons but different mass numbers are sometimes referred to as *isotones*. For example, $_{14}Si^{30}$, $_{15}P^{31}$, and $_{16}S^{32}$ are isotones because they all contain 16 neutrons per nucleus.

Among the natural radioactive bodies discussed in chapter 1 there are two, UX_2 and UZ, which have the same mass number as well as the same atomic number but differ in their radioactive properties. This is an example of nuclear *isomerism*. Although UX_2 and UZ had been known for several years, the phenomenon of nuclear isomerism did not receive much attention until another pair of isomers, Br^{80}, was discovered among artificially produced radioactive species in 1937. About 250 cases of isomerism are now known. Nuclear isomers are different energy states of the same nucleus, each having a different measurable lifetime (except that the ground state may be stable).[3] In a number of cases more than two isomeric states have been found for a given A and Z. For example, three radioactive species of half-lives 60 days, 1.5 minutes, and 21 minutes have been assigned to Sb^{124}. The notation which has become fairly standard for representing isomeric states other than the ground state is a superscript m (for metastable) following the mass number; if there are two or more excited isomeric states, they are labeled m_1, m_2, etc., in order of increasing excitation energy. Thus the isomers of Sb^{124} are denoted as Sb^{124} (60 day), Sb^{124m_1} (1.5 minutes) and Sb^{124m_2} (21 minutes). We shall not consider each isomeric state as an individual nuclide.

[3] The question what constitutes a "measurable lifetime" has become complicated by continued extension of lifetime-measurement techniques to shorter and shorter time scales. Cf. discussions of isomerism in section E4 of this chapter and in chapter 8, section D.

C. NUCLEAR PROPERTIES

Mass and Energy.   Masses of atomic nuclei are so small when stated in ordinary units (less than 10^{-21} g) that they are generally expressed on a different scale.   The scale used for many years was one in which the mass of an atom of O^{16} was taken as the standard and assigned a mass of exactly 16.00000 units.   This was known as the physical atomic-weight scale.   It should be noted that it is not identical with the chemical atomic-weight scale which is used for expressing atomic weights in chemical calculations; in the chemical scale the atomic weight of the natural isotopic mixture of oxygen (containing small amounts of O^{17} and O^{18}) was, until recently, assigned the value of exactly 16.00000.   The unit used was therefore larger in the chemical than in the physical scale, and the numerical value of any atomic weight was smaller when expressed on the chemical scale.   The conversion factor between the two is 1.000272 ± 0.000005, the uncertainty being caused by the uncertainty in the isotopic composition of normal oxygen.

Partly in order to eliminate this duality of scales and the uncertainty in chemical atomic weights caused by the variable isotopic composition of natural oxygen, a new scale was adopted in 1960–1961 by the International Unions of Pure and Applied Physics and Chemistry.   This is based on the mass of an atom of C^{12} taken as exactly 12.00000 units.   On this scale the mass of O^{16} is 15.994915 and all other atomic masses are therefore about 0.0318 per cent smaller than on the old physical atomic-weight scale.   On the other hand, the difference between the new C^{12} scale and the old chemical atomic-weight scale is only about 0.005 per cent, so that for all practical purposes chemical atomic weights remain unchanged.

The values of nuclidic masses given in this book and in recent publications on nuclear physics and chemistry are on the new C^{12} scale.   Caution will be required for some time in the use of the literature because all older publications quote masses on the O^{16} scale.[4]   Whichever scale is used, it should be noted that atomic rather than nuclear masses are given; in other words, the masses quoted include the masses and the binding energies of the extranuclear electrons in the neutral atoms.   This convention, as we shall see, turns out to have some advantages in the treatment of nuclear reactions and energy relations. More importantly, however, it arises from the fact that it is always *atomic* masses or differences between *atomic* masses that are measured experimentally.

The experimental determination of exact atomic masses involves the use of a mass spectrograph or mass spectrometer.   In most of these instruments the charge-to-mass ratio of positive ions is determined from the amount of deflection in a combination of magnetic and electric fields; different arrangements

---

[4] A complete set of atomic masses on both scales may be found in reference E1.   A more up-to-date mass table, but on the C^{12} scale only, is given in reference K2.

are used for bringing about velocity focusing or directional focusing, or both, for ions of a given e/M. Instruments that use photographic plates for recording the mass spectra are called mass spectrographs; those that make use of collection and measurement of ion currents are referred to as mass spectrometers. More recently the fact that ions of the same kinetic energy and different masses require different times to traverse a given path length has been utilized in the design of several types of so-called time-of-flight mass spectrometers. These devices have proved particularly useful for the determination of accurate mass values.

Mass determinations throughout the mass range from hydrogen to bismuth have been made with precisions varying between about 0.02 and 10 parts per million. For precision mass determinations the method generally used is the so-called doublet method. This substitutes the measurement of the difference between two almost identical masses for the direct measurement of absolute masses. All measurements must, of course, eventually be related to the standard C^{12}. But for convenience the masses of H^1, H^2, and O^{16} have been adopted as substandards and for this purpose have been carefully measured by determinations of the fundamental doublets:

$(C^{12}H^1{}_4)^+$ and $(O^{16})^+$ at mass-to-charge ratio 16,
$(H^2{}_3)^+$ and $(C^{12})^{2+}$ at mass-to-charge ratio 6,
$(H^2)^+$ and $(H^1{}_2)^+$ at mass-to-charge ratio 2.

On the C^{12} scale the mass of a hydrogen atom (sometimes loosely called the proton mass) is 1.0078252, the mass of a neutron 1.0086654, and that of an electron 0.0005486 mass units. One mass unit equals 1.660×10^{-24} g.

One of the important consequences of Einstein's special theory of relativity[5] is the equivalence of mass and energy. The total energy content E of a system of mass M is given by the relation

$$E = Mc^2,$$

where c is the velocity of light (2.99792×10^{10} cm per second). Therefore the mass of a nucleus is a direct measure of its energy content. The measured mass of a nucleus is always smaller than the combined masses of its constituent nucleons, and the difference between the two is called the binding energy of the nucleus.

To find the energy equivalent to 1 mass unit we put $M = 1.660 \times 10^{-24}$ g and $c = 2.998 \times 10^{10}$ cm per second and find $E = Mc^2 = 1.492 \times 10^{-3}$ erg. However, energy units much more useful in nuclear work than the erg are the electron volt (eV), the kiloelectron volt (keV; 1 keV = 1000 eV), and the million electron volt (MeV; 1 MeV = 10^6 eV). The electron volt is defined

[5] A summary of the most frequently used relativistic equations may be found in appendix B.

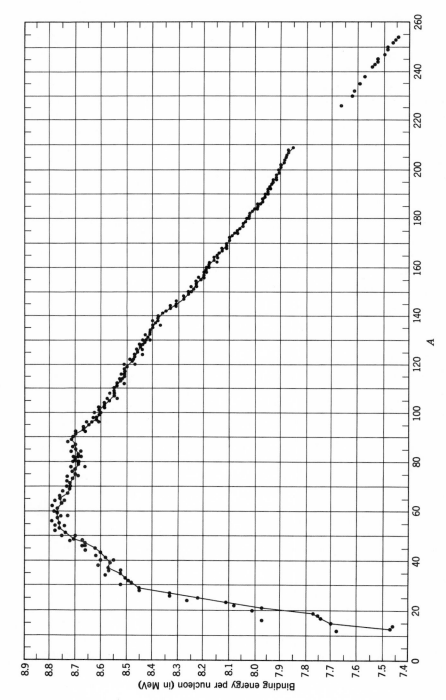

Fig. 2-1a   The average binding energy per nucleon as a function of A.   The line drawn connects the odd-A points.

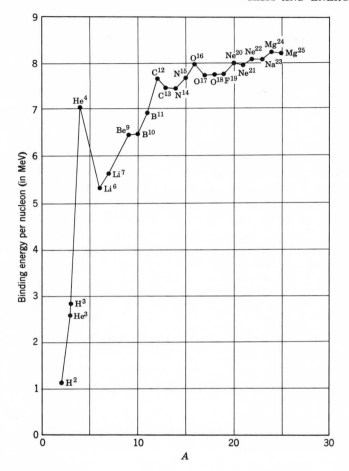

Fig. 2-1b   The average binding energy per nucleon in the region of the lightest nuclei.

as the energy necessary to raise one electron through a potential difference of 1 volt.

$$1 \text{ eV} = 1.602 \times 10^{-12} \text{ erg}; \qquad 1 \text{ MeV} = 1.602 \times 10^{-6} \text{ erg}.$$

Using these new units, we find

$$1 \text{ mass unit} = 931.5 \text{ MeV},$$

and

$$1 \text{ electron mass} = 0.5110 \text{ MeV}.$$

As an example, we shall calculate the binding energy of He^4.   The mass of He^4 is 4.002604; the combined mass of two hydrogen atoms[6] and two neutrons

---

[6] Since the mass of He^4 includes the mass of two electrons it is clear that it is the *atomic* mass of H^1 that must be used.

is $2 \times 1.0078252 + 2 \times 1.0086654 = 4.032981$.   Thus the binding energy of He^4 is $4.032981 - 4.002604 = 0.030377$ mass unit or $0.030377 \times 931.5 = 28.30$ MeV.   The binding energy per nucleon in He^4 is therefore approximately 7.1 MeV.

The binding energy of the deuteron calculated by the same method is found to be 2.225 MeV.   Actually this value was determined experimentally from the threshold for the photodisintegration of the deuteron into proton and neutron and combined with the mass-spectrographically measured masses of proton and deuteron to calculate the neutron mass.   No accurate method for a direct measurement of the neutron mass is known.

The average binding energy per nucleon is remarkably constant in all nuclei except for a few of the lightest ones.   For $A > 11$ it ranges between 7.4 and 8.8 MeV throughout the table of elements, with the maximum values (near 8.8 MeV) occurring in the vicinity of $A = 60$ (for iron and nickel nuclei).   In figure 2-1 the average binding energy per nucleon is plotted as a function of A for all stable and some of the heavy radioactive nuclei.   From the maximum in the iron region the values are seen to decrease more slowly towards the high-A than towards the low-A side.   Despite the near-constancy of the average binding energy, some interesting details can be discerned.   Among the lighter nuclei the value for a nucleus of even A is generally higher than the average of the values for the adjacent odd-A nuclei.   The same is true at higher mass numbers when the most stable nucleus of a given even A (there are often two and occasionally three) is compared with the neighboring odd-A nuclei.   The slight deviations from a completely smooth curve (e.g., at $A \cong 88$) are real and well established and are discussed later in connection with nuclear shell structure.

A number of irregularities occur among the lightest nuclei; in particular, the binding energies of $_2He^4$, $_6C^{12}$, and $_8O^{16}$ are very high.   These trends have some important consequences.   The sun's radiant energy is believed to result from a series of nuclear transformations whose net effect is the building up of helium atoms from hydrogen atoms, which is a very exoergic process.   The energy released in the fission of the heaviest nuclei is large because nuclei near the middle of the periodic table have higher binding energies per nucleon.   The abnormally high natural abundance of iron and nickel is presumably connected with the maximum in the nuclear stability curve in the region of these elements (cf. chapter 15, section B).

Quantities related to the binding energy are the mass defect and the packing fraction.   These are, in fact, more frequently tabulated in the older literature than are the binding energies.   The mass defect Δ is the difference between the atomic mass M and the mass number A: $\Delta = M - A$.   (Some authors call this the mass excess.)   The packing fraction f is the mass defect divided by the mass number: $f = \Delta/A$.   The packing fraction goes through a minimum in the region of iron and, in general, shows an inverse correlation with the average binding energy.

Although the *average* binding energy per nucleon is a rather slowly varying function, the contribution to the binding energy from the addition of *one* more proton or neutron shows large fluctuations from one nucleus to the next. (Chemists may enjoy thinking of the binding energy of one additional nucleon as a sort of partial molal binding energy.)   The quantity may be defined here as the mass of the nucleus plus the mass of the additional nucleon minus the mass of the resulting nucleus, expressed in energy units.   As an illustration of the fluctuations in this quantity, consider the binding energies for an additional neutron to Ti^{45}, Ti^{46}, Ti^{47}, Ti^{48}, Ti^{49}, and Ti^{50}, which have values of 13.19, 8.88, 11.61, 8.15, 10.93, and 6.36 MeV; the even-odd effect is much more pronounced here than with the *average* binding energy per nucleon. Similar trends are found in the binding energies for additional protons; for example, proton addition to the nuclei $_{50}Sn^{122}$, $_{51}Sb^{123}$, $_{52}Te^{124}$, $_{53}I^{125}$, and $_{54}Xe^{126}$ involves the liberation of 6.5, 8.6, 5.6, 7.6, and 4.4 MeV, respectively.

For some purposes it is convenient to consider the binding energies of nuclear aggregates, such as α particles, in particular nuclei.   The binding energy of an α particle (He^4, mass 4.00260) in U^{235} (mass 235.04393) may be obtained from these masses and the mass of Th^{231} (231.03635): $231.03635 + 4.00260 - 235.04393 = -0.00498$ mass unit, or -4.64 MeV.   The negative binding energy means that the U^{235} nucleus is thermodynamically unstable with respect to decomposition into Th^{231} and He^4.   Alpha-particle binding energies are negative in all "stable" nuclei with $A \gtrsim 140$ (see chapter 8, section A).

The masses of some radioactive nuclei can be determined from an accurate knowledge of the energy balance in nuclear reactions involving these nuclei and from their disintegration energies.   This subject is discussed in section F.

Radius (E2).   In the discussion of Rutherford scattering (p. 21) we have already seen that early α-particle scattering experiments led to approximate values of nuclear dimensions for light nuclei and to upper limits of those for heavier nuclei.   When beams of protons and α particles of higher energies than are available from natural sources were produced in accelerators, it became possible to observe deviations from Coulomb's law in scattering experiments with even the heaviest elements and thus to obtain a measure of their nuclear sizes (found to be in the neighborhood of 10^{-12} cm).   The nuclear radii determined in this manner by charged-particle scattering are those distances from the centers of nuclei within which, in addition to the Coulombic (repulsive) forces, specific nuclear (attractive) forces act.   As a result of the interplay of these two forces, we may picture (in somewhat idealized representation) the potential energy between a nucleus and a positively charged particle, as shown in figure 2-2, as a function of the distance r between centers.

The potential well within which the short-range nuclear forces are dominant, does not necessarily have vertically rising walls as shown in figure 2-2, and the

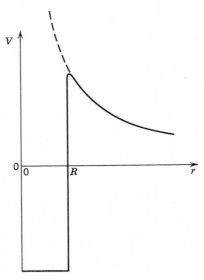

Fig. 2-2   Potential energy in the neighborhood of a nucleus. R is the radius of the potential well.

nuclear radius R is therefore not really as well defined as indicated there. Although all available experimental data show that the radius of a nucleus is a much more definite quantity than the radius of an atom, we should not be surprised to find that different methods for determining nuclear radii give slightly different results.   Perhaps the degree of *agreement* obtained with different probes should, in fact, occasion some surprise.

The cross-sectional "area" which a nucleus presents to a beam of fast neutrons can be determined from experiments on fast-neutron absorption and scattering.   The neutrons must be of sufficiently high energy ($\gtrsim 10$ MeV) to have de Broglie wavelengths that are small compared to the nuclear radii to be determined.   If a nucleus can be considered a completely opaque sphere of radius R, it will present a cross-sectional area of πR^2 for absorption of such a fast-neutron beam.[7]   In this manner nuclear radii ranging from about 3.8×10^{-13} cm for carbon to about 8×10^{-13} cm for bismuth have been

---

[7] In most experiments of this type, in which the transmitted neutron beam is observed some distance behind a target, the total cross section actually measured is $2\pi R^2$ because it includes, in addition to the cross section for absorption, that for "shadow scattering," which is also πR^2.   This phenomenon, the scattering of radiation of wavelength λ by a black object of radius R, which causes a shadow to extend only a finite distance behind the object, is a general result of wave optics, but it is not observed with light waves incident on macroscopic objects because the scattering is confined to angles $\lesssim \lambda/2\pi R$.   For a further discussion of total cross sections, cf. chapter 10, section E.

obtained.   Since nuclear dimensions are of the order of 10^{-13} cm, it has been found convenient to define a new unit of length, the Fermi (F): 1 F $= 10^{-13}$ cm.

The emission of α particles from heavy nuclei involves a transition from a state in which the α particle is inside the nuclear potential to one in which it is outside the range of nuclear forces.   The probability for this process and therefore the lifetime for α decay depends very strongly on the height of the potential barrier which the α particle must penetrate and therefore on the nuclear radius R beyond which the repulsive Coulomb potential is not compensated by the attractive nuclear potential.   The quantum-mechanical theory of α decay accounts very well for the relation between lifetimes for α decay and α-particle energies (see chapter 8, section A) and allows nuclear radii to be deduced from the experimental half-lives and decay energies.   Values between 8.4 and 9.8 $\times$ 10^{-13} cm are obtained in this way for the radii of α-emitting nuclei with $A > 208$.   The penetration of potential barriers by charged particles, although in the opposite direction from that encountered in α decay, also governs the probability of reactions between nuclei and charged particles; from the measured probabilities for such reactions, radii of nuclei over the entire range of A values can therefore be deduced via the theory of barrier penetration.

All the methods mentioned lead to the determination of the distance from the nuclear center within which nuclear forces act.   The results can be represented approximately by the empirical formula

$$R = r_0 A^{\frac{1}{3}}, \tag{2-2}$$

where r_0 is a constant independent of A.   The values of r_0 obtained differ slightly for the different methods, clustering around 1.4 F for the fast-neutron data, around 1.3 F for α-decay measurements, and tending to be somewhat higher ($\cong$1.6 F) and not quite so independent of A for charged-particle reaction results.   The important conclusion is that nuclear volumes are very nearly proportional to nuclear masses or, in other words, that all nuclei have approximately the same density.

A method for the determination of nuclear radii which differs basically from the others mentioned is that of electron scattering.   Electrons are not appreciably affected by the specific nuclear forces, and therefore the scattering of electrons by a nucleus is sensitive only to the extent and distribution of electric charge in the nucleus.   Scattering data obtained with electrons of moderate energies ($<$100 MeV) are compatible with nuclei being spheres of uniformly distributed charges but with radii distinctly smaller than indicated by the other methods mentioned.   Equation 2-2 is not quite an adequate representation of these electron-scattering results, since they indicate r_0 values varying from about 1.4 F for light nuclei to about 1.2 F for heavy ones.

As the electron energy is increased above $\cong$100 MeV, the wavelength becomes so short that one can hope to see some details of the charge distribution in nuclei.   Experiments with such high-energy electrons, particularly by the Stanford group, have led to the definite conclusion that nuclei do not

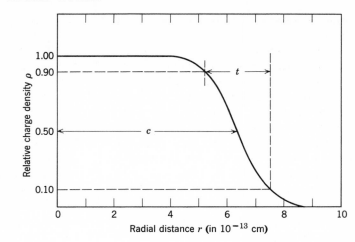

Fig. 2-3   Typical charge distribution in a nucleus, as determined by electron scattering experiments.   The half-density radius c and the skin thickness t are indicated.   The particular distribution shown is that of the gold nucleus.   (Data from R. Hofstadter, ref. H1.)

possess a uniform charge distribution out to a sharp boundary (H1, E2). Rather the charge distribution found, which is shown in figure 2-3, has a central region of approximately uniform density and a "skin" with tapering density.   The charge distributions can be characterized by two parameters, the half-density radius c (distance from the center to where the charge density has reached one half the central value) and the skin thickness t (radial distance between the 90 and 10% density points).   For all nuclei with $Z > 10$, the value of t was found to be constant at $\cong 2.4$ F, whereas $c = 1.07 \times A^{1/3}$ F. Even for the lightest nuclei the skin thickness stays nearly the same, so that they have almost no constant-density core.   Charge density distributions for some representative nuclei are shown in figure 2-4.

The distribution of total mass in a nucleus is not so readily measured as the distribution of charge.   Available experimental evidence favors the view that neutrons and protons are distributed in approximately the same manner, and the charge density distributions deduced from the electron-scattering data are therefore commonly used to represent mass density distributions also.   The result that the radius of the charge (and mass) distribution in a nucleus turns out to be smaller than the radius of action of the nuclear forces is not unexpected in view of the finite range ($\cong 2 \times 10^{-13}$ cm) of the nuclear forces.

In all that has been said it was implicit that nuclei are spherical.   This is essentially true for most nuclei.   However, as we shall see in the next section, definite deviations from the spherical shape have been established for some types of nuclei.

Spin and Nuclear Moments (R3).   Since both neutron and proton have an intrinsic angular momentum of $\frac{1}{2}(h/2\pi)$ and since the neutrons and protons in a nucleus may also move in orbits with respect to one another, it is not unexpected that nuclei possess angular momenta.   The intrinsic angular momentum of a nucleus is always expressed as an integral or half-integral multiple, I, of $h/2\pi$.   This quantity, I, is referred to as the spin of the nucleus. According to quantum mechanics, angular momentum associated with the orbital motion of the nucleons can only be an integral multiple of $h/2\pi$.   Since each nucleon can only add or subtract its intrinsic spin $\frac{1}{2}$, it follows that the spin of any nucleus of even A must be zero or integral and that of any odd-A nucleus must be half-integral.   All spin measurements have confirmed this

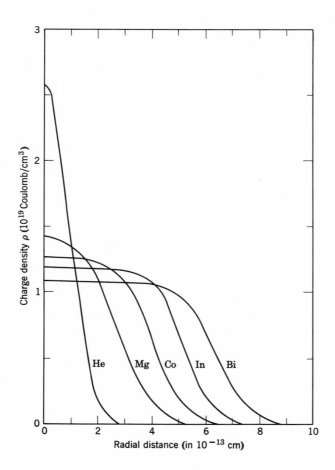

Fig. 2-4   Nuclear-charge distributions for a number of elements, as determined by electron scattering.   (From R. Hofstadter, ref. H1.)

rule; furthermore, it appears that all nuclei of even A and even Z have $I = 0$ in their normal or ground states (cf. appendix E).

The observation that all odd-A nuclei have half-integral and all even-A nuclei have integral spins was one of the major obstacles to the proton-electron hypothesis of nuclear structure.   On this hypothesis, for example, a nucleus of odd Z and even A (such as N^{14}) would have an odd number of constituent particles, all of intrinsic spin $\frac{1}{2}$, and would thus have to have half-integral spin; but the measured spin of N^{14} is 1.

Since the rotation of a charged particle produces a magnetic moment, nuclei with nonzero angular momenta have magnetic moments.   The prediction of Dirac's theory for the magnetic moment of an electron (charge e, mass M_e), namely $eh/4\pi M_e c = 0.927 \times 10^{-20}$ erg/gauss $\equiv 1$ Bohr magneton, agrees so well with the experimentally determined value that similar success might be expected in the case of the proton (charge e, mass $M_p = 1836 M_e$).   However, the magnetic moment of the proton is not equal to $1/1836$ Bohr magneton, but about 2.79 times this value.   Nevertheless, $1/1836$ Bohr magneton is used as the unit of nuclear magnetic moments and called a nuclear magneton ($\cong 5.05 \times 10^{-24}$ erg/gauss).

The fact that the proton has a magnetic moment very different from that expected from the theory for a simple structureless charged particle indicates that the proton is, in fact, not such a simple entity.   Perhaps even more startling is the observation that the neutron has a magnetic moment of -1.91 nuclear magnetons (the negative sign indicates that spin and magnetic moment are in opposite directions).   This magnetic moment must result from a distribution of charges in the neutron, with negative charge (perhaps due to negative mesons) concentrated near the periphery and overbalancing the effect of an equal positive charge nearer the center.

In general, the magnetic moments of nuclei differ from values calculated by any simple theory.   Magnetic moments are often expressed in terms of gyromagnetic ratios (nuclear g factors); the magnetic moment is then $g \cdot I$ nuclear magnetons, with g positive or negative, depending on whether spin and magnetic moment are in the same or opposite directions.

Nuclear spins and magnetic moments can sometimes be determined from hyperfine structure in atomic spectra.   Hyperfine structure derives from the fact that the energy of an atom is slightly different for different (quantized) orientations between nuclear spin and angular momentum of the electrons because of the interaction between the nuclear magnetic moment and the magnetic field of the electrons.   From the number of lines in a spectroscopic "hypermultiplet," the nuclear spin I can be determined under suitable conditions.   Many nuclear spins have been measured by this method.

Although the number of components in a hypermultiplet is determined by the nuclear spin, the amount of the splitting (which is usually of the order of 1 Å or less) depends on the value of the nuclear magnetic moment.   This dependence is understood well enough to allow a calculation of the magnetic moment from the magnitude of the splitting, provided that the nuclear spin is known.   Thus the nuclear magnetic moments of Bi^{209} and Na^{23} have been determined by this method to be $+4.080$ and $+2.217$ nuclear magnetons, respectively.

A second method for the determination of nuclear magnetic moments (and for the measurement of nuclear spin) is the atomic-beam method of I. I. Rabi and co-workers.   In this method (which is an extension of the Stern-Gerlach experiment for the determination of magnetic moments of atoms) a beam of atoms is sent through an inhomogeneous magnetic field.   The nuclear spin I, uncoupled from the electron angular momentum J by the external field, orients itself with respect to the field.   This orientation is governed by the usual quantum conditions, and the beam is therefore split into $2I + 1$ components whose separations are dependent on the nuclear magnetic moment.   The energies of these splittings may be found in terms of characteristic alternating magnetic-field frequencies which induce transitions between components. Various modifications of this method, notably the addition of focusing devices and the adaptation to molecular rather than atomic beams, have greatly improved the accuracy of the results obtainable.   The magnetic moment of the neutron was directly determined by a suitable (and rather drastic) modification of this principle.

A more recent technique for the study of nuclear spins, and especially of magnetic moments, uses nuclear resonance absorption.   The magnetic dipoles of nuclei of spin I can align themselves with a strong external magnetic field in $2I + 1$ different orientations.   The energy differences between the resulting $2I + 1$ energy states correspond to the radio-frequency region, and their magnitude depends on the gyromagnetic ratio.   Resonance absorption of radio-frequency radiation will, therefore, take place at a frequency corresponding to these transitions; the resonance frequency is a measure of the gyromagnetic ratio and, if I is known, of the magnetic moment.   In some cases I can be determined separately because the intensity of absorption is a function of I (and of other factors).

A very powerful tool for nuclear spin determinations is microwave spectroscopy.   Many transitions, particularly between rotational states of molecules, occur in the microwave region (wavelengths of the order of 1 cm).   The hyperfine structure of such transitions is quite analogous to that in atomic spectra and has been used for a number of nuclear spin determinations.   Still another type of hyperfine structure measurement which has found some application in this field uses the paramagnetic resonance method.   What is observed here is the resonance absorption frequency for a paramagnetic substance in a radio-frequency field and the splitting of this frequency caused by the interaction between the nuclear spin and the electronic angular momentum of the molecule or ion.

Information on spins of radioactive nuclei has been inferred from detailed studies of β- and γ-decay processes.   This subject is discussed in chapter 8.

In addition to its magnetic dipole moment a nucleus may have an electric quadrupole moment. This property may be thought of as arising from an elliptic charge distribution in the nucleus. The quadrupole moment q is given by the equation $q = \frac{2}{5}(a^2 - b^2)$, where a is the semiaxis of rotation of the ellipsoid and b is the semiaxis perpendicular to a; q has the dimensions of area. For the deuteron $q = +2.74 \times 10^{-27}$ cm^2, and the charge distribution is cigar-shaped. Quadrupole moments, including both positive and negative values, have been determined for quite a number of nuclei with $I > \frac{1}{2}$. (Nuclei with $I = 0$ or $I = \frac{1}{2}$ cannot have quadrupole moments.) The interactions of nuclear quadrupole moments with the electric fields produced by electrons in atoms and molecules give rise to abnormal hyperfine splittings in spectra, and the methods for quadrupole-moment measurements are therefore the ones already discussed: optical spectroscopy, microwave spectroscopy, nuclear resonance absorption, and some modified molecular-beam techniques.

Statistics (R3). This is a quantum-mechanical property of particles which becomes important when large numbers of them occur together in a system. For detailed discussions of the concept the reader is referred to other works (B1, B2). Here we merely indicate the nature of this property and give some useful results.

All nuclei and elementary particles are known to obey one of two kinds of statistics: Bose-Einstein or Fermi-Dirac. If all the coordinates describing a particle in a system (including three space coordinates and the spin) are interchanged with those describing another identical particle in the system, the absolute magnitude of the wave function representing the system must remain the same; but the wave function may or may not change sign. If it does not change sign (the wave function is then called symmetrical), Bose statistics applies. If the sign of the wave function does change sign with the interchange of coordinates (antisymmetrical wave function), the particles obey Fermi statistics. In Fermi statistics each completely specified quantum state can be occupied by only one particle; that is, the Pauli exclusion principle applies to all particles obeying Fermi statistics. For particles obeying Bose statistics no such restriction exists. Protons, neutrons, electrons (and some other elementary particles such as positrons, neutrinos, and some types of mesons) all obey Fermi statistics. A nucleus will obey Bose or Fermi statistics, depending on whether it contains an even or odd number of nucleons.

The statistics of nuclei can be deduced from the alternating intensities in rotational bands of the spectra of diatomic homonuclear molecules. With Bose statistics the even-rotational states and with Fermi statistics the odd-rotational states are more populated. This can be illustrated by the rotational spectra of hydrogen and deuterium. In normal hydrogen, H$_2$, the ratio of the populations in the states of odd- and even-rotational quantum numbers is 3:1 corresponding to spin $\frac{1}{2}$ and Fermi statistics; in deuterium, D$_2$, the ratio is 1:2 corresponding to spin 1 and Bose statistics.

Table 2-1   Properties of some elementary particles

Symbol	Name	Charge[1]	Rest Mass[2]	Spin[3]	Magnetic Moment[4]	Statistics[5]
e^- or β^-	electron	-1	0.0005486	$\frac{1}{2}$	-1836	F
e^+ or β^+	positron	$+1$	0.0005486	$\frac{1}{2}$	$+1836$	F
γ	photon	0	0	1	0	B
ν	neutrino	0	$<2 \times 10^{-7}$	$\frac{1}{2}$	<0.3	F
n	neutron	0	1.008665	$\frac{1}{2}$	-1.913	F
μ	mu-meson	$+1, -1$	0.113	$\frac{1}{2}$		F
$\pi^\pm$	pi-meson	$+1, -1$	0.150	0		B
π^0	pi-meson	0	0.145	0		B
H^1 or p	proton	$+1$	1.007825	$\frac{1}{2}$	$+2.793$	F

[1] In units of $e = 4.8030 \times 10^{-10}$ esu.
[2] For the proton the mass of the neutral atom is listed.   The unit is the atomic mass unit ($C^{12} = 12,00000$).
[3] In units of $h/2\pi$.
[4] In units of the nuclear magneton ($eh/4\pi Mc$), where M is the proton mass. Positive values indicate moment orientations with respect to spin orientations that would result from spinning positive charges.
[5] F means Fermi and B mean Bose statistics.

The determination of the statistics of nuclei provides an important test of nuclear models.   The demonstration in 1929 that N^{14} nuclei obey Bose statistics was one of the first clear-cut contradictions to the proton-electron model. According to that model, the N^{14} nucleus would consist of 14 protons and 7 electrons, a total of 21 Fermi particles, and it should therefore obey Fermi statistics.   On the proton-neutron model this nucleus is composed of 14 Fermi particles, which is consistent with the Bose statistics.   All the information on statistics of nuclei is compatible with the proton-neutron rather than the proton-electron model.

Parity.   Another nuclear property connected with symmetry properties of wave functions is parity.   A system is said to have odd or even parity according to whether or not the wave function for the system changes sign when the signs of all the space coordinates are changed.   We shall make some use of the concept of parity in our discussions of nuclear reactions and radioactive decay processes because the parity of an isolated system, like its total energy, momentum, angular momentum, and statistics, is conserved.[8]   We shall

[8] As postulated in 1956 by T. D. Lee and C. N. Yang and subsequently verified by many experiments, parity is *not* conserved in the so-called weak interactions (e.g., β decay). Although this discovery was very important in the development of modern physics, it is of little consequence for most of the discussions of nuclear phenomena in this book (see chapter 8, section C).

require merely the very simple rules of combination for parity. Two particles in states of even parity or two particles in states of odd parity can combine to form a state of even parity only. A particle of even parity and one of odd parity result in a system of odd parity. We may illustrate this by an example from atomic spectroscopy: allowed transitions in atoms occur only between an atomic state of even and one of odd parity, not between two even or two odd states, because the quanta of ordinary dipole radiation are characterized by odd parity.

In discussing nuclear energy states we shall make use of the fact that parity is connected with the angular-momentum quantum number l. States with even l (s, d, g . . . states) have even parity, those with odd l (p, f, h . . . states) have odd parity.

Now that we have briefly discussed the principal properties by which nuclei are characterized, we list in table 2-1 some of the values of these properties for elementary particles.

D. NUCLEAR SYSTEMATICS

Binding Energies. We have seen in preceding sections that both the volumes and the total binding energies of nuclei are very nearly proportional to the numbers of nucleons present. From the first of these observations we can conclude that nuclear matter is quite incompressible, from the second that the nuclear forces must have a saturation character; that is, a nucleon in a nucleus can apparently interact with only a small number of other nucleons, just as an atom in a liquid or solid is strongly bound to only a small number of neighboring atoms. These characteristics of nuclei suggest a similarity with drops of liquid and have prompted attempts to account for the binding energies of nuclei in terms of a model in which nuclei are considered as liquid drops. By a semiempirical approach it has thus been possible to arrive at a very useful equation[9] relating the binding energy (or total mass) of any nucleus to its composition (Z and A). Depending on the manner in which they fitted experimental data, different authors have given somewhat different coefficients for the various terms in the equation, but the terms are always essentially the same in their functional form and we shall discuss how they come about. The following form of the binding-energy equation gives agreement within about 1 per cent with measured values for $A > 40$:

$$E_B = 14.0A - 13.1A^{2/3} - 0.585Z(Z - 1)A^{-1/3} - 18.1(A - 2Z)^2 A^{-1} + \delta A^{-1},$$
$$(2\text{-}3)$$

where the binding energy E_B (the energy required to dissociate the nucleus into its constituent nucleons) is expressed in million electron volts (MeV).

[9] An equation of this form was first derived by C. F. von Weizsäcker (W1). A critical review of the semiempirical mass equation and of the values of the coefficients was given by E. Feenberg (F1). For discussions of additional refinements see C2 and G1.

The first and dominant term in (2-3) expresses the fact, already discussed, that the binding energy is nearly proportional to the number of nucleons.   It is a direct consequence of the short range and saturation character of the nuclear forces.   The saturation of these forces is almost (but certainly not entirely) complete when four nucleons, two protons and two neutrons, interact, as is indicated by the large observed binding energies of He^4, C^{12}, O^{16} (see figure 2-1).

The nucleons at the surface of a nucleus can be expected to have unsaturated forces, and consequently a reduction in the binding energy proportional to the nuclear surface should be taken into account.   This effect gives rise to the second (negative) term; it contains $A^{2/3}$ which is a measure of the surface (since A is proportional to the volume).   With increasing nuclear size, the surface-to-volume ratio decreases, and therefore this term becomes relatively less important.

The Coulombic repulsive force between protons is, of course, not of the saturation type and is of sufficient range to be effective for all the protons in a nucleus.   Therefore each of the Z protons interacts with the other $Z - 1$ protons to reduce the binding energy as shown in the third term.   The average separation distance between protons is proportional to the nuclear radius, hence the factor $A^{-1/3}$ appears in the Coulomb energy term.   This term becomes more and more important as protons are added and accounts for the fact that all stable nuclei with $Z > 20$ contain more neutrons than protons (see figure 2-5).

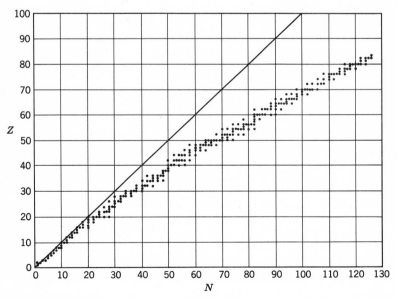

Fig. 2-5   The known stable nuclei on a plot of Z versus N.   Note the gradual increase in the neutron-proton ratio; the 45° line corresponds to a neutron-proton ratio of unity.

As shown graphically in figure 2-5, the stablest nuclear species among the light elements tend to have equal numbers of neutrons and protons.[10] [The deviation from this trend among heavier nuclei can be completely accounted for by the Coulomb effect just discussed.]   Thus the binding-energy equation must contain a negative term whose magnitude increases with increasing value of $|N - Z|$ (or $|A - 2Z|$).   Furthermore, the binding energies of unstable light nuclei on both sides of the stable region indicate that the effect is symmetrical around $N = Z$.   The simplest functional form expressing these empirical facts is the one chosen for the fourth or symmetry term in (2-3), namely $(A - 2Z)^2$. The A^{-1} dependence comes about because the binding-energy contribution per neutron-proton pair is proportional to the probability of having such a pair within a certain volume (determined by the range of nuclear forces), and this probability in turn is inversely proportional to the nuclear volume.   The fact that neutron-proton pairs appear to contribute more to nuclear binding than do pairs of like nucleons is discussed in terms of the character of nuclear forces in chapter 9, section A.

The final term in (2-3) is a quantitive expression of the fact (already noted on p. 28) that binding energies for a given A depend somewhat on whether N and Z are even or odd.   So-called even-even nuclei (Z and N even) are the stablest, and for them δ in (2-3) may be taken as $+132$; for even-odd (Z even, N odd) and odd-even (Z odd, N even) nuclei, $\delta = 0$; for odd-odd nuclei, $\delta = -132$.[11]   The difference in the stabilities of these four types of nuclei is manifested in the distribution of the known stable nuclides among them: 163 even-even, 55 even-odd, 50 odd-even, and 4 odd-odd.   The striking preponderance of even-even nuclei and complete absence of odd-odd nuclei outside the region of the lightest elements (the four odd-odd nuclei are $_1H^2$, $_3Li^6$, $_5B^{10}$, and $_7N^{14}$) can be explained in terms of a tendency of two like particles to complete an energy level by pairing opposite spins.   The δ term in the binding-energy equation is often called the pairing term.[12]

The greater stability of nuclei with filled energy states is apparent not only in the larger number of even-even nuclei but also in their greater abundance

---

[10] This effect becomes even more pronounced when the natural abundances of the nuclides are taken into account.

[11] The quantity δ is actually not a constant but varies rather irregularly with A.   The value given is an average for $A > 80$.   For $A < 60$ a lower value, perhaps $\delta \sim \pm 65$, should be used, but in that region (2-3) does not give very reliable results anyway.   There is evidence that in some regions of A and Z, δ shows a slight systematic difference in odd-A nuclei depending on whether the odd nucleon is a proton or a neutron (cf. C2).   We disregard this small effect.

[12] The pairing energy for a neutron-proton pair in the same energy state is actually larger than that for a pair of like nucleons because of the spin-dependent character of nuclear forces (see chapter 9, section A): these forces are stronger between two nucleons of parallel spin than between two nucleons of opposite spin, and the Pauli principle prevents two like

relative to the other types of nuclei.   On the average, elements of even Z are much more abundant than those of odd Z (by a factor of about 10).   For elements of even Z the isotopes of even mass (even N) account in general for about 70 to 100 per cent of the element (beryllium, xenon, and dysprosium being exceptions).   The general shape of the binding-energy curve (figure 2-1) with the maximum at $A \cong 60$ comes about through the opposing trends with mass number of the relative contributions of surface energy (decreasing with A) and Coulomb and symmetry energies (increasing with A).

Nuclear Energy Surface.   The binding energies of all nuclei can be represented as a function of A and Z by means of a three-dimensional plot of an equation such as (2-3).   Without attempting to construct this nuclear energy surface in three dimensions, we can obtain useful information about some of its features.   For this purpose it is more convenient to consider the total atomic mass M rather than the binding energy E_B.   According to the definition of binding energy, we can write

$$M = ZM_H + (A - Z)M_N - E_B, \qquad (2\text{-}4)$$

where M_H and M_N are the masses of the hydrogen atom (938.77 MeV) and the neutron (939.55 MeV), respectively.   By combining (2-3) and (2-4) we obtain the semiempirical mass equation:

$$M = 925.55A - 0.78Z + 13.1A^{2/3} + 0.585Z(Z - 1)A^{-1/3}$$
$$+ 18.1(A - 2Z)^2 A^{-1} - \delta A^{-1}. \quad (2\text{-}5)$$

Equation 2-5 is quadratic in Z and can be rewritten as follows:

$$M = aZ^2 + bZ + c - \delta A^{-1}, \qquad (2\text{-}6)$$

where $a = 0.585A^{-1/3} + 72.4A^{-1}$,
$\quad b = -0.585A^{-1/3} - 73.18$,
$\quad c = 943.65A + 13.1A^{2/3}$.

Thus we see that for constant A the coefficients a, b, and c are constants, and (2-6) therefore represents a parabola for a given value of δ.   A section through

---

nucleons with parallel spins from being in the same energy state.   It is this large neutron-proton pairing energy that stabilizes the odd-odd nuclei H^2, Li^6, B^{10}, and N^{14} relative to their even-even isobars, the di-neutron, He^6, Be^{10}, and C^{14}.   With increasing Z the Coulomb effect keeps increasing and soon prevents the most loosely bound protons from occupying the same energy state as the most loosely bound neutrons; thus the neutron-proton pairing energy is no longer observable in heavier nuclei, whereas the pairing of like nucleons (expressed by the δ term in the binding energy equation) is manifest throughout the table of nuclides.   (Cf. ref. B1, pp. 211–225.)

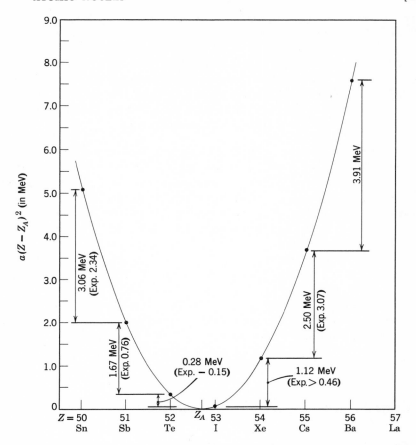

Fig. 2-6   Mass parabola for $A = 125$ as calculated from (2-8) with the constants given in the text.   The calculated mass differences (β-decay energies) between neighboring isobars are indicated, with experimentally determined values shown in parentheses for comparison.

the nuclear energy surface at any odd value of A ($\delta = 0$) is a single parabola. A section at an even value of A ($\delta = \pm 132$) results in two parabolas, displaced from each other by $2\delta/A$ along the energy axis but otherwise identical.

These mass (or energy) parabolas are useful in β-decay systematics because approximate values of the energy available for β decay between neighboring isobars can be read directly from them.   Parabolas for $A = 125$ and $A = 128$, calculated from (2-6), are shown in figures 2-6 and 2-7.

The vertex of each energy parabola gives, for the given value of A, the minimum mass or maximum binding energy.   To find the nuclear charge Z_A corresponding to this point, we differentiate (2-6) with respect to Z, consider-

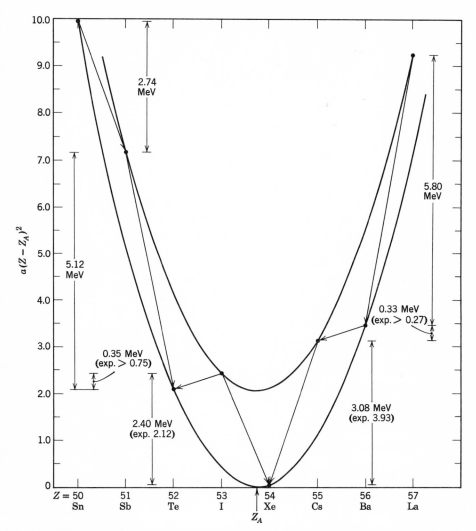

Fig. 2-7   Mass parabolas for $A = 128$ as calculated from (2-8) with the constants given in the text.   Calculated mass differences (β-decay energies) between neighboring isobars are shown, with experimental values given in parentheses.

ing A constant, and set the derivative equal to zero:

$$\frac{\partial M}{\partial Z} = 2aZ_A + b = 0,$$

$$Z_A = -\frac{b}{2a} = \frac{0.585A^{\frac{2}{3}} + 73.18A}{2(0.585A^{\frac{2}{3}} + 72.4)}. \qquad (2\text{-}7)$$

Since we have treated Z as a continuous function, we must, of course, expect to find nonintegral values for Z_A. For $A = 125$, for example, we get $Z_A = 52.7$; for $A = 128$, $Z_A = 53.7$.

For the purpose of plotting energy parabolas we can now use the expression (2-7) for Z_A to write (2-6) in a more convenient form often found in the literature:

$$M = a(Z - Z_A)^2 - \delta A^{-1} + \left(c - \frac{b^2}{4a}\right). \qquad (2\text{-}8)$$

The last term, $c - b^2/4a$, is a function of A only and does not usually need to be evaluated. In figures 2-6 and 2-7, for example, only $a(Z - Z_A)^2 - \delta A^{-1}$ is plotted as the ordinate; that is, the zero of the ordinate scale is the atomic mass corresponding to Z_A (for even A, the mass of the even-even nucleus occurring at Z_A is used as the zero mass).

The widths of the energy parabolas are determined by the values of the constant a, which decrease with increasing A. It is thus evident that the "stability valley" in the nuclear energy surface broadens with increasing nuclear size. The position of the valley floor (Z_A) and the width of the valley in various mass regions are, in fact, among the important input data used to adjust the coefficients in (2-3) and (2-5).

By considering the parabolic curves for a set of isobars we can draw several important conclusions about nuclear stability. For example, it is immediately clear that for any given odd A there can be only one β-stable nuclide,[13] that nearest the minimum of the parabola. For even A there are usually two and sometimes three possible β-stable isobars, all of the even-even type. In figure 2-7 both Te^{128} and Xe^{128} are stable. Strictly speaking, one of these, the figure indicates Te^{128}, is not really stable because it has a lower binding energy than the other (Xe^{128}). However, the transition requires a so-called double β-decay process (involving simultaneous emission of two β particles or simultaneous capture of two electrons); such processes are expected to have exceedingly long half-lives, and none has been established with certainty (see D1, p. 619).

In figures 2-6 and 2-7 the experimentally determined energy differences between neighboring isobars have been included for comparison with those obtained from the binding-energy equation. The agreement is seen to be within 1 MeV in the particular mass region considered. Considerably better agreement may be obtained (C2, G1) by adjustment of a and Z_A to fit known points in the particular region of Z and A. For example, according to our equations, Z_A for $A = 125$ is 52.7, and consequently I^{125} would be expected to be the stable nuclide of mass 125. Actually Te^{125} is stable and I^{125} is unstable by about 0.13 MeV with respect to electron capture. To obtain agreement

[13] An apparent exception to this rule still exists at $A = 113$, with Cd^{113} and In^{113}. One of the members of this pair of seemingly stable isobars is undoubtedly radioactive but with a very long half-life. The same is presumably true for the odd-odd nuclide Ta^{180}, which occurs in nature in small abundance.

with this experimental fact, we could shift the parabola empirically by 0.3 unit along the Z axis, using $Z_A = 52.4$; this would improve the agreement with the other experimentally known decay energies and would allow a more reliable prediction of the still unmeasured decay energies of Ba^{125} and La^{125}. The maximum error in Z_A values as determined from our equations is approximately one unit.

Nuclear Shell Structure. In the preceding discussion we have treated nuclei essentially as statistical assemblies of neutrons and protons. This model, in which a nucleus is considered analogous to a liquid drop, is successful in accounting for many properties of nuclei. However, there is also strong experimental evidence for some sort of shell structure in nuclei analogous to the electron-shell structure in atoms, although not nearly so pronounced. That certain numbers of neutrons and protons seem to form particularly stable configurations was pointed out by W. Elsasser in 1934. However, except for attempts to account for the well-known special stability of light nuclei with N and Z values of 2, 8, and 20 ($_2He^4$, $_8O^{16}$, and $_{20}Ca^{40}$), the subject of nuclear shell structure received little attention until 1948 when M. G. Mayer pointed out the strong evidence for the additional "magic numbers" 50 and 82 for protons and 50, 82, and 126 for neutrons (M2). Since then 28 has been recognized as an additional magic number. We shall briefly summarize some of the evidence.

Above $Z = 28$, the only nuclides of even Z which have isotopic abundances exceeding 60 per cent are Sr^{88} (with $N = 50$), Ba^{138} ($N = 82$), and Ce^{140} ($N = 82$). No more than five isotones occur in nature for any N except $N = 50$, where there are six, and $N = 82$, where there are seven. Similarly, the largest number of stable isotopes (10) occurs at tin, $Z = 50$, and both in calcium ($Z = 20$) and tin the stable isotopes span an unusually large mass range. The fact that all the heavy natural radioactive chains end in lead ($Z = 82$) is significant, as is the neutron number 126 of the two heaviest stable nuclides, Pb^{208} and Bi^{209}.

The particularly weak binding of the first nucleon outside a closed shell (analogous to the low ionization potential for the valence electron in an alkali atom) is shown by the unusually low probabilities for the capture of neutrons by nuclides having $N = 50$, 80, and 126. Also, in the nuclei Kr^{87} ($N = 51$) and Xe^{137} ($N = 83$) one neutron is bound so loosely that it is emitted spontaneously when these nuclei are formed by β decay from Br^{87} and I^{137}, respectively.

The major discrepancies between the semiempirical binding-energy equation and experimental decay energy and mass data occur in the regions of the magic numbers. Attempts to fit the isobaric parabolas to experimental β-decay data in these regions fail unless discontinuities of as much as 2 MeV in neutron and proton binding energies at the shell edges are taken into account (C2). Similarly, careful mass measurements for nuclides in the neighborhood of $N = 50$, 82, and 126 and $Z = 28$, 50, and 82 have shown that the first nucleon outside one of these closed shells is always weakly bound (K2); the binding energy curve (figure 2-1) is therefore not really perfectly smooth in these regions but has slight breaks.   A great deal of evidence for the $N = 126$ shell has been accumulated from α-decay systematics.   Alpha-decay energies are rather smooth functions of A for a given Z but show striking discontinuities at $N = 126$.   Finally, there are some interesting correlations between the occurrence of nuclear isomerism and the magic numbers: so-called islands of isomerism occur for N and Z values just below 50, 82, and 126.

In chapter 9 we shall return to a discussion of nuclear shell structure and show how the magic numbers can be interpreted in terms of the filling of proton and neutron energy levels.

E. DECAY OF UNSTABLE NUCLEI

1. Sources of Instability

In discussing the nuclear energy surface we have seen that for each mass number there can be only one nuclide that is stable with respect to its isobars. For even A as many as three even-even isobars may be stable with respect to ordinary β decay into their odd-odd neighbors, and, since double β decay is an exceedingly slow process ($t_{1/2} > 10^{15}$ years) not yet experimentally established, we shall consider such nuclides as β stable.

If we consider all nuclei rather than isobars only, it is clear from the binding-energy curve (figure 2-1) that there is only one nuclide (probably Fe^{56}) that represents the thermodynamically stable equilibrium form of nuclear matter. At absolute zero all other nuclei would eventually (but at negligible rates) transform into Fe^{56}.   As we shall discuss in chapter 15, relatively rapid transformations of lighter nuclei into nuclei in the Fe region are indeed believed to occur under extreme conditions of temperature and pressure in the interiors of certain stars.

We have already mentioned (p. 29) that nuclei with $A \gtrsim 140$ are unstable with respect to α-particle emission.   This comes about because the emission of an α particle lowers the Coulomb energy, the principal negative contribution to the binding energy of heavy nuclei, but changes the nuclear binding very little, since the α particle itself is almost as tightly bound as a heavy nucleus. Alpha radioactivity, both in naturally occurring and artificially produced nuclides, is confined to the mass region above $A \cong 140$, and the observed half-lives for α emission vary from small fractions of a second to about 10^{15} years; the so-called stable nuclides in the high-A region are, strictly speaking, α unstable but have half-lives that are so long that their decay is unobservable. Proton decay does not occur because, although it would result in a reduction in Coulomb energy, it would also lead to substantial reduction in nuclear binding except far on the proton-excess side of β stability; even in that region proton decay is not likely to compete successfully with β^+ emission (cf. chapter 8, section B).

Since the binding energy per nucleon decreases with increasing A above $A \cong 60$, all nuclides with $A \gtrsim 100$ are, in fact, unstable with respect to spontaneous fission (breakup into two fragments of approximately equal masses). Because of the high Coulomb barriers for the emission of fission fragments, measurable rates of spontaneous fission have been observed among the heaviest elements ($A > 230$) only. The measured rates range from about 4×10^{-5} fission g^{-1} sec^{-1} for Th^{232} to about 2×10^{17} fissions g^{-1} sec^{-1} for Fm^{256}. The general trends of spontaneous-fission rates (see chapter 8) indicate that at sufficiently high Z this mode of decay may well become the limiting factor for the synthesis of new elements and new isotopes, although α-decay rates also become very high in the same region.

2. Alpha Decay

The α particles from a given nuclide either all have the same energy or are distributed among a few monoenergetic groups. When a single α-particle energy occurs, for example in the decays of Po^{215} (AcA) and Rn^{222}, the transition evidently takes place to a single energy level (generally the ground state) of the product nucleus. The emission of α particles of several different energies by one nuclide occurs when the product nucleus can be left in different states of excitation which subsequently transform to the ground state by γ emission (see p. 56). Known α-particle energies in radioactive decay range from about 1.5 MeV (Ce^{142}) to about 11.7 MeV (Po^{212m}).

From the total disintegration energy E_d associated with an α emission to the ground state of the product nuclide and from the known atomic mass of He^4 (M_α), the difference ΔM between the masses of emitting and product atoms can be computed: $\Delta M = M_\alpha + E_d/c^2$. Note that E_d exceeds the kinetic energy E_α of the α particle by the recoil energy of the product nucleus which is of the order of 0.1 MeV for heavy α emitters.[14] The disintegration energy E_d, strictly speaking, also includes the binding energy of the atomic electrons in He^4 which is usually not included in the experimental measurements, but this is a negligible quantity.

In discussing nuclear radii (p. 33), we have already mentioned that the quantum-mechanical theory of α decay relates the half-life for α emission to the kinetic energy of the α particle and the height and width of the potential barrier. We defer further discussion of the quantitative treatment until

---

[14] Since, in nonrelativistic mechanics, the momentum p and kinetic energy E of a particle of mass M are related by the equation $p^2 = 2ME$, it follows from conservation of momentum that $M_\alpha E_\alpha = M_{\text{nucleus}} E_{\text{nucleus}}$. For example, for the 6.05-MeV α particles of Bi^{212} the recoil energy is $6.05 \times 4/208 = 0.116$ MeV.

chapter 8.   Here we merely point out that the observed range of α-particle energies (1.5 $-$ 11 MeV) is associated with a range in half-lives extending over a factor of about 10^{27}.   Increasing decay energy, of course, leads to decreasing half-life.

3. Beta Decay

Any radioactive decay process in which the mass number of the nucleus remains unchanged, but the atomic number changes, is classed as a β decay. The only type of β decay occurring among the natural radioactive series is what is now called negatron (β^-) decay; the particles emitted in this decay mode were quickly recognized as electrons, the ultimate units of negative electricity discovered before the discovery of radioactivity.   In β^- decay the atomic number increases by one unit and, on present views, this comes about by the transition of a nucleon from its neutron to its proton state.   Negatron decay occurs on the neutron-excess side of the stability valley.   The β^- decay of a species of atomic number Z to its isobar of atomic number $Z + 1$ is energetically possible when $M_Z - M_{Z+1} > 0$, where M_Z and M_{Z+1} are the atomic masses; the numerical value of this mass difference is the available decay energy.

Positrons.   Positron (β^+) decay, arising from the transition of a proton into a neutron and accompanied by a decrease in Z by one unit, takes place in nuclei that are proton-rich in relation to their most stable isobars.   This mode of decay was discovered some years after the existence of positrons had been postulated by P. A. M. Dirac on purely theoretical grounds.   Dirac had found that his relativistic wave equations for electrons had solutions corresponding to electrons in negative as well as positive energy states, but with the magnitude of the energy always greater than mc^2 (where m is the electron mass). As to the physical meaning of the unobserved negative-energy states of electrons, Dirac suggested that normally all the negative-energy states are filled. The raising of an electron from a negative- to a positive-energy state (by the addition of an amount of energy necessarily greater than $2mc^2$) should then be observable not only in the appearance of an ordinary electron but also in the simultaneous appearance of a "hole" in the infinite "sea" of electrons of negative energy.   This hole would have the properties of a positively charged particle, otherwise identical with an ordinary electron.   The subsequent discovery of positrons, first in cosmic rays and then in radioactive disintegrations, was soon followed by discoveries of the processes of pair production and positron-electron annihilation, which may be regarded as experimental verifications of Dirac's theory.

Pair production is the name for a process which involves the creation of a positron-electron pair by a photon of at least 1.02 MeV ($2mc^2$).   It can be shown that in this process both momentum and energy cannot be conserved in empty space; however, the pair production may take place in the field of a nucleus which can then carry off some momentum and energy.   The cross section for pair production goes up with increasing Z and with increasing photon energy.   Pair production may be thought of as the lifting of an electron from a negative- to a positive-energy state.   The reverse process, the falling of an ordinary electron into a hole in the sea of electrons of negative energy, with the simultaneous emission of the corresponding amount of energy in the form of radiation, is observed in the so-called positron-electron annihilation process. This process accounts for the very short lifetime of positrons; whenever a hole in the sea of electrons is created, it is quickly filled again by an electron.   The energy corresponding to the annihilation of a positron and electron is released either in the form of two γ quanta emitted in nearly opposite directions (to conserve momentum) or, much more rarely, in the form of a single quantum if the electron involved in the annihilation is strongly bound (say, in an inner shell of an atom) so that a nucleus is available to carry off the excess momentum.   The two-quantum annihilation occurs principally with very slow positrons, that is, positrons which have almost come to rest by ionization processes.   It is then accompanied by the emission of two γ quanta, each of energy equal to mc^2 (0.51 MeV); this radiation is often referred to as annihilation radiation.

Electron Capture.   In the sense of the Dirac theory, positron emission is the capture of an electron from the continuum of negative-energy states.   This suggests that a possible alternative way for a nucleus to accomplish the result of decreasing Z by one unit without changing A is by the capture of any convenient electron (of positive energy).   As the K electrons in an atom are, on the average, closest to the nucleus (quantum-mechanically the wave functions of the K electrons have larger amplitudes at the nucleus than those of the L, M, etc., electrons), the capture probability should ordinarily be greatest for a K electron.   In 1938 L. Alvarez reported experimental evidence for this mode of decay.   Since then electron capture has been established as a very common mode of decay among neutron-deficient nuclides.   It is, in fact, the only β-decay mode possible for such a nuclide when the decay energy (the mass difference between decaying and product atoms) is less than $2mc^2$.   Only when the decay energy exceeds this value can positron emission take place,[15] and the higher the decay energy, the more effectively positron emission competes with electron capture (EC).   The EC/β^+ ratio depends on atomic number also,

---

[15] Let the *nuclear* masses of decaying and product nucleus be m_Z and m_{Z-1}, the corresponding atomic masses, M_Z and M_{Z-1}, and the decay energy Q.   Then the energetic condition for β^+ emission (assuming the neutrino rest mass to be zero) is

$$m_Z = m_{Z-1} + m_e + Q \text{ or } M_Z = M_{Z-1} + 2m_e + Q.$$

The corresponding condition for electron capture is

$$m_Z + m_e = m_{Z-1} + Q \text{ or } M_Z = M_{Z-1} + Q.$$

increasing with increasing Z for a given decay energy. Among the heaviest elements β^+ emission is therefore very rare.

Extranuclear Effects of Electron Capture. The electron-capture process is more difficult to observe than positron or negatron emission because it is not accompanied by the emission of detectable nuclear radiation except in cases in which the product nuclei are left in excited states and γ rays are emitted.[16] The most characteristic radiations accompanying electron capture are the X rays emitted as a consequence of the vacancy created in the K (or L, etc.) shell. The atomic rearrangement processes following electron capture can be quite extensive. The initial vacancy is filled most frequently by an electron from the next higher shell.

If an L electron falls into the K shell, the difference between the K- and L-binding energies may be emitted as a characteristic X ray or may be used in an internal photoelectric process in which an additional extranuclear electron from the L, or M, etc., shell is emitted with a kinetic energy equal to the characteristic X-ray energy minus its own binding energy. Such electrons are called Auger electrons. The whole process of readjustment in a heavy atom may involve many X-ray emissions and Auger processes in successively higher shells. The fraction of vacancies in a given shell that is filled with accompanying X-ray emission is called the fluorescence yield, and the fraction that is filled by Auger processes is the Auger yield. The K-shell fluorescence yields are plotted as a function of Z in figure 2-8. The L-shell fluorescence yield varies with Z in a similar manner but is several times smaller than the K yield for a given Z or about the same as the K fluorescence yield for a given electron binding energy. Knowledge of the fluorescence yield is important in the measurement of disintegration rates of electron-capture nuclides, since the radiation most frequently detected are the X rays.

The Neutrino. The β particles[17] from a given radioactive species are emitted with a continuous energy distribution extending from zero up to a maximum value. Beta-ray spectra have been studied in some detail by magnetic-

---

[16] A continuous spectrum of electromagnetic radiation of very low intensity is found to be emitted in electron-capture processes and in fact in all β-decay processes. The quanta of this so-called inner bremsstrahlung have part or all of the energy ordinarily carried away by the neutrino. The total number of quanta per electron-capture disintegration is approximately $7.4 \times 10^{-4} E_0{}^2$, where E_0 is the transition energy (in MeV) of the electron-capture process. When nuclear γ rays are emitted, the inner bremsstrahlung usually escapes detection because of its low intensity. However, for electron-capture transitions not accompanied by γ emission measurement of the upper energy limit of the inner-bremsstrahlung spectrum is a very useful method for the determination of the transition energy.

[17] By β particles we shall mean any electrons, positive or negative, emitted from nuclei. Whenever necessary we shall distinguish between negative β particles or negatrons (β^-) and positive β particles or positrons (β^+). Electrons originating in the extranuclear shells (see later) should not be referred to as β particles; they are often represented by the symbol e^-. In the early literature any electrons emitted in radioactive processes were usually called β particles.

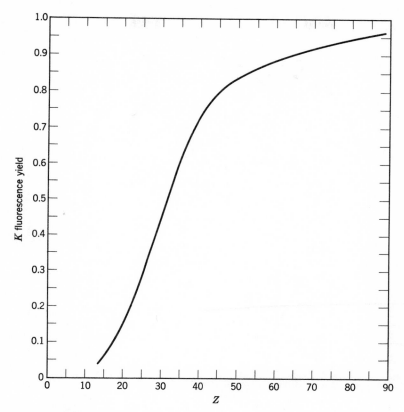

Fig. 2-8   *K*-shell fluorescence yields as a function of atomic number.   [From C. D. Broyles, D. A. Thomas, and S. K. Haynes, *Phys. Rev.* **89**, 715 (1953).]

deflection methods.   Some typical shapes of β spectra are shown in figure 2-9. The average energy is about one third the maximum energy.   A β transition is usually characterized in terms of the upper limit of the β-particle spectrum. Maximum energies ranging approximately from 15 keV to 15 MeV occur among known β emitters.   Since its discovery by Chadwick in 1914, the continuous spectrum of β rays has presented a puzzling problem.   Studies of the α- and γ-ray spectra have revealed that nuclei exist in definite energy states. Yet in every known β-decay process the transition from one such definite energy state to another takes place with the emission of β particles of variable kinetic energy.   It was proved by calorimetric measurements that when all the β particles are absorbed in a calorimeter the measured energy per β particle is the average and not the maximum energy of the β spectrum.   Thus the law of the conservation of energy might appear to be violated in β decay.

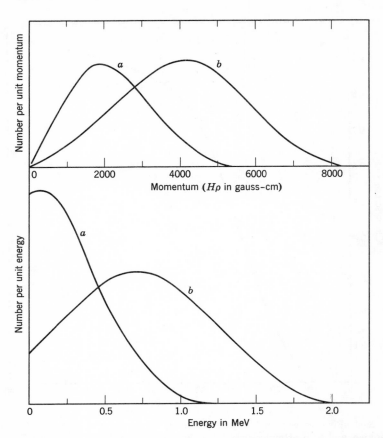

Fig. 2-9 Typical shapes of β-ray spectra; the same data are plotted as a function of β-ray momentum (top) and as a function of β-ray energy (bottom). The momentum is given in terms of the quantity most frequently measured experimentally, the "magnetic rigidity" $H\rho$ where H is magnetic field strength and ρ the radius of curvature of the electron trajectory in that field. Curves a and b represent characteristic negatron and positron spectra, respectively.

Furthermore, the observations show discrepancies with other conservation laws also. As we have seen in section C, all nuclei of even mass number have integral spins and obey Bose statistics; all nuclei of odd mass number have half-integral spins and obey Fermi statistics. Since the mass number remains unchanged in β decay the spins of initial and final nuclei should belong to the same class, either integral or half-integral, and the statistics should remain the same. Yet electrons (and positrons) have one half unit of spin and obey Fermi statistics. Thus angular momentum and statistics appear not to be conserved in β decay.   Finally, experiments in which the recoil momenta of nuclei as well as the corresponding β-particle momenta were measured seem to indicate that conservation of linear momentum also is violated in β decay.

To avoid the necessity of abandoning all these conservation laws for the case of β-decay processes, W. Pauli postulated that in each β disintegration an additional unobserved particle is emitted.   The properties attributed to this hypothetical particle, which has come to be known as the neutrino, are such that the conservation difficulties are eliminated.   The neutrino is supposed to have zero charge, spin $\frac{1}{2}$, and Fermi statistics, and it is thought to carry away the appropriate amount of energy and momentum in each β process to conserve these quantities.   To account for the fact that neutrinos are almost undetectable, it is in addition necessary to assume that they have a very small or zero rest mass and a very small or zero magnetic moment.   By careful measurements of the maximum energy of a β spectrum and determination of the masses of the corresponding β emitter and product atom an upper limit can be obtained for the rest mass of the neutrino.   Best suited for such measurements is the β decay of H^3, and here the most accurate data give an upper limit of about 200 eV (0.0004 times the electron rest mass) for the neutrino rest mass.

In recent years the existence of neutrinos has, in fact, been proved by the observation of their capture by protons to give neutrons and positrons.   This is an example of so-called inverse β processes which take place with extremely small probability and are therefore exceedingly hard to observe.   Yet it has now been possible, by investigation of inverse processes, to establish with certainty that the neutrinos emitted in β^+ decay are *not* identical with those emitted in β^- decay; the latter are called antineutrinos.   This nonidentity of neutrino and antineutrino also follows from the presently accepted theories of β decay that include nonconservation of parity.   However, the properties of neutrinos and antineutrinos are indistinguishable (except in the capture reactions just mentioned), and we shall sometimes use "neutrinos" as a generic term for both.

In electron capture, as in other β-decay processes, conservation of momentum, angular momentum, and statistics require that a neutrino be emitted. However, since the electron is captured from a definite energy state, the neutrinos emitted in this process are presumably monoenergetic.

Half-lives.   Known half-lives for β-decay processes vary over a wide range, from milliseconds to 10^{15} years.   The relation between decay energy and half-life is not nearly so simple as in the case of α decay.   As we shall see in chapter 8, $\dot{\beta}$-decay half-lives depend strongly on spin and parity changes as well as on available energy, and β decays are grouped into allowed, first-forbidden, second-forbidden, etc., transitions, depending on spin and parity changes. Within each of these groups, half-lives and decay energies are related in a relatively simple manner.

4. Gamma Transitions and Isomerism

An α- or β-decay process may leave the product nucleus either in its ground state or, more frequently, in an excited state.

De-excitation Processes. A nucleus in an excited state may give up its excitation energy and return to the ground state in a variety of ways. The most obvious, and the most common, transition is by the emission of electromagnetic radiation. Such radiation is called γ radiation; the γ rays have a frequency determined by their energy $E = h\nu$. Frequently the transition does not proceed directly from an upper state to the ground state but may go in several steps involving intermediate excited states. Gamma rays with energies between about 10 keV and about 7 MeV have been observed in radioactive processes.

Gamma-ray emission may be accompanied, or even replaced, by another process, the emission of internal-conversion electrons. Internal conversion comes about by the (purely electromagnetic) interaction between nucleus and extranuclear electrons which can lead to the emission of an electron with a kinetic energy equal to the difference between the energy of the nuclear transition involved and the binding energy of the electron in the atom.

A third process for the de-excitation of a nucleus is possible if the available energy exceeds 1.02 MeV. This energy is equivalent to the mass of two electrons. It is possible for the excited nucleus to create simultaneously one new electron and one positron and to emit them with kinetic energies that total the excitation energy minus 1.02 MeV. This is an uncommon mode of de-excitation; it is the mode of decay of the first excited state of O^{16}, $E = 6.05$ MeV, $t_{1/2} = 7 \times 10^{-11}$ sec.

All the processes just described we call γ transitions, although only in the first is a γ ray emitted by the nucleus. All are characterized by a change in energy without change in Z and A.

In a number of instances a nuclide in an excited state decays predominantly by α- or β-decay.[18] It is even possible for such a β decay to be followed by another β process leading back to the original nucleus in its ground state. One instance of such a sequence (which we would not like to designate as a γ transition) has been observed: the isomer Sr^{87m} decays partially by electron capture to Rb^{87}, a β^- emitter which decays to Sr^{87}.

---

[18] The so-called long-range α particles of Po^{212} (ThC') and Po^{214} (RaC') arise from the decay of excited states in these nuclei so unstable with respect to α emission that α decay can compete with γ emission. Usually the lifetime of an excited state for γ emission is much shorter than that for β or α decay, although this is, of course, not true for some metastable states (see discussion of isomerism below). As the excitation energy of a nucleus exceeds the binding energy of the most loosely bound nucleon (most often a neutron), emission of this nucleon rapidly becomes a much more probable process than γ emission.

Nuclear Isomerism.   Gamma transitions, like the other spontaneous nuclear changes, proceed in accordance with the exponential decay law, although in most cases the γ decays occur so quickly that the half-lives have not been measured.   Indirect evidence from competition between γ and α decay and from the energy widths of γ-emitting levels indicates that many of the lifetimes are as short as 10^{-13} second.   By modern electronic techniques the exponential decay has been measured for γ transitions with half-lives as short as 10^{-11} second.   At the other extreme, long-lived excited states are known; Ir^{192m_2} has a half-life of about 600 years.

The theoretical expectations and the experimental evidence are in agreement that the γ-transition lifetimes depend on the energy of the transition (E), on the nuclear spin change $(|I_i - I_f| = \Delta I)$, and on the mass number of the nucleus (A).   The total lifetime for de-excitation (by γ emission and internal conversion) increases with decreasing E, increasing ΔI, and decreasing A. The selection rules which govern γ decay and lead to these results are discussed in chapter 8.

Excited states with measurable lifetimes are known as metastable or isomeric states.   The phenomenon of nuclear isomerism was discovered and several pairs of nuclear isomers (including $UZ - UX_2$ and $Br^{80} - Br^{80m}$) were known before the principles governing γ-transition lifetimes were understood.   The γ transitions between isomeric pairs are now known as isomeric transitions.   One member of each pair of isomers must be an excited state of the particular nucleus, and for its lifetime to be measurably long the γ transitions producing de-excitation must be sufficiently restricted by the selection rules.   An important aspect of nuclear shell structure has been the correlation of nuclear spins and isomer lifetimes.

A few remarks may be in order on the usual definition of nuclear isomers, namely, two or more energy levels of a nucleus with half-lives not too short to be measured.   Directly measurable half-lives now extend into the 10^{-11}-second region.   The much shorter, indirectly measured half-lives have been excluded, but borderline measuring techniques, based on decay in flight (for O^{16m}) and on Doppler broadening (for Li^{7m}, $t_{1/2} = 5 \times 10^{-14}$ sec), are tending to make this definition either arbitrary or trivial.   Some workers are now using the terms "long-lived isomers" $(t_{1/2} > \sim 1$ sec) and "short-lived isomers" $(t_{1/2} \ll \sim 1$ sec). (Lifetimes in the millisecond range appear to be extremely rare for γ transitions.)   Chemists with some chemical experiment in mind are likely to think mostly of the "long-lived" isomers.

Internal Conversion.   As mentioned earlier, this is an alternative to γ-ray emission.   The ratio of the rate of the internal conversion process to the rate of γ emission (or the ratio of the number of internal conversion electrons to the number of γ quanta emitted) is known as the internal conversion coefficient α; it may have any value between 0 and ∞.   Separate coefficients for internal conversion in the K, L, M shell, etc. (α_K, α_L, α_M, etc.) and even in the subshells (α_{LI}, α_{LII}, α_{LIII}, etc.) may be measured as well as computed.   In general, the coefficients for any shell increase with decreasing energy, increasing ΔI, and increasing Z.   Accurately computed internal conversion coefficients for various shells are available in tabular form.   Comparison of measured conversion coefficients with these computed values is very helpful for the determination of spin changes in γ transitions.

Internal conversion electrons, examined in an electron spectrograph, show a line spectrum with lines corresponding to the γ-transition energy minus the binding energies of the K, L, M . . . shells in which conversion occurs.   The differences in energy between successive lines serve to identify Z and to classify groups of lines resulting from different γ transitions.   The ratios of the intensities of the lines measure the ratios of the conversion coefficients α_K, α_L, α_M, etc. Experimentally these ratios can be determined more accurately than any individual coefficient.

An internal conversion process leaves the atom with a vacancy in one of its shells.   The subsequent atomic rearrangement processes are essentially the same as those following electron capture (p. 52).

F. NUCLEAR REACTIONS

1. Nature and Energetics of Nuclear Reactions

A nuclear reaction is a process in which a nucleus reacts with another nucleus, an elementary particle, or a photon to produce, in a time of the order of 10^{-12} second or less, one or more other nuclei (and possibly other particles).   Most of the nuclear reactions studied to date are of the type in which a nucleus reacts with a light particle (neutron, proton, deuteron, triton, helium ion, electron, meson, photon), and the products are a nucleus of a different species and again one or more light particles.   The chief exceptions to this description are the fission reaction and the more recently investigated reactions induced by heavy ions, such as lithium, beryllium, boron, and carbon.

Notation.   As an example of a nuclear reaction, we may cite the first such process discovered, the disintegration of nitrogen by α particles.   When Rutherford bombarded nitrogen with α particles from RaC′, he could observe scintillations on a zinc sulfide screen even when enough material to absorb all the α particles was interposed between the nitrogen and the screen.   Further experiments proved the long-range particles causing the scintillations to be protons, and the results were interpreted in terms of a nuclear reaction between nitrogen and α particles to give oxygen and protons, or, in the usual notation,

$$_7\text{N}^{14} + {_2}\text{He}^4 \rightarrow {_8}\text{O}^{17} + {_1}\text{H}^1.$$

Since 1919 thousands of different nuclear reactions have been studied.   The recognition of reaction products is greatly facilitated when they are unstable because characteristic radioactive radiations can then be observed. The positron emitter P^{30} discovered by Joliot and Curie in the α-particle bombardment of aluminum is produced by the reaction

$$_{13}Al^{27} + {}_2He^4 \rightarrow {}_{15}P^{30} + {}_0n^1.$$

The notation used for nuclear reactions is analogous to that in chemical reactions, with the reactants on the left- and the reaction products on the right-hand side of the equation. In all reactions so far observed (except those involving creation or annihilation of antinucleons) the total number of nucleons (total A) is conserved. In addition, other properties such as charge, energy, momentum, angular momentum, statistics, and parity are conserved in nuclear reactions.

A short-hand notation is often used for the representation of nuclear reactions. The light bombarding particle and the light fragments (in that order) are written in parentheses between the initial and final nucleus; in this notation the two reactions mentioned above would read

$$N^{14}(\alpha, p)O^{17} \quad \text{and} \quad Al^{27}(\alpha, n)P^{30}.$$

The symbols n, p, d, α, e, γ, x, π, $\bar{p}$ are used in this notation to represent neutron, proton, deuteron, alpha particle, electron, gamma ray, X ray, pi meson, and antiproton.

Comparison of Nuclear and Chemical Reactions. Nuclear reactions, like chemical reactions, are always accompanied by a release or absorption of energy, and this is expressed by adding the term Q to the right-hand side of the equation. Thus a more complete statement of Rutherford's first transmutation reaction reads

$$_7N^{14} + {}_2He^4 \rightarrow {}_8O^{17} + {}_1H^1 + Q.$$

The quantity Q is called the energy of the reaction or more frequently just "the Q of the reaction." Positive Q corresponds to energy release (exoergic reaction); negative Q to energy absorption (endoergic reaction).

Here an important difference between chemical and nuclear reactions must be pointed out. In treating chemical reactions, we always consider macroscopic amounts of material undergoing reactions, and, consequently, heats of reaction are usually given per mole or occasionally per gram of one of the reactants. In the case of nuclear reactions we usually consider single processes, and the Q values are therefore given per nucleus transformed. If the two are calculated on the same basis, the energy release in a representative nuclear reaction is found to be many orders of magnitude larger than that in any chemical reaction. For example, the reaction $N^{14}(\alpha, p)O^{17}$ has a Q value of -1.19 MeV or $-1.19 \times 1.602 \times 10^{-6}$ erg or $-1.19 \times 1.602 \times 10^{-6} \times 2.390 \times 10^{-8}$ cal $= -4.56 \times 10^{-14}$ cal for a single process. Thus, to convert 1 g atom of N^{14} to O^{17}, the energy required would be $6.02 \times 10^{23} \times 4.56 \times 10^{-14}$ cal $= 2.74 \times 10^{10}$ cal. This is about 10^5 times as large as the largest values observed for heats of chemical reactions. On the other hand, we must keep in mind that nuclear reactions are extremely rare events compared with chemical reactions; one reason is that the small sizes of nuclei make effective nuclear collisions quite improbable.

Q Values.   It is clear from the foregoing discussion that the energy changes involved in nuclear reactions are of such magnitude that the corresponding mass changes must be observable. (The mass changes accompanying chemical reactions are too small to be observable with the most sensitive balances available.) If the masses of all the particles participating in a nuclear reaction are known from mass-spectrographic data, as is the case for the $N^{14}(\alpha, p)O^{17}$ reaction, the Q of the reaction can be calculated. The sum of the N^{14} and He^4 masses is 18.005678 mass units, and the sum of the O^{17} and H^1 masses is 18.006958 mass units; thus an amount of energy equivalent to 0.001280 mass unit has to be supplied to make the reaction energetically possible, or $Q = -0.001280 \times 931.5$ MeV $= -1.192$ MeV. When the Q value is known experimentally (from the kinetic energies of the bombarding particle and the reaction products), it is sometimes possible to compute the unknown mass of one of the participating nuclei. By this method the masses of a number of radioactive nuclei have been determined (see exercise 17).

It is often possible to calculate the Q value of a reaction even if the masses of the nuclei involved are not known, provided that the product nucleus is radioactive and decays back to the initial nucleus with known decay energy. Consider, for example, the reaction $Pd^{106}(n, p)Rh^{106}$. The product Rh^{106} decays with a 30-second half-life and the emission of 3.54-MeV β particles to the ground state of Pd^{106}. We can write this sequence of events as follows:

$$_{46}Pd^{106} + {}_0n^1 \rightarrow {}_{45}Rh^{106} + {}_1H^1 + Q;$$

$$_{45}Rh^{106} \rightarrow {}_{46}Pd^{106} + \beta^- + \bar{\nu} + 3.54 \text{ MeV}.$$

Adding the two equations, we see that the net change is just the transformation of a neutron into a proton, an electron, and an antineutrino, with accompanying energy change; or, symbolically,

$$_0n^1 \rightarrow {}_1H^1 + \beta^- + \bar{\nu} + Q + 3.54 \text{ MeV}.$$

Note that the symbol $_1H^1$ must here stand for a bare proton (evident from the charge conservation), whereas the listed "proton mass" includes the mass of one orbital electron.[19]   For energy balance we therefore write

$$M_n = M_{H^1} + Q + 3.54 \text{ MeV},$$

---

[19] In general, for negative-β-particle emission and electron-capture processes, the masses of electrons never have to be included in calculations when atomic masses are used. However, whenever emission of a positron is involved, two electron masses have to be taken into account: one for the positron and one for the extra electron that has to leave the electron shells to preserve electrical neutrality.

where $M_n = 1.008665$ and $M_{H^1} = 1.007825$ mass units. Then
$$Q = (1.008665 - 1.007825) \times 931.5 - 3.54 = -2.76 \text{ MeV.}$$

In the first example calculated we found the Q value of the reaction $N^{14}(\alpha, p)O^{17}$ to be -1.19 MeV. Does that mean that this reaction can actually be produced by α particles whose kinetic energies are just over 1.19 MeV? The answer is no, for two reasons. First, in the collision between the α particle and the N^{14} nucleus conservation of momentum requires that at least $\frac{4}{18}$ of the kinetic energy of the α particle must be retained by the products as kinetic energy; thus only $\frac{14}{18}$ of the α particle's kinetic energy is available for the reaction. The threshold energy of α particles for the $N^{14}(\alpha, p)O^{17}$ reaction, that is, the kinetic energy of α particles just capable of making the reaction energetically *possible*, is $\frac{18}{14} \times 1.19$ MeV $= 1.53$ MeV. The fraction of the bombarding particle's kinetic energy which is retained as kinetic energy of the products becomes smaller with increasing mass of the target nucleus (see exercise 19).

Barriers for Charged Particles. The second reason why the α particles must have higher energies than is evident from the Q value to produce the reaction $N^{14}(\alpha, p)O^{17}$ in *good yield* is the Coulomb repulsion between the α particle and the N^{14} nucleus. The repulsion increases with decreasing distance of separation until the α particle comes within the range of the nuclear forces of the N^{14} nucleus. This Coulomb repulsion gives rise to the potential barrier already mentioned in section C. The height V of the potential barrier around a nucleus of charge $Z_1 e$ and radius R_1 for a particle of positive charge $Z_2 e$ and radius R_2 may be estimated as the energy of Coulomb repulsion when the two particles are just in contact:

$$V = \frac{Z_1 Z_2 e^2}{(R_1 + R_2)}. \tag{2-9}$$

If R_1 and R_2 are expressed in fermis (units of 10^{-13} cm),

$$V = 1.44 \frac{Z_1 Z_2}{R_1 + R_2} \text{ MeV.} \tag{2-10}$$

Obtaining the nuclear radii from the formula[20] (cf. p. 33),

$$R = 1.6 \times 10^{-13} A^{1/3} \text{ cm,}$$

we get for the barrier height between N^{14} and an α particle a value of about 3.2 MeV. According to classical theory, an α particle must thus have at least $\frac{18}{14} \times 3.2 = 4.0$ MeV kinetic energy to enter a N^{14} nucleus and produce the α, p reaction, even though the energetic threshold for the reaction is only 1.53 MeV. In the quantum-mechanical treatment of the problem there exists

[20] For the lightest nuclei this formula for nuclear radii is actually a poor approximation; but for a rough estimate of barrier heights it is adequate. In using (2-10) to estimate proton barriers of medium and heavy nuclei, one commonly considers the proton as a point charge ($R_2 = 0$).

a finite probability for "tunneling through the barrier" by lower-energy particles, but this probability drops rapidly as the energy of the particle decreases. (The penetration of potential barriers is discussed in connection with α decay in chapter 8, section A.) Rutherford actually used α particles of more than 7 MeV in his experiments.

It follows from (2-9) that the Coulomb barrier around a given nucleus is about half as high for protons and for deuterons as it is for α particles. The height of the barrier increases with increasing Z of the target nucleus; it is roughly proportional to $Z^{2/3}$ (not to Z because the nuclear radius R increases approximately as $Z^{1/3}$). For the heaviest elements the potential barriers are about 12 MeV for protons and deuterons and about 25 MeV for α particles. In order to study nuclear reactions induced by charged particles, especially reactions involving heavy elements, it was therefore necessary to develop machines capable of accelerating charged particles to energies of many millions of electron volts.

Neutrons. It is apparent that the entry of a neutron into a nucleus is not opposed by any Coulomb barrier, and even neutrons of very low energy react readily with even the heaviest nuclei. In fact, the so-called thermal neutrons, that is, neutrons whose energy distribution is approximately that of gas molecules in thermal equilibrium at ordinary temperatures, have particularly high probabilities for reaction with target nuclei. This important effect was discovered at the University of Rome by E. Fermi, E. Amaldi, B. Pontecorvo, F. Rasetti, and E. Segrè in 1934 in experiments on the neutron irradiation of silver; they found that the neutron-induced radioactivity was much greater when a bulk of hydrogen-containing material such as paraffin was present to modify the neutron beam. Fermi reasoned correctly that fast neutrons would lose energy in collisions with protons, that repeated collisions might reduce the energy to the thermal range, and that such slow neutrons could show large capture cross sections. Other workers found the effect to be sensitive to the temperature of the paraffin, thus demonstrating that the neutrons were actually slowed to approximately thermal energies.

Centrifugal Barriers. In the foregoing discussion of the Coulomb barrier we have assumed that the incident particle collides head-on with the nucleus. Interactions do occur in which the original direction of motion of the particle does not pass through the center of the target nucleus; such systems of projectile and target have angular momentum, and this angular momentum is quantized in integral multiples of the universal unit $h/2\pi$. In general the interactions are classified as s, p, d, etc., in the notation derived from atomic structure: s interactions correspond to $l = 0$, p interactions to $l = 1$, d interactions to $l = 2$, etc., where l is the value of the orbital angular momentum in units of $h/2\pi$. The resulting hindrance to the close approach of projectile to target is usually described as a centrifugal barrier, which is, of course, zero for s-wave interactions but appreciable for higher l values. In crude approximation this barrier keeps any projectile just $l/2\pi$ de Broglie wavelengths from the target center; the barrier is overcome at incident kinetic energies large enough to make this length as small as the sum of the nuclear radii of target and projectile. This approximation gives for the centrifugal barrier: $V = h^2 l^2 / [8\pi^2 M (R_1 + R_2)^2]$, where M is the mass of the projectile. In the correct expression derived from wave mechanics l^2 is replaced by $l(l + 1)$; also M should be replaced by μ, the reduced mass of the system (see chapter 10, section D):

$$V = \frac{h^2 l(l + 1)}{8\pi^2 \mu (R_1 + R_2)^2}.$$

For the reaction of an α particle and N^{14} the centrifugal barrier alone is $0.145l(l + 1)$ MeV, which is 0.29 MeV for p-wave and 0.87 MeV for d-wave interactions. The total barrier height is the sum of the Coulomb and the centrifugal barriers. Keep in mind, however, that, of the two, only the Coulomb barrier contributes to the minimum energy requirement for producing a nuclear reaction except in very special cases in which s collisions may not lead to the reaction of interest.

It must be emphasized that potential barriers have an effect not only for particles entering but also for particles leaving nuclei. For this reason a charged particle has to be excited to a rather high energy inside the nucleus before it can either go over the top of the Coulomb barrier or, according to the quantum-mechanical picture, leak through the barrier with appreciable probability. Therefore, charged particles are usually emitted from nuclei with considerable energies (more than 1 MeV).

2. Cross Sections

Definitions. We now turn to a more quantitative consideration of reaction probabilities. The probability of a nuclear process is generally expressed in terms of a cross section σ which has the dimensions of an area. This originates from the simple picture that the probability for the reaction between a nucleus and an impinging particle is proportional to the cross-sectional target area presented by the nucleus. Although this classical picture does not hold for reactions with charged particles which have to overcome Coulomb barriers or for slow neutrons (it does hold fairly well for the total probability of a fast neutron interacting with a nucleus), the cross section is a useful measure of the probability for any nuclear reaction. For a beam of particles striking a thin target, that is, a target in which the beam is attenuated only infinitesimally, the cross section for a particular process is defined by the equation

$$R_i = \mathbf{I} n \sigma_i x, \tag{2-11}$$

where R_i is the number of processes of the type under consideration occurring in the target per unit time,

I is the number of incident particles per unit time,

n is the number of target nuclei per cubic centimeter of target,

σ_i is the cross section for the specified process, expressed in square centimeters, and

x is the target thickness in centimeters.

The total cross section for collision with a fast particle is never greater than twice[21] the geometrical cross-sectional area of the nucleus, and therefore fast-particle cross sections are rarely much larger than 10^{-24} cm^2 (radii of the heaviest nuclei are about 10^{-12} cm). Hence a cross section of 10^{-24} cm^2 is considered "as big as a barn," and 10^{-24} cm^2 has been named the barn, a unit generally used in expressing cross sections and often abbreviated b. The millibarn (mb, 10^{-27} cm^2) and the microbarn (μb, 10^{-30} cm^2) are also used.

If instead of a thin target we consider a thick target, that is, one in which the intensity of the incident particle beam is attenuated, the attenuation $-d\mathbf{I}$ in the infinitesimal thickness dx is given by the equation

$$-d\mathbf{I} = \mathbf{I}n\sigma_t\,dx,$$

where σ_t is the total cross section. If we are able to neglect the variation in σ_t as the incident particles traverse the target, which is often the case for neutron reactions, we obtain, by integration,

$$\mathbf{I} = \mathbf{I}_0 e^{-n\sigma_t x},$$

$$\mathbf{I}_0 - \mathbf{I} = \mathbf{I}_0(1 - e^{-n\sigma_t x}), \tag{2-12}$$

where **I** is the intensity of the beam after traversing a target thickness x, $\mathbf{I}_0$ is the incident intensity, and $\mathbf{I}_0 - \mathbf{I}$ is the number of reactions per unit time.

As an illustration, we shall calculate the number of radioactive Au^{198} nuclei produced per second in a sheet of gold 0.3 mm thick and 5 cm^2 in area exposed to a thermal neutron flux of 10^7 neutrons per cm^2 per second. The capture cross section of Au^{197} for thermal neutrons is 99 barns, and we neglect any other reactions of neutrons with gold. The density of gold is 19.3 g cm^{-3}, and its atomic weight is 197.2; therefore

$$n = \frac{19.3}{197.2} \times 6.02 \times 10^{23} = 5.89 \times 10^{22} \text{ Au}^{197} \text{ nuclei/cm}^3;$$

$$x = 0.03 \text{ cm};$$

$$\mathbf{I}_0 = 5 \times 10^7 \text{ incident neutrons per second.}$$

Therefore, from (2-12),

$$\mathbf{I}_0 - \mathbf{I} = 5 \times 10^7(1 - e^{-5.89 \times 10^{22} \times 99 \times 10^{-24} \times 0.03})$$

$$= 8.0 \times 10^6 \text{ Au}^{198} \text{ nuclei formed per second.}$$

[21] The reason why total cross sections may be as large as $2\pi R^2$ was briefly mentioned on p. 32 and is further discussed in chapter 10, section E.

Partial and Total Cross Sections. A cross section may be given for any particular nuclear process. For example, the total cross section for interaction of 10-MeV neutrons with a particular nuclear species may be measured by means of a transmission experiment in a collimated neutron beam, with the detector subtending a small solid angle with respect to the sample. In this case only those neutrons that have been neither absorbed nor scattered by the sample will be detected in the transmitted beam. The total cross section σ_t measured in this way is the sum of the cross sections for all the possible processes. Also we may define and measure cross sections for particular processes, such as n, γ or n, α reactions. We have already indicated that in (2-12) total cross sections must be used, and only the total number of reactions may be obtained directly. This may be multiplied by the ratio of a partial cross section to the total cross section to obtain the number of reactions of a particular kind. The partial cross section might be for an effect in one component of a mixture or compound or in a single isotope, but the total cross section must be that for the target substance; in such cases it is the product of cross section times number of effective nuclei that must be compared.

As an example, consider the irradiation of carbon tetrachloride to produce S^{35} by the reaction $Cl^{35}(n, p)S^{35}$. The sample is a 1-cm cube (1.46 g), and the thermal neutron flux normal to one face is 10^9 cm^{-2} sec^{-1}. How many S^{35} atoms are formed in 24 hours? According to appendix C the total absorption cross section for chlorine is 33.8 barns, that for carbon is only 0.0037 barn and may be neglected, and the isotopic cross section for the reaction $Cl^{35}(n, p)S^{35}$ is 0.19 barn. The number of chlorine atoms per cubic centimeter of sample is $(1.46/153.8) \times 4 \times 6.02 \times 10^{23} = 2.28 \times 10^{22}$. Therefore, according to (2-12), the total number of neutrons absorbed in the sample in 24 hours is $24 \times 60 \times 60 \times 10^9[1 - \exp(-2.28 \times 10^{22} \times 33.8 \times 10^{-24}) = 4.64 \times 10^{13}$. Since the isotopic abundance of Cl^{35} is 75.4 per cent, the fraction of neutrons absorbed that lead to n, p reactions in Cl^{35} is $(0.754 \times 0.19)/33.8 = 4.24 \times 10^{-3}$, and the number of S^{35} atoms formed is $4.24 \times 10^{-3} \times 4.64 \times 10^{13} = 2.0 \times 10^{11}$.

Sometimes the angular distribution of particles resulting from a particular process is of interest. In this case it is convenient to define a differential cross section $d\sigma/d\Omega$; this is the cross section for that part of the process in which the particles are emitted into unit solid angle at a particular angle Ω. Then the cross section for the over-all process under consideration is $\sigma = \int (d\sigma/d\Omega)\, d\Omega$.

Excitation Functions. Frequently the variation of a particular reaction cross section with incident energy is of interest; the relation between the two is called an excitation function. Examples of excitation functions are shown in figures 10-1 and 10-6. The determination of an absolute cross section requires a knowledge of the incident beam intensity and a measurement of the absolute number of reactions in the target. It is usually much simpler, and still frequently of considerable interest, to determine merely the relative cross sections of a reaction at different incident energies (most often done by exposure of several target foils in the same beam, with energy-degrading foils interposed—the method of "stacked foils") or the relative cross sections of different reactions at one energy.

A detailed discussion of the factors determining reaction cross sections is deferred to chapter 10. In general, the higher the energy of a bombarding particle, the more complex the possible reactions. For example, with a few exceptions among the lightest nuclides, thermal neutrons can induce (n, γ) reactions only. With neutrons of several MeV kinetic energy, (n, p) reactions become possible and prevalent; at still higher energies $(n, 2n)$, (n, α), and (n, np) reactions set in. In the energy range of primary interest to most radiochemists (in contrast to many nuclear chemists), namely up to about 50 MeV bombarding energy, the cross section of a given reaction rises with increasing bombarding energy from threshold to some maximum value which is usually reached about 10 MeV above threshold and then drops again to some low value; the drop is accompanied by the rise of cross sections for other more complex reactions. This behavior is illustrated by the excitation functions shown in figure 10-6.

REFERENCES

*B1   J. M. Blatt and V. F. Weisskopf, *Theoretical Nuclear Physics*, Wiley, New York, 1952.

*B2   H. A. Bethe and P. Morrison, *Elementary Nuclear Theory*, second edition, Wiley, New York, 1956.

C1   J. Chadwick, "The Existence of a Neutron," *Proc. Roy. Soc. (London)* **A136,** 692 (1932).

C2   C. D. Coryell, "Beta-Decay Energetics," *Ann. Rev. Nuclear Sci.* **2,** 305–34 (1953).

D1   M. Deutsch and O Kofoed-Hansen, "Beta-Rays," in *Experimental Nuclear Physics* (E. Segrè, Editor), Vol. III, Wiley, New York, 1959, pp. 426–638.

E1   F. Everling, L. A. König, J. H. E. Mattauch, and A. H. Wapstra, "Relative Nuclidic Masses," *Nucl. Phys.* **18,** 529 (1960).

*E2   L. R. B. Elton, *Nuclear Sizes*, Oxford University Press, 1961.

*F1   E. Feenberg, "Semiempirical Theory of the Nuclear Energy Surface," *Revs. Mod. Phys.* **19,** 239 (1947).

*F2   N. Feather, *Nuclear Stability Rules*, Cambridge University Press, 1952.

G1   A. E. S. Green and D. F. Edwards, "Discontinuities in the Nuclear Mass Surface," *Phys. Rev.* **91,** 46 (1953).

H1   R. Hofstadter, "Nuclear and Nucleon Scattering of High-Energy Electrons," *Ann. Rev. Nuclear Sci.,* **7,** 231–316 (1957).

*H2   D. Halliday, *Introductory Nuclear Physics*, second edition, Wiley, New York, 1955.

*K1   I. Kaplan, *Nuclear Physics*, second edition, Addison-Wesley, Cambridge, Mass., 1963.

K2   L. A. König, J. H. E. Mattauch, and A. H. Wapstra, "1961 Nuclidic Mass Table," *Nucl. Phys.,* **31,** 18 (1962).

M1   H. G. J. Moseley, "The High-Frequency Spectra of the Elements," *Phil. Mag.* **26,** 1024 (1913) and **27,** 703 (1914).

M2   M. G. Mayer, "On Closed Shells in Nuclei," *Phys. Rev.* **74,** 235 (1948).

*M3   P. Morrison, "A Survey of Nuclear Reactions," in *Experimental Nuclear Physics* (E. Segrè, Editor), Vol. II, Wiley, New York, 1953, pp. 1–207.

R1   E. Rutherford, "The Scattering of α and β Particles by Matter and the Structure of the Atom," *Phil. Mag.* **21,** 669 (1911).

R2   E. Rutherford, J. Chadwick, and C. D. Ellis, *Radiations from Radioactive Substances,* Cambridge University Press, 1930.

*R3   N. F. Ramsey, "Nuclear Moments and Statistics," in *Experimental Nuclear Physics* (E. Segrè, Editor), Vol. I, Wiley, New York, 1953, pp. 358–467.

W1   C. F. v. Weizsäcker, "Zur Theorie der Kernmassen," *Z. Physik* **96,** 431 (1935).

EXERCISES

1. Show that h ($=6.626 \times 10^{-27}$ erg-sec) has the dimensions of angular momentum.

2. Calculate the binding energy per nucleon for Li^6, P^{31}, Ni^{60}, Pd^{108}, Pt^{195}, and U^{238} both from masses given in appendix E and from the semiempirical binding-energy equation.

3. Estimate the level width (in electron volts) of a quantum state whose lifetime is 10^{-15} second.

4. From the masses given in appendix E, find (a) the binding energy for an additional neutron to O^{16}, V^{50}, Pu^{239}; (b) the binding energy for an additional proton to B^{10}, Mn^{52}, $Th^{234}(UX_1)$.   *Answers:* (a) Pu^{239} 6.39 MeV; (b) Mn^{52} 7.27 MeV.

5. What is the kinetic energy of (a) an electron, (b) a proton, (c) a π meson with a de Broglie wavelength of 1.5×10^{-13} cm?   (You may want to refer to the relativistic relations in appendix B.)   *Answer: (a):* 826 MeV.

6. Estimate what might be the heaviest element for which deviations from the predictions of the Rutherford scattering formula would be observed with 12-MeV He^4 ions.   *Answer:* In the vicinity of Zr.

7. The three fundamental mass doublets have been found to have the following separations.

$$(C^{12}H_4)^+ - (O^{16})^+ = 36.385 \text{ millimass units,}$$

$$H_2^+ - D^+ = 1.548 \text{ millimass units,}$$

$$D_3^+ - (C^{12})^{2+} = 42.307 \text{ millimass units.}$$

Calculate the atomic masses of H, D, and O^{16}.

8. With the aid of the semiempirical mass or binding-energy equations, estimate (a) the energy liberated when one additional neutron is added to U^{235}, (b) the energy liberated when one additional neutron is added to U^{238}, (c) the amount of energy by which I^{129} is unstable with respect to β decay to Xe^{129}.
        Answer: (a) 7.2 MeV.

9. Determine from the semiempirical mass equations the atomic number Z_A corresponding to maximum stability for $A = 27$, $A = 131$, and $A = 204$.   Compare your results with the experimental data as listed, for example, in appendix E.

10. What can you say about the intensities of alternate lines in the rotational spectrum of iodine?

11. Estimate as best you can the energy available for (a) α decay, (b) spontaneous fission into equal fragments, of Pt^{192}, U^{238}, and Cf^{252}.

12. Show that the production of a positron-electron pair by a photon in vacuum is impossible. (*Note.* Set up the conditions for momentum and energy conservation, using relativistic expressions, and show that they lead to a contradiction, for example, to the inequality $\cos \theta > 1$, where θ is the angle between the directions of motion of positron and electron.)

13. In the α decay of AcC (Bi^{211}) to AcC''(Tl^{207}) two groups of α particles, with kinetic energies 6.617 and 6.273 MeV, are observed. They populate the ground state and first excited state of Tl^{207}, what is the energy difference between these two states?   *Answer:* 0.351 MeV.

14. Estimate the rate of energy deposition (in calories per hour) in a large calorimeter when each of the following samples if placed in it; (a) C^{14}, (b) Sm^{146}, (c) Cf^{254}, each undergoing 10^4 disintegrations per minute.

15. Compute, from masses in appendix E, the Q values for the following reactions: (a) Mg^{24}(d, p)Mg^{25}; (b) B^{10}(n, α)Li^7.

16. Would the Ca^{43}(n, α) reaction proceed in good yield with thermal neutrons? Justify your answer.   *Answer:* No.

17. The reaction S^{33}(n, p)P^{33} is exoergic by 0.53 MeV. What is the mass of P^{33}?

18. Estimate the Coulomb barrier and the centrifugal barrier for p-wave interaction between protons and Be^9.   *Answer:* 1.2 MeV, 1.7 MeV.

19. Show from the conservation of momentum that, at the threshold of an endothermic nuclear reaction $A(a, b)B$, the fraction of the kinetic energy of the bombarding particle a which goes into kinetic energy of the products is $M_a/(M_a + M_A)$, where M_a and M_A are the masses of a and A, respectively.

20. Calculate the approximate heights of the Coulomb barriers around $_{13}$Al^{27}, $_{26}$Fe^{56}, $_{47}$Ag^{107}, $_{73}$Ta^{181}, and $_{92}$U^{238} for protons.   *Answers to first two:* ~3, ~5 MeV.

21. Calculate the de Broglie wavelength of a neutron of (a) 1 eV, (b) 1 keV, (c) 1 MeV kinetic energy.   *Answer:* (b) 0.9×10^{-10} cm.

22. Approximately what thickness of cadmium is necessary to reduce a beam of thermal neutrons to 0.1 per cent of its intensity?

23. The nuclide Cl^{33} can be produced by the reaction S^{33}(p, n)Cl^{33}. The Cl^{33} emits positrons with an upper energy limit of 4.5 MeV. What is the Q value of the reaction? What is the height of the potential barrier around the S^{33} nucleus for the proton? Estimate the minimum proton energy required to produce the reaction.   *Answer:* Minimum proton energy $\cong$ 6.5 MeV.

24. (a) Calculate the number of Co^{60} atoms produced in a 10-mg sample of cobalt metal exposed for two minutes to a thermal-neutron flux of 2×10^{13} cm^{-2} sec^{-1} in a reactor. (b) What is the disintegration rate of the cobalt sample a few hours after the irradiation?   *Answer:* (b) 2.2×10^6 disintegrations per minute.

25. A copper foil of surface density 20 mg/cm^2 is exposed for five minutes to a beam of 14-MeV neutrons. The beam intensity is 10^7 per second. At the end of the irradiation the foil is found to contain 2.0×10^5 atoms of Cu^{64}. Assuming that they have been formed entirely by the reaction Cu^{65}(n, $2n$) and neglecting any decay of the 12.8-hour nuclide during the short bombardment, estimate the cross section for the n, $2n$ reaction. Would you expect the n, γ reaction on Cu^{63} to have an important effect on the actual result of the experiment? *Answer:* $\sigma_{n,2n}$ = 1.1b.

26. Estimate the total energy release and (from Coulomb barrier considerations) the minimum total kinetic energy (in MeV) of the two fission fragments when a U^{235} nucleus captures a thermal neutron and splits into (a) Kr^{90} and Ba^{143}, (b) Rh^{113} and Ag^{121}.

Equations of Radioactive Decay and Growth

A. EXPONENTIAL DECAY

Half-life. We have seen (in chapter 1) that a given radioactive species decays according to an exponential law: $N = N_0 e^{-\lambda t}$ or $\mathbf{A} = \mathbf{A}_0 e^{-\lambda t}$, where N and $\mathbf{A}$ represent the number of atoms and the measured activity, respectively, at time t, and N_0 and $\mathbf{A}_0$ the corresponding quantities when $t = 0$, and λ is the characteristic decay constant for the species. The half-life $t_{1/2}$ is the time interval required for N or $\mathbf{A}$ to fall from any particular value to one half that value. The half-life is conveniently determined from a plot of log $\mathbf{A}$ versus t when the necessary data are available and is related to the decay constant:

$$t_{1/2} = \frac{\ln 2}{\lambda} = \frac{0.69315}{\lambda}.$$

Average Life. We may determine the average life expectancy of the atoms of a radioactive species. This average life is found from the sum of the times of existence of all the atoms divided by the initial number; if we consider N to be a very large number, we may approximate this sum by an equivalent integral, finding for the average life τ

$$\tau = -\frac{1}{N_0} \int_{t=0}^{t=\infty} t \, dN = \frac{1}{N_0} \int_0^\infty t\lambda N \, dt = \lambda \int_0^\infty t e^{-\lambda t} \, dt$$

$$= -\left[\frac{\lambda t + 1}{\lambda} e^{-\lambda t} \right]_0^\infty = \frac{1}{\lambda}.$$

We see that the average life is greater than the half-life by the factor $1/0.693$; the difference arises because of the weight given in the averaging process to the fraction of atoms that by chance survive for a long time. It may be seen that during the time $1/\lambda$ an activity will be reduced to just $1/e$ of its initial value.

Mixtures of Independently Decaying Activities. If two radioactive species, denoted by subscripts 1 and 2, are mixed together, the observed total activity is the sum of the two separate activities: $\mathbf{A} = \mathbf{A}_1 + \mathbf{A}_2 = c_1\lambda_1 N_1 +$

$c_2\lambda_2N_2$. The detection coefficients c_1 and c_2 are by no means necessarily the same and often are very different in magnitude. In general, $\mathbf{A} = \mathbf{A}_1 + \mathbf{A}_2 + \cdots + \mathbf{A}_n$ for mixtures of n species.

For a mixture of several *independent* activities the result of plotting log $\mathbf{A}$ versus t is always a curve concave upward (convex toward the origin). This curvature results because the shorter-lived components become relatively less significant as time passes. In fact, after sufficient time the longest-lived activity will entirely predominate, and its half-life may be read from this late portion of the decay curve. Now, if this last portion, which is a straight line, is extrapolated back to $t = 0$ and the extrapolated line subtracted from the original curve, the residual curve represents the decay of all components except the

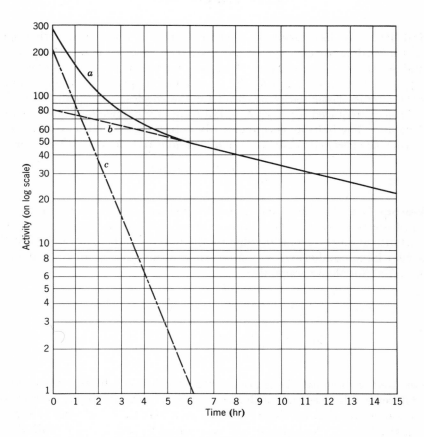

Fig. 3-1 Analysis of composite decay curve:

    (a) composite decay curve;
    (b) longer-lived component ($t_{1/2} = 8.0$ hr);
    (c) shorter-lived component ($t_{1/2} = 0.8$ hr).

longest-lived.   This curve may be treated again in the same way, and in principle any complex decay curve may be analyzed into its components.   In actual
practice experimental uncertainties in the observed data may be expected to
make it difficult to handle systems of more than three components, and even
two-component curves may not be satisfactorily resolved if the two half-lives
differ by less than about a factor of two.   The curve shown in figure 3-1 is for
two components with half-lives differing by a factor of 10.

The resolution of a decay curve consisting of two components of known but
not very different half-lives is greatly facilitated by the following approach.
The total activity at time t is

$$A = A_1{}^0 e^{-\lambda_1 t} + A_2{}^0 e^{-\lambda_2 t}.$$

By multiplying both sides by $e^{\lambda_1 t}$ we obtain

$$Ae^{\lambda_1 t} = A_1{}^0 + A_2{}^0 e^{(\lambda_1 - \lambda_2)t}.$$

Since λ_1 and λ_2 are known and A has been measured as a function of t, we can
construct a plot of $Ae^{\lambda_1 t}$ versus $e^{(\lambda_1 - \lambda_2)t}$; this will be a straight line with intercept $A_1{}^0$ and slope $A_2{}^0$.

Least-squares analysis is a more objective method for the resolution of complex decay curves than the graphical analysis described.   Computer programs
for this analysis have been developed (C1) which give values of A^0 and its
standard deviation for each of the components.   Some of the programs can
also be used to search for the "best values" of the decay constants.

B. GROWTH OF RADIOACTIVE PRODUCTS

General Equation.   In chapter 1 we considered briefly a special case in
which a radioactive daughter substance was formed in the decay of the parent.
Let us take up the general case for the decay of a radioactive species, denoted
by subscript 1, to produce another radioactive species, denoted by subscript 2.
The behavior of N_1 is just as has been derived; that is, $-(dN_1/dt) = \lambda_1 N_1$,
and $N_1 = N_1{}^0 e^{-\lambda_1 t}$, where we use the symbol $N_1{}^0$ to represent the value of
N_1 at $t = 0$.   Now the second species is formed at the rate at which the first
decays, $\lambda_1 N_1$, and itself decays at the rate $\lambda_2 N_2$.   Thus

$$\frac{dN_2}{dt} = \lambda_1 N_1 - \lambda_2 N_2$$

or

$$\frac{dN_2}{dt} + \lambda_2 N_2 - \lambda_1 N_1{}^0 e^{-\lambda_1 t} = 0. \tag{3-1}$$

The solution of this linear differential equation of the first order may be obtained by standard methods and gives

$$N_2 = \frac{\lambda_1}{\lambda_2 - \lambda_1} N_1{}^0(e^{-\lambda_1 t} - e^{-\lambda_2 t}) + N_2{}^0 e^{-\lambda_2 t}, \tag{3-2}$$

where $N_2{}^0$ is the value of N_2 at $t = 0$. Notice that the first group of terms shows the growth of daughter from the parent and the decay of these daughter atoms; the last term gives the contribution at any time from the daughter atoms present initially.

Transient Equilibrium. In applying (3-2) to considerations of radioactive (parent and daughter) pairs, we can distinguish two general cases, depending on which of the two substances has the longer half-life. If the parent is longer-lived than the daughter ($\lambda_1 < \lambda_2$), a state of so-called radioactive equilibrium is reached; that is, after a certain time the ratio of the numbers of atoms and, consequently, the ratio of the disintegration rates of parent and daughter become constant. This can be readily seen from (3-2); after t becomes sufficiently large, $e^{-\lambda_2 t}$ is negligible compared with $e^{-\lambda_1 t}$, and $N_2{}^0 e^{-\lambda_2 t}$ also becomes negligible; then

$$N_2 = \frac{\lambda_1}{\lambda_2 - \lambda_1} N_1{}^0 e^{-\lambda_1 t},$$

and, since $N_1 = N_1{}^0 e^{-\lambda_1 t}$,

$$\frac{N_1}{N_2} = \frac{\lambda_2 - \lambda_1}{\lambda_1}. \tag{3-3}$$

The relation of the two measured activities is found, from $\mathbf{A}_1 = c_1 \lambda_1 N_1$, $\mathbf{A}_2 = c_2 \lambda_2 N_2$, to be

$$\frac{\mathbf{A}_1}{\mathbf{A}_2} = \frac{c_1(\lambda_2 - \lambda_1)}{c_2 \lambda_2}. \tag{3-4}$$

In the special case of equal detection coefficients ($c_1 = c_2$) the ratio of the two activities, $\mathbf{A}_1/\mathbf{A}_2 = 1 - (\lambda_1/\lambda_2)$, may have any value between 0 and 1, depending on the ratio of λ_1 to λ_2; that is, in equilibrium the daughter activity will be greater than the parent activity by the factor $\lambda_2/(\lambda_2 - \lambda_1)$. In equilibrium both activities decay with the parent's half-life.

As a consequence of the condition of transient equilibrium ($\lambda_2 > \lambda_1$), the sum of the parent and daughter disintegration rates in an initially pure parent fraction goes through a maximum before transient equilibrium is achieved. This situation is illustrated in figure 3-2. The more general condition for the total *measured activity* ($A_1 + A_2$) of an initially pure parent fraction to exhibit a maximum is found to be $c_2/c_1 > \lambda_1/\lambda_2$. This condition holds regardless of the relative magnitudes of λ_1 and λ_2. The condition $(\lambda_1 - \lambda_2)/\lambda_2 \leq c_2/c_1 \leq$

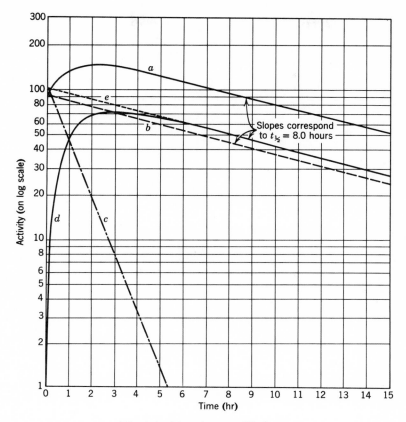

Fig. 3-2   Transient equilibrium:

(a) total activity of an initially pure parent fraction;
(b) activity due to parent ($t_{1/2} = 8.0$ hr);
(c) decay of freshly isolated daughter fraction ($t_{1/2} = 0.80$ hr);
(d) daughter activity growing in freshly purified parent fraction;
(e) total daughter activity in parent-plus-daughter fractions.

λ_1/λ_2 will give a maximum in the total measured activity which occurs at a negative time.

Secular Equilibrium.   A limiting case of radioactive equilibrium in which $\lambda_1 \ll \lambda_2$ and in which the parent activity does not decrease measurably during many daughter half-lives is known as secular equilibrium.   We illustrated this situation in chapter 1 and now may derive the equation presented there as a useful approximation of (3-3):

$$\frac{N_1}{N_2} = \frac{\lambda_2}{\lambda_1}, \quad \text{or} \quad \lambda_1 N_1 = \lambda_2 N_2.$$

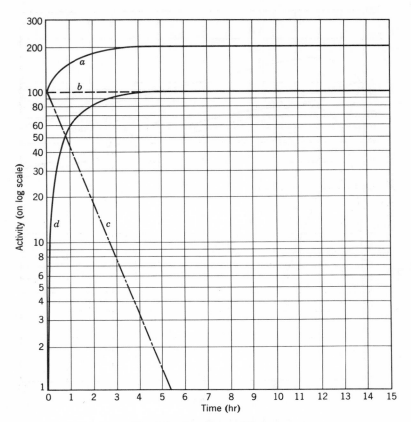

Fig. 3-3   Secular equilibrium:

(*a*) total activity of an initially pure parent fraction;
(*b*) activity due to parent ($t_{1/2} = \infty$); this is also the total daughter activity in parent-plus-daughter fractions;
(*c*) decay of freshly isolated daughter fraction ($t_{1/2} = 0.80$ hr);
(*d*) daughter activity growing in freshly purified parent fraction.

In the same way (3-4) reduces to

$$\frac{\mathbf{A}_1}{\mathbf{A}_2} = \frac{c_1}{c_2},$$

and the measured activities are equal if $c_1 = c_2$.

Figure 3-2 presents an example of transient equilibrium with $\lambda_1 < \lambda_2$ (actually with $\lambda_1/\lambda_2 = \frac{1}{10}$); the curves represent variations with time of the parent activity and the activity of a freshly isolated daughter fraction, the growth of daughter activity in a freshly purified parent fraction, and other relations; in preparing the figure we have taken $c_1 = c_2$. Figure 3-3 is a similar plot for secular equilibrium; it is apparent that as λ_1 becomes smaller compared to λ_2 the curves for transient equilibrium shift to approach more and more closely the limiting case shown in figure 3-3.

The Case of No Equilibrium. If the parent is shorter-lived than the daughter ($\lambda_1 > \lambda_2$), it is evident that no equilibrium is attained at any time. If the parent is made initially free of the daughter, then as the parent decays the amount of daughter will rise, pass through a maximum, and eventually decay with the characteristic half-life of the daughter. This is illustrated in figure 3-4; for this plot we have taken $\lambda_1/\lambda_2 = 10$, and $c_1 = c_2$. In the figure

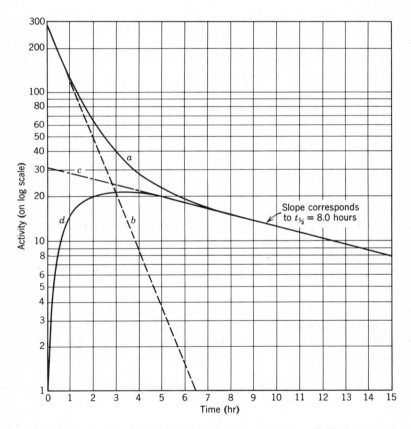

Fig. 3-4   The case of no equilibrium:

(a) total activity;
(b) activity due to parent ($t_{1/2} = 0.80$ hr);
(c) extrapolation of final decay curve to time zero;
(d) daughter activity in initially pure parent.

the final exponential decay of the daughter is extrapolated back to $t = 0$. This method of analysis is useful if $\lambda_1 \gg \lambda_2$, for then this intercept measures the activity $c_2\lambda_2N_1{}^0$: the $N_1{}^0$ atoms give rise to N_2 atoms so early that $N_1{}^0$ may be set equal to the extrapolated value of N_2 at $t = 0$. The ratio of the initial activity, $c_1\lambda_1N_1{}^0$, to this extrapolated activity gives the ratios of the half-lives if the relation between c_1 and c_2 is known:

$$\frac{c_1\lambda_1N_1{}^0}{c_2\lambda_2N_1{}^0} = \frac{c_1}{c_2} \times \frac{\lambda_1}{\lambda_2} = \frac{c_1}{c_2} \times \frac{(t_{1/2})_2}{(t_{1/2})_1}.$$

If λ_2 is not negligible compared to λ_1, it can be shown that the ratio λ_1/λ_2 in this equation should be replaced by $(\lambda_1 - \lambda_2)/\lambda_2$ and the expression involving the half-lives changed accordingly.

Both the transient-equilibrium and the no-equilibrium cases are sometimes analyzed in terms of the time t_m for the daughter to reach its maximum activity when growing in a freshly separated parent fraction. This time we find from the general equation (3-2) by differentiating,

$$\frac{dN_2}{dt} = -\frac{\lambda_1{}^2}{\lambda_2 - \lambda_1} N_1{}^0 e^{-\lambda_1 t} + \frac{\lambda_1\lambda_2}{\lambda_2 - \lambda_1} N_1{}^0 e^{-\lambda_2 t},$$

and setting $dN_2/dt = 0$ when $t = t_m$:

$$\frac{\lambda_2}{\lambda_1} = e^{(\lambda_2-\lambda_1)t_m} \quad \text{or} \quad t_m = \frac{2.303}{\lambda_2 - \lambda_1} \log \frac{\lambda_2}{\lambda_1}.$$

At this time the daughter decay rate, λ_2N_2, is just equal to the rate of formation, λ_1N_1 [this is obvious from (3-1)]; in figures 3-2, 3-3, and 3-4, in which we assumed $c_1 = c_2$, we have the parent activity $\mathbf{A}_1$ intersecting the daughter growth curve d at the time t_m. (The time t_m is infinite for secular equilibrium.)

Many Successive Decays. If we consider a chain of three or more radioactive products, it is clear that the equations already derived for N_1 and N_2 as functions of time are valid, and N_3 may be found by solving the new differential equation:

$$\frac{dN_3}{dt} = \lambda_2N_2 - \lambda_3N_3. \tag{3-5}$$

This is entirely analogous to the equation for dN_2/dt, but the solution calls for more labor, since N_2 is a much more complicated function than N_1. The next solution for N_4 is still more tedious. H. Bateman (B1) has given the solution for a chain of n members with the special assumption that at $t = 0$ the parent substance alone is present, that is, that $N_2{}^0 = N_3{}^0 = \cdots N_n{}^0 = 0$. This

solution is

$$N_n = C_1 e^{-\lambda_1 t} + C_2 e^{-\lambda_2 t} + \quad \cdot \cdot \; C_n e^{-\lambda_n t},$$

$$C_1 = \frac{\lambda_1 \lambda_2 \, \cdot \cdot \cdot \, \lambda_{n-1}}{(\lambda_2 - \lambda_1)(\lambda_3 - \lambda_1) \, \cdot \cdot \cdot \, (\lambda_n - \lambda_1)} \, N_1{}^0,$$

$$C_2 = \frac{\lambda_1 \lambda_2 \, \cdot \cdot \cdot \, \lambda_{n-1}}{(\lambda_1 - \lambda_2)(\lambda_3 - \lambda_2) \, \cdot \cdot \cdot \, (\lambda_n - \lambda_2)} \, N_1{}^0, \quad \text{etc.}$$

If we do require a solution to the more general case with $N_2{}^0, N_3{}^0 \cdot \cdot \cdot N_n{}^0 \neq 0$, we may construct it by adding to the Bateman solution for N_n in an n-membered chain, a Bateman solution for N_n in an $(n-1)$-membered chain with substance 2 as the parent, and, therefore, $N_2 = N_2{}^0$ at $t = 0$, and a Bateman solution for N_n in an $(n-2)$-membered chain, etc.

Branching Decay.   Another variant that is met in general decay schemes is the branching decay, illustrated by

Here the two partial decay constants, λ_b and λ_c, must be considered when the general relations in either branch are studied because, for example, the substance B is formed at the rate $\lambda_b N_A$, but A is consumed at the rate $(\lambda_b + \lambda_c)N_A$. Notice that A can have but one half-life, in this case $t_{1/2} = 0.693/(\lambda_b + \lambda_c)$.   By definition the half-life is related to the total rate of disappearance of a substance, regardless of the mechanism by which it disappears.

If the Bateman solution is to be applied to a decay chain containing branching decays, the λ's in the numerators of the equations defining C_1, C_2, etc., should be replaced by the partial decay constants; that is, λ_i in the numerators should be replaced by $\lambda_i{}^*$, where $\lambda_i{}^*$ is the decay constant for the transformation of the ith chain member to the $(i+1)$th member.   If a decay chain branches, and subsequently the two branches are rejoined as in the natural radioactive series, the two branches are treated by this method as separate chains; the production of a common member beyond the branch point is the sum of the numbers of atoms formed by the two paths.

C. EQUATIONS OF TRANSFORMATION DURING NUCLEAR REACTIONS

Stable Targets.   When a target is irradiated by particles that induce nuclear reactions, a steady state can be reached in which radioactive products disintegrate at just the rate at which they are formed; the situation is analogous to that of secular equilibrium.   If the irradiation is terminated before the steady state is achieved, then the disintegration rate of a particular active nuclide is less than its rate of formation R.   The differential equation that governs the number of product atoms, N, present at time t during the irradiation is

$$\frac{dN}{dt} = R - \lambda N,$$

the solution to which is

$$R = \frac{N\lambda}{1 - e^{-\lambda t}}. \tag{3-6}$$

For very large irradiation times $(t \gg 1/\lambda)$ the disintegration rate, λN, approaches the "saturation value," R.   The factor $(1 - e^{-\lambda t})$ is often called the "saturation factor."   If the disintegration rate of a particular radioactive product at the end of a steady bombardment of known duration is divided by this saturation factor, the rate at which the product was formed during the bombardment is obtained.[1]

Occasionally a product is formed during irradiation both directly by a nuclear reaction and by the decay of an active parent that is produced by another reaction [e.g., the product of a (p, pn) reaction, if unstable, may decay by positron emission or electron capture into the product of the $(p, 2p)$ reaction on the same target].   Under these circumstances the number of atoms of the product of interest present at a time t_s after the end of a bombardment of duration t_b has three sources:

1. Those formed directly in nuclear reactions.
2. Those formed by the decay of the parent during bombardment.
3. Those formed by the decay of the parent during the interval t_s (which may, for example, be the time between the end of bombardment and the chemical separation of daughter from parent).

---

[1] If the rate of formation R is not constant during the irradiation (beam current varies during the bombardment), the bombardment may be divided into time intervals Δt_i during each of which the rate R_i is approximately steady.   Under these circumstances the number of atoms present after a bombardment of total length t comes directly from an obvious modification of (3-6):

$$N = \frac{1}{\lambda} \sum_{i=1}^{n} R_i(1 - e^{-\lambda \Delta t_i})e^{-\lambda(t - t_i)},$$

where t_i is the time at the end of the ith interval.   In the event that the time intervals are short compared to the half-life of the product $(\lambda \Delta t_i \ll 1)$, expansion of the exponential gives

$$N = \sum_{i=1}^{n} R_i \Delta t_i e^{-\lambda(t - t_i)}.$$

If R_1 and R_2 are the rates of the nuclear reactions that directly form the parent and daughter products, respectively, then the number of daughter atoms (characterized by subscript 2) arising from each of the three sources is

$$N_2' = \frac{R_2}{\lambda_2}(1 - e^{-\lambda_2 t_b})e^{-\lambda_2 t_s},$$ (3-7)

$$N_2'' = \left[\frac{R_1}{\lambda_2}(1 - e^{-\lambda_2 t_b}) + \frac{R_1}{\lambda_1 - \lambda_2}(e^{-\lambda_1 t_b} - e^{-\lambda_2 t_b})\right]e^{-\lambda_2 t_s},$$ (3-8)

$$N_2''' = \frac{R_1(1 - e^{-\lambda_1 t_b})(e^{-\lambda_1 t_s} - e^{-\lambda_2 t_s})}{\lambda_2 - \lambda_1}.$$ (3-9)

Experimentally, it is, of course, only the totality of the daughter atoms $(N_2' + N_2'' + N_2''')$ that is observed; but from a knowledge of the times t_b and t_s, the decay constants λ_1 and λ_2, and the rate of formation of the parent, R_1 (which can be determined in a separate experiment) it is possible to calculate R_2.

Radioactive Targets in a High-Flux Reactor. When nuclear reactions are induced in a radioactive nuclide, the rate of disappearance of the substance is no longer governed by the law of radioactive transformation alone but by a modified law which takes into account the disappearance by transmutation reactions also. Under most practical bombardment conditions the rate of transformation of radioactive species by nuclear reactions is negligible compared to the rate of radioactive decay. However, in the case of long-lived nuclides, and with the large neutron fluxes available in nuclear reactors, transformations by both mechanisms sometimes have to be considered. We shall state the modified transformation equations for the case of a neutron flux; they are equally applicable for any other bombarding particle. The treatment given here follows that developed by W. Rubinson (R1).

Consider N atoms of a single radioactive species of decay constant λ (in reciprocal seconds) and total neutron reaction cross section σ (in square centimeters) in a constant neutron flux nv (neutrons $\text{cm}^{-2}\ \text{sec}^{-1}$). The rate of radioactive transformation is λN, the rate of transformation by neutron reactions is $nv\sigma N$, and the total rate of disappearance is

$$-\frac{dN}{dt} = (\lambda + nv\sigma)N = \Lambda N,$$ (3-10)

where Λ may be considered as a modified decay constant. Equation (3-10) has the same form as the standard differential equation of radioactive decay and is integrated to give

$$N = N_0 e^{-\Lambda t}.$$ (3-11)

If we consider a parent-daughter pair, the parent disappears by both transmutation and decay: $-dN_1/dt = (\lambda_1 + nv\sigma_1)N_1 = \Lambda_1 N_1$; but the daughter grows by decay of the mother only and disappears by both processes: $dN_2/dt =$

$\lambda_1 N_1 - \Lambda_2 N_2$, or, in more general notation,

$$\frac{dN_{i+1}}{dt} = \lambda_i N_i - \Lambda_{i+1} N_{i+1}.$$

Actually we may want to consider chains in which the transformation from one member to the next may occur by nuclear reaction as well as by radioactive decay. Then λ_i must be replaced by a modified decay constant, $\Lambda_i{}^* = \lambda_i{}^* + nv\sigma_i{}^*$, where the asterisks serve as a reminder that if either the decay or reaction of the parent does not always lead to the next chain member then $\lambda_i{}^*$ must be the partial decay constant and $\sigma_i{}^*$ must be the partial reaction cross section leading from the ith member to the $(i + 1)$th member of the chain. With this notation the general solution is written, as in the Bateman equations, for $N_2{}^0 = N_3{}^0 = \cdots N_n{}^0 = 0$:

$$N_n = C_1 e^{-\Lambda_1 t} + C_2 e^{-\Lambda_2 t} + \cdots C_n e^{-\Lambda_n t}, \tag{3-12}$$

where

$$C_1 = \frac{\Lambda_1{}^* \Lambda_2{}^* \cdots \Lambda_{n-1}{}^*}{(\Lambda_2 - \Lambda_1)(\Lambda_3 - \Lambda_1) \cdots (\Lambda_n - \Lambda_1)} N_1{}^0,$$

$$C_2 = \frac{\Lambda_1{}^* \Lambda_2{}^* \cdots \Lambda_{n-1}{}^*}{(\Lambda_1 - \Lambda_2)(\Lambda_3 - \Lambda_2) \cdots (\Lambda_n - \Lambda_2)} N_1{}^0, \quad \text{etc.}$$

As an illustration, we compute the amount of 3.15-day Au^{199} formed by two successive n, γ reactions when 1 g Au^{197} is exposed for 30 hr in a neutron flux of 1×10^{14} cm^{-2} sec^{-1}. The chain of reactions is

$$Au^{197} \xrightarrow[n,\,\gamma]{\sigma = 99b} Au^{198} \xrightarrow[n,\,\gamma]{\sigma = 26{,}000b} Au^{199}$$

$$\beta^- \downarrow t_{1/2} = 2.7\text{d} \qquad\qquad \beta^- \downarrow t_{1/2} = 3.15\text{d}$$

We use (3-12) for this three-membered chain:

$$N_{199} = \Lambda_{197}{}^* \Lambda_{198}{}^* N_{197}{}^0 \left[\frac{e^{-\Lambda_{197} t}}{(\Lambda_{198} - \Lambda_{197})(\Lambda_{199} - \Lambda_{197})} \right.$$

$$\left. + \frac{e^{-\Lambda_{198} t}}{(\Lambda_{197} - \Lambda_{198})(\Lambda_{199} - \Lambda_{198})} + \frac{e^{-\Lambda_{199} t}}{(\Lambda_{197} - \Lambda_{199})(\Lambda_{198} - \Lambda_{199})} \right].$$

The numerical values to be substituted are

$$t = 1.08 \times 10^5 \text{ sec,}$$

$$nv = 10^{14} \text{ cm}^{-2} \text{ sec}^{-1},$$

$$\sigma_{197} = 9.9 \times 10^{-23} \text{ cm}^2,$$

$$\sigma_{198} = 2.6 \times 10^{-20} \text{ cm}^2$$

$$N_{197}{}^0 = \frac{6.02 \times 10^{23}}{197} = 3.05 \times 10^{21},$$

$$\Lambda_{197}{}^* = \Lambda_{197} = nv\sigma_{197} = 9.9 \times 10^{-9}\ \text{sec}^{-1},$$

$$\Lambda_{198} = \lambda_{198} + nv\sigma_{198} = 3.0 \times 10^{-6} + 2.6 \times 10^{-6}$$

$$= 5.6 \times 10^{-6}\ \text{sec}^{-1},$$

$$\Lambda_{198}{}^* = nv\sigma_{198} = 2.6 \times 10^{-6}\ \text{sec}^{-1},$$

and

$$\Lambda_{199} = \lambda_{199} = 2.55 \times 10^{-6}\ \text{sec}^{-1}.$$

Using these values, we get

$$N_{199} = 7.85 \times 10^7 \left(\frac{e^{-0.00107}}{5.6 \times 10^{-6} \times 2.55 \times 10^{-6}} \right.$$

$$\left. + \frac{e^{-0.605}}{5.6 \times 10^{-6} \times 3.05 \times 10^{-6}} - \frac{e^{-0.275}}{2.55 \times 10^{-6} \times 3.05 \times 10^{-6}} \right)$$

$$= 7.85 \times 10^7 (6.99 \times 10^{10} + 3.20 \times 10^{10} - 9.77 \times 10^{10})$$

$$= 3.3 \times 10^{17}.$$

The disintegration rate of Au^{199} at the end of the irradiation is $\lambda_{199}N_{199} = 0.84 \times 10^{12}\ \text{sec}^{-1}$. For comparison we compute the disintegration rate of Au^{198} in the sample [again from (3-12) for a two-membered chain]:

$$\lambda_{198}N_{198} = \lambda_{198}nv\sigma_{197}N_{197}{}^0 \left(\frac{e^{-\Lambda_{197}t}}{\Lambda_{198} - \Lambda_{197}} + \frac{e^{-\Lambda_{198}t}}{\Lambda_{197} - \Lambda_{198}} \right)$$

$$= 9.06 \times 10^7 \frac{0.999 - 0.546}{5.6 \times 10^{-6}} = 7.33 \times 10^{12}\ \text{sec}^{-1}.$$

Thus about 10 per cent of the radioactive disintegrations in the sample occur in Au^{199}.

D. UNITS OF RADIOACTIVITY

A familiar unit of radioactivity is the curie. Originally the term referred to the quantity of radon in equilibrium with one gram of radium. Later it came to be used as a unit of disintegration rate for any radioactive preparation, defined as that quantity of the preparation which undergoes the same number of disintegrations per second as one gram of pure radium. With this definition the value of the curie varied with successive refinements in the measurement of the decay constant or atomic weight of radium. In 1950 a Joint Commission of the International Union of Pure and Applied Chemistry and the International Union of Pure and Applied Physics adopted the following definition: "The curie is a unit of radioactivity defined as the quantity of any radioactive nuclide in which the number of disintegrations per second is 3.700×10^{10}." The millicurie (mc) and the microcurie (μc) are practical units also in common use, and the megacurie finds use in reactor technology.

As an illustration, we calculate the weight in grams W of 1.00 mc of C^{14} from its half-life of 5720 years:

$$\lambda = \frac{0.693}{5720 \times 3.156 \times 10^7} = 3.83 \times 10^{-12} \text{ sec}^{-1};$$

$$-\frac{dN}{dt} = \lambda N = \lambda \frac{W}{14} \times 6.02 \times 10^{23} = 1.65W \times 10^{11} \text{ sec}^{-1}.$$

With $-\dfrac{dN}{dt} = 3.700 \times 10^7$ disintegrations per second (1 mc),

$$W = \frac{3.700 \times 10^7}{1.65 \times 10^{11}} = 0.224 \times 10^{-3} \text{ g}.$$

E. DETERMINATION OF HALF-LIVES

Decay Curves. Half-lives in the range from several seconds to several years are usually determined experimentally by measurements of the activity with an appropriate instrument at a number of suitable successive times. Then log **A** is plotted versus time and the half-life is found by inspection, provided that the activity is sufficiently free of other radioactivities that a straight line (exponential decay) is found, preferably extending over several half-life intervals.   As we have already discussed, the decay curve resulting from a mixture of independent activities may often be analyzed to yield the half-lives of the various components.   When difficulties arise in this analysis, it is often adequate to measure decay curves separately through several different thicknesses of absorbing material to obtain curves with some components relatively suppressed; even better is the use of selective detection equipment which will measure separately the radiations from the several activities in the sample.   Our treatments of the more general equations have already suggested methods of finding half-lives from more complicated growth and decay curves.

The manipulations necessary for activity measurements become difficult as the time scale to be investigated becomes short.   The use of electronic and photographic recording devices can extend the working region to half-lives well below 0.1 second.   With short-lived gaseous products, or products in solution, a method that has been particularly useful for fission products with half-lives of the order of a few seconds is to measure the activity at different points along a tube through which the fluid flows at a measured rate.   The ordinary decay curve is then found on a plot of log **A** versus distance along the tube. A method based on a similar principle, using a rapidly rotating wheel, has been employed for solid samples; the half-life (0.020 second) of B^{12} was determined in this way. In these procedures the limitation usually arises not in the activity measurements but in the rapid preparation, and possibly isolation, of the short-lived sample. The possible use of a modulated source, such as a betatron or synchrotron, or a cyclotron modified to produce periodic pulses of accelerated ions, may be mentioned. Appropriate electric circuits will divide the time between pulses into an arbitrary number of intervals and measure the average resulting activity in each interval.

Variable-Delay Coincidences (S1). When a body of very short half-life results from a radioactive decay with moderate or long half-life, the method of variable-delay coincidences can be used. In one form of apparatus the electric pulse produced in a detection instrument by a ray from the parent is electrically delayed by a time t and then recorded in coincidence with any ray from the daughter that may produce a pulse in a detector at that time or, more correctly, at that time within the limits of the resolving time τ of the coincidence equipment. Now, as t is varied by electrical means, the coincidence counting rate—coincidences per unit τ—will vary; the effect is essentially to record disintegrations over the period τ at a time t after formation of the short-lived nucleus. The very short half-life is determined from the typical decay curve with the logarithm of the coincidence rate plotted against t. So long as Geiger counters were used as the detection instruments the lower limit of measurable half-lives was a few tenths of a microsecond because of inherent response times in these counters. With scintillation detectors (described in chapter 5, section B) the method has more recently been extended down to about 10^{-11} second.

Specific Radioactivity. If the half-life, or disintegration constant, is to be determined for a substance of very long half-life (very small λ), the activity $\mathbf{A} = c\lambda N$ may not change measurably in the time available for observation. In such cases λ may be found from the relation $\lambda N = -dN/dt = \mathbf{A}/c$, provided that N is known and $-dN/dt$ may be determined in an absolute way (through knowledge of the detection coefficient c). This method is most accurate for α emitters, and the absolute rates of emission of α particles from uranium samples have been investigated with great care to measure the half-life of U^{238}. In an accurate determination of the half-life of Pu^{239} the value of $-dN/dt$ was established in a calorimetric measurement of the heating effect, with the α-particle energy known from the α-particle range.

In some instances the disintegration rate is better obtained from a measurement of the equal disintegration rate of a daughter in secular equilibrium. Early determinations of the half-life of U^{235} were based on the α-particle counting rate of Pa^{231} obtained in known yield from old uranium ores; the U^{235} α particles were not measurable in a direct way because of the much larger number of α disintegrations occurring in the U^{238} and U^{234}.

REFERENCES

B1 H. Bateman, "Solution of a System of Differential Equations Occurring in the Theory of Radio-active Transformations," *Proc. Cambridge Phil. Soc.* **15,** 423 (1910).
C1 J. B. Cumming, "CLSQ, The Brookhaven Decay-Curve Analysis Program," *Application of Computers to Nuclear- and Radiochemistry* (G. D. O'Kelly, Editor), Washington, NAS-NRC, 1963, p. 25.
R1 W. Rubinson, "The Equations of Radioactive Transformation in a Neutron Flux," *J. Chem. Phys.* **17,** 542 (1949).
*R2 E. Rutherford, J. Chadwick, and C. D. Ellis, "Radiations from Radioactive Substances," Cambridge University Press, 1930.
S1 A. Schwarzschild, "A Survey of the Latest Developments in Delayed Coincidence Measurements," *Nucl. Inst. Methods,* **21,** 1 (1963).

EXERCISES

1. The following experimental data were obtained when the activity of a certain β-active sample was measured at the intervals shown.

Time (in hours)	Activity (in counts/min)	Time (in hours)	Activity (in counts/min)
0	7300	4.0	481
0.5	4680	5.0	371
1.0	2982	6.0	317
1.5	1958	7.0	280
2.0	1341	8.0	254
2.5	965	10.0	214
3.0	729	12.0	181
3.5	580	14.0	153

Plot the decay curve on semilog paper and analyze it into its components. What are the half-lives and the initial activities of the component activities?

2. Compute (a) the weight of 1 curie of Rn^{222}; (b) the weight of 1 curie of P^{32} (see appendix E for the half-life); (c) the disintegration rate of 1 cm^3 of tritium (H^3) at STP.   *Answer:* (a) 6.49 μg.

3. What was the rate of production, in atoms per second, of I^{128} during a constant 1-hour cyclotron (neutron) irradiation of an iodine sample if the sample is found to contain 2.00 mc of I^{128} activity at 15 minutes after the end of the irradiation?

4. From data in appendix E calculate the total rate of emission of α particles from 1 mg of ordinary uranium. Calculate this answer also for the case of 1 mg of very old uranium in secular equilibrium with all its decay products.
                                          Answer to first part: 24.0/sec.

5. A 0.100 mg sample of pure $_{94}Pu^{239}$ (an α-particle emitter) was found to undergo 1.40×10^7 disintegrations per minute. Calculate the half-life of this isotope. Pu^{239} is formed by the β decay of Np^{239}. How many curies of Np^{239} would be required to produce a 0.100-mg sample of Pu^{239}?
                                          Answer to second part: 23.2 curies.

6. To determine the thermal-neutron capture cross section of 31-hour Os^{193}, a 100-mg sample of osmium metal is placed in a thermal-neutron flux of 2×10^{12} cm^{-2} sec^{-1} for 30 days. The amount of 700-day Os^{194} activity formed is found from subsequent decay measurements to be 210 disintegrations per second at the end of irradiation. From the known activation cross section of Os^{192} (1.6b) and the half-lives of Os^{193} and Os^{194} compute the capture cross section of Os^{193}.

Answer: See appendix C.

7. A sample of 1.00×10^{-10} g of RaE is freshly purified at time $t = 0$. (a) If this sample is left without further treatment, when will the amount of Po^{210} in it be a maximum? (b) At that time of maximum growth what will be the weight of Po^{210} present, the α activity in disintegrations per second, the beta activity of the sample in disintegrations per second, the number of microcuries of Po^{210} present? (c) Sketch on semilog paper a graph of α activity and β activity versus time.

8. (a) Show that the number of radioactive daughter atoms present in a sample at time t is given by

$$(\lambda N_1{}^0 t + N_2{}^0)e^{-\lambda t}$$

if the parent and daughter atoms have the same half-life and $N_1{}^0$ and $N_2{}^0$ are the numbers of parent and daughter atoms, respectively, that were initially present. (b) Show that the sum of the disintegration rates of parent and daughter decays with the half-life of the daughter if the daughter substance has twice the half-life of the parent. (c) Derive the condition $c_2/c_1 > \lambda_1/\lambda_2$ (given on p. 72) for the occurrence of a maximum in the total counting rate of an initially pure parent fraction.

9. In the slow-neutron activation of a sample of separated Mo^{100} isotope some 14.6-min Mo^{101} is produced; this decays to 14.0-min Tc^{101}. A sample of Mo^{101} is chemically freed of technetium and then immediately placed under a counter. Sketch the activity as a function of time, assuming the detection coefficient to be the same for the Tc^{101} as for the Mo^{101} radiation.

10. Carry out the solution of the differential equation 3-5. Compare your result with the Bateman solution for this case with $N_2{}^0$ and $N_3{}^0$ not equal to 0.

11. A sample of an activity whose half-life is known to be 7.50 min was measured from 10:03 to 10:13. The total number of counts recorded in this 10-minute interval was 34,650. What was the activity of the sample (in counts per minute) at 10:00?

Answer: 7012.

12. What will be the disintegration rate of Au^{198} in a 10-mg sample of gold that has been irradiated in a nuclear reactor at a flux of 10^{12} neutrons cm^{-2} sec^{-1} for one day? Take the capture cross section of Au^{197} as 98.8 barns (appendix C).

Answer: 6.8×10^8 sec^{-1}.

13. A 1.0-mg cm^{-2} target of Zn^{64} enriched to 100 per cent is bombarded for 38 minutes in an α-particle beam of intensity 6×10^{12} alphas per second. The energy of the α particles is such that there is production of Ge^{67} and Ga^{67} by the (α, n) and (α, p) reactions, respectively, during the bombardment. From Ga and Ge samples isolated from the target it was found that the disintegration rate of the Ge^{76} one hour after the irradiation was 1.34×10^6 disintegrations per second, whereas that for Ga^{67} was 2.50×10^5 disintegrations per second. The Ga sample from which the disintegration rate of Ga^{67} was determined was separated from the Ge in the target 0.5 hour after the end of the irradiation. From this information calculate the rates of formation of Ge^{67} and Ga^{67} directly by nuclear reactions during the irradiation and their production cross sections.

Answers: Ge^{67}—1.60×10^7 sec^{-1},
$\sigma = 283$ mb.
Ga^{67}—3.19×10^7 sec^{-1},
$\sigma = 564$ mb.

4

Interaction of Radiations with Matter

Nuclear radiations, both corpuscular and electromagnetic, are detectable only through their interactions with matter.   If this interaction is sufficiently small, as in the case of the neutrino, the radiation remains undetected.   For an understanding of the methods and instruments used for the detection, measurement, and characterization of nuclear radiations it is necessary to consider the manner in which these radiations interact with matter.

Up to about 1950 absorption studies of the radiation from radioactive substances were an important technique for energy determinations.   They no longer are; most serious energy determinations are now performed by the measurement of deflections in electric and magnetic fields, by the use of detectors whose output is sensitive to energy, or by the use of crystal diffraction. Nevertheless, the slowing down and absorption of radiation in matter remain important for the reduction in energy of beams of high-energy particles and, when feasible, for the detection of a particular radiation in the presence of others with sufficiently different absorption characteristics.

A. ALPHA PARTICLES AND OTHER IONS

Processes Responsible for Energy Loss.   In passing through matter, α particles lose energy chiefly by interaction with electrons.[1]   This interaction may lead to the dissociation of molecules or to the excitation or ionization of atoms and molecules.   The effect most easily measured and most often used for the detection of α particles is ionization.   The details of the ionization processes and other effects associated with α-particle passage are more readily investigated in gases than in liquids or solids, although the processes are pre-

---

[1] The interaction of α particles with nuclei (scattering and nuclear reactions) is discussed in chapter 10.   The contribution of these processes to energy loss of helium ions in passing through matter is entirely negligible except at very high energies (>100 MeV).

86

Table 4-1   Average energy lost by alpha particles in producing one ion pair in various gases

Gas	Energy per Ion Pair W (eV)	First Ionization Potential I (eV)	Fraction of the Energy Used in Ionization (I/W)
H_2	36.3	15.6	0.43
He (very pure)	43	24.5	0.58
He (tank)	30		
N_2	36.5	15.5	0.42
O_2	32.5	12.5	0.38
Air	35.0		
Ne (very pure)	36.8	21.5	0.58
Ne (tank)	28		
Ar	26.4	15.7	0.59
Kr	24.1	13.9	0.58
Xe	21.9	12.1	0.55
CH_4	30	14.5	0.48
C_2H_4	29	10.5	0.36
CO	34	14.3	0.42
CO_2	34		
CS_2	26	10.4	0.40
CCl_4	27		
NH_3	39	10.8	0.28

sumably about the same.   We shall therefore speak mostly of phenomena observed in the passage of α particles through gases.

Because α particles travel only short distances in matter before being reduced to thermal energies, a known number of α particles of known initial energy can be made to spend their entire energy inside an ionization chamber and thus the total ionization produced per α particle is readily measured.   These experiments show that on the average approximately 35 eV of energy are dissipated for each ion pair formed in air.   The energy expended in the formation of an ion pair is listed for a number of other gases[2] in table 4-1, together with the first ionization potentials of these gases.   In the noble gases the fraction of the α-particle energy spent in ionization processes is larger than it is in the diatomic and polyatomic gases where dissociation of molecules is also possible.   Most of the values tabulated were determined for α particles from Po^{210}, but the energy per ion pair is surprisingly insensitive to the energy and the nature of the

[2] It is to be noted that the energy expended in producing an ion pair (or rather an electron-hole pair) in semiconductors is about an order of magnitude less than in gases.   This fact is discussed further in chapter 5, section A3.

radiation; almost identical values for several of the gases have been obtained with 340-MeV protons.   A very good way to measure the energy of an α particle or similar ion is to measure electrically the total number of ions produced when it is stopped in a gas-filled ionization chamber.

A large part of the energy loss of α particles and other ions is accounted for by the kinetic energy given to the electrons removed from atoms or molecules in close collisions with the α particle.   It can easily be shown from conservation of momentum that the maximum velocity which an α particle of velocity v can impart to an electron is about $2v$; therefore the maximum energy which an electron can receive from the impact of a 6-MeV α particle, for example, is about 3 keV.   The average energy imparted to electrons by α particles in their passage through matter is of the order of 100 to 200 eV.   Many of these secondary electrons or δ rays are energetic enough to ionize other atoms.   In fact about 60 to 80 per cent of the ionization produced by α particles is due to secondary ionization; the exact ratio of primary to secondary ionization is difficult to determine.   Delta-ray tracks are often seen in cloud-chamber pictures of α-particle tracks.

When the velocity of the α particles has been reduced to a point at which it is comparable to the velocity of the valence electron in an atom of the stopping material, a new phenomenon becomes important: the α particle starts making elastic collisions with the atoms rather than exciting the atomic electrons. These ion-atom collisions give rise to what is known as "nuclear stopping," as compared to the "electronic stopping" which occurs at the higher velocities. In addition, when the velocity of the α particles becomes comparable to that of an electron in a He^+ ion, the α particles will start picking up electrons from the atom in the stopping material, and the average charge of the particles will change from $+2$ to $+1$.   On the average, heavy ions passing through matter will be stripped of all orbital electrons whose orbital velocity is less than that of the heavy ion.

In sum, there are three important phenomena attendant upon the passage of a heavy ion through matter:

1. At sufficiently high velocities the ion is stripped of all of its electrons and the energy loss is essentially all through electronic excitation and ionization of the stopping material.

2. At velocities comparable to the velocities of its K-shell electrons the heavy ion starts to pick up electrons from the stopping material.   The mechanism of energy loss is still essentially all electronic.

3. At velocities comparable to those of the valence electrons of the stopping material the mechanism of energy loss essentially becomes one of elastic collisions between the heavy ion, even if it still has a charge, and the atoms of the stopping material.

There is no sharp gradation between (2) and (3); there is an energy region in which the heavy ion makes both elastic and inelastic atomic collisions.

Range. Because an α particle loses only a very small fraction of its energy in a single collision with an electron and is not appreciably deflected in the collision, α-particle paths are very nearly straight lines. Furthermore, because of the very large number of collisions (of the order of 10^5) necessary to bring an α particle of a few million electron volts initial energy to rest, the ranges of all α particles of the same initial energy are the same within narrow limits. Ranges of α particles are generally determined by absorption methods, either with solid absorbers or, more accurately, with a gaseous absorber at variable pressures. Ranges may be determined with a precision of about one part in 5000.

In figure 4-1 the number of α particles found in a gas at a distance r from the source is plotted against r for the case of a source which emits α particles of a single energy. It is seen that the ranges of all the particles in a given medium are not exactly the same but show a small spread of about 3 or 4 per cent. This phenomenon, called the straggling of α-particle ranges, is caused by the statistical fluctuations in the number of collisions and in the energy loss per collision. The dotted curve in figure 4-1 is obtained by differentiation of the other (integral) curve and represents the distribution of ranges or the amount of straggling; it is approximately a Gaussian curve. The distance r corresponding to the maximum of the differential curve (point of inflection of the integral curve) is called the mean range R of the α particles. The distance r obtained by extrapolating to the abscissa the approximately straight portion of the integral curve is the extrapolated range R_{ex}. The mean range

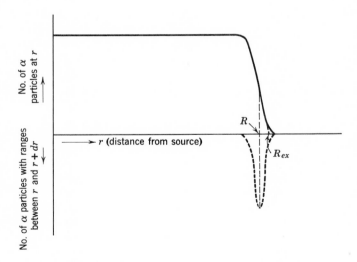

Fig. 4-1   Number of α particles from a point source as a function of distance from the source (full curve). The derivative of that function is also shown (dotted curve); the latter represents the distribution in ranges.

is now generally used in range tables and in range-energy relations.   Extrapolated ranges were often given in the older literature and are more easily determined experimentally.   Relations between the two ranges are available; the difference is approximately 1.1 per cent for α particles of ordinary energies.

Range-Energy Relation and Specific Ionization.   The relationship between the energy of a heavy ion (mass greater than that of an electron) and its range may be more clearly seen in terms of dE/dx, the rate at which the charged particle loses energy in passing through matter.   In a given medium the quantity dE/dx, often called the specific ionization, is a function of the energy, charge, and mass of the heavy ion.   The general properties of this function may be seen from the data in table 4-2, which gives the differential energy loss of protons and of α particles in air, and from figure 4-2, a classical Bragg curve that gives the specific ionization as a function of the distance of an α particle from the end of its path.   From these data it is seen that there is a maximum rate of energy loss that occurs at rather low energies and a decrease toward higher energy that is approximately given by an inverse dependence on the energy. This behavior can be rather easily understood qualitatively.

Table 4-2   Ionizing effects of protons and alpha particles in air

Energy (MeV)	Ion Pairs per 1.00 mg cm^{-2} Produced by	
	Protons	α Particles
0.025	16,700	. . . . . .
0.1	18,000	. . . . . .
0.2	16,700	40,000
0.5	11,200	56,000
1.0	6,800	54,000
2.0	4,200	41,000
3.0	3,100	32,000
4.0	2,400	26,000
5.0	2,100	22,000
6.0	1,800	20,000
8.0	1,400	16,000
10.0	1,100	13,000
14.0	900	10,300
25.0	560	6,500
70.	240	2,900
100.	190	2,200
1000.	55	400
10000.	61	210

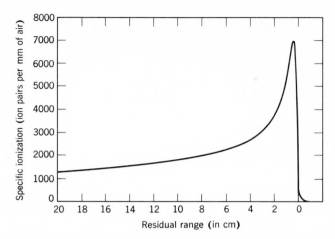

Fig. 4-2   Bragg curve for initially homogeneous α particles.

The interaction between the charged particle and the atomic electrons is nearly completely described as a Coulomb interaction between the electrons and a positive point charge. Thus, if the heavy ion diminishes its charge by picking up electrons in its passage through matter, the Coulomb interaction and the rate of energy loss will diminish. This is what occurs on the low-energy side of the maximum and ultimately changes the mechanism of energy loss to one of elastic collisions between atoms. As a rough rule, a heavy ion will pick up an electron whose orbital velocity is no greater than its own speed. Actually, a heavy ion will pick up and then lose electrons many times near the end of its trajectory. For α particles the fluctuations in charge occur in the last few millimeters of their path.

The diminution in the rate of energy loss with increasing energy on the high-energy side of the maximum is a consequence of the diminished "time of inter-action" between the charged particle and the atomic electrons. If the velocity of the ion is v, the time that the ion spends within a given distance from the atom is proportional to $1/v$; hence the impulse on the electrons in the atom, or the momentum transfer, is also proportional to $1/v$. Since the energy transfer to the electrons in the atom is equal to the energy loss by the ion and is proportional to the square of the momentum transfer, it is evident that the rate of energy loss must have a term inversely proportional to v^2 or inversely proportional to E, the energy of the particle.

Derivation of Stopping-Power Formula. We now proceed to use the simple model of momentum transfer via a Coulomb interaction just outlined as the basis for a classical treatment of dE/dx, which yields many of the important features that result from a quantum-mechanical analysis of the problem. The central problem of the classical calculation is that of computing the

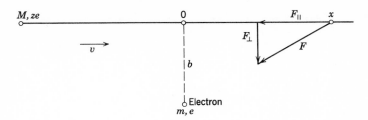

Fig. 4-3   Ion of mass M and charge Ze losing energy to an electron (mass m and charge e) in an interaction with impact parameter b.

momentum transfer to an electron from a heavy ion of charge z and mass M moving with a velocity v along a linear trajectory characterized by a distance of closest approach to the electron (impact parameter) b.   The model is illustrated in figure 4-3.   The assumption that the ion moves undeflected is justified because the mass of the electron is negligible compared to that of the heavy ion.   The electron is considered to be free and stationary; this is acceptable as long as the velocity of the heavy ion is much greater than that of the electron, or $(m/M)E \gg I$ where I is approximately the ionization potential of the electron and m is the electron mass.

The force exerted on the electron by the heavy ion at point x (origin of coordinates taken at the point of closest approach) can be resolved into two components, that normal to and that parallel to the trajectory:

$$F_{\perp}(x) = \frac{ze^2}{(x^2 + b^2)} \frac{b}{(x^2 + b^2)^{\frac{1}{2}}}, \tag{4-1a}$$

$$F_{\parallel}(x) = \frac{ze^2}{(x^2 + b^2)} \frac{x}{(x^2 + b^2)^{\frac{1}{2}}}. \tag{4-1b}$$

From the symmetry of the problem it is evident that the average value of the parallel component will vanish.   The momentum transfer to the electron will then be given by the usual impulse integral:

$$\Delta p = \int_0^{\infty} F_{\perp}(x) \, dt. \tag{4-2}$$

The integration over time may be transformed to an integration over x through the function $dx/dt = v$:

$$\Delta p = \frac{1}{v} \int_{-\infty}^{+\infty} F_{\perp}(x) \, dx \tag{4-2a}$$

$$= \frac{ze^2 b}{v} \int_{-\infty}^{+\infty} \frac{dx}{(x^2 + b^2)^{\frac{3}{2}}} \tag{4-2b}$$

$$= \frac{2ze^2}{bv}. \tag{4-2c}$$

The energy of the accelerated electron is $(\Delta p)^2/2m$; thus the energy transfer in a "collision" at impact parameter b is

$$\Delta E = \frac{2z^2e^4}{mb^2v^2}. \tag{4-3}$$

The number of electrons with impact parameters between b and $b + db$ in a small thickness dx of the absorber is

$$2\pi bn\, db\, dx,$$

where n is the number of electrons per unit volume in the absorber. Thus the energy transfer to electrons at impact parameters between b and $b + db$ is $4\pi z^2e^4n\, db/mbv^2$ and the total rate of energy loss, often called the stopping power, is obtained by integrating over all allowable values of b:

$$-\frac{dE}{dx} = \frac{4\pi z^2e^4n}{mv^2} \int_{b_{min}}^{b_{max}} \frac{db}{b} \tag{4-4a}$$

$$= \frac{4\pi z^2e^4n}{mv^2} \ln \frac{b_{max}}{b_{min}}. \tag{4-4b}$$

The negative sign of the differential arises because (4-4b) represents the rate of energy *loss* by the heavy ion. The allowable extremes of the impact parameter that appear as the limits of integration in (4-4a) are determined by the assumptions made. The maximum impact parameter is limited by the assumption that the electron is both free and stationary during the time of the interaction. Although, strictly speaking, the range of the Coulomb interaction is infinite, the major effect, as seen from (4-1), occurs when $|x| < b$; thus the time of the interaction is approximately $2b/v$. We demand that this time be less than $1/\omega$ where ω is the classical frequency of motion of the electron in the atom. The upper limit on the impact parameter is then

$$b_{max} = \frac{v}{2\omega}. \tag{4-4c}$$

The minimum impact parameter is determined by the maximum amount of energy that can be transferred in a single collision. As stated earlier, the maximum velocity that may be imparted to an electron by collision with a heavy ion is $2v$, which corresponds to a maximum energy transfer of $2mv^2$. From (4-3) the minimum impact parameter is

$$b_{min} = \frac{ze^2}{mv^2} \tag{4-4d}$$

The substitution of (4-4c) and (4-4d) into (4-4b) gives

$$-\frac{dE}{dx} = \frac{4\pi z^2e^4n}{mv^2} \ln \frac{mv^3}{2ze^2\omega}, \tag{4-4e}$$

a formula first derived by Bohr (B1).

A more exact analysis that proceeds through a quantum-mechanical and relativistic statement of the conditions required by the assumptions leads to the more accurate expression

$$-\frac{dE}{dx} = \frac{4\pi z^2 e^4 n}{mv^2}\left[\ln\frac{2mv^2}{I} - \ln(1-\beta^2) - \beta^2\right]. \tag{4-5}$$

The quantity I is the effective ionization potential of the atoms in the absorber, and $\beta = v/c$ where c is the velocity of light.

For kinetic energies of heavy ions small compared to their rest-mass energy ($\beta \ll 1$), (4-5) reduces to

$$-\frac{dE}{dx} = \frac{4\pi z^2 e^4 n}{mv^2}\ln\frac{2mv^2}{I}. \tag{4-6}$$

Equation 4-6 gives a continuous decrease in the rate of energy loss as the energy of the ion increases.

On the other hand, when the kinetic energy of the ion becomes larger than its rest-mass energy ($\beta \to 1$), the term $\ln(1-\beta^2)$ in (4-5) is the most rapidly varying one and the rate of energy loss increases with increasing energy. Thus the rate of energy loss goes through a broad minimum which occurs at approximately twice the rest-mass energy of the ion. The specific ionization of a singly charged particle at the minimum is about 1.8 MeV g^{-1} cm^{-2} in carbon and 1.1 MeV g^{-1} cm^{-2} in lead. Physically, the specific ionization increases with energy in the relativistic region because the Lorentz contraction shortens the time of a collision and thereby allows a larger $b_{\max}$.

An important feature of the equations just given is immediately evident: the rate of energy loss of all charged particles moving with the same velocity in a given absorber is proportional to the square of their charges. Thus the rates of energy loss of protons of energy E, deuterons of energy $2E$, and tritons of energy $3E$ are all the same and are one quarter as large as those of a He^3 of energy $3E$ or an α particle of energy $4E$. From data of the kind given in table 4-2, the rate of energy loss of any other charged particle may be simply computed as long as the energies are within the range in which energy loss is caused mainly by excitation and ionization of atoms.

Range-Energy Relations. The range of an ion may be immediately computed by integration of the energy-loss expression:

$$R = \int_{E_0}^{0}\frac{1}{dE/dx}\,dE. \tag{4-7}$$

Figure 4-4 gives the ranges in dry air of protons and of helium ions (α particles) as functions of their kinetic energy. These curves, taken from the literature (see references), are based more on theoretical calculations than on experiment because the calculated values are believed to be more accurate over most of the energy region. The only energy losses considered were those covered by

(4-5).   At very high energies other mechanisms may become relatively important; for example, 2-GeV protons are appreciably (approximately 15 per cent) attenuated in intensity by nuclear reactions in a lead absorber 1 in. thick but lose relatively little (less than 3 per cent) of their energy by ionization processes.

The ranges in figure 4-4 are expressed in milligrams of air per square centimeter in the absorption path.   The range can be expressed as the length of the absorption path in centimeters, provided that the density of the air is specified. Dry air at 15°C and 760 mm pressure (the standard for range values) has a density of 1.226 mg cm^{-3}, and for such air the range in centimeters is given by the range in milligrams per square centimeter divided by 1.226.   Obviously the range in centimeters for air at other temperatures and pressures is readily computed by use of the corresponding density.   Even in liquid air the ranges in milligrams per square centimeter are virtually unchanged from those given in figure 4-4.

The ranges of α particles and other ions in absorbing materials other than air are often wanted.   Data can hardly be accumulated for all substances and again theoretical calculations based on (4-5) and (4-6) are resorted to.   These computations are time-consuming and have been carried through for relatively few elementary substances (B2, B3, N1, R1, S1).   Several approximate rules, now chiefly of historical significance, have been given for estimating ranges in an absorber of atomic number Z and mass number A; one that is easy to remember is that the range in milligrams per square centimeter is proportional to $A^{\frac{1}{2}}$ (Bragg's rule).   Actually, the range in an absorber, compared to the range in air for the same ray, is a complicated function of Z, A, and the radiation energy E.   The following equation approximates this function well enough to be useful, within the limits indicated below, for protons, deuterons, and helium ions with initial energies from about 0.1 to about 1000 MeV and for any elementary absorber:

$$\frac{R_Z}{R_a} = 0.90 + 0.0275Z + (0.06 - 0.0086Z) \log \frac{E}{M} \qquad (4\text{-}8)$$

Here R_Z is the range in element Z in milligrams per square centimeter, R_a is the range of the same ray in air in the same unit, M is the mass number of the particle (1 for protons, 4 for α particles, etc.), and E is the initial particle energy in million electron volts.   As written, the equation is applicable to absorbers with $Z > 10$.   For lighter elements, replace $(0.90 + 0.0275Z)$ by 1.00, except for helium and hydrogen for which 0.82 and 0.30, respectively, are used.   (For deuterium as absorber double the range computed for hydrogen.)   For elements heavier than air the results are improved by replacement of R_Z by $R_Z + (0.01Z/z)$, where z is the atomic number of the particle; this correction is quite negligible except for very large Z or very low E.   With these modifications, (4-8) fits available calculated range-energy curves for the light-element absorbers and for aluminum, copper, silver, and lead to within a few per cent, at least from 1 to 100 MeV.

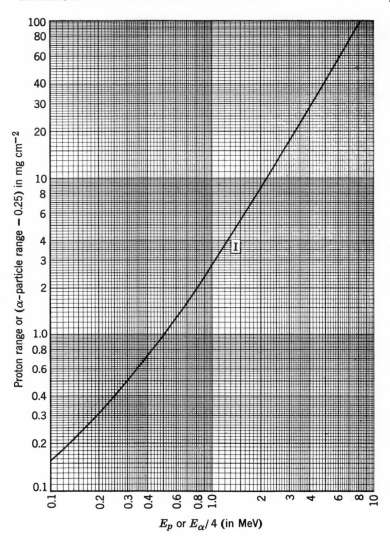

Fig. 4-4   Range-energy relations for protons and helium ions in air.

In many actual cases, indeed for air, the stopping substance is not a single element but rather a compound or mixture of elements.   For practical purposes we make the further approximation that the stopping effect of a molecule or of a mixture of atoms or molecules is given by the sum of the stopping effects of all the component atoms (another rule due to Bragg).[3]   Therefore, if R_1, R_2,

[3] In view of the fact that a considerable fraction of the α-particle energy is expended in molecular excitation and dissociation processes, this simple additivity relation is somewhat

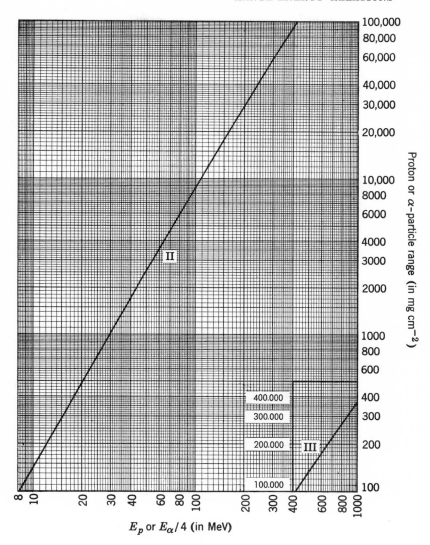

Fig. 4-4 *(continued)*

R_3 . . . denote the ranges (in milligrams per square centimeter) of a particular ray in each of several elements, the range R_t of that ray in a compound or in an essentially homogeneous mixture of these elements with respective weight

---

surprising. The stopping power of water vapor has been measured to be about 3 per cent less than that of the equivalent mixture of hydrogen and oxygen; range measurements in a number of organic isomers show that ranges in them are the same within less than 1 per cent.

fractions w_1, w_2, w_3 . . . is approximately given by

$$\frac{1}{R_t} = \frac{w_1}{R_1} + \frac{w_2}{R_2} + \frac{w_3}{R_3} + \cdots \qquad (4\text{-}9)$$

Because the relative stopping effects of various elements are functions of the energy of the α particle or other ion, (4-9) is not to be applied to grossly heterogeneous absorbers, which contain separate phase regions large enough to produce serious changes in the particle energy.

As an example, we shall determine the range of 20-MeV He^4 ions in polyethylene, $(CH_2)_x$. From figure 4-4 the range in air is 41.3 mg cm^{-2}. Using (4-8) (modified), we get for the range in hydrogen

$$R_H = 41.3(0.30 + 0.051 \log \tfrac{20}{4}) = 13.9 \text{ mg cm}^{-2},$$

and for the range in carbon

$$R_C = 41.3(1.00 + 0.012 \log \tfrac{20}{4}) = 41.6 \text{ mg cm}^{-2}.$$

Polyethylene is 85.6 per cent by weight carbon and 14.4 per cent hydrogen; hence by (4-9)

$$\frac{1}{R_{CH_2}} = \frac{0.856}{41.6} + \frac{0.144}{13.9}$$

and $R_{CH_2} = 32.3$ mg cm^{-2}.

Suppose now that a beam of 20-MeV He^4 ions is incident on a polyethylene absorber 15 mg cm^{-2} thick. At what energy E' will the He^4 ions emerge and how much farther will they travel in air? The emerging He^4 ions would have a residual range in polyethylene of $32.3 - 15.0 = 17.3$ mg cm^{-2}. Therefore their range in air R_a' is given by

$$\frac{1}{17.3} = \frac{0.856}{R_a'[1.00 + 0.012 \log (E'/4)]} + \frac{0.144}{R_a'[0.30 + 0.051 \log (E'/4)]}.$$

As a first approximation we neglect the small log terms in the denominators and get $R_a' = 23.1$ mg cm^{-2}. According to figure 4-4, this corresponds to $E' = 16$ MeV, and, using this value, we obtain as a second (and evidently sufficiently accurate) approximation $R_a' = 22.2$ mg cm^{-2} and $E' = 15.6$ MeV.

Ranges of Other Ions. Any ion moving at high speed through matter loses its energy by essentially the same mechanism. We have already stated that at a given velocity the energy loss (measured by the specific ionization) is proportional to z^2 (z is the *net* charge on the ion). Therefore α particles and protons of the same velocity have almost equal ranges because the α particle loses energy at four times the rate for the proton and has just four times as much energy to lose. In general, for an ion of charge z, mass number M, and

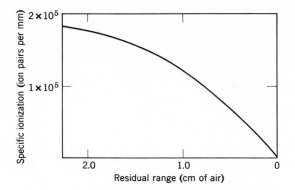

Fig. 4-5 "Bragg curve" for a fission fragment.

energy E the range is approximately

$$R_{z,M,E} = \frac{M}{z^2} R_{p,E/M}, \tag{4-10}$$

where $R_{p,E/M}$ is the range in the same absorber of a proton of energy E/M. Equation 4-10 gives only the lower limit to the range; the pickup of electrons by the heavy ion as it is slowed down will lower the net charge, z, and increase the range.

Fission fragments have $M \cong 100$, $Z \cong 45$, and initially $E \cong 100$ MeV. If the fragments were completely stripped of electrons, the large value of $z^2 = Z^2$ would lead to ranges of the order of 0.14 mg cm^{-2}, roughly 1 mm in air. Actually a fission fragment probably will be stripped only of electrons with orbital velocities smaller than its own velocity and so will hold all electrons with binding energies greater than about 1 keV.[4] The fragment starts with net $z \cong 20$ and gains electrons as its velocity decreases until $z \cong 0$ at about 1 MeV (approximately 3 mm before the end of the range).   Measured fission fragment ranges in slow-neutron fission are about 1.9 to 2.9 cm in air; a curve of specific ionization versus residual range looks about as shown in figure 4-5. The relative stopping powers of various substances are very nearly the same for fission fragments as for α particles.

In figures 4-6a and b, taken from the work of Alexander and Gazdik (A1), the energy dependence of the ranges of typical light and heavy fission fragments in several elements is shown.

[4] The kinetic energy of a bound electron (which is equal to its binding energy) with velocity equal to that of a fission fragment is in a nonrelativistic approximation the energy of the fission fragment multiplied by the ratio of the electron-to-fragment masses.   For a 100-MeV fragment with mass 100 the binding energy of the corresponding electron is $100 \times 10^6 \times (0.00055/100) = 550$ eV.

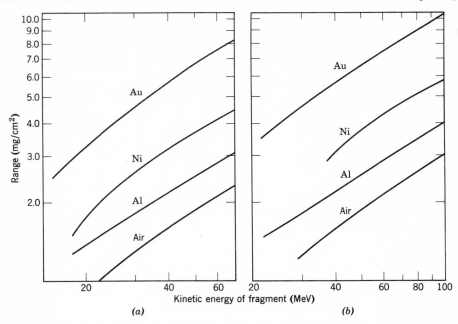

Fig. 4-6   Range-energy curves for fission fragments: (*a*) median heavy fragment (A ≅ 139); (*b*) median light fragment (A ≅ 95). [From J. Alexander and M. F. Gazdik, *Phys. Rev.* **120**, 874 (1960).]

The ranges of other ions, heavier than α particles but lighter than fission fragments, are given in figure 4-7.   These curves are based on the work of Northcliffe (N1).   Again, the relative stopping powers of other materials for these ions are similar to those for alpha particles.   A critical review of the theories and experimental results for the ranges of ions heavier than α particles is found in ref. N2.   Ranges of heavy ions with initial energies so low that most of the energy loss occurs through elastic atomic collisions (nuclear stopping) present special problems which have been treated in detail in reference L1.

Straggling.   The rate of energy loss, as given in (4-5), is only an average quantity; there are fluctuations in the energy lost by an ion in each collision as well as fluctuations in the number of collisions per unit path length.   These fluctuations in the fractional energy loss per collision become even larger at low energies where the nuclear stopping is dominant as well as at slightly higher energies where the fluctuations in the charge of the ion occur.   Further, the ions, largely through nuclear stopping, will undergo scattering, and thus the distance traveled by the ion along its original direction of motion is less than the actual distance traversed.   As a consequence of all of these effects, an initially monoenergetic beam of ions of a given type does not have a unique range in an absorber. As pointed out in the discussion of figure 4-1, there is a distribution of ranges. These phenomena all come under the heading of straggling.

Quantitatively, the straggling, S, is defined as the difference between the mean and extrapolated ranges, as defined in figure 4-1. For protons moving

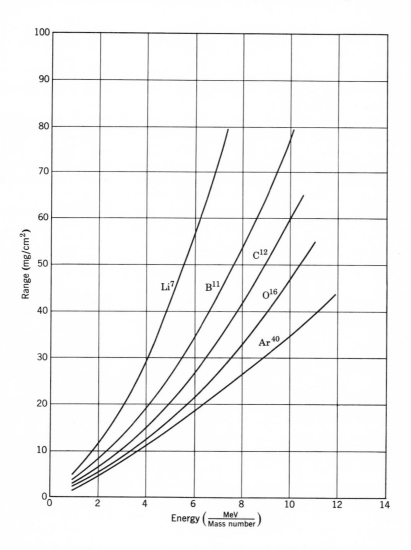

Fig. 4-7 Range-energy relation for several heavy ions in aluminum. [Data from L. Northcliffe, *Phys. Rev.* **120**, 1744 (1960).]

through air the straggling expressed in terms of the mean range varies from about 1.9 to 1.1 per cent as the initial energy varies from about 8 to 500 MeV; the percentage of straggling diminishes by about 0.3 for each fourfold increase in energy. The straggling of any other particle of charge z and M may be approximated from that of protons of the *same initial velocity if its energy is sufficiently high so that all but a negligible part of its range is spent in the completely ionized form:*

$$S_{z,M} = \frac{\sqrt{M}}{z^2} S_{1,1}.$$

(4-11)

This expression, for example, would be useful for 40-MeV α particles but useless for fission fragments.

At the other extreme, where most of the stopping of the heavy ions is largely nuclear stopping (below about 1 MeV for ions of atomic number greater than about 35 being stopped in an absorber of the same atomic number), the mean square deviation of the range divided by the square of the range is approximately $\frac{2}{3}[M_1 M_2/(M_1 + M_2)^2]$ where M_1 and M_2 are the atomic weights of the heavy ion and of the atoms in the absorber, respectively (L1).

B. ELECTRONS

Processes Responsible for Energy Loss. The interaction of electrons with matter is in many ways fundamentally similar to that of α particles and other ions. The processes responsible for the energy loss are qualitatively the same in both cases. In fact, the average energy loss per ion pair formed is very closely the same for electrons as for α particles (35 eV for electrons in air). The primary ionization by electrons accounts for only about 20 to 30 per cent of the total ionization; the remainder is due to secondary ionization.

There are a number of differences between the interactions of the two types of particles with matter. First, for a given energy the velocity of an electron is much larger than that of an α particle, and therefore the specific ionization is less for electrons. In table 4-3 the specific ionization in air is given for electrons of different energies. The largest specific ionization, 5950 ion pairs per milligram per square centimeter, occurs at 146 eV (velocity = 0.024c), which is a much lower energy but somewhat higher velocity than corresponds to the peak in the Bragg curve for α particles. In air, ionization stops when the electron energy has been reduced to 12.5 eV (the ionization potential of oxygen molecules). On the higher-energy side of the maximum the specific ionization reaches a flat minimum at about 1.4 MeV. The increase beyond this energy is a relativistic effect, as discussed in connection with (4-5). The Lorentz

Table 4-3   Specific ionization and velocity for electrons of various energies in air

Velocity (in units of the velocity of light, c)	Energy (MeV)	Ion Pairs per 1.00 mg cm^{-2}
0.001979	10^{-6}	0
0.006257	10^{-5}	0
0.0240	1.46×10^{-4}	5950 (maximum)
0.1950	10^{-2}	~850
0.4127	0.05	154
0.5483	0.10	116
0.8629	0.50	50
0.9068	0.70	47
0.9411	1.0	46
0.9791	2.0	46
0.9893	3.0	47
0.9934	4.0	48
0.9957	5.0	49
0.9988	10	53
0.99969	20	57
0.999949	50	63
0.9999871	100	66

contraction of lengths enables the fast electron to ionize atoms at greater distances, even at distances of several molecular diameters.[5]

An electron may lose a large fraction of its energy in one collision; therefore a statistical treatment of the energy-loss processes is less justified than for α particles, and straggling is much more pronounced. In the passage of an initially homogeneous beam of electrons through matter the apparent straggling is further increased by the pronounced scattering of the electrons into different directions, which makes possible widely different path lengths for electrons traversing the same thickness of absorber. Nuclear scattering is responsible for most of the large-angle deflections, although energy loss is caused almost entirely by interactions with electrons.

For electrons of high energy an additional mechanism for losing energy must be taken into account: the emission of radiation (bremsstrahlung) when an

[5] This has the perhaps unexpected consequence of making the physical state of the absorber of importance. For example, in liquid rather than gaseous air the dielectric polarization of the medium probably reduces the specific ionization from the values in table 4-3 by about 10 per cent at 10 MeV and about 20 per cent at 100 MeV, if W remains 35 eV per ion pair in liquid air.

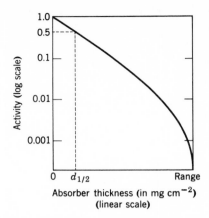

Fig. 4-8   Idealized β-ray absorption curve (semilog plot).

electron is accelerated in the electric field of a nucleus.   The ratio of energy loss by this radiation to energy loss by ionization in an element of atomic number Z is approximately equal to $EZ/800$, where E is the electron energy in millions of electron volts.   Thus in heavy materials such as lead the radiation loss becomes appreciable even at 1 MeV, whereas in light materials (air, aluminum) it is unimportant, at least for the energies available from β emitters.

Finally, the additional fact that β particles are emitted with a continuous energy spectrum further complicates any attempt at detailed analysis of their absorption in matter.

Absorption of Beta Particles.   The combined effects of continuous spectrum and scattering lead—quite fortuitously—to an approximately exponential absorption law for β particles of a given maximum energy.   Absorption curves, that is, curves of activity versus thickness of absorber traversed, are for this reason usually plotted on semilogarithmic paper.   The nearly exponential decrease applies both to numbers and specific ionizations of β particles, although absorption curves taken with counters and ionization chambers cannot be expected to be completely identical.   The exact shape of an absorption curve depends also on the shape of the β-ray spectrum and, because of scattering effects, on the geometrical arrangement of active sample, absorber, and detector.   If sample and absorber are as close as possible to the detector, the semilog absorption curve becomes most nearly a straight line; otherwise, some curvature toward the axes is generally found.   When β particles belonging to two spectra of widely different maximum energies are present in a source, this is apparent from the change of slope in the absorption curve; such an absorption curve is roughly analogous to the semilog decay curve of an activity containing two different half-life periods.

Determination of Beta-Particle Ranges. It is generally the purpose of absorption measurements to determine the upper energy limit of a β-ray spectrum.   We should say at the outset that precision determinations of upper energy limits can be made only with electron spectrographs; yet for many purposes absorption measurements are useful.

To get a measure of the upper energy limit of a β-ray spectrum one must find the range in the absorber of the most energetic β particles.   The fact that a range exists for a given β-ray spectrum means that the absorption curve cannot continue as an approximate exponential but must eventually turn downward toward − ∞ on a semilog plot (see figure 4-8).   The ratio of range to initial half-thickness in aluminum is generally between 5 and 10.   In practice, a β-ray absorption curve is never found to reach − ∞ on a semilog plot and may not even turn in that direction because of the presence of more penetrating radiation beyond the range of the β rays.   Even if neither nuclear γ radiation nor characteristic X rays are present, there is always some background of bremsstrahlung from the deceleration of the β particles in the sample itself and in the absorbers.   If elements of low Z are used as absorbers, the difference in slopes between β-ray and γ- or X-ray absorption curves is particularly marked; the absorption curve then exhibits a fairly sharp break where the β-ray component turns over into the photon "tail" (see figure 4-9).   For this

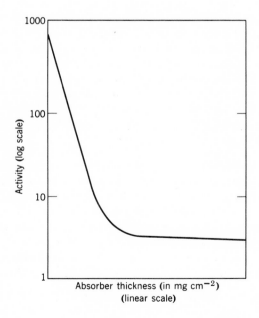

Fig. 4-9   Typical β-ray absorption curve in aluminum (semilog plot).   A γ-ray component is present.

reason β-ray absorption curves are always taken with absorbers of low atomic number; aluminum or plastic absorbers are most commonly used, and for differentiating β particles from soft X rays beryllium absorbers are particularly useful. (Beta-ray ranges expressed in milligrams per square centimeter are nearly independent of the absorber material, whereas the absorption of electromagnetic radiation increases rapidly with Z as discussed in section C.)

The relative efficiencies of measuring instruments for β and γ rays vary with the energies of the radiations and with different instruments; most gas-filled counters and ionization chambers are about 100 times as efficient for β particles as for 1-MeV γ quanta, and the efficiency for γ rays is roughly proportional to the γ-ray energy from a few hundred kiloelectron volts to a few million electron volts. Thus a typical β-ray absorption curve for a β spectrum accompanied by γ rays will have a γ-ray "tail" with intensity of the order of 1 per cent of the initial β activity. A pure bremsstrahlung tail is usually at least an order of magnitude smaller, about 0.05 per cent or less of the initial β activity. In β^+ absorption curves there is, in addition to other electromagnetic radiation that may be present, always a background of annihilation radiation, about 1 per cent of the initial β^+ intensity in a typical detector.

The maximum range for a β-particle spectrum may be obtained from an experimental absorption curve in various ways. Visual inspection gives a rough value (usually too small) for the point at which the β activity ceases to be detectable above the γ- or X-ray background; the lower the γ- or X-ray background, the better the visual method. Better results can sometimes be obtained by subtraction of the penetrating background radiation from the total absorption curve, which should result in a curve similar to the one in figure 4-8.

More reliable methods for the determination of beta-ray ranges from absorption curves depend on comparisons with absorption curves for "standard" substances measured with the same experimental arrangement. The first method of this sort, which has been of great value in the earlier determinations of beta-ray energies, was suggested by N. Feather (F1); a more elaborate method was suggested by E. Bleuler and W. Zünti (B4).

Range-Energy Relations. Once the range of β particles or conversion electrons is known, a range-energy relation can be used to deduce the maximum energy. Many empirical relations have been proposed. One given by Feather for energies above 0.6 MeV has been most widely used: $R = 0.543E - 0.160$, where E is the maximum β energy in million electron volts and R is the range in aluminum in grams per square centimeter. This relation is useful up to at least 15 MeV. In the lower energy region (below about 0.7 MeV) it is best to use a range-energy curve such as that plotted in figure 4-10.

Back-Scattering. As already mentioned, scattering of electrons, both by nuclei and by electrons, is much more pronounced than scattering of heavy particles. A significant fraction of the number of electrons striking a piece of material may be reflected as a result of single and multiple scattering processes.

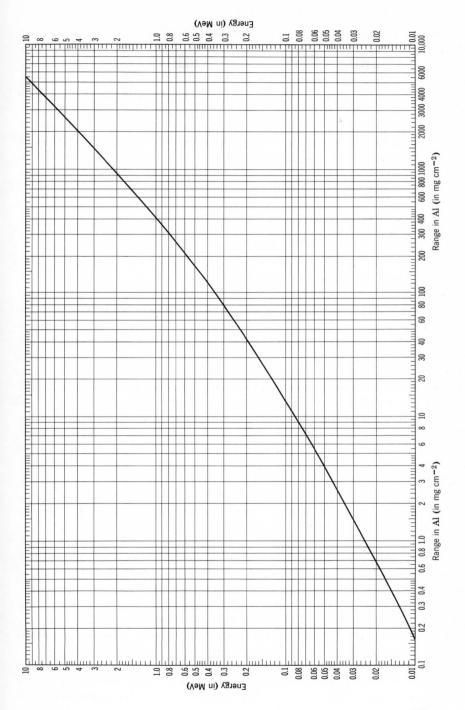

Fig. 4-10 Range-energy relation for β particles and electrons in aluminum.

107

The reflected intensity increases with increasing thickness of reflector, except that for thicknesses greater than about one third of the range of the electrons saturation is achieved and further increase in thickness does not add to the reflected intensity. The ratio of the measured activity of a β source with reflector to that without reflector is known as the back-scattering factor. The saturation back-scattering factor is essentially independent of the maximum beta-particle energy for energies above about 0.6 MeV and varies from about 1.3 for aluminum to about 1.8 for lead. These factors, though, are dependent on the particular counting arrangement used and should be determined in each configuration. There may also be a small difference between the back-scattering of positrons and negatrons of the same energy.

C. ELECTROMAGNETIC RADIATION

Processes Responsible for Energy Loss (H1). The average specific ionization caused by a γ ray is perhaps one tenth to one hundredth of that caused by an electron of the same energy. The practical ranges of γ rays are therefore much greater than those of β particles. The ionization observed for γ rays is almost entirely secondary in nature, as we shall see from a discussion of the three processes by which γ rays (and X rays) lose their energy. The average energy loss per ion pair formed is the same as for β rays, namely, 35 eV in air.

At low energies the most important process is the photoelectric effect. In this process the electromagnetic quantum of energy $h\nu$ ejects a bound electron from an atom or molecule and imparts to it an energy $h\nu - b$, where b is the energy with which the electron was bound. The quantum of radiation completely disappears in this process, and momentum conservation is possible only because the remainder of the atom can receive some momentum. For any photon energy greater than the K-binding energy of the absorber, photoelectric absorption takes place primarily in the K shell, with the L shell contributing only of the order of 20 per cent and outer shells even less. For this reason the probability for photoelectric absorption has sharp discontinuities at energies equal to the binding energies of the K, L, etc., electrons. For photon energies well above the K-binding energy of the absorber the photoelectric absorption first falls off rapidly (about as $E_\gamma^{-7/2}$), then more slowly (eventually as E_γ^{-1}) with increasing energy. It is also approximately proportional to Z^5. The γ-ray energy at which the photoelectric contribution to the total γ-ray absorption is about 5 per cent is 0.15 MeV for aluminum, 0.4 MeV for copper, 1.2 MeV for tin, and 4.7 MeV for lead. Except in the heaviest elements, photoelectric absorption is relatively unimportant for energies above 1 MeV.

The ionization produced by photoelectrons accounts largely for the ionization effect of low-energy photons. The photoelectric effect is frequently used to determine γ-ray energies. This may be accomplished by measurement of

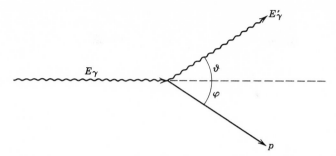

Fig. 4-11   Schematic diagram of Compton scattering of a gamma ray by an electron.

the total ionization due to the photoelectrons in a proportional or scintillation counter.   In another method a thin foil of high atomic number, called a "radiator" or "converter," is placed over the γ-active sample, and the energies of the ejected photoelectrons are measured in an electron spectrograph.

Instead of giving up its entire energy to a bound electron, a photon may transfer only a part of its energy to an electron, which in this case may be either bound or free; the photon is not only degraded in energy but also deflected from its original path.   This process is called the Compton effect or Compton scattering.   The relation between energy loss and scattering angle can be derived from the relativistic conditions for conservation of momentum and energy.   The important relativistic expression required relates the *total* energy of a particle E to its momentum p (cf. appendix B):

$$E = (E_0{}^2 + c^2 p^2)^{1/2}. \tag{4-12}$$

The quantity E_0 is the total energy of the particle when it is at rest and is given by mc^2 where m is the rest mass of the particle.   It must also be recalled that the rest mass of a photon is zero.   The scattering event is depicted in figure 4-11.   We define E_γ as the energy of the initial gamma ray, E_γ' as its energy after being scattered through an angle ϑ, E_0 as the rest energy of the electron (511 keV), and p as the magnitude of the momentum of the originally stationary electron after being struck by the incident gamma ray and projected at an angle φ with respect to the incident direction.   The conservation of total energy requires that

$$E_\gamma + E_0 = E_\gamma' + (E_0{}^2 + c^2 p^2)^{1/2}; \tag{4-13}$$

the conservation of momentum along the direction of the incident gamma-ray requires that

$$\frac{E_\gamma}{c} = \frac{E_\gamma'}{c} \cos \vartheta + p \cos \varphi; \tag{4-14}$$

and the conservation of momentum normal to the incident direction requires that

$$\frac{E_\gamma'}{c} \sin \vartheta = p \sin \varphi. \tag{4-15}$$

The angle φ may be eliminated between (4-14) and (4-15) through the relation $\sin^2 \varphi + \cos^2 \varphi = 1$; this gives

$$E_\gamma^2 - 2E_\gamma E_\gamma' \cos \vartheta + (E_\gamma')^2 = c^2 p^2. \tag{4-16}$$

Equation 4-16 may be substituted into (4-13) which yields after some simple manipulations

$$\frac{1}{E_\gamma'} - \frac{1}{E_\gamma} = \frac{1 - \cos \vartheta}{E_0}. \tag{4-17}$$

Through the relation between the energy of a photon and its wavelength, $E = hc/\lambda$, where c is the velocity of light, (4-17) takes the more familiar form

$$\lambda' - \lambda = \frac{h}{m_e c} (1 - \cos \vartheta), \tag{4-18}$$

where m_e is the rest mass of the electron. The quantity $h/m_e c = 2.426 \times 10^{-10}$ cm is the well-known Compton wavelength of the electron.

Equation 4-18 shows that for a given incident energy there is a minimum energy (maximum wavelength) for the scattered γ ray and that this occurs for scattering in the backward direction ($\cos \vartheta = -1$). This minimum energy is readily obtained from (4-17):

$$(E_\gamma')_{\min} = \frac{E_0}{2} \frac{1}{1 + E_0/2E_\gamma}. \tag{4-19}$$

For large incident γ-ray energies ($E_\gamma \gg \frac{1}{2}E_0$) the minimum energy of the scattered γ rays approaches $\frac{1}{2}E_0 = 250$ keV. For this reason scintillation spectra of high-energy γ rays always show a "back-scattering peak" at ≤ 250 keV which is caused by Compton scattering in *surrounding material* (cf. chapter 12, section F) and a valley between photopeak and Compton continuum whose width corresponds to the minimum energy (≤ 250 keV) carried off by γ rays Compton-scattered in the *crystal*.

The Compton scattering per electron is independent of Z, and therefore the scattering coefficient per atom is proportional to Z. For energies in excess of 0.5 MeV it is also approximately proportional to E_γ^{-1}. Thus Compton scattering falls off much more slowly with increasing energy than photoelectric absorption, at least at moderate energies (up to 1 or 2 MeV), and even in lead it is the predominant process in the energy region from about 0.6 to 4 MeV.

The third mechanism by which electromagnetic radiation can be absorbed is the pair-production process (discussed in chapter 2, p. 50). Pair production cannot occur when $E_\gamma < 1.02$ MeV. Above this energy the atomic cross

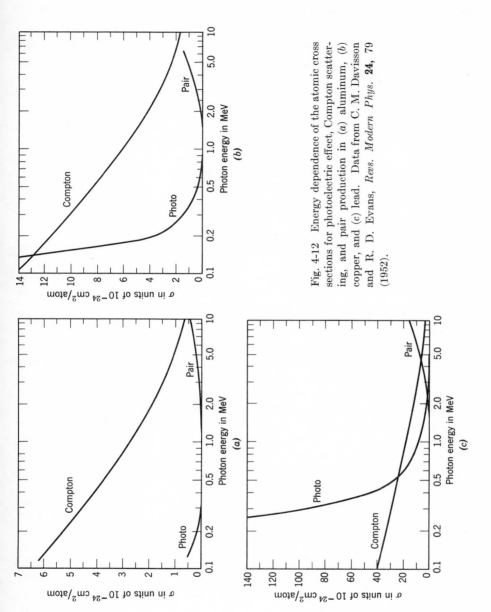

Fig. 4-12 Energy dependence of the atomic cross sections for photoelectric effect, Compton scattering, and pair production in (a) aluminum, (b) copper, and (c) lead. Data from C. M. Davisson and R. D. Evans, *Revs. Modern Phys.* **24,** 79 (1952).

section for pair production first increases slowly with increasing energy and above about 4 MeV becomes nearly proportional to log E_γ. It is also proportional to Z^2. At high energies, where pair production is the predominant process, γ-ray energies can best be determined by measurements of the total energies of positron-electron pairs. Pair production is always followed by annihilation of the positron, usually with the simultaneous emission of two 0.51-MeV photons. The absorption of quanta by the pair-production process is therefore always complicated by the appearance of this low-energy secondary radiation.

The atomic cross sections for all three processes discussed increase with increasing Z, except for the photoelectric effect at very low energies. For this reason heavy elements, atom for atom, are much more effective absorbers for electromagnetic radiation than light elements, and lead is most commonly used as an absorber. Because photoelectric effect and Compton effect decrease and pair production increases with increasing energy, the total absorption in any one element has a minimum at some energy. For lead this minimum absorption, or maximum transparency, occurs at about 3 MeV; for copper at about 10 MeV; and for aluminum at about 22 MeV. The energy dependence of the three processes is shown for Al, Cu, and Pb in figure 4-12.

Determination of Photon Energies by Absorption. If only photons of the incident energy are considered, all the processes by which γ rays interact with matter lead to exponential attenuation; that is, the intensity $\mathbf{I}_d$ transmitted through a thickness d is given by $\mathbf{I}_d = \mathbf{I}_0 e^{-\mu d}$, where $\mathbf{I}_0$ is the incident intensity and μ is called the absorption coefficient. Separate absorption coefficients for photoelectric effect, Compton scattering, and pair production are sometimes quoted, and the total absorption coefficient μ is the sum of the three. The half-thickness $d_{1/2}$ is defined as the thickness which makes $\mathbf{I}_d = \frac{1}{2}\mathbf{I}_0$; $d_{1/2} = 0.693/\mu$. Absorber thicknesses are frequently given in terms of surface density (ρd, expressed in grams per square centimeter). Then $\mathbf{I}_d = \mathbf{I}_0 e^{-(\mu/\rho)\rho d}$, and μ/ρ is called the mass absorption coefficient.

Unless absorption is entirely by the photoelectric process, the condition that only photons of the incident energy be measured is not always easy to meet experimentally. It requires either a very "good" geometry (large distances between source and absorber and between absorber and detector) or a detector which responds over a narrow energy range only. Often the detector receives and records some of the degraded radiation produced in Compton and pair-production processes. This tends to cause some deviations from true exponential absorption unless the absorber is thick enough for equilibrium with the secondary radiations to be established.

The approximately exponential nature of γ-ray absorption may be used for the determination of γ-ray energies. Various thicknesses of absorber are placed between γ emitter and detector, and the measured activity is plotted versus absorber thickness (usually on semilog paper) to yield an absorption curve.   For a single incident photon energy a line results which is straight over a factor of 10 or 20 in intensity if an appropriate experimental arrangement is used.   When two components differing sufficiently (perhaps a factor of 2) in energy are present, the absorption curve can often be resolved into two straight lines in the same manner as a decay curve is resolved.   Resolution into more than two components with any precision is generally not possible.

In an adequate, simple experimental arrangement for absorption measurements the active source and absorbers are as far as practicable from the detector to prevent most of the scattered quanta and secondary electrons from falling on the detector.   An additional absorber of low Z near the detector stops a large fraction of the secondary electrons (as well as any β particles emitted by the source which otherwise might enter the detector when little or no lead absorber is used).

The quantity most conveniently obtained from an absorption curve for electromagnetic radiation is the half-thickness, and this may be translated into photon energy.   Curves of half-thickness in various absorbers versus photon energy are reproduced in figure 4-13 for energies of 0.2 to 6 MeV and in figure

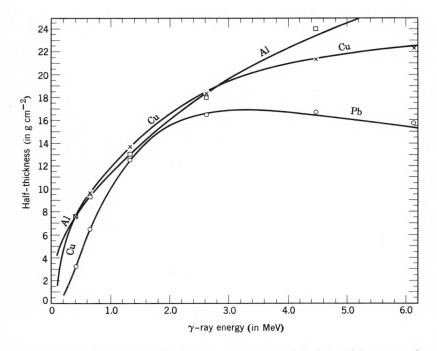

Fig. 4-13   Half-thickness values in aluminum, copper, and lead for high-energy photons. The curves are based on the calculated absorption coefficients of C. M. Davisson and R. D. Evans, *Revs. Modern Phys.* **24,** 79 (1952).   Some experimental points ($\square$ Al, $\times$ Cu, $\bigcirc$ Pb) are taken from S. A. Colgate, *Phys. Rev.* **87,** 592 (1952).

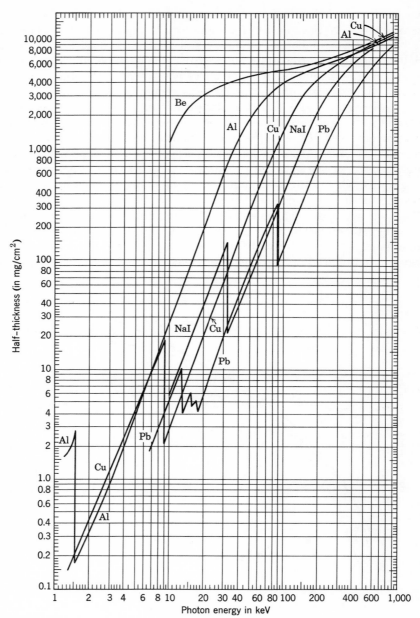

Fig. 4-14   Half-thickness values in beryllium, aluminum, copper, lead, and sodium iodide for low-energy photons.   The K-absorption edges of aluminum, copper, iodine and lead, as well as the L_I, L_{II}, and L_{III} edges of lead are shown.   (Data from G. W. Grodstein, *National Bureau of Standards Circular* 583, April 1957, and from *Handbook of Chemistry and Physics*, 44th ed., 1962.)

4-14 for energies of about 1 to 800 keV.   It should be remembered that a given half-thickness may correspond to two different energies because of the minimum in the curve of absorption versus energy.   For example, a half-thickness of 15.5 g cm^{-2} in lead may correspond either to an energy of 2.0 MeV or to an energy of about 5.9 MeV.   This ambiguity can always be eliminated by taking absorption curves in two different materials.

Sometimes the absorption of low-energy X rays in aluminum may simulate that of β particles.   This source of confusion can be eliminated by the use of a second absorber material because in the X-ray region the photon mass absorption coefficients vary quite rapidly with Z.

Critical Absorption of X Rays.   We have already mentioned the discontinuities in absorption coefficients at photon energies corresponding to the electron binding energies.   These absorption edges and their variation from element to element may be used to identify the energies of characteristic X rays. To understand this method of critical absorption we recall that the *emission* of an X ray from an atom is due to the transition of an electron from one of the outer shells to a vacancy in a shell farther in, say, from the L to the K shell.[6]   Photoelectric *absorption* in a given electron shell, on the other hand, can occur only if the photon has enough energy to promote an electron from that shell to a vacant level (which means very nearly enough energy to remove the electron from the atom).   It follows that an element is a poor absorber for its own characteristic X rays.   The K_α X rays of an element have an energy equal to the difference between the K and L shells and so cannot lift a K electron to one of the outer vacant shells in the same element.   However, the binding energy of electrons decreases with decreasing Z; therefore the K_α emission line of an element Z has an energy rather close to but slightly greater than the K absorption edge of some element of slightly lower Z and is strongly absorbed by that element but not by the next higher one.   These two neighboring elements will thus have very different absorption coefficients for the particular rays, and the one that absorbs more strongly is called the critical absorber for these X rays.   Critical absorption can also be applied to L-emission lines, especially of heavy elements.

As an example, consider the K_α X rays of zinc ($Z = 30$) which have a wavelength of 1.43 Å (energy 8.7 keV).   The K absorption edges of $_{29}$Cu and $_{28}$Ni are at 1.38 Å (9.0 keV) and 1.48 Å (8.4 keV), respectively.   Therefore, nickel is a good absorber for zinc K_α X rays, and copper is not (figure 4-15).   The K_α X rays of gallium ($Z = 31$), on the other hand, are strongly absorbed both in nickel and copper because their wavelength is 1.34 Å (9.3 keV), but they are not absorbed well in zinc whose K absorption edge is at 1.28 Å (9.7 keV).

---

[6] In X-ray terminology, X rays due to transitions from the L to the K shell are called K_α X rays ($K_{\alpha1}$ and $K_{\alpha2}$ corresponding to the electron originating in different sublevels of the L shell); X rays due to transitions from the M to the K shell are called K_β, etc.   Similarly, there are L_α, L_β, etc., X rays.

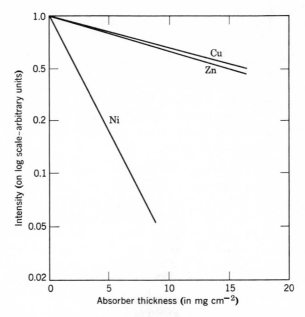

Fig. 4-15 Absorption of zinc K_α X rays in zinc, copper, and nickel. (These absorption curves were calculated from data given in reference C1.)

Both the X-ray emission lines (C1) and the absorption edges (H2) of the elements can be found in tables, and suitable elements can be chosen as absorbers to decide the origin of a set of X rays accompanying a nuclear decay process. The K_α X rays are usually the most prominent lines, but occasionally, especially with very heavy elements, the absorption of other lines (K_β and L) also must be taken into account. Absorption curves are taken with each of two or three neighboring elements, and their comparison usually brackets the energy of the emission line sufficiently to determine the corresponding atomic number. It should be emphasized that it is not always necessary to use pure elements as absorbers (this would be difficult for some elements). Compounds of the desired element can be used, provided that other elements in the compounds do not appreciably absorb the X rays under investigation. Light elements are poor absorbers for energetic X rays, and oxides, hydroxides, or carbonates are therefore usually suitable as absorbers.

Critical absorption measurements are sometimes used to aid in the identification of a decay process. The X radiation is characteristic of the particular value of Z at the time of emission of the ray; X rays following β^- decay of a nucleus of charge Z correspond to atomic number $Z + 1$; those following β^+ or electron-capture decay correspond to atomic number $Z - 1$; those following internally converted isomeric transitions correspond to atomic number Z.

Even with energy-sensitive detectors of moderate resolution, such as proportional counters with pulse height analyzers, critical absorbers can sometimes be used to advantage, for example, to suppress one X-ray line so that the spectrum of a neighboring one can be measured cleanly.

D. NEUTRONS

Because neutrons carry no charge, their interaction with electrons is exceedingly small, and primary ionization by neutrons is a completely negligible effect. The interaction of neutrons with matter is confined to nuclear effects, which include elastic and inelastic scattering and nuclear reactions such as (n, γ), (n, p), (n, α), $(n, 2n)$, and fission. These nuclear interactions have been briefly mentioned in chapter 2 and are further discussed in chapter 10; here we shall merely indicate how each of these types of interaction may be applicable to the detection and measurement of neutrons.

The recoil protons produced by the elastic scattering of fast neutrons in hydrogenous material are often used for the detection of such neutrons. About seven protons leave a thick paraffin layer per 10^4 incident neutrons of 1 MeV energy, and for other energies the ratio of protons to neutrons is roughly proportional to neutron energy. The energy of the fastest recoil protons equals the neutron energy.

The ionization produced by protons or α particles created in n, p or n, α reactions can also be used for neutron detection. Ionization chambers or proportional counters (described in chapter 5) may be lined with boron or filled with gaseous BF_3, and the particles from the $B^{10}(n, \alpha)Li^7$ reaction may be detected. The separated isotope B^{10} is particularly effective. Fission fragments may be detected in an ionization chamber lined with fissionable material and exposed to a neutron source. Neutron-capture reactions leading to radioactive products are frequently used for detection of neutrons by means of the induced activity.

Slowing Down of Neutrons. From the description of nuclear reactions given in chapter 2, section F, it will be recalled that thermal neutrons, neutrons whose energy distribution is approximately that of gas molecules at ordinary temperatures, are very efficient at producing nuclear reactions. Because of this fact processes for reducing the energy of high-energy neutrons produced in nuclear reactions to a thermal energy distribution have received much theoretical and experimental study.

Fast neutrons may lose large amounts of energy in inelastic collisions, especially with heavy nuclei. This process ceases to be effective after intermediate energies are reached and does not produce slow neutrons. Most slowing down is accomplished through a process of many successive elastic collisions with nuclei. Because of the conservation of momentum a neutron of energy E_0 making an elastic collision with a heavy nucleus bounces off with most of its original energy, giving up no more energy than $4AE_0/(A + 1)^2$ to the recoil nucleus, where A is the mass number of the target nucleus. The lighter the nucleus with which a neutron collides, the greater the fraction of the neutron's kinetic energy that can be transferred in the elastic collision. For this reason hydrogen-containing substances such as paraffin or water are the most effective slowing-down media for neutrons.

In the elastic scattering of neutrons with energies below about 10 MeV all energy transfers between zero and the upper limit, $4AE_0/(A + 1)^2$, are equally probable. Thus the probability that a neutron of energy E_0 has a residual energy between E and $E + dE$ is

$$P(E)\,dE = \frac{dE}{4AE_0/(A + 1)^2};$$

and the average energy retained by the neutron is

$$\begin{aligned}
\bar{E} &= \int_{E_0[1-4A/(A+1)^2]}^{E_0} P(E)E\,dE \\
&= \frac{(A + 1)^2}{4AE_0} \int_{E_0[1-4A/(A+1)^2]}^{E_0} E\,dE \\
&= E_0\left[1 - \frac{2A}{(A + 1)^2}\right].
\end{aligned} \tag{4-20}$$

From this result it is seen that the average value of E/E_0 is independent of E_0; therefore the average value of E/E_0 after n collisions is simply

$$\frac{\overline{E_n}}{E_0} = \left[1 - \frac{2A}{(A + 1)^2}\right]^n. \tag{4-21}$$

The average value after n collisions is a rather misleading quantity, as the distribution of energies is strongly skewed. The probability that a neutron of initial energy E_0 has an energy between E_n and $E_n + dE_n$ after n elastic collisions with hydrogen nuclei may be obtained from the recursion relation[7]

$$P_n(E_n)\,dE_n = \int_{E_n}^{E_0} [dE_{n-1}\,P_{n-1}(E_{n-1})]\left[\frac{dE_n}{E_{n-1}}\right]; \tag{4-22}$$

where the first bracketed term is the probability of obtaining energy between E_{n-1} and $E_{n-1} + dE_{n-1}$ in $n - 1$ collisions, and the second bracketed term is the probability of going from the interval $E_{n-1} \rightarrow E_{n-1} + dE_{n-1}$ to the interval $E_n \rightarrow E_n + dE_n$ in the nth collision. The integration is, of course, performed over the variable E_{n-1}. Equation 4-22 has the solution (ignoring

---

[7] The recursion relations for heavier nuclei are more complicated than (4-22), since it is not possible for neutrons of all values of E_{n-1} to go to E_n in a single collision.

thermal motion):

$$P_n(E_n) = \frac{1}{(n-1)!E_0}\left(\ln\frac{E_0}{E_n}\right)^{n-1}. \qquad (4\text{-}23)$$

Another question of interest is the average number of collisions required to slow a neutron of energy E_0 down to an energy E. We may write for the energy after n collisions

$$E = E_0 f_1 f_2 \cdots f_i \cdots f_n; \qquad (4\text{-}24)$$

where

$$f_i = \frac{E_i}{E_{i-1}}. \qquad (4\text{-}25)$$

As we stated before, f_i has equal probability for all values between 1 and $1 - 4A/(A+1)^2$. It is clear that (4-24) has an infinite number of possible solutions for any value of n above a certain minimum value determined by the mass of the scattering nucleus. It is tempting to estimate the average value of n by putting the average value of f from (4-21) into (4-24); but this would be wrong. It would be wrong for the same reason that the average square of a set of random numbers is, in general, not equal to the square of the average. The solution to the problem, however, can be immediately obtained by a simple transformation of (4-24), which turns the problem into a more familiar one whose answer is well known.

Take the logarithm of both sides of (4-24):

$$\ln\left(\frac{E}{E_0}\right) = \ln\left(f_1 f_2 \cdots f_i \cdots f_n\right) = \sum_{i=1}^{n}\ln f_i;$$

define $x_i = -\ln f_i$ so that

$$\ln\left(\frac{E_0}{E}\right) = \sum_{i=1}^{n} x_i. \qquad (4\text{-}26)$$

Again, an infinite number of values of n will satisfy (4-26); but now to ask for the average value of n is equivalent to asking for the average number of collisions made by a gas molecule when it travels a "distance" $\ln(E_0/E)$. The answer to this problem is well known; it is just the distance traveled divided by the mean free path.[8] For our problem this means

$$\bar{n} = \frac{\ln(E_0/E)}{\bar{x}} = \frac{\ln(E_0/E)}{\ln(E_{i-1}/E_i)}, \qquad (4\text{-}27)$$

[8] The probabilities of the various values of n are given by the Poisson distribution which is discussed in section D of chapter 6.

where the quantities with bars over them denote mean values. The mean value of $\ln (E_{i-1}/E_i)$ may be obtained in the same manner as $\bar{E}$ in (4-20) [cf. (6-6)]; the result is

$$\overline{\ln \left(\frac{E_{i-1}}{E_i}\right)} = 1 - \frac{(A-1)^2}{2A} \ln \left(\frac{A+1}{A-1}\right). \tag{4-28}$$

Substituting (4-28) into (4-27) gives:

$$\bar{n} = \frac{\ln (E_0/E)}{1 - [(A-1)^2/2A] \ln [(A+1)/(A-1)]}. \tag{4-29}$$

For collisions with protons ($A = 1$) the denominator in (4-29) becomes unity, hence $E_n = E_0 e^{-\bar{n}}$; approximately 20 collisions are therefore necessary to reduce neutrons from a few million electron volts to thermal energies (about 0.04 eV at ordinary room temperature). Paraffin about 8 in. thick surrounding a neutron source is adequate for reducing most neutrons to the thermal energy distribution. The whole slowing-down process requires less than 10^{-3} second.

The probable eventual fate of a thermal neutron in a hydrogenous medium like water or paraffin is capture by a proton to form a deuteron; but, since the cross section for this reaction is quite small compared with the cross section for scattering, a neutron after reaching thermal energies makes about 150 further collisions before being captured. Paraffin and water are good substances to use for the slowing down of neutrons because the capture cross sections of oxygen and carbon are even much smaller than the hydrogen capture cross section. Heavy water is better than ordinary water because of the low probability of neutron capture by deuterium. Carbon (graphite) is also useful as a slowing-down medium; many more (about 120) collisions are necessary to reduce neutrons to thermal energies in carbon than in hydrogen, but after reaching thermal energies the neutrons can exist longer in carbon. In either substance the lifetime of a neutron before capture is only a fraction of a second.

Even if neutrons could be kept in a medium in which they would not eventually be captured, they would not exist very long. The systematics of β radioactivity predict that free neutrons are unstable and should decay rather quickly into protons and electrons. This decay was observed by A. H. Snell and by J. M. Robson in 1950. The neutrons, from nuclear reactors, were in free flight in vacuum. Robson measured the energy released in the disintegration, 0.78 MeV, and the half-life, 13 minutes.

Thermal Distribution. It should be apparent that not all thermal neutrons have the same energy. After neutrons are slowed to energies comparable to thermal agitation energies they may either lose or gain energy in collisions, and the result is a Maxwellian distribution of velocities in which the fraction of the total number of neutrons with velocity between v and $v + dv$ is given by

$$F(v)\, dv = 4\pi^{-\frac{1}{2}} \left(\frac{M}{2kT}\right)^{\frac{3}{2}} v^2 e^{-Mv^2/2kT}\, dv. \qquad (4\text{-}30)$$

Here the fraction is denoted by $F(v)\, dv$, M is the neutron mass, T is the absolute temperature, and k is the Boltzmann constant. Some properties of this distribution, usually derived in books on the kinetic theory of gases, are that the most probable velocity is

$$v_m = \left(\frac{2kT}{M}\right)^{\frac{1}{2}},$$

the average velocity is

$$\bar{v} = \left(\frac{8kT}{\pi M}\right)^{\frac{1}{2}} = \frac{2v_m}{\pi^{\frac{1}{2}}},$$

and the average kinetic energy is $\bar{E} = \frac{3}{2}kT$. The average energy of the neutrons depends on the temperature of the slowing-down medium. At very low temperatures the Maxwellian distribution function becomes a poor approximation because of the discrete energy levels of the bound atoms of the medium. At all temperatures the approximation can be poor if the neutron path in the medium is too short or if the distribution is seriously altered by neutron absorption or leakage from the surface.

A significant point is the distinction between the velocity distribution present in a medium and that felt by a sample placed in the medium. The two distributions are different because the probability that a particular neutron will strike the sample in a given time is proportional to v. It is this altered or weighted distribution, denoted here by $F'(v)\, dv$, that is significant in any transmutation or cross section computation:

$$F'(v)\, dv = 2\left(\frac{M}{2kT}\right)^2 v^3 e^{-Mv^2/2kT}\, dv. \qquad (4\text{-}31)$$

E. RADIATION CHEMISTRY

The study of the chemical effects caused by ionizing and other nuclear radiations in their passage through matter is called radiation chemistry. Although any detailed discussion of this rather large field is beyond the scope of this book, we shall mention briefly a few of its important aspects. Radiation-chemical effects are of great practical significance in nuclear-reactor technology and in the interaction of radiations with biological systems.

In gases, liquids, and covalently bonded solids the chemical effects of ionizing radiations can be ascribed almost entirely to ionization, excitation, and dissociation of molecules.   In gases the chemical reactions observed appear to be rather independent of the type and energy of ionizing radiation used, and the magnitude of the effects in a particular system is determined by the total energy absorbed.   In condensed systems the chemical effects for a given total ionization may vary somewhat with ionization density as manifested, for example, in differences between α- and β-ray effects.

Radiation Dosimetry.   In the study of chemical and biological effects of radiation a quantitative measure of radiation energy absorption (usually called the dose) is required.   The unit of measure that is now used is called the rad. A dose of one rad deposits 100 ergs per gram of material.   The most frequently used unit of radiation dose in the past was the roentgen or r unit.   It is defined as "that quantity of X or γ radiation such that the associated corpuscular emission per 0.001293 g[9] of air produces, in air, ions carrying 1 esu of quantity of electricity of either sign."   This means that 1 r produces 1.61×10^{12} ion pairs per gram of air, which corresponds to the absorption of 84 ergs of energy per gram of air.   In water the energy absorption corresponding to 1 r is about 93 ergs per gram or 0.93 rad for all X- or γ-ray energies above about 50 keV.

A common quantitative measure of the efficiency of a radiation-chemical effect is the number of molecules destroyed or produced for each 100 eV of energy absorbed, called the G value for the reaction.

The possible use of radiation-induced chemical reactions for the measurement of integrated radiation doses has been extensively investigated.   To be suitable for this purpose, the reaction should have a constant G value over a wide range of intensities and types of radiation; the extent of reaction should be proportional to dose over a wide range, and easily measurable, and the reagents employed should be convenient to prepare and store.   The chemical dosimeter (or actinometer) that appears most nearly to fulfill these conditions uses the oxidation of Fe(II) in air-saturated, dilute sulfuric acid solution.   The G value for this reaction has been carefully determined to be 15.5 ± 0.4. Chemical dosimeters require rather sizable radiation doses to produce a measurable reaction.   Since 1 rad corresponds to an energy loss of 100 ergs per gram of water, we see that the energy deposited per rad in 1 ml of solution is about $100/(1.60 \times 10^{-12})$ eV and therefore the number of moles of Fe(II) oxidized per milliliter is

$$\frac{100 \times 15.5 \times 10^{-2}}{1.60 \times 10^{-12} \times 6.02 \times 10^{23}} = 1.6 \times 10^{-11}.$$

Various other reactions have been suggested for use as actinometers, among them some that are accompanied by readily measurable color changes.   The bleaching by radiation of very dilute solutions of dyes (such as methylene blue) in aqueous gelatin looks promising in this regard.   The use of chain reactions,

---

[9] This is the weight of 1 cc of dry air at 0°C and 760 mm pressure.

although desirable because of enhanced sensitivity, is probably precluded by the lack of a linear relation between response and dose rate.

Mechanisms (A2).   Much of the work of radiation chemists has been directed toward gaining a detailed understanding of the energy absorption mechanism and of the reaction paths of the unstable intermediates (excited molecules, radicals, and ions).   In general, the processes of importance in radiation chemistry include ionization, formation of excited electronic states, transfer of electronic excitation from one molecule to another, dissociation of excited vibrational states, electron capture, neutralization, and radical reactions.

It is apparent from the energy requirement for ion pair formation that only about half the energy dissipated in the passage of radiation through matter is used for ionization; the other half presumably goes into molecular excitation. Both the initial ionization and excitation may be followed by dissociation, and in either case free-radical mechanisms are believed to play an important role. For example, the radiation decomposition of water to H_2, O_2, and H_2O_2 is thought to involve H and OH radicals as intermediates.   A great deal of work has been done to elucidate the reaction mechanisms in the radiation chemistry of aqueous solutions; studies of the effects of various solutes on the products of water decomposition have been particularly helpful.

As an example of the mechanism deduced for a radiation-induced reaction, we consider the ferrous sulfate oxidation in air-saturated acid solution mentioned earlier.   The primary reaction is thought to be the dissociation of water into radicals:

$$H_2O \rightarrow H + OH.$$

The subsequent steps are believed to be as follows:

(a) $$OH + Fe^{2+} \rightarrow Fe^{3+} + OH^-,$$

(b) $$H + O_2 \rightarrow HO_2,$$

(c) $$H^+ + HO_2 + Fe^{2+} \rightarrow Fe^{3+} + H_2O_2,$$

(d) $$H_2O_2 + Fe^{2+} \rightarrow Fe(OH)^{2+} + OH.$$

The OH radical produced in step (d) reacts according to (a).   Thus, for every H_2O molecule decomposed, four Fe^{2+} ions are oxidized, and this agrees with experimental observations.

Radiolysis of Organic Compounds.   Most organic systems are much too complex to encourage any detailed studies of the mechanism of their radiation decomposition.   However, a few generalizations may be made about the types of net reactions observed.   A great variety of products can usually be found: gases such as H_2, CO, and CO_2; fragments smaller than the irradiated molecules; and polymerization products.   Radiation-induced polymerization of acetylene to benzene and of styrene to polystyrene have been studied and found to involve chain reactions induced by free radicals. Radiation-induced changes in the mechanical properties of some polymers such as polyethylene can be traced to the formation of cross linkages between polymer chains, and these effects have found some practical applications.

One very striking general observation is that aromatic compounds are much more stable toward radiation decomposition than are aliphatic compounds. This is explained by the resonance stabilization of even the excited states of the benzene ring; as a consequence of this extra stability these excited states presumably do not dissociate readily but may be de-activated by collisions or by emission of radiation. Aromatic compounds with aliphatic side chains (e.g., ethyl benzene) exhibit about the same stability as purely aromatic substances, which indicates that excitation energy is readily transferred from the side chain to the ring before dissociation can occur. Even in mixtures this "protective" influence of benzene rings has been demonstrated; for example, the radiation decomposition of cyclohexane in benzene solution is much less than that of pure cyclohexane.

The relative stability of the first excited electronic states of aromatic molecules toward dissociation is closely connected with the fact that these states have large probabilities for de-excitation by the emission of fluorescent radiation. It is this de-excitation mechanism that accounts for the scintillations of organic compounds such as anthracene, naphthalene, stilbene, and terphenyl exposed to ionizing radiations. The fluorescence properties of these organic substances are characteristic of their molecular structure, and they scintillate in solutions as well as in the solid state.

The differences in the effects of radiation on aromatic and aliphatic compounds indicate that, even in a given compound, radiations should not be expected to affect different bonds equally. Some specificity in the attack of radiation on certain groups has indeed been established. Yet a statistical approach to the problem of the probabilities for various bond ruptures is often useful, particularly when the radiation-induced reactions of a homologous series are considered. On a statistical basis one might expect that in such a series the relative yields of the products should vary from compound to compound linearly with the relative abundances of parent groups. This is experimentally found; for example, in the radiolysis of straight-chain alkanes the ratio of H_2 to CH_4 yields varies almost linearly with the ratio of H to CH_3 groups in the molecules.

Radiation Effects in Solids (G1). The irradiation of ionic crystals and of other insulators such as glass often results in intense coloration. This phenomenon is ascribed to absorption bands produced when electrons are trapped by lattice imperfections or impurity atoms. Various bands (designated as F, F', V, etc.) are distinguished according to the type of electron-trapping center involved. Some energy levels caused by lattice defects or impurity ions are so-called luminescence centers: the electrons brought into these levels as a result of radiation effects can return to the filled (ground-state) band with the emission of photons in the visible or near-ultraviolet part of the spectrum. The scintillations in inorganic phosphors (thallium-activated sodium iodide, silver-activated zinc sulfide, etc.) produced by this mechanism are finding increasing use in radiation detection devices.   The transparency of a phosphor to its own luminescence radiation is very important in this connection and is due to the fact that the luminescence centers lie in energy below the conduction band to which electrons can be raised by photon absorption.

In metals and semiconductors the ionization effects produced by radiation are of relatively little importance because the ionized electrons find themselves in the conduction bands, and their energy is quickly transformed into heat. In these solids the dominant effects produced by neutrons, protons, and heavier ions are caused by displacement of atoms, which in other materials are quite negligible compared with the ionization effects.

Irradiation with fast neutrons or ions is found to alter many properties of solids, including thermal and electrical conductivity, hardness, other mechanical properties, and crystal lattice parameters.   Many of the observed changes are similar to those induced by quite different means such as cold-working, and in most cases the original properties can be restored by appropriate heat treatment (provided, of course, that the chemical composition has not been appreciably altered by nuclear transmutations).   Changes caused by nuclear transmutations are of some significance in the irradiation of semiconductors; the production of impurities by nuclear reactions, for example the formation of gallium from germanium, can be used to bring about permanent changes in the electrical properties of the semiconductor.

F. BIOLOGICALLY PERMISSIBLE DOSES

Radiation chemistry has an important application in the field of biology. The biological effects of radiation are brought about through chemical changes in the cells caused by ionizations, excitations, dissociations, and atom displacements.   In determining radiation effects on living organisms, whether from external radiation or from ingested or inhaled radioactive material, one has to take into consideration not only the total dosages of ionization produced in the organism but also such factors as the density of the ionization, the dosage rate, the localization of the effect, and the rates of administration and elimination of radioactive material.

The unit of radiation dosage that is currently used in measuring biological effects is the rem (roentgen equivalent man).   The dosage in rems is equal to that in rads, multiplied by the relative biological effectiveness (RBE) of the radiation, a factor that reflects the effects of different densities of ionization along the path of the radiation.   For example, a dose of one rad of 2-MeV neutrons (ionization caused by the recoil protons) does about 10 times the damage caused by one rad of X rays; thus one rad of neutron dosage is about 10 rem.   In this connection it should be noted that a flux of 20 neutrons of 2-MeV per cm^2 per second will deliver a dose of 20 millirem in a period of 8 hours.   Table 4-4 lists the RBE values adopted by the National Committee on Radiation Protection and by the International Commission on Radiological Protection for various types of radiation (M1).   The same organizations have adopted maximum permissible doses for human beings occupationally exposed to external radiation.   For whole-body exposure this maximum has been set at 3 rem over any period of 13 consecutive weeks for people over the age of 18, with the additional provision that the total accumulated dose should not exceed $5(N - 18)$ rem, where N is the age of the dosee in years.   When the irradiation is only on hands and forearms or feet and ankles, the corresponding figures are 25 rem and $75(N - 18)$ rem (P1).   Most presently available radiation-survey instruments (chapter 5, section D) give the dosage rate in milliroentgens (mr) per unit time.   For practical purposes the dosage in milliroentgens from γ-emitting radioactive sources is essentially the same as that in units of millirems.

As an example of the practical application of some of the concepts discussed we estimate the dosage rate in rads per hour to be expected at a distance of 50 cm from a 100-mc Co^{60} source.   Each disintegration of Co^{60} is accompanied by two γ quanta with energies 1.17 and 1.33 MeV; for simplicity we use for each an average energy of 1.25 MeV.   The source emits $2 \times 100 \times 3.7 \times 10^7 = 7.4 \times 10^9$ quanta per second.   At a distance of 50 cm the γ flux is $7.4 \times 10^9/(4\pi \times 2500) = 2.3 \times 10^5$ photons cm^{-2} sec^{-1} or $2.3 \times 10^5 \times 1.25 \times 10^6 = 2.9 \times 10^{11}$ eV cm^{-2} sec^{-1}.   Since at an energy of 1.25 MeV the mass absorption coefficients in air and aluminum are about the same, we read the half-thickness from figure 4-13 as 12.5 g cm^{-2}.   Then the fractional energy loss for the γ rays per g cm^{-2} of air is given by $\mu/\rho = 0.693/12.5 = 0.055$, and the energy lost by the γ rays in going through 1 g cm^{-2} of air is $0.055 \times 2.9 \times$

Table 4-4   RBE values for various types of radiation

Radiation	RBE
X and γ rays	1
Beta rays and electrons	1
Thermal neutrons	2.5
Fast neutrons	10
Alpha particles	10
Protons	10
Heavy ions	20

Table 4-5   *Biologically permissible levels of radionuclides*[1]

Nuclide	In Critical Organ (μc)	In Water ($\mu c/cm^3$)	In Air[2] ($\mu c/cm^3$)
Pu^{239}	0.04 (bone)	10^{-4}	2×10^{-12}
U^{238}	0.005	10^{-3}	7×10^{-11}
Ra^{226}	0.1 (bone)	4×10^{-7}	3×10^{-11}
Po^{210}	0.03 (spleen)	2×10^{-5}	2×10^{-10}
$Sr^{90}(+ Y^{90})$	2.0 (bone)	4×10^{-6}	3×10^{-10}
Co^{60}	10.0 (G.I. tract)	10^{-3}	3×10^{-7}
S^{35}	90.0 (testes)	2×10^{-3}	3×10^{-7}
P^{32}	6.0 (bone)	5×10^{-4}	7×10^{-8}
Na^{24}	7.0 (G.I. tract)	8×10^{-4}	10^{-7}
$C^{14}(CO_2)$	300 (fat)	2×10^{-2}	4×10^{-6}
$H^3(H_2O)$	1000	10^{-1}	2×10^{-5}
I^{131}	0.7 (thyroid)		9×10^{-9}
Cs^{137}	30 (whole body)		6×10^{-8}

[1] Maximum Permissible Body Burdens and Maximum Permissible Concentrations of Radionuclides in Air and in Water for Occupational Exposure, *Nat. Bur. Std.* (*U. S.*) *Handbook* **69,** 1959, available from Superintendent of Documents, Washington 25, D. C. (35 cents).

[2] Occupational exposure, 40 hours per week for 50 years.

$10^{11} \times 3600 = 5.7 \times 10^{13}$ eV hr^{-1} or 92 erg hr^{-1}.   Setting the energy absorbed per gram of air equal to this energy loss,[10] we get $92/100 = 0.92$ rad/hr^{-1}.

Before leaving the subject of maximum allowable doses of radiation we should note that the body may receive excessive irradiations from internal as well as external sources.   Many radioactive nuclides when ingested or inhaled become fixed in the body for varying lengths of time.   Care must therefore be taken to avoid intake of radioactive materials.   Table 4-5 lists the maximum allowable concentrations of a few nuclides in inhaled air and in ingested liquids and also the maximum permissible amounts in the body.

REFERENCES

A1   J. Alexander and M. F. Gazdik, "Recoil Properties of Fission Products," *Phys. Rev.* **120,** 874 (1960).

[10] This procedure leads to an overestimate (in the present case, by about a factor 2) of the energy absorption in air because a fraction of the energy loss occurs by Compton scattering, and some of the secondary quanta leave the local region of interest.   The method used applies when the primary radiation is in equilibrium with secondaries; inside a mass of tissue this condition obtains more nearly than it does in air.

*A2   A. O. Allen, *"The Radiation Chemistry of Water and Aqueous Solutions,"* Van Nostrand, Princeton, New Jersey, 1961.

B1   N. Bohr, "On the Theory of the Decrease of Velocity of Moving Electrified Particles on Passing through Matter," *Phil. Mag.* **25,** 10 (1913).

B2   H. Bichsel, R. F. Mozley, and W. A. Aron, "Range of 6- to 18-MeV Protons in Be, Al, Cu, Ag, and Au," *Phys. Rev.* **105,** 1788 (1957).

*B3   H. A. Bethe and J. Ashkin, "Passage of Radiations through Matter," *Experimental Nuclear Physics* (E. Segré, Editor) Vol. 1, J. Wiley, New York, 1953.

B4   E. Bleuler and W. Zünti, "On the Absorption Method for the Determination of Beta and Gamma Energies," *Helv. Phys. Acta* **19,** 375 (1946).

C1   A. H. Compton and S. K. Allison, *X-rays in Theory and Experiment,* Van Nostrand, Princeton, New Jersey, 1935.

F1   N. Feather, "Further Possibilities for the Absorption Method of Investigating the Primary β-particles from Radioactive Substances," *Proc. Cambridge Phil. Soc.* **34,** 599 (1938).

G1   A. N. Goland, "Atomic Displacements in Solids by Nuclear Radiation," *Ann. Rev. Nuclear Sci.,* **12,** 243 (1962).

H1   W. Heitler, *The Quantum Theory of Radiation,* Oxford, Clarendon Press, 1944.

H2   R. D. Hill, E. L. Church, and J. W. Mihelich, "The Determination of Gamma-Ray Energies from Beta-Ray Spectroscopy and a Table of Critical X-Ray Absorption Energies," *Rev. Sci. Instr.* **23,** 523 (1952).

L1   J. Lindhard, M. Scharff, and H. E. Schiøtt, "Range Concepts and Heavy Ion Ranges," Kgl. Danske Videnskab. Selskab, Mat.-Fys. Medd **33,** No. 14 (1963).

M1   K. Z. Morgan, "Techniques of Personnel Monitoring and Radiation Surveying," *Nuclear Instruments and Their Uses* (A. H. Snell, Editor), Wiley New York, 1962.

N1   L. Northcliffe, "Energy Loss and Effective Charge of Heavy Ions in Aluminum," *Phys. Rev.* **120,** 1744 (1960).

N2   L. C. Northcliffe, "Passage of Heavy Ions through Matter," *Ann. Rev. Nuclear Sci.,* **13,** 67–102 (1963).

P1   "Permissible Dose from External Sources of Ionizing Radiation," *National Bureau of Standards Handbook* **59** (1954 and 1958 addendum), available from Superintendent of Documents, Washington 25, D. C. (35 cents).

R1   M. Rich and R. Madey, "Range-Energy Tables," University of California Radiation Laboratory Report UCRL-2301 (1954).

R2   B. Rossi, *High Energy Particles,* Prentice-Hall, Englewood Cliffs, New Jersey, 1952.

S1   R. M. Sternheimer, "Range-Energy Relations for Protons in Be, C, Al, Cu, Pb, and Air," *Phys. Rev.,* **115,** 137 (1959).

*W1   W. Whaling, "The Energy Loss of Charged Particles in Matter," *Encyclopedia of Physics,* Vol. 34 (E. Flügge, Editor), Springer-Verlag, Berlin, 1958.

EXERCISES

1. Show that the maximum velocity an electron can receive in an impact with an α particle of velocity v is approximately $2v$.

2. Estimate the ranges in air of (a) 10-Mev H^3 ions, (b) double charged 10-Mev He^3 ions.   Assume STP.   *Answer:* (a) 51 cm.

3. (a) What thickness of aluminum foil will reduce the energy of 40-Mev He^4 ions to 32 Mev?   (b) What energy loss will a 20-Mev deuteron beam suffer in passing through the same foil?   *Answer:* (a) 55 mg cm^{-2}; (b) $\Delta E \cong 2$ Mev.

4. An absorption curve of a sample emitting β and γ rays was taken, using a gas-flow proportional counter, with aluminum absorbers.   The data obtained were:

Absorber Thickness (g/cm^2)	Activity (counts/min)	Absorber Thickness (g/cm^2)	Activity (counts/min)
0	5800	0.700	101
0.070	3500	0.800	100
0.130	2200	1.00	98
0.200	1300	2.00	92
0.300	600	4.00	80
0.400	280	7.00	65
0.500	120	10.00	53
0.600	103	14.00	40

(a) Estimate the maximum energy of the β spectrum (in million electron volts). (b) Find the energy of the γ ray. (c) What would be the absorption coefficient of that γ ray in lead? *Answers:* (b) 0.8 MeV; (c) 1.0 cm^{-1}.

5. Estimate the straggling of 32-MeV α particles in air. *Answer:* $\cong 1$ mg cm^{-2}.

6. Estimate the range of Po^{210} α particles in (a) air, (b) methane, (c) argon, (d) uranium.

7. At 1.00 meter from 1.00 g radium (in equilibrium with its decay products and enclosed in 0.5 mm of platinum) the γ-ray dosage rate is 0.84 rad per hour. What is the minimum safe working distance from a 20 mg radium source for 13 consecutive 40-hour weeks? *Answer:* 5.6 ft.

8. From data on X-ray spectra and X-ray absorption coefficients (in the Chemical Rubber Company Handbook, for example) locate the critical absorbers for the identification of the X rays following K capture (a) in Ar^{37}, (b) in Pd^{101}.

9. At a distance of 2 ft from a Cs^{137} source the dosage rate due to the γ rays from this source is found to be 127 mr per hour. The decay of Cs^{137} is accompanied by the emission of a 0.66-Mev γ ray in about 82 per cent of the disintegrations; no other γ rays are emitted. (a) Estimate the strength of the Cs^{137} source in millicuries. (b) What thickness of lead shielding (in inches) is required around the source to reduce the dosage rate at 2 ft to 3 mr per hour? *Answer:* (a) 60 mc; (b) $\sim 1\frac{1}{4}$ in.

10. The radiations emitted by a radioactive strontium sample were examined by aluminum absorption measurements. The following data were obtained with an argon-filled proportional counter. The decay of the sample during the course of the measurements was negligible.

Absorber Thickness (mg/cm^2)	Activity (counts/min)	Absorber Thickness (mg/cm^2)	Activity (counts/min)
0	32,100	200	525
10	20,900	250	445
20	13,600	300	397
40	6,050	350	368
60	2,950	400	355
80	1,610	500	342
100	1,080	700	330
120	810	1200	316
150	670	1800	300

Some check measurements with beryllium instead of aluminum absorbers gave the following results:

Absorber Thickness (mg/cm^2)	Activity (counts/min)	Absorber Thickness (mg/cm^2)	Activity (counts/min)
0	32,100	100	2220
10	21,000	125	2050
30	9,450	250	1940
50	4,850	400	1820
75	2,800	650	1650

Finally, the following lead absorption measurements were taken, with 150 mg cm^{-2} of beryllium always covering the counter window, in addition to the lead absorbers listed.

Lead Absorber (g/cm^2)	Activity (counts/min)	Lead Absorber (g/cm^2)	Activity (counts/min)
0	2000	0.100	343
0.005	1110	1.20	291
0.010	680	2.50	240
0.015	490	4.30	184
0.020	405	7.00	122
0.030	350	9.90	80

What conclusions can you draw about the mode of decay of the (hypothetical) strontium isotope present in the sample and about the radiations and their energies and relative intensities?

11. Derive (4-22) for $n = 2$.

12. Calculate the average number of elastic collisions required to slow a neutron from 10^6 to 10^{-2} eV in (a) U^{238}, (b) C^{12}, and (c) H^1.   *Answer:* (a) 2.2×10^3.

13. (a) A beam of 160-MeV O^{16} ions stripped of all its electrons enters an aluminum absorber. Ignoring the decrease in charge on the ion caused by electron pickup, estimate its range in aluminum by (4-8) and (4-10). Compare this estimate with the range given in figure 4-7. (b) Make the same comparison under the same assumption for 48-MeV O^{16} ions. Explain why the percentage discrepancy is larger at the lower incident energy. (c) Estimate the energy at which the totally stripped O^{16} ion picks up the first electron and the energy at which the O^{16} ion is neutralized (take ionization potential of O^{7+} as 871 eV and of the oxygen atom as 13.5 eV).   *Answers:* (a) 43 mg cm^{-2} at 160 MeV.
                     (b) 5.3 mg cm^{-2} at 48 MeV.

5

Radiation Detection and Measurement

In the preceding chapter we saw that the principal interactions of the radio-active radiations with matter result in the production of ions with a reduction in energy of the radiation of about 35 eV per ion pair formed. All methods for detection of radioactivity are based on interactions of the charged particles or electromagnetic rays with matter traversed. The uncharged neutron is detected only indirectly, through recoil protons (from fast neutrons) or through nuclear transmutations or induced radioactivities (from fast or slow neutrons). Neutrinos have neither charge nor rest mass and therefore do not interact measurably with matter to produce either ions or recoil particles. As mentioned on p. 55, neutrinos are expected to be capable of causing nuclear transmutations which are the inverse of β-decay processes; observation of such reactions has been reported, but the cross sections are extremely small—of the order of 10^{-40} cm^2 or less (R1).

A. ION COLLECTION METHODS

1. Recombination, Saturation Collection, Multiplication

Many common radiation detectors make use of the electric conductivity of a gas resulting from the ionization produced in it. This conductivity is somewhat analogous to the electric conductivity of solutions caused by the presence of electrolyte ions. In gas conduction, as produced by radiation, the ion current first increases with applied voltage; with increasing voltage the current eventually reaches a constant value which is a direct measure of the rate of production of charged ions in the gas volume. This constant value of the current is called the saturation current. A schematic representation of a gas volume and collecting electrodes, with potential difference V and meter to measure the ionization current I, is shown in figure 5-1, along with a plot of I versus V that might be obtained.

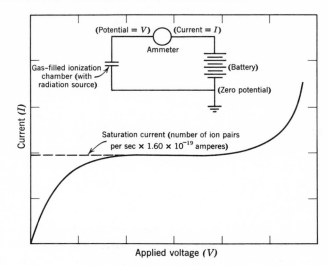

Fig. 5-1   Ionization current.

In the region of applied voltage below that necessary for the saturation current, recombination of positive and negative ions reduces the current collected. As the applied voltage is increased beyond the upper limit for saturation collection, the current increases again, and finally the gap breaks down into a glowing discharge or arc, with a very sharp rise in the current. In the measurement of gas ionization it is obviously of some advantage to measure the saturation current: the current is easily interpreted in terms of the rate of gas ionization, and the measured current does not depend critically on the applied voltage. The range of voltage over which the saturation current is obtained depends on the geometry of the electrodes and their spacing, the nature and pressure of the gas, and the general and local density and spatial distribution of the ionization produced in the gas. In air, for many practical cases, this range may extend from $\sim 10^2$ to $\sim 10^4$ volts per centimeter of distance between the electrodes.

We may classify detection systems (of the ion-collection type) according to whether saturation collection is employed or whether the multiplicative collection region is used. In the multiplicative region, where V is above the maximum value for saturation collection, the additional current is caused by secondary ionization processes which result from the high velocities reached by the ions (particularly electrons) moving in the high potential gradient. The use of this current amplification makes multiplicative collection methods inherently sensitive but unfortunately also inherently critical to many experimental variables.

2. Saturation Collection in Gases

Current Collection versus Pulsed Operation. We shall call the gas-filled electrode systems designed for saturation collection ionization chambers. Saturation current instruments consist of the ionization chamber, in which ions produced are collected with as little recombination and as little multiplication as possible, and of an electric system for measurement of the very small currents obtained.   The essential differences between the various instruments of this sort are in the nature of the systems for the measurement of the collected charges.   These may be grouped into two categories, those that measure steady-state currents and those that detect pulses resulting from individual ionizing events.   The difference lies in the magnitude of the so-called time constant of the device relative to the average time between ionizing events in the ionization chamber.   The charge in a capacitor of capacitance C short-circuited with a high resistance R will be dissipated exponentially; the time required for the charge to be reduced to $1/e$ of its value is RC, and this is known as the time constant of the circuit.   If RC is long compared to the time between ionizing events, a steady state is reached, and a direct current (or a voltage developed across a known resistor through which this current flows) may be measured.   On the other hand, if RC is small compared to the time between ionizing events, the charges collected during individual events (or the corresponding voltages) may be measured by means of appropriate a-c circuitry.

Direct-Current Instruments. The simplest and least expensive instruments of the d-c type, at one time widely used but now considered outmoded, are the various types of electroscopes.   Here a quartz-fiber or gold-leaf electrode system is initially charged to a voltage V, and the change in voltage (ΔV) produced by collection of the ionization charge is measured as a function of time. For a collected charge q the resulting $\Delta V = q/C$, where C is the (approximately constant) capacity of the electrode system.   The capacity C in centimeters or micromicrofarads is of the same order of magnitude as the dimensions of the electrode system in centimeters.   With a Lauritsen-type quartz fiber electroscope, intensities in the range of 10^3 to 10^6 β particles per minute entering the chamber are readily measurable.

In another type of instrument the ionization chamber is connected to an electronic d-c amplifier.   The ionization current I is caused to flow through a very high resistance R, and the voltage developed, $V = IR$, is applied to the control grid of a vacuum tube and measured in terms of the plate current of the tube (for example, by a galvanometer).   To measure the smallest currents, the vacuum tube must be chosen for low inherent grid current.   High stability of the circuit is essential, and some balancing feature to reduce effects of battery voltage variations is generally employed.

The ionization chamber may contain air at atmospheric pressure, which permits the use of an exceedingly thin aluminum leaf window for particles of low penetration but makes the response to penetrating radiations proportional to the barometric pressure. Sealed chambers are also used, and in special applications low-energy β emitters such as C^{14} and H^3 may be introduced as gaseous compounds directly into the chamber volume. For γ and X rays the ionization chamber may be filled to a pressure of several atmospheres with Freon (a chlorofluoromethane) or methyl bromide.

An instrument of this type with $R = 10^{11}$ Ω is sensitive to 500 β particles per minute entering the ionization chamber. With R switched to 10^8 Ω, about 10^7 β particles per minute may be measured.

As an example we estimate the ionization current I and the voltage drop IR for an ionization chamber under these assumptions: $R = 10^{11}$ Ω; the sample is an emitter of moderately energetic β rays with 1000 disintegrations per minute; the geometry is such that 50 per cent of the β particles enter and spend an average 8-cm path length in the effective volume of an air-filled chamber. The number of ion pairs to be expected[1] is about $1000 \times 0.50 \times 80 \times 10 = 4 \times 10^5$ per minute, or 6.7×10^3 per second. The current I will be the corresponding charge per second:

$$I = 6.7 \times 10^3 \times 1.6 \times 10^{-19} = 1.1 \times 10^{-15} \text{ A}.$$

$$IR = 1.1 \times 10^{-15} \times 10^{11} = 1.1 \times 10^{-4} \text{ V}.$$

If the use of the number of ion pairs rather than the total number of ions in this calculation is not entirely clear, remember that only half the ions—those with the proper sign of charge—are collected at either electrode.

Vibrating-Reed Electrometer. Even though special tubes and balanced circuits may be used with ionization chambers, d-c amplifiers are more susceptible to disturbance and drift and more difficult to arrange with several successive stages of amplification than those designed for amplifying alternating currents. An important development is the use of a continuously vibrating reed, which, through its oscillating electrostatic capacitance to a fixed electrode, converts the IR voltage to an approximately sinusoidal alternating potential; the a-c signal is then amplified in a highly stable audio-frequency amplifier. This instrument, the vibrating-reed electrometer, is easily sensitive to 10^{-15} A, thus to 1000 β disintegrations per minute with a typical ionization chamber, and is unusually free from troublesome zero drift and external disturbance.

Because the a-c voltage generated by the reed is affected by the amplitude of vibration, this amplifier is almost always used as a null instrument. The voltage IR developed by the ionization chamber current (with R perhaps as large as 10^{12} Ω) is applied to one side of the vibrating-reed condenser and an

---

[1] The estimate of 10 ion pairs per millimeter over the 80-mm path is taken from the information on β-ray ionization in chapter 4.

accurately known d-c reference voltage from a potentiometer is applied to the other. When these two voltages are the same, there is no a-c signal output. Commercial instruments find this null point electronically, and a conventional voltmeter connected to the reference voltage indicates the ionization current.

For most applications requiring high sensitivity and stability the vibrating-reed electrometer is usually the instrument of choice. The reeds must be constructed with great care, and the entire set of equipment is not inexpensive.

Ion Chambers with Pulse Amplifiers (F1, W1, W2). An ionization chamber may be directly connected to an a-c amplifier for measurements of individual ionization pulses. A short burst of intense ionization, such as results from the passage of an α particle through the chamber, will give a sudden change of voltage on the grid of the first amplifier tube; this grid voltage will return to normal in a time of the order of RC, where C represents the distributed capacitance of the grid and collecting-electrode system and R is the effective resistance to ground. With sufficient amplification, a large pulse will appear at the amplifier output terminal; the shape in time of this voltage pulse will depend on several factors, including the value of RC and the frequency-response characteristics of the amplifier. It is ordinarily desirable to have the height of the output pulse proportional to the amount of ionization produced by the particle in the chamber; thus the name linear amplifier or linear pulse amplifier is often applied to this instrument.

Since the instrument is used for counting single α particles, we may estimate the voltage amplification factor (gain) needed. A fast α particle traveling 1 cm in the chamber would give in air about 25,000 ion pairs, and the collected ion charge, $q \cong 25,000 \times 1.6 \times 10^{-19} \cong 4 \times 10^{-15}$ Coulomb. Guessing $C \cong 10\mu\mu$F, we have then $V = q/C \cong 4 \times 10^{-4}$ V. If an output pulse of 10 V is wanted (for oscillographic observation and for driving a scaling circuit), the required gain is 2.5×10^4. Three amplifier stages might be used, each with gain of roughly 30.

A practical ionization chamber may have a background rate of the order of 0.1 to 1 α per minute; the lower limit of sample strength easily detectable we may take as $\cong 1$ α disintegration per minute (with 50 per cent geometry). With appropriate amplifier and recording equipment the maximum usable rate is limited by the duration ($\cong RC$) of the voltage pulse, for if the average rate of arrival of pulses is such that there is an appreciable chance of one following another within the time RC appreciable counting error results. With $R = 10^8\Omega$, $RC \cong 10^{-3}$ sec, and a few thousand counts per minute would be the useful upper limit. Of course R is easily made smaller, but the full voltage q/C is achieved only if RC is long compared to the time of collection of ions in the chamber. The velocity v of ions in air under a voltage gradient E volts per centimeter is about (perhaps 1.5 times) E centimeters per second; with 1000 V applied to a 0.4-cm chamber, $v \cong 4000$ cm/sec, and the ion collection time is $\cong 0.4/4000 \cong 10^{-4}$ sec. In practice, RC is usually made somewhat longer than this time; to waste much of the voltage pulse is not advisable because with higher amplifier gains much trouble would be caused by tube "noise" and by microphonic effects (sensitivity of the chamber and amplifier to vibration).

In a closed ionization chamber filled with pure argon or nitrogen, or with one of several other gases, the electrons formed in ionization processes do not become attached to gas molecules to form negative ions but remain principally as free electrons. The drift velocity of electrons in the field is much greater than that of ions, and they reach the collector electrode in about 10^{-6} sec. In this case not only may much higher counting rates be used but also, if the amplifier is responsive to the higher frequencies only, the microphonic disturbances are very much reduced. For these reasons most ion chamber work nowadays employs electron collection only.

One of the most valuable applications of ion chambers is the accurate measurement of the energies of α particles and other ions with discrete spectra. In this application it is necessary for the entire particle range to be contained in the chamber volume. In addition, some provision has to be made to ensure that the size of the voltage pulse is independent of the location of the ionizing event in the chamber. In an ordinary parallel-plate chamber with electron collection the positive ion cloud left behind when the electrons are collected induces a charge on the collecting electrode, and the magnitude of this so-called image charge depends on the distance between the ions and the collector. Thus α particles of a given energy but moving in different trajectories give rise to different pulse heights. To overcome this effect, a negatively charged grid is usually placed between the collector and the ionization region; this shields the collector from the positive ion cloud, yet allows most of the electrons to pass through to the collector. In such grid ion chambers the pulse height is independent of track location, and excellent energy resolution is obtainable (0.6 per cent width at half maximum for 5-MeV α particles).

Ion chambers with linear amplifiers may be used to count fissions; because a fission fragment has approximately 10 or 20 times the energy of an α particle they are easily distinguished. Exceedingly low fission rates (of the order of 1 count per day) can be measured in the presence of high α-particle fluxes. This was important in the determination of some low spontaneous-fission rates. Fission chambers are also useful as neutron detectors.

Beta particles and γ rays do not produce enough ionization in the usual chambers to give pulses detectable above the background noise.

3. Solid-State Detectors (M1)

Insulating Crystals as Counters. Ionization-chamber operation is not limited to gas-filled ion chambers. The use of denser ionizing media has obvious advantages: ions of higher energy can be stopped completely within the chamber without recourse to unreasonably large volumes, and detectable pulses can be obtained from the passage of individual electrons or γ quanta in spite of their low specific ionization.   Ionization chambers filled with liquid argon have been tested.   Somewhat more promising have been the so-called crystal counters, which are essentially ion chambers using solid dielectrics between parallel-plate electrodes.   The ionizing radiation lifts electrons into the conduction band—a process analogous to ionization in an atom or molecule—and these electrons then travel toward the positive electrode with fairly high mobilities.   The positive charges (electron "holes") travel in the opposite direction, not by bodily movement of the ions through the crystal but by successive exchanges of electrons between neighboring lattice sites.   Diamond and cadmium sulfide are among the crystals which have been used successfully at room temperature.   Other solids, such as the halides of silver and thallium, which are ionic conductors at room temperature, can be employed at low temperatures.   The average energy required to lift an electron into the conduction band in one of these solid dielectrics is typically about 10 eV, compared with the average energy of about 30 eV expended in formation of an ion pair in a gas.

Semiconductors as Ion Chambers.   The application of crystal counters is seriously limited by the tendency of the charge carriers (electrons and positive "holes") to get trapped by impurities and crystal imperfections; these trapped charges distort the applied field and lead to polarization of the device and to nonuniform pulse sizes for mono-energetic ionizing particles.   Also, reproducibility from crystal to crystal is rather poor.   These disadvantages are largely obviated in the more recently developed semiconductor devices.   Mobilities and lifetimes of the charge carriers are much greater in semiconductors than in insulators, and trapping is therefore much less of a problem.   Also, the energy gap between the highest filled band and the conduction band is typically only about 1 eV in semiconductors; thus the energy required to produce an electron-hole pair is relatively small, and therefore semiconductor detectors have potentially good energy resolution.

So far the most widely used semiconductor material for radiation detectors has been silicon, with a band gap of 1.1 eV and an average energy ϵ for electron-hole pair production of 3.5 eV.[2]   In germanium ϵ is even smaller (2.9 eV), but its small band gap of 0.66 eV leads to excessive thermal excitation of electrons across the gap at room temperature; germanium detectors must therefore be cooled to eliminate this thermal noise.   In spite of this inconvenient feature, germanium detectors are coming into use as γ-ray spectrometers because the higher Z of germanium makes them much more sensitive to γ rays than silicon crystals.

[2] It appears that ϵ in silicon is independent of the energy and type of ionizing particle (electrons, protons, α particles, and heavier ions).   The value of ϵ always exceeds the energy gap between the bands because some energy is used in coupling electrons to lattice vibrations. This is analogous to the situation in gases in which the energy required to form an ion pair always exceeds the ionization potential because of "losses" to excitation and dissociation processes.

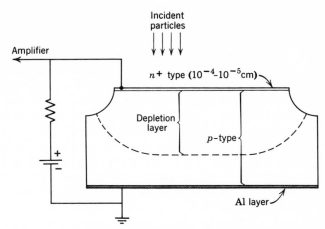

Fig. 5-2   Schematic diagram of a *p-n* junction detector.

It is in the very nature of a semiconductor that, with an electric field applied across it, it will pass a leakage current.[3]   To make semiconductors useful as particle detectors, one has to be able to apply appreciable electric fields without such excessive leakage currents.   The two basic techniques that have been employed are the controlled introduction of certain impurities which greatly increase resistivity and the use of reverse-biased diodes.   We shall discuss the second type of device first.

In the reverse-bias type of detector a barrier to the flow of leakage current is created either by appropriate chemical treatment of one of the surfaces ("surface barrier detector") or by providing a *p-n* junction[4] ("diffused junction detector").   Figure 5-2 schematically shows a *p-n* junction detector with *p*-type silicon as the base material.   A thin (10^{-4} to 10^{-5} cm) layer of n^+-type silicon has been produced at one surface by diffusion of phosphorus (e.g., by painting P_2O_5 on the surface and heating the slab).   On the opposite face a thin layer of aluminum is evaporated to facilitate electrical contact, the edges of the device are etched to reduce surface leakage, and a reverse bias (+charge on *n*-type, −charge on *p*-type side) is applied, providing a field of the order of 10^3 volts per centimeter.   In the presence of this field the positive holes in the *p*-type silicon are pulled toward the negative electrode, the electrons toward

---

[3] A slab of the highest-resistivity silicon available ($\cong 10^4$ Ω-cm), 0.5 cm thick and 1 cm^2 in area, would pass a current of 0.1 A at an electric field of 10^3 volts per centimeter.

[4] *p*- and *n*-type semiconductors are characterized by the presence of excess positive (hole) and negative (electron) charge carriers, respectively.   Small impurity concentrations can be used to produce *p*- and *n*-type semiconductors.   In silicon and germanium, phosphorus and arsenic serve as donors of electrons and produce *n*-type; boron and gallium are electron acceptors and produce *p*-type.   Semiconductor materials *heavily* doped with acceptor and donor impurities are sometimes designated as p^+ and n^+.

the p-n junction, and thus a "depletion layer" with a very small concentration of free charge carriers is created in the p-type material. The result is an extremely low leakage current. The thickness of the depletion layer depends on the magnitude of the applied field. When ionizing particles enter the device (usually through the p-n-junction side), the depletion layer serves as the sensitive volume, and electron-hole pairs created there will be quickly collected (carrier velocities are $10^6 - 10^7$ centimeters per second). One interesting feature of solid-state detectors is that hole mobilities are only about three times smaller than electron mobilities (in contrast to positive-ion mobilities in gases), so that all charge carriers are usually collected and the current produced is therefore independent of the location of the ionizing event within the sensitive volume.

The general features of the p-n^+-junction detector described apply equally well to a n-p^+ junction device made from n-type silicon with a p^+ layer produced by diffusion of boron or gallium. Also, surface barrier detectors, made, for example, by evaporation of a thin layer of gold or nickel onto n-type silicon or by surface oxidation of an etched n-type silicon surface operate in an essentially similar way, although the mechanism of surface barrier junctions does not seem to be fully understood.

The depth w of the depletion layer is approximately proportional to $\sqrt{\rho V}$ where ρ is the resistivity of the silicon in the main body of the detector and V is the applied potential. For p-type silicon

$$w \cong 3 \times 10^{-5} \sqrt{\rho V} \text{ cm,} \qquad (5\text{-}1)$$

if ρ is in Ω-cm and V in volts. With p-type silicon of 10^4 Ω-cm and with a bias voltage of $\cong 500$ V, depletion layers approaching 0.1 cm in thickness have been achieved. Silicon of sufficient purity to have even higher resistivity is very difficult to obtain, and bias voltages are limited by the onset of breakdown across the junction.

For some applications sensitive volumes larger than can be achieved with diffused-junction detectors are desirable. A promising approach is the attainment of very high resistivity by controlled compensation of the excess acceptors in ordinary p-type material (typically 10–100 Ω-cm) with donor atoms, usually of lithium. The technique is to allow the lithium to diffuse into the silicon at an elevated temperature (120–150°C) and under reverse bias. Under these conditions the mobility of Li^+ ions is large, and they drift under the influence of the bias voltage toward the negative electrode. The current carried by Li^+ ions (and electrons) will drop nearly to zero when almost perfect compensation of the acceptor concentration by the lithium donor concentration has been accomplished; resistivities of $\cong 10^5$ Ω-cm have been achieved this way, and depletion layers extending throughout the compensated region (many millimeters thick) are readily attained with modest bias voltages in such lithium-drift devices. Equation 5-1 does not apply to lithium-drift detectors.

Lithium-drifted germanium detectors are of particular interest as γ-ray counters.

Semiconductor detectors have many advantages over other particle detectors; chief among them are the good energy resolution attainable with them (roughly three and 10 times as good as for gas-filled ion chambers and scintillation counters, respectively), their linearity as a function of energy for a variety of particles, and their short pulse-rise times. Their principal applications to date have been in spectroscopy of protons, deuterons, α particles, fission fragments, and other heavy ions; but with the advent of the lithium-drift technique their usefulness for electron and γ-ray spectroscopy has increased substantially. The compactness of solid-state detectors and their insensitivity to low temperatures and magnetic fields suggests their utility in various special applications.

The output pulses from semiconductor detectors are of the order of milli-volts. Amplifiers of fairly high gain are therefore required, and, if full advantage is to be taken of the intrinsic resolving power of these counters, special care must be taken to achieve low noise levels in the electronic circuitry.

4. Multiplicative Ion Collection (S1, R2, W2)

In the preceding section we discussed detection techniques that utilize saturation collection of ions in ionization chambers. As already mentioned, at sufficiently high applied voltages the ion current or pulse height in an ion-chamber device increases above the saturation value because electrons moving in these high fields acquire enough energy to cause secondary ionization. In practice, multiplicative collection is always coupled with pulsed operation (small RC values), and the devices employing this scheme are referred to as "counters." Multiplicative ion collection has been confined to gas-filled electrode systems, although the idea might conceivably be applicable to solid or liquid media.

To obtain multiplicative collection of the type desired, we might at first consider simply increasing the voltage applied to an ordinary parallel-plate ionization chamber. This is ordinarily not practical for several reasons which are suggested by the following discussion.

Voltage Gradients and Electrode Shapes. Figure 5-3 shows the electrostatic lines of force between parallel-plate electrodes. The density of the lines of force is a measure of the voltage gradient (field) E in any region. The voltage gradient is the same everywhere between the plates, except for effects near the edge, and is given by the applied voltage difference ΔV divided by the plate separation. Lines of force converge on a curved electrode such as a sphere or wire or point, and indeed unless the parallel plates are perfectly smooth high local gradients will exist at surface irregularities.

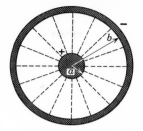

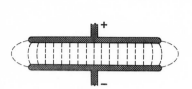

Fig. 5-3 Electrostatic lines of force between parallel-plate electrodes.

Fig. 5-4 Electrostatic lines of force between coaxial cylindrical electrodes.

In counters one electrode is usually a cylinder, the other an axial wire. Figure 5-4 shows a cross-sectional view with the wire radius exaggerated; the lines of force are sketched in.   It is readily seen that the density of these lines is inversely proportional to the radial distance r; that is,

$$E = \frac{k}{r} \qquad (5\text{-}2)$$

By definition, $E = dV/dr$, and we may represent the voltage difference between the electrodes of radii a and b:

$$\Delta V = \int_{r=a}^{r=b} dV = \int_{a}^{b} E\, dr = \int_{a}^{b} \frac{k}{r}\, dr = k \ln\left(\frac{b}{a}\right). \qquad (5\text{-}3)$$

In a practical case we might have $b = 1$ cm, $a = 4 \times 10^{-3}$ cm, and $\Delta V = 1000$ V.   Then

$$1000 = k \ln\left(\frac{1}{4 \times 10^{-3}}\right) = 5.5k; \qquad k = 180.$$

The voltage gradients at wall and wire are

$$E_b = 180 \text{ V cm}^{-1};$$

$$E_a = \frac{180}{4 \times 10^{-3}} = 4.5 \times 10^4 \text{ V cm}^{-1}.$$

The field at the wire and for a small space around it is above the maximum value for saturation collection (say $\sim 10^3$ V cm^{-1} in a practical counter gas). The voltage difference ΔV is always applied with the wall (cathode) negative with respect to the wire (anode); in this way free electrons and negative ions move to the wire.

Regions of Multiplicative Operation.   With an electrode system of the kind just described (cylindrical cathode with central wire anode), filled with a suitable gas, and connected to a high-gain amplifier and oscilloscope, the pulse

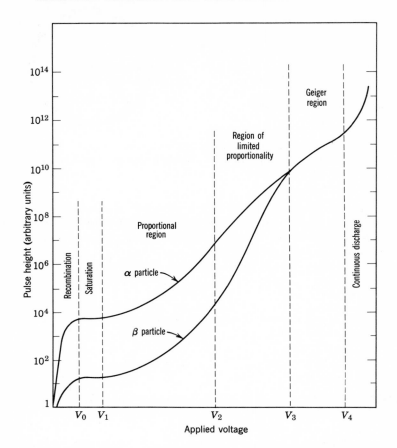

Fig. 5-5   Variation of pulse height with applied voltage in a counter.

heights obtained as a function of applied voltage would be approximately as shown in figure 5-5.   Curves are drawn for two types of ionizing particles, one losing several hundred times as much energy in the chamber volume as the other (they might be typical α and β particles, respectively).

In the region of saturation collection, the voltage pulses caused by β particles are, as already discussed, generally too small to be detected with practical amplifiers, whereas the α pulses are measurable with a sensitive pulse amplifier. Once the voltage is raised above the limit of the saturation region the pulse heights increase as a result of secondary ionization by the electrons accelerated in the high field gradient near the wire.   For a considerable voltage range (V_1 to V_2 in figure 5-5, typically of the order of several hundred to a thousand volts) the *ratio* of pulse heights for different ionizing events remains independent of applied voltage, or, in other words, the pulse height remains proportional to the amount of energy lost in the chamber by the primary ionizing particle.   In this voltage region the apparatus operates as a proportional counter.   Multiplication factors (number of electrons collected per initial ion pair) in actual proportional counters may vary from $\cong 10$ to $\cong 10^4$.   The gain required in the external amplifier depends on the multiplication factor used and on the radiation to be detected.

If the voltage is increased further (above V_2), pulse heights continue to increase.   From V_2 to V_3 pulse heights are still related, but no longer proportional, to initial ionization intensity.   This is sometimes referred to as the region of limited proportionality.   Finally, at V_3 the pulse height becomes independent of initial ionization—a pulse caused by a single ion pair becomes indistinguishable from one due to a fission fragment depositing all its energy. A device operated in this region is called a Geiger-Müller counter.   Pulse heights are typically of the order of volts and very little if any additional amplification is needed.   At a still higher voltage (V_4) the Geiger-Müller action is terminated by the onset of self-excitation, and eventually the counter goes into continuous discharge.

Gas Multiplication.   The gas multiplication factor obtained in a given counter tube depends on the nature and pressure of the gas, on the tube dimensions, particularly the wire diameter, and on the applied voltage.   From (5-2) it is clear that as the wire diameter is decreased the field strength in the immediate vicinity of the wire goes up and that with increasing applied voltage the region around the wire in which the gradient exceeds the minimum value necessary for multiplication will extend farther and farther out.   We must also expect the critical field gradient to be different for different gases, generally higher for the polyatomic than for the rare gases.   Finally, since an electron must gain sufficient energy between collisions to produce ionization, the multiplication factor decreases under otherwise identical conditions with decreasing mean free path and therefore with increasing gas pressure.   It has been shown (R2) that in a given gas the functional dependence of the multiplication factor M on wire radius a, cathode radius b, pressure P, and voltage V is of the form

$$M = f\left[\frac{V}{\ln (b/a)}, (Pa)\right] \tag{5-4}$$

As an illustration, figure 5-6 (based on data in R2) shows the variation of M with voltage in argon and methane at two different pressures.   As these data suggest, it is found that relatively small admixtures of argon lower the threshold and operating voltages of methane-filled proportional counters considerably.   On the other hand, the presence of methane or other polyatomic gases in argon-filled counters decreases the dependence of M on applied voltage and thus improves the stability of operation with respect to voltage variations.

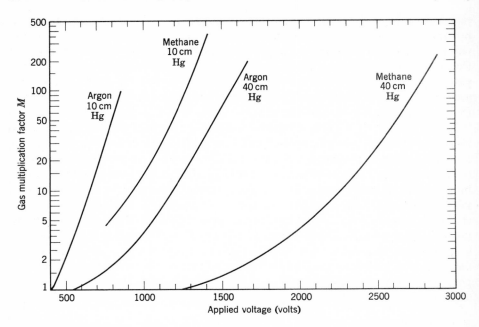

Fig. 5-6 Multiplication factors in argon and methane as a function of applied voltage. Wire radius a = 0.005 in., cathode radius b = 0.435 in.   Data from reference R2.

Proportional Counters (C1, C6, F1).   True proportionality between pulse height and primary ionization requires that the avalanches produced by individual primary electrons in an ionization track be essentially independent of one another; thus each avalanche must be confined to a very small region of the central wire.   In the course of an avalanche excitation of molecules can lead to the emission of ultraviolet photons, which in turn are capable of producing photoelectrons at the cathode or in some constituent of the gas.   At sufficiently high voltages the number of photons per primary electron becomes so great that a given avalanche is likely to spread along the entire tube by these photoionization phenomena; the final pulse size is then no longer dependent on primary ionization and the Geiger region has been reached.

In the discussion of ionization chambers we noted that the pulse height for a given type of ionizing event depends on the location of the track.   In the proportional counter this is not the case, for with any reasonable value of M the collection of primary ions and electrons contributes only negligibly to the final pulse size and the amount of avalanche ionization is independent of the location of the primary track, provided it does not lie too close to the anode wire.   Since the multiplicative region is confined to a small distance from the wire, the electrons travel only through a small part of the potential difference applied to the counter and therefore contribute relatively little to the pulse size.   Most of the pulse height is due to the motion of the positive ions away from the central wire (an effect analogous to that of the image charge in ion chambers).   However, although the total time for the positive ions to reach the cathode is typically of the order of 10^{-3} second, the variation of field gradient with radius is such that initially the pulse rises very rapidly and then approaches its final height extremely slowly.   If the total collection time is T, the time at which half the final pulse height is reached is about $(a/b)T$, or typically of the order of 10^{-6} second.   Amplifier circuits with time constants of this order are therefore used to "clip" the pulses from proportional counters.

With a clipping circuit of the type just mentioned, a proportional counter is ready to record a new ionizing event within 1 to 2 μsec after a count.   Thus proportional counters can be used at counting rates up to $\sim 10^{6}$/min with very small dead-time losses (cf. chapter 6, section F).   This is one of their chief advantages over the much slower Geiger-Müller (GM) counters, even when proportionality of response is not important.   It should be noted, however, that even with sharp clipping of pulses proportionality is preserved because the pulse *shape* is independent of pulse *height*.   In general, proportional counters are also more stable and more reproducible in their operation than are GM counters, and they tend to have longer and flatter voltage plateaus.

A voltage plateau is any region in which the counting rate caused by a given constant radiation source is independent of applied voltage.   Note that we are here focusing on the behavior of *counting rate*, not (as in figure 5-5) of *pulse height* as a function of voltage.   Throughout the proportional region, the pulse height for a given amount of energy loss in the counter increases with voltage; yet the counting rate observed with a sample is expected to exhibit a plateau.   With a particular amplifier gain, the position of the plateau will depend on the amount of primary ionization deposited in the counter by each event.   It is thus possible to operate a proportional counter in a voltage region in which it exhibits a plateau for α particles (which might produce 2×10^{5} primary ion pairs in the counter) but does not respond to β particles (producing about 100 times less ionization); in a higher voltage range the same counter will generally have another plateau on which both types of radiation are recorded.   Figure 5-7 shows the response of such a counter to a mixed source of α and β particles as a function of voltage.   Voltage plateaus of several hundred volts with slopes of 1 per cent or less per hundred volts are quite common.   The characteristics of the plateau are by no means independent of the associated electronic circuits; in particular, the onset and length of the plateau are affected by the discriminator setting.   A discriminator is a device used to prevent pulses below a certain size from being registered and is generally needed to cut out pulses due to electronic noise.

Several types of proportional counters are available.   Very popular are the flow-type counters, in which the argon-methane or other gas mixture flows at atmospheric pressure from a compressed-gas tank through the counter at a

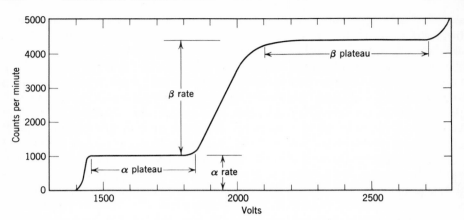

Fig. 5-7   Counting rate as a function of applied voltage for a proportional counter exposed to a source emitting both α and β particles.

slow rate.   This procedure avoids deterioration of the gas and minimizes the effects of gas leaks.   In some of these designs the mounted sample is introduced into the counter gas volume through an air lock; others are equipped with a thin window of aluminum or of a plastic film (such as Mylar) with an aquadag or evaporated metal coating.   Thin beryllium windows are sometimes used for X rays of very low energy.

The proportional response of the counter can be a great advantage in several applications.   If the counter tube is large enough to include all the ionization produced by a particular radiation, the pulse height (observed after linear amplification) measures the energy of the ray.   Thus the proportional counter can serve as a very useful spectrometer, particularly for β rays and conversion electrons below about 0.15 MeV and for low-energy electromagnetic radiation such as X rays.   For X-ray spectrometry the counter tubes have to be relatively large (several inches in diameter) and sometimes filled to more than atmospheric pressure with an appropriate gas so that virtually complete absorption of the X rays in the counter gas can be assured.

Geiger-Müller Counters (C1, W2).   As we have already mentioned, the proportional region of counter operation is limited at the upper voltage end by the onset of photoionization; this process spreads the intense ionization produced by a single primary electron along the anode, thus causing interaction with the avalanches produced by other primaries, and destroying proportionality of response.   As the voltage is further increased, a condition is finally reached (V_3 in figure 5-5) in which each ionizing event is spread along the entire length of the wire and the final pulse size becomes independent of primary ionization—the tube now operates as a Geiger counter.

In a Geiger counter, as in a proportional counter, the negative ions formed (mostly free electrons) reach the wire very quickly, typically in about 5×10^{-7} second; but now the entire wire is surrounded by a sheath of positive ions, which has the effect of reducing the voltage gradient below the value necessary for ion multiplication. Thus, in contrast to a proportional counter, the GM counter is left insensitive after a pulse and must recover before another event can be counted. Recovery is effected through migration of the positive ions to the cathode which takes about 100 to 500 μsec. This is the inherent dead-time of a GM counter, which limits its use to counting rates below a few tens of thousands per minute.

When positive ions reach the cathode, secondary electrons might be emitted from the surface; this would produce a new counter discharge a few hundred microseconds after the first and quite independent of the source of radiation that the counter is intended to measure. To prevent these self-perpetuating discharges, two schemes have been employed. One is to use an external quench circuit which keeps the voltage across the counter below the threshold voltage for Geiger action until after the positive ions have reached the cathode. Much more common are the so-called self-quenching GM counters in which the emission of secondary electrons from the cathode is suppressed by the presence of a polyatomic vapor or gas, such as alcohol, ether, or methane, along with the usual filling gas, such as argon. These mixtures seem to be effective because the positive ions are by electron transfers all converted to organic ions while moving to the cathode, and these polyatomic ions may dissipate energy by predissociation and reduce enormously the probability of secondary-electron emission. Also they may serve to quench metastable states of the argon atoms. It is significant that the organic additive is considerably consumed after 10^8 or 10^9 counts and that a polyatomic filling gas, tetramethyl lead, requires no additive. Counter-filling mixtures containing small amounts of halogens as quench gas do not appear to deteriorate with use. However, halogen-quenched GM counters usually have poorer plateau characteristics than the other types.

Properly working Geiger counters exhibit voltage plateaus; but these plateaus are usually shorter and have more of a slope than those obtained with proportional counters. As the upper voltage limit of a Geiger counter plateau is exceeded, double, triple, and other multiple pulses occur with increasing frequency. Their time spacing is a few hundred microseconds, suggesting that a break-down of the self-quenching mechanism is responsible for the end of the plateau. At still higher voltages the counters break into continuous glow discharge.

The principal advantage of GM counters lies in the large pulse sizes they produce, which obviate the need for electronic amplifiers. The resulting simplicity of operation led to the preponderant use of GM counters in preference to proportional counters over a period of many years. The advantages of proportional counting and the commercial availability of reliable high-gain amplifiers have reversed this trend in recent years.

Counter Backgrounds. Since both GM and proportional counters register individual ionizing events, their sensitivities are limited purely by background counting rates. Even in a laboratory not contaminated by radiochemical work small amounts of activity are present as impurities in construction materials. Also the air contains an appreciable and variable amount of radon and thoron and their decay products. In free air at the earth's surface most of the ionization comes from these two causes, with the cosmic radiation contributing a smaller part. However, because the counter is itself closed and enclosed in a building, it is not accessible to most of the radioactive α, β, and even γ radiation, and the cosmic-ray effect is the most significant. A β-sensitive counter with a diameter of 2.5 cm and length 6 cm may have a background rate of about 30 counts per minute; this may be reduced to less than 15 counts per minute by the usual lead shield of a few centimeters thickness. We may take about 10 β disintegrations per minute as a minimum sample strength easily detected by such a counter (with the 50 per cent geometry estimate as before). Several techniques are sometimes employed to reduce these backgrounds, including special shielding and anticoincidence circuits which reject those counts occurring simultaneously with counts in nearby auxiliary counters (chapter 15, section A). These techniques are employed in a number of commercially available "low-level" counting systems which achieve background rates of less than 0.5 per minute with counters of approximately the dimensions mentioned above. In the case of proportional counters it is possible to obtain further reduction by the use of circuits which reject pulses of sizes that do not correspond to the radiation being measured (see section C). Proportional counters operating on their α plateaus require no special shielding and may have background rates of about 0.1 per minute if constructed of selected materials.

B. METHODS NOT BASED ON ION COLLECTION

1. Image-Forming Devices

Photographic Film (B1). The historical method for the detection of radioactivity was the general blackening or fogging of photographic negatives, apparent on chemical development in the usual way. This method was soon supplanted by ionization measurements but has reappeared more recently in the "film badge" for personnel exposure control (see section D) and in the γ raying (analogous to X raying) of castings and other heavy metal parts for hidden flaws. Also, in the radioautograph technique the distribution of a radioactive tracer (preferably an α or soft-β emitter) is revealed when a thin section, perhaps of biological material, is kept in contact with a photographic plate.

Special photographic emulsions known as nuclear emulsions exposed to ionizing radiations, such as α rays, protons, mesons, and electrons, on development show blackened grains along the path of each particle; these tracks, some of which may be quite short, are observed and their characteristics (length, point of origin, direction, ionization density, and scattering) are measured under a microscope. The number of developed grains along a track is smaller by several orders of magnitude than the number of ion pairs produced.

Nuclear emulsions differ from ordinary light-sensitive plates in that they have considerably higher silver halide content and smaller grain sizes (0.1 to 0.6 μ). The smaller the grain size, the less sensitive the emulsion to anything but the most densely ionizing particles. Thus different commercially available emulsions, differing chiefly in grain size, can be used to discriminate between different particles. Further differentiation can be obtained by variations in development. The least sensitive emulsions will show only fission fragment tracks, whereas the most sensitive ones reveal the tracks of singly charged minimum-ionizing particles.

The complete identification of a particle from its emulsion track may involve a variety of measurements. If the nature of the particle is known and if it spends its entire range in the emulsion, the energy may, of course, be deduced from the track length and the range-energy relation applicable to the particular emulsion used. Particles of moderate velocities ($v \ll c$) whose tracks end in the emulsion can be identified by measurements of track length and grain density, since grain density is proportional to specific ionization. Grain densities range from about 3000 grains/cm for minimum ionizing particles to approximately 500 times that value, at which point saturation sets in. The energies of very heavily ionizing particles can be estimated from the density of δ rays (cf. chapter 4, p. 88). For high-energy particles ($v \cong c$), measurement of the average scattering per unit path length is one of the most valuable techniques; the average multiple scattering for a particle of momentum p is inversely proportional to pv and therefore to the total energy E. One of the major shortcomings of the nuclear emulsion technique is the difficulty of using magnetic fields to deflect particles traveling in emulsions—the tracks are generally too short for this very useful identification method to be applicable. The extremely high magnetic fields made possible by superconducting coils may change this picture in the future.

The chief advantages of nuclear emulsions (as compared, for example, with the devices to be discussed next) are their high densities, their good spatial resolution, and, for some applications, their continuous sensitivity. The technique has been particularly useful for the recording of very rare events, for example those of interest in cosmic-ray studies. A number of elementary particles were first discovered by means of emulsions, among them the positron and the pi meson.

Cloud Chamber (F2). A pictorial representation of the paths of ionizing particles similar to the photographic track but capable of finer detail is given by the cloud chamber (Wilson chamber). In this instrument, invented in 1911 by C. T. R. Wilson, the particle track through a gas is made visible by the condensation of liquid droplets on the ions produced. To accomplish this, an enclosed gas saturated with vapor (water, alcohol, and the like) is suddenly cooled by adiabatic expansion to produce supersaturation. Ordinarily a fog would be formed but, if conditions are right and the gas is free of dust, scattered ions, and so on, the supersaturation is maintained except for local condensation along the track where the ions serve as condensation centers. The piston or diaphragm causing the expansion is operated in a cyclic way, and a small electrostatic gradient is provided to sweep out ions between expansions. There is usually an arrangement of lights, camera, and mirrors to make stereoscopic photographs of the fresh tracks at each expansion. The supersaturated vapor for cloud chamber operation can be achieved in other ways, notably by the diffusion of a saturated organic vapor into a colder region. In the diffusion cloud chamber the working volume is continuously sensitive rather than intermittently so, and the whole instrument is considerably less complicated than the conventional Wilson chamber. Expansion chambers were therefore largely replaced by diffusion chambers in the early 1950's until the latter were themselves made obsolete by the advent of the bubble chamber. Particularly useful to high-energy physics research were diffusion chambers filled with hydrogen, deuterium, or helium to pressures of about 25 atm.

In cloud chambers α tracks appear as straight lines of dense fog droplets, with thousands of droplets per centimeter. The β tracks are much less dense, with discrete droplets visible, several per centimeter along the path. In general, track density is related to specific ionization and therefore depends on mass, charge, and energy of the ionizing particle. Cloud chambers are often operated in a magnetic field, and momenta of particles can then be deduced from the track curvatures. As in photographic emulsions, measurements of scattering, straggling, and δ-ray density may also aid in the characterization of particles.

Bubble Chamber (B2, F2, B3). Advantages of photographic emulsions and of cloud chambers are combined in the bubble chamber. This device, invented in 1952 by D. A. Glaser, makes use of the well-known fact that liquids can be heated for finite though short times above their boiling points without actually boiling. In the bubble chamber charged particles traveling through a superheated liquid cause vapor bubbles to form along their tracks, probably because of local heating; these bubble tracks can be photographed in strong illumination. Since superheated liquids are not stable for long periods of time, bubble chambers are always operated in pulsed fashion, with the liquid normally under such a pressure that at the operating temperature it is below its boiling point; it is made sensitive by sudden reduction of the pressure and remains sensitive for about 10^{-3} to 10^{-2} second.   The best operating temperature appears to be roughly two thirds of the way from the normal boiling point to the critical temperature, and the corresponding pressure is in the neighborhood of half the critical pressure.

Bubble chambers are principally useful in conjunction with high-energy accelerators and are indeed among the most important research tools at these machines.   The greater densities of liquids compared to gases make bubble chambers superior to cloud chambers for the study of energetic particles. Even so, the dimensions of bubble chambers used at the biggest accelerators are gigantic.   A chamber 80 in. deep and filled with 1500 liters of liquid hydrogen was put into operation at the Brookhaven National Laboratory. Hydrogen is probably the most important bubble-chamber liquid because it makes possible the unambiguous identification of interactions with individual protons.   However, construction and operation of large hydrogen chambers pose formidable cryogenic and safety problems.   Chambers filled with propane, deuterium, helium, xenon, and a variety of other liquids have been very successfully used.   As with cloud chambers, magnetic fields are common aids to particle identification.

The scanning of bubble-chamber photographs for particular events of interest and the subsequent detailed analysis of these events is exceedingly complex and time-consuming.   Special computer methods are being developed constantly to solve these problems (R4).

Spark Chamber (C2).   One of the chief limitations of the bubble chamber lies in the limited number of tracks that can be tolerated in the chamber during one expansion cycle.   This makes detection and study of extremely rare events in the presence of many unwanted interactions excessively time-consuming.   The spark chamber, developed since about 1957 but based on previously known ideas, is suitable for this type of application.   It consists essentially of a series of parallel metal plates, spaced a few millimeters apart, and with neon or argon between them.   If a short ($\cong 2 \times 10^{-7}$ second) high-voltage pulse ($\cong 10$ kV) is applied to alternate plates, with every other plate left at ground potential, the chamber becomes momentarily sensitive to charged particles in the sense that a visible spark develops between adjacent plates wherever an ionizing particle crosses the gap.   The path of an ionizing particle can thus be photographed.   The sensitive time of the chamber can be held down to $< 5 \times 10^{-7}$ second by the application of a sweeping field of $\cong 100$ V to sweep out the ions formed.   The efficiency, even for minimum ionizing particles, is almost 100 per cent, and the high-voltage pulse can be triggered by any desired arrangement of coincidence or anticoincidence counters so that almost none but the desired events are recorded in the chamber, even though it is placed in a very large flux of ionizing particles.

2. Counters Based on Light Emission

Scintillation Counting (B4, R3).   When α particles strike a prepared fluorescent screen of zinc sulfide, discrete flashes of light may be seen by the dark-accustomed eye.   The counting of α rays by this scintillation method in a device called the spinthariscope was of great value in the early studies of radioactivity.   Although it is no longer used in this way, there is a modern adaptation of scintillation counting, especially for β and γ rays.   The rays produce light in an anthracene crystal or other suitable scintillator; the light can produce photoelectrons from the first photosensitive electrode of a photomultiplier tube, and the output pulse may be recorded.   The theory of the scintillation process, the role of impurities, etc., are not discussed here.   Excellent treatments of these matters are available in books and review articles (e.g., R3, B4, C3, S2).

Active experimentation in recent years has developed a variety of practical scintillators, each with particular advantages.   Of the organic phosphors, anthracene gives the highest yield of photons, about 15 for each 1000 eV of energy dissipated in the crystal.   Sizable anthracene crystals are commercially available.   Stilbene crystals give about half as big a light yield as anthracene but are useful in coincidence experiments because the pulse has a short decay time, of the order of 10^{-9} second.   Liquid solutions, such as p-terphenyl or stilbene in xylene or toluene, give good yields and short time constants and are easily prepared in large volumes.   If loaded with boron or cadmium compounds, such liquid scintillators become efficient neutron detectors.   (A liquid scintillation counter of 300-liter sensitive volume was developed for the neutrino experiment mentioned parenthetically on p. 55.)   Scintillators incorporated in plastics also are available, and some success has been reported with the use of scintillations in rare gases.

For γ-ray measurements NaI crystals (activated with about 1 per cent TlI) are most widely used.   Their high density and the high Z of iodine make them especially suitable.   Commercially available crystals of this phosphor, machined to specified shapes up to several inches in each dimension, can give high γ-ray counting efficiencies, close to 100 per cent for energies up to about 200 keV, perhaps 20 per cent at 1 MeV.   It is somewhat inconvenient that these (expensive) crystals are hygroscopic and must be protected from moisture.   The NaI (Tl) scintillator has a high light yield (about twice that of anthracene), but its longer decay time prevents its use in high-speed coincidence counting.   For the production of scintillations from α particles conventional ZnS phosphors are still used in conjunction with photomultiplier tubes.   The addition of boron to a ZnS phosphor permits the detection of neutrons through the α particles produced in the $B^{10}(n, \alpha)Li^7$ reaction.   A ZnS phosphor in plastic will give light pulses from the proton recoils produced by fast neutrons.   Also, europium-activated LiI phosphor crystals are useful for neutron detection.

The developments in scintillators have been accompanied by improvements in the photomultiplier tubes. Modern tubes may have large, almost uniformly sensitive photocathodes at which about 1 photoelectron results per 10 photons of the typical phosphorescence wavelength (near 4400 Å). The electrons are accelerated by a potential of 100 V or more to the first electrode (dynode) where each one produces several (n) secondary electrons; these secondary electrons are similarly accelerated to and increased n-fold at the second dynode and so on. With 10 dynodes the charge of the original photoelectrons is multiplied n^{10} times, which may be of the order of 10^5. Thus a 0.1-MeV electron, absorbed in an anthracene scintillator, might produce 1500 photons, giving of the order of 100 photoelectrons and leading eventually to an output pulse of about 10^7 electrons or 1.6 $\mu\mu$C. (In an output circuit of 160-$\mu\mu$F capacity this pulse would amount to 0.01 V, and further amplification in a conventional linear pulse amplifier would be provided.) Because the multiplication in a photomultiplier is sensitive to the applied voltage, good high-voltage stabilization (to ±0.1 per cent or better) must be provided for reproducible operation. For optimum performance the photomultiplier tube must be carefully shielded from magnetic fields, including the earth's field. So-called mu-metal and other shields are available for this purpose.

In typical scintillation counters the crystal is fixed to the photosensitive face of the multiplier tube, either directly or through a short lucite connecting section, with balsam or oil to provide good optical contact at the interface. A light-tight enclosure of aluminum foil serves also as a reflector to assist in directing light to the photocathode. For β or γ rays of more than about 50 keV very efficient light transfer may not be needed to give output pulses greater than the background of electrical "noise." However, for many applications it is important to have the pulse height accurately proportional to the energy dissipated in the crystal. This result is obtained only if light produced by rays in all parts of the crystal can reach the cathode with a uniform attenuation. The crystal should be quite transparent, and nearly total reflections at surfaces other than the photocathode should be the goal; roughening of the crystal faces and the use of a diffuse reflector such as MgO are recommended practices. Integrally canned tube-crystal assemblies incorporating all of these features are commercially available. Well-type scintillation detectors for measurement of liquid samples in vials are very popular.

The condition of proportional response can be achieved within about 2 per cent by good design and construction, with careful regulation of dynode voltages and other electronic variables. This proportionality is most useful if the crystal is large enough so that it can contain the entire path of the ray and so measure its total energy. For β radiation this requires a crystal with dimensions considerably larger than the range of the particles and some geometrical provisions to minimize scattering out of the crystal. For γ radiation the crystal must be large enough to contain the ranges of a good share of the photoelectrons produced in it; the resulting pulses will correspond to the total γ energy, since the energy used to overcome the electron's binding energy will also contribute to the scintillation through X-ray and Auger-electron emission. In addition to this "photoelectric peak" there will appear a continuous distribution of smaller energies corresponding to Compton-recoil electrons, to the extent that the crystal is not large enough to contain the entire sequence of processes that consume the initial γ-ray energy. In practice, β- and γ-ray spectrometry based on scintillation response offers the best sensitivity but less than the best resolution, and the technique is widely used for energies from about 0.01 to 2 or 3 MeV. For most γ-ray measurements, scintillation spectrometry is without doubt the method of choice (cf. chapter 12, section F).

Background rates in NaI(Tl) scintillation counters are rather high (of the order of 10^3 counts per minute in a 1-in.^{3} scintillator in an unshielded room) and approximately proportional to scintillator volume. Massive shielding can be effective in reducing the background effects due to cosmic rays and to γ rays from surrounding material, and further background reduction can be achieved with anticoincidence arrangements. Pulse height selection helps, of course, to obtain improved ratios of sample to background rates. In certain low-level activity measurements it is worthwhile to use phototubes with quartz rather than glass envelopes in order to avoid the background contribution of the K^{40} γ rays originating in the glass.

Čerenkov Counters (L1, M2). These devices which are now among the major tools of the high-energy physicist are based on the then rather surprising discovery (C4) by P. A. Čerenkov (1934) that a beam of γ rays in water was accompanied by the emission of light at a definite angle to the beam direction ($\cong 40°$). The effect was subsequently explained by I. Frank and I. Tamm (F3) in terms of the electromagnetic shock wave produced when a charged particle travels through a transparent medium at a speed exceeding the velocity of light in the same medium. Thus, if n is the refractive index of the medium and βc the velocity of the particle, the condition for emission of Čerenkov light is

$$n\beta > 1. \tag{5-5}$$

The angle θ between the particle trajectory and the direction of light emission is given by the relation

$$\cos \theta = \frac{1}{n\beta}. \tag{5-6}$$

Although the theory of Čerenkov radiation, including intensity and frequency relationships, was worked out in 1937, a practical Čerenkov counter was not built until 10 years later when the development of photomultiplier tubes made efficient light collection feasible. The intensity of Čerenkov light is weak compared with the light output of a scintillator, but the directional properties can be used to advantage to improve light collection, and many types of highly efficient Čerenkov counters for relativistic particles have been constructed.

The most important applications of Čerenkov counters are based on their velocity-selecting properties. From condition (5-5) it is evident that a Čerenkov counter will always act as a threshold detector, recording only those particles with $\beta > 1/n$. For example, with water ($n = 1.332$) the threshold is at $\beta = 0.751$, which corresponds to 500-MeV protons, 73-MeV π mesons, or 265 keV electrons. Liquid nitrogen ($n = 1.205$, $\beta_{min} = 0.830$) can be used (by direct coupling of a Dewar flask to a photomultiplier) to detect protons above 760 MeV or pions above 112 MeV, etc. For still lower refractive indices (down to about 1.01) various compressed gases have been used.

The angular definition of Čerenkov light (5-6) is used in many ingenious ways to achieve selectivity in detecting particles in certain velocity intervals. An upper limit to β can be set by taking advantage of internal reflections at the exit face of the Čerenkov medium for all light arriving at less than a critical angle; blackening of other faces aids in the absorption of the internally reflected photons. In other types of arrangements focusing of light emitted in a certain angular interval is accomplished by systems of mirrors and lenses.

Čerenkov counters, because of their velocity selectivity, are particularly useful for experiments with particle beams at high-energy accelerators. They may serve as elements in counter telescopes in which, together with momentum selection by magnetic deflection, they can be used to obtain good energy resolution for specific particles; they are also ideally suited as triggering devices for cloud and spark chambers which are to be sensitive for incident particles of specified properties only.

C. AUXILIARY INSTRUMENTATION

Scalers and Recorders. The pulses produced by the various types of counters and pulse-ionization chambers, after suitable amplification, may be recorded or measured in a variety of ways. A cathode-ray oscilloscope on which individual pulses may be displayed is almost indispensable if the behavior of counters and associated electronic equipment is to be checked, but it is, of course, not very useful for ordinary counting-rate measurements. At low counting rates (perhaps up to 10 per second) pulses may be used directly to actuate a mechanical recorder. Almost universally, however, the mechanical register is preceded by a scaling circuit which reduces the rate electronically by some factor, usually a power of 2 or, more conveniently, a power of 10. Most of these circuits employ multivibrator[5] scaling pairs, each pair reducing

---

[5] In this application both tubes of the basic multivibrator pair (square-waveform oscillator) are biased to prevent oscillation; each incoming pulse triggers the pair through a half-cycle; completion of each cycle provides an output pulse to trigger the next pair. One of the most dependable circuits is that employing direct rather than capacitor coupling, devised by W. A. Higinbotham. A scale factor of 10 may be obtained by modification of a scale of 16. A

the rate by a factor of 2.   The scaled impulses are recorded mechanically, with electric power supplied by a suitable output circuit.   Many types of scaling circuits are commercially available, and most of these instruments have provisions for selection among several possible scaling factors.   A commercial circuit may or may not include an amplifier, the mechanical register, and the stabilized high-voltage supply (perhaps 1500 or 3000 V maximum, with 0.1 per cent stability for a 10-V change in line voltage).   Depending on the accessories included, the price might be somewhere between 300 and 2500 dollars. The trend is away from vacuum-tube electronics and toward the more reliable, more compact, less power-consuming, and more expensive transistorized circuitry.

The particular choice of circuit depends on the counter used.   For GM tubes no amplifier is required, high scaling factors (>100) are usually not needed (because of the dead-time limitation on counting rates), and good voltage stabilization is relatively unimportant.   For scintillation counters, excellent voltage regulation is desirable, and for both scintillation and proportional counting stable nonoverloading amplifiers are wanted with gains somewhere between 10^2 and 10^5.   The amplifiers need not be linear unless the counters are used for pulse-height analysis.

Many useful variations of the basic scaling-circuit arrangements are available, differing principally in the output stages, and with many of the commerical units one of several modes of operation can be selected.   The scaler may be automatically turned off after a preset number of counts (with the time recorded on an electric clock or printed out) or after a preset amount of time.   The accumulated register counts may be printed out at selected time intervals by a printing register, or scaled pulses may be recorded by a galvanometer on paper tape moving at constant known speed.   These modifications are particularly useful for determinations of decay curves.   In a large laboratory with many routine counting instruments it may be advantageous to use even more sophisticated output arrangements, such as recording of data on punched paper tape or magnetic tape for processing on an electronic computer. Refinements such as automatic changing of samples after a certain time or after a certain number of counts are available in some commercial instruments. In the counting-rate meter the pulses are integrated electrically, and their average rate of arrival (averaged over a suitably long time interval) is indicated directly by a meter deflection.

Pulse Height Analysis (C5).   Whenever the output pulse from a counter or ionization chamber is proportional to the energy dissipation in the detector, the measurement of pulse heights may be a useful tool for energy determi-

---

more recent development is the glow-transfer tube which provides a scale of 10 in a single gas-filled tube by means of a glow discharge that is moved to 10 different positions by 10 successive pulses.   These tubes are limited to slower repetition rates than are the vacuum-tube or transistor scalers.

nations. Some pulse height selection is used even in the simplest scalers in the form of a "discriminator," which allows only pulses above a certain minimum size to be recorded. If a calibrated adjustable discriminator is provided, counting rates may be measured as a function of discriminator setting; the resulting curve is called an integral bias curve. The derivative of this curve (a differential bias curve) gives the distribution of pulse heights.

Pulse height distributions may be obtained more directly and much more accurately with a single-channel analyzer. In this instrument there are two discriminators, and usually an anticoincidence arrangement is used to pass only pulses of such a height that they fall between the two discriminator settings. The two discriminators may be separately controlled or, more frequently, they may be moved up and down the voltage scale together, with a constant "channel" or "window" width between them. The usual pulse height range is 0 to 50 V or 0 to 100 V for vacuum tube analyzers, 0 to 10 V for transistorized analyzers. Different parts of a pulse height spectrum may be brought into this range by the choice of suitable amplifier gains. Channel widths of 0.5 to 10 per cent of the full range are often used. High stability is required in the discriminators, the amplifier, and the high-voltage supply.

A typical pulse height spectrum obtained with a NaI scintillation counter and single-channel analyzer is shown in figure 5-8. The γ-ray source used is Cs^{137} with the single γ-ray energy 0.662 MeV. The full-energy peak is due to photoelectric absorption and to those Compton processes followed by absorption of the scattered photon in the crystal. The broad Compton distribution with its spectrum of lower pulse heights is also seen. The pulse height scale is usually calibrated by use of one or more radiations of known energy. Although curves of efficiency versus energy have been published for a few "standard" arrangements (see, for example, J1, p. 115), the efficiency (area under photo peak) as a function of γ-ray energy is best determined by means of calibrated standard sources for each particular crystal and geometrical arrangement used.

Both vacuum-tube and transistorized single-channel analyzers are commercially available. A complete set-up including high-voltage supply with 0.01 per cent regulation, amplifier, analyzer, and scaler might cost 2000 to 3000 dollars.

Pulse height analysis is made much more versatile and rapid if, instead of a single-channel analyzer, a multichannel analyzer is used in which the pulses are sorted according to size and simultaneously recorded in many consecutive channels. A variety of such instruments has been developed; the most widely used ones employ conversion from pulse height to time and sorting of the different time durations by means of an oscillator and scaler. Data storage may be handled in a variety of ways, the most common scheme being a magnetic-core memory very much like those used in high-speed computers. Here the number of pulses in each channel is stored in an array of ferrite elements, each serving as a binary digit (being either magnetized or unmagnetized and thus either 0 or 1). If the number of cores for each channel is m, $(2^m - 1)$

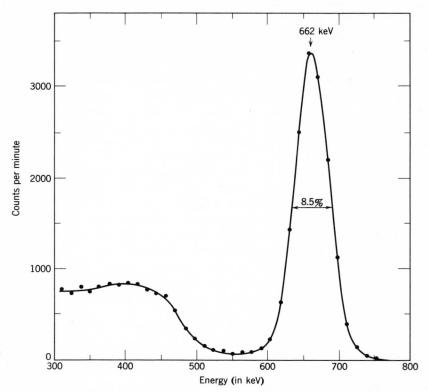

Fig. 5-8   Pulse height spectrum obtained with a NaI (Tl) scintillation counter for a Cs[137] source.   The broad peak at the left is the Compton distribution; at the right is the photo-peak due to absorption of the full energy (662 keV) of the γ ray in the crystal.   The width of the photopeak at half maximum is 56 keV or 8.5 per cent of 662 keV.   (Courtesy J. P. Welker.)

pulses per channel can be stored.   When a pulse "addressed" to a particular channel or memory location arrives, the number stored in that location is read out into a scaler, one is added to it, and the new number is stored in the same location.   The time required for sorting and storing a pulse is appreciable (typically $15 + 0.25n$ μsec, if n is the channel number to which the pulse is addressed) and the analyzer cannot process a new pulse during this "dead time."   Provision is often made to record automatically the fraction of a counting interval during which the analyzer is dead.   Even though the dead-time depends on channel number, spectrum shapes are not distorted.

A multichannel analyzer must, of course, include circuitry for obtaining in useful form the information stored in the memory.   This may include display of the spectrum on an oscilloscope, conversion from binary to decimal numbers, and the facility to drive an automatic printer, plotter, or magnetic tape unit so that the content of each channel can be appropriately recorded. Commercial multichannel analyzers with as many as 4000 channels are available, almost all of them now completely transistorized. For ordinary spectral analyses with pulsed ion chambers, scintillation counters, proportional counters, and solid-state detectors, models with 100 to 500 channels are most popular. Their cost is in the range of 6000 to 20,000 dollars.

Coincidence Techniques (D2). Studies of the time relations between various radiations emitted from one nucleus may be made by means of coincidence techniques and are very useful in decay scheme studies.

Whether a β ray goes to the ground state of the product nucleus or is followed by γ emission can be established by a coincidence experiment in which the sample is placed between a β and a γ counter and time-coincident pulses in the two counters are recorded. Similarly, with appropriately chosen detectors, and possibly with the aid of absorbers or pulse height discrimination to make detection more selective, γ-γ, α-γ, X-γ, β-e^-, e^--γ, etc., coincidences may be studied. Coincidence measurements with pulse height analysis at one or both detectors offer a particularly powerful tool for detailed decay scheme studies. A number of the available multichannel analyzers are of the so-called two-dimensional or two-parameter variety; that is, they can be used to record simultaneously the coincidences between each of n pulse height groups from one detector with each of m groups from the other. Thus a 4096-channel machine might be used to give arrays of 64×64 or of 32×128. The display of the output from these two-parameter analyzers can take various forms, and the analysis of the data obtained with these devices often requires computer methods.

In most coincidence measurements rather strong samples are used. This is because the number of coincidence counts recorded is proportional to the product of the solid angles subtended by the two counters at the sample, and frequently the sample-to-counter distances have to be rather large (inches) to minimize scattered radiation from one counter entering the other. Since the coincidence rates are often quite low, background rates are a problem. Apart from a very small true coincidence background (e.g., due to a cosmic ray striking both detectors), there is always a certain chance or accidental background that comes about because sometimes two rays not originating from the same nucleus happen to arrive at the two counters within the resolving time of the coincidence circuit. If the single counting rates in the two counters are R_1 and R_2 per second and if the coincidence resolving time (the time within which the two counters have to be tripped for a coincidence to be recorded) is τ second, then the accidental coincidence rate is $2R_1R_2\tau$ per second. To reduce the chance rate it is desirable to make the resolving time as short as possible. Coincidence resolving times of 10^{-6} to 10^{-8} sec are common, and for delayed-coincidence measurements of very short half-lives resolving times of less than 10^{-9} sec have been achieved. To be used with coincidence circuits, the detectors must have pulse-rise times not much longer than the coincidence resolving time, and for this reason GM tubes are not very useful for fast coincidence work. Scintillation counters and solid-state detectors are most commonly used.

Spectrographs and Spectrometers. In addition to the scintillation- and proportional-counter spectrometers there is a variety of devices for energy measurements on α, β, γ, and X rays and conversion electrons (Y1, S3). We can make only the briefest mention of these. In the 180° magnetic spectrograph (or spectrometer) use is made of the fact that identical charged particles emerging from a point source with equal momenta but at slightly divergent angles (say within 20°) are brought to an approximate focus after traveling about 180° in a plane perpendicular to a uniform magnetic field (figure 5-9). If a constant magnetic field is used, electrons or α particles of different momenta are detected in different positions, either on a photographic film or by a movable counter. Instruments of this type are particularly useful for line spectra such as conversion-electron or α spectra. For the continuous β spectra it is more practical to leave the detector (GM, proportional, or scintillation counter) fixed and vary the field to bring β particles of different energies into focus at the detector. By slight shaping of the magnetic field greatly improved focusing has been achieved in 180° spectrometers, and much precision work in β-ray spectrometry has been done with such instruments.

Another type of β-ray spectrometer is the lens spectrometer. Here the source and the detector are located on the axis of an axially symmetric magnetic field. It is a property of such a system that all electrons emitted with a large spread of angles but with a given momentum will, after traveling along spiral paths, be focused at some other point on the axis. Very high geometric efficiencies (several per cent of 4π) and good momentum resolution ($\sim$1 per

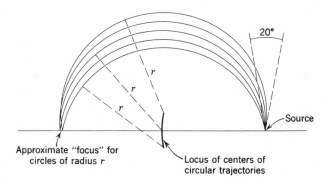

Fig. 5-9 Principle of 180° focusing. The magnetic field is perpendicular to the plane of the paper. The angular divergence of the trajectories shown is 20° at the source.

cent) can be attained, although in the latter respect the 180° instruments are superior.

In all electron spectrometers source preparation is a problem because sources must be extremely thin (often <0.1 mg cm^{-2}) and mounted on equally thin backings to minimize self-absorption and back-scattering. Also the counters used must have very thin windows.

Electron spectrometers are used not only for conversion-electron and β-ray measurements but also for γ-ray spectra. The energies of photoelectrons ejected from a radiator are measured in this case. For certain detailed studies of complex decay schemes two lens-type spectrometers have been used end-to-end; coincidences between two electron lines or between one electron line and a β spectrum can thus be studied. More common are coincidence arrangements with one scintillation counter and one electron spectrograph as the detectors.

An instrument for the precision measurement of X rays and low-energy γ rays is the curved-crystal spectrograph (D1). This is analogous to an optical spectrograph of the grating type, with the atomic planes of a bent crystal replacing the ruled lines of the curved grating. The detector may be a photographic plate, or a counter in the curved-crystal spectrometer.

D. HEALTH PHYSICS INSTRUMENTS

By the term health physics instruments we refer to detection and measuring instruments designed for the monitoring of personnel radiation exposures and for the surveying of laboratories, equipment, clothing, hands, and the like for biologically harmful radioactive contaminations. Many types are in use (M3). Those that are generally available commercially may be divided into a few categories, and are all derived from the instrumentation principles already discussed in this chapter.

Film Badges. Personnel in almost every nuclear or radiochemical project routinely wear badge-type holders containing photographic film, which gives a permanent record of general body exposure to radiation integrated over a period of time—usually one week. Ordinary dental X-ray film is generally used for β and γ dosimetry. To obtain information about the type and energy of radiation, various filters (plastic, aluminum, and cadmium foils) are placed over certain areas of the film. Development, calibration, and dosage evaluation for the badges can be obtained on a subscription basis. Film holders can be worn on the wrist or even arranged as finger rings, thus serving as more accurate monitors of hand exposure. Because thermal-neutron flux is not well indicated on ordinary films, special boron-loaded films are also available on a routine basis. Fast neutrons and radiations of very high energies (from synchrocyclotrons, proton synchrotrons, etc.) may be monitored with nuclear emulsions; track counting must then be resorted to.

Pocket Ion Chambers. Another widely used radiation monitor is the pocket ionization chamber. This is an ordinary ionization chamber in most respects, made small enough to be worn clipped in the pocket like a fountain pen. The charging potential is applied through a temporary connection, and at the end of the day or end of exposure the residual charge is read on an electrometer. A pocket ionization chamber with a built-in electrometer and scale is more convenient but also more expensive. This style of chamber is initially charged on an external device and then may be read directly at any time without auxiliary apparatus and without effect on the indication.

These pocket meters are calibrated in roentgen units, with full scale corresponding most commonly to 0.1 or 0.2 r, so that they easily detect general radiation dosage below tolerance levels. They may not give a measure of local exposure (say, of the hands, while other parts of the body are shielded by a lead screen) and, of course, are not sensitive to soft radiations that do not penetrate the chamber wall. Some pocket ion chambers produce audible signals whose frequency is proportional to dose rate.

Portable Counters and Survey Meters. More sensitive detection instruments are used to determine the rate at which exposure is being received in a given radiation field. These may be larger ionization chambers with compact d-c amplifiers operated from self-contained batteries, so that a readily portable survey meter weighing perhaps 3 lb is achieved. Models of this type ordinarily have several calibrated scale ranges, from about 0 to 20 mr per hour to about 0 to 3000 mr per hour or have a logarithmic response which compresses a wide range into a single scale. Battery-operated portable Geiger counter sets of about the same size and weight can be used for the same purpose. They are usually arranged as counting-rate meters, with full-scale readings calibrated at about 0.2 to 20 mr per hour. Although the counter type of meter is much more sensitive than the ionization chamber, the chamber is usually sensitive enough and may be expected to give a response more nearly proportional to the biological effects of the radiation. Both types are ordinarily provided with a movable shield to permit a distinction between hard and soft radiations. These two types of instruments are also useful for surveying the laboratory and its apparatus for radioactive contamination. The GM counter instrument with its higher sensitivity, especially when used with earphones so that each count may be heard, is more convenient in rapid surveys for small amounts of activity, but in its usual form it is not useful for very soft β rays such as those from C^{14}. The ionization chamber instrument is more easily fitted with a window thin enough for this purpose (not more than a few milligrams per square centimeter), and some available models have very thin windows (or simply open screens) that will pass even α particles. The most sensitive γ-ray monitor is the portable scintillation counter.

Other Procedures. A number of other more specialized instruments have been devised. Geiger counters and atmospheric-pressure proportional counters may be arranged particularly to detect β and α contaminations on the hands. The monitoring of air-borne contamination requires special instruments and may be particularly important in laboratories handling long-lived α activities. One method for air-borne dusts is to filter a large volume of the air and to assay the activity left on the filter paper with a standard counter or linear-amplifier instrument. (The radon decay products ordinarily present in air can be detected in this way.) A very simple and widely applicable semi-quantitative method of contamination monitoring which requires no special instrumentation is worth mentioning here; this is the so-called swipe method. A small piece of clean filter paper of a standard size is wiped over a roughly uniform path length on the suspected desk top, floor, wall, laboratory ware, or almost anywhere, and then measured for α, β, or γ activity on a standard instrument. Even air-borne contamination may be checked in a rough way by swipe samples of accumulated dust from an electric-light fixture or some such place exposed only to contamination from the air.

REFERENCES

B1   M. Blau, "Photographic Emulsions," in ref. Y1, pp. 208–264.

B2   D. V. Bugg, "The Bubble Chamber," *Progress in Nuclear Physics* (O. R. Frisch, Editor), Vol. 7, Pergamon, London and New York, 1959, pp. 1–52.

B3   H. Bradner, "Bubble Chambers," *Ann. Rev. Nuclear Sci.* **10**, 109–160 (1960).

*B4 J. B. Birks, *Scintillation Counters*, McGraw-Hill, New York, 1953.

*C1 S. C. Curran and J. D. Craggs, *Counting Tubes*, Academic, New York, 1949.

C2   B. Cork, "Spark Chambers," in ref. Y1, pp. 281–288.

*C3 S. C. Curran, *Luminescence and the Scintillation Counter*, Academic, New York, 1953.

C4   P. A. Cerenkov, *Compt. Rend. Acad. Sci. URSS* **2**, 451 (1934).

*C5 R. L. Chase, *Nuclear Pulse Spectrometry*, McGraw-Hill, New York, 1961.

C6   S. C. Curran, "Proportional Counter Spectrometry," in *Beta- and Gamma-Ray Spectroscopy* (K. Siegbahn, Editor) North-Holland, Amsterdam, 1955, pp. 165–183.

D1   J. W. M. DuMond, "Gamma-Ray Spectroscopy by Direct Crystal Diffraction," *Ann. Rev. Nuclear Sci.* **8**, 163–180 (1958).

D2   S. De Benedetti and R. W. Findley, "The Coincidence Method," in ref. F4, p. 222.

F1   W. Franzen and L. W. Cochran, "Pulse Ionization Chambers and Proportional Counters," in *Nuclear Instruments and Their Uses* (A. H. Snell, Editor), Vol. I, Wiley, New York, 1962, pp. 3–81.

F2   W. B. Fretter, "Nuclear Particle Detection (Cloud Chambers and Bubble Chambers)," *Ann. Rev. Nuclear Sci.* **5**, 145–178 (1955).

F3   I. Frank and I. Tamm, *Compt. Rend. Acad. Sci. URSS* **14**, 109 (1937).

*F4 S. Flügge and E. Creutz (Editors), *Encyclopedia of Physics*, XLV, *Nuclear Instrumentation II*, Springer Verlag, Berlin, 1958.

*J1 N. F. Johnson, E. Eichler, and G. D. O'Kelley, *Nuclear Chemistry*, Chapter VI, "Detection and Measurement of Nuclear Radiation," Interscience, New York, 1963.

L1   S. J. Lindenbaum and L. C. L. Yuan, "Čerenkov Counters," in ref. Y1, pp. 162–194.

*M1 G. L. Miller, W. M. Gibson, and P. F. Donovan, "Semiconductor Particle Detectors," *Ann. Rev. Nuclear Sci.* **12**, 189–220 (1962).

M2   B. J. Moyer, "A Survey of Cerenkov Counter Technique," in *Nuclear Instruments and Their Uses* (A. H. Snell, Editor), Vol. I, Wiley, New York, 1962, pp. 166–193.

*M3   K. Z. Morgan, "Techniques for Personnel Monitoring and Radiation Surveying," in *Nuclear Instruments and Their Uses* (A. H. Snell, Editor), Vol. I, Wiley, New York, 1962, pp. 391–469.

R1   F. Reines, "Neutrino Interactions," *Ann. Rev. Nuclear Sci.* **10**, 1–26 (1960).

*R2   B. B. Rossi and H. H. Staub, *Ionization Chambers and Counters*, National Nuclear Energy Series, Div. V, Vol. 2, McGraw-Hill, New York, 1949.

R3   G. T. Reynolds and F. Reines, "Scintillation Counters and Luminescent Chambers," in ref. Y1, pp. 120–162.

R4   A. H. Rosenfeld and W. E. Humphrey, "Analysis of Bubble-Chamber Data," *Ann. Rev. Nuclear Sci.* **13**, 103–144 (1963).

*S1   H. H. Staub, "Detection Methods," in *Experimental Nuclear Physics* (E. Segré, Editor) Vol. I, Wiley, New York, 1953, pp. 1–165.

S2   R. K. Swank, "Characteristics of Scintillators," *Ann. Rev. Nuclear Sci.* **4**, 111–140, (1954).

S3   K. Siegbahn, "β-Ray-Spectrometer Theory and Design. High Resolution Spectroscopy," in *Beta- and Gamma-Ray Spectroscopy* (K. Siegbahn, Editor), North-Holland Amsterdam, 1955, pp. 52–99.

W1   R. W. Williams, "Ionization Chambers," in ref. Y1, pp. 89–110.

*W2   D. H. Wilkinson, *Ionization Chambers and Counters*, Cambridge University Press, 1950.

*Y1   L. C. L. Yuan and C. S. Wu (Editors), *Methods of Experimental Physics*, Vol. 5A, *Nuclear Physics*, Academic, New York, 1961.

EXERCISES

1. Estimate roughly the voltage (IR) applied to the grid of a d-c amplifier tube; use these assumptions: $R = 10^{11}$ ohms; the sample emits 1-Mev γ rays at the rate of 10^6 per minute; the geometry is such that 30 per cent of the γ's spend an average 8-cm path length in the ionization chamber, which is filled with CF_2Cl_2 at 2 atm total pressure.

2. Calculate the time required for a positive ion to move from the wire to the wall of a Geiger counter; take 0.005 in. for the wire diameter, 1⅛ in. for the cathode diameter, 1000 volts as the applied voltage, 10 cm Hg as the gas pressure, and 1.5 cm sec^{-1} for the mobility of the ion at 1 volt cm^{-1} gradient and at 76 cm pressure.
   Answer: 490 μsec.

3. What type of instrument would you use for each of the following (a) Detection of 0.1 μc of P^{32}? (b) Detection of 10^{-3} μc of H^3? (c) Detection of 10^{-5} μc of At^{211}? (d) Following the decay of a sample of Cu^{64} (initially 3×10^5 disintegrations per second) over a period of 8 days? (e) Determining the relative amounts of Co^{57} and Co^{60} in a sample (without decay measurements over a long period of time)? State briefly the reason for each choice.

4. An anthracene crystal and a 10-stage photomultiplier tube are to be used as a scintillation spectrometer for β rays. It is desired that a 10-keV β particle incident on the scintillator produce a 2-millivolt pulse in the photomultiplier output circuit, which has a capacity of 120 μμF. What average electron multiplication per stage is required in the photomultiplier? Assume a light collection efficiency of unity and a photocathode efficiency (number of photoelectrons emitted per incident photon) of 0.1.
   Answer: 3.3.

5. A certain nuclide decays predominantly by emission of β^- particles of 1.25 MeV maximum energy to a 25-second isomeric state, which in turn decays to the product ground state by emission of 0.45-MeV γ rays. A rare β^- branch (0.1 per cent abundance, 0.81 MeV maximum energy) leads to a state which decays by γ emis-

sion to the 25-second isomeric level.   To establish this branch by a β-γ coincidence measurement with scintillation detectors, a true coincidence rate at least three times the accidental coincidence rate is desired.   The sample strength available is 1000 disintegrations per second, and background effects in the counters may be neglected.   Assume equal counting efficiency of the β counter for the two β groups. What coincidence resolving time is required?   *Answer:* $\leq 0.17\,\mu$sec.

6. Consider the relative merits of a proportional counter, a NaI(Tl) scintillation counter, and a lithium-drifted semiconductor detector for the measurement of 12-keV X rays in the presence of a 1-MeV β spectrum and of 0.5-MeV γ rays. Discuss details such as counter dimensions and, in the case of the proportional counter, the gas filling conducive to optimizing the measurement of the X rays. Also comment on the energy resolution attainable with each instrument (assuming that a pulse height analyzer is available).

7. A beam of 210-MeV π^- mesons has been momentum-analyzed by magnetic deflection but is contaminated with μ^- mesons of the same momentum.   How could two Čerenkov counters and an anticoincidence circuit be used to detect the π mesons only?   What are the requirements for the refractive indices of the substances used in the two Čerenkov counters?   Could the same type of system be used to discriminate against electron contamination in the π beam?

8. An n^+p-type diffused-junction detector made from p-type silicon of 5×10^3 Ω-cm resistivity is to be used to measure kinetic-energy spectra of protons incident on its face.   Estimate the maximum proton energy for which the detector will be useful if a bias voltage of 200 V is applied.   (Take 2.4 g cm^{-3} for the density of silicon.)
Answer: 6.1 MeV.

9. A proportional counter of 2 cm radius and with a center wire of 4×10^{-3} cm diameter is filled to 1 atm with methane.   It is operated with an applied voltage of 4000 V to achieve a certain gas multiplication.   Under what conditions of pressure and voltage would the same gas multiplication be obtained in methane-filled counters of (a) 1 cm radius and 4×10^{-3} cm wire diameter, (b) 2 cm radius and 8×10^{-3} cm wire diameter?

10. How would you propose to measure (a) the ratio of K- to L-conversion coefficients for a 150-keV transition in a germanium isotope, (b) the relative intensities of 560-keV and 820-keV γ rays in a sample, (c) the relative amounts of U^{236} and U^{233} present in approximately 1μg of a mixture of the two isotopes?

11. What gas multiplication is required in a methane-flow proportional counter if a minimum-ionizing electron that spends a 2 cm path length in the active counter volume is to result in a 5×10^{-4} V pulse?   Assume $\sim 5\mu\mu$F for the capacity of the counter.   Take the ionization loss of minimum-ionizing electrons (MeV/mg cm^{-2}) in methane the same as in air.   *Answer:* Approximately 210.

6

Statistical Considerations

in Radioactivity Measurements

The radioactive-decay law discussed in chapter 3 describes the average behavior of a sample of radioactive atoms. In measurements of radioactive decay we are concerned with observations which show fluctuations about the average behavior predicted by the decay law. Therefore we shall discuss in this chapter the applications of statistical methods to the treatment of radioactivity measurements.

A. DATA WITH RANDOM FLUCTUATIONS

Consider the set of data actually obtained with a Geiger counter measuring a "steady" source, as given in table 6-1. The number of counts recorded per minute (the counting rate) is clearly not uniform. Which minute gave the most accurate result? The best thing we can do is to compute the arithmetic mean (the average value) and consider it as representing the proper counting rate. What we are trying to do is to estimate from a finite number of observations the results of an essentially infinite number of observations. In particular, we wish to estimate the average value that we would find and the distribution of the observed values about that average.

Average Value. If the determinations, minute by minute, are denoted by $x_1, x_2, \ldots x_i$ for the first, second, $\ldots$ ith minute, then the arithmetic mean value $\bar{x}$ is, by definition,

$$\bar{x} = \frac{1}{N_0} \sum_{i=1}^{i=N_0} x_i, \tag{6-1}$$

where N_0 is the number of values of x to be averaged. For the counting rates in the table $\bar{x} = 990/10 = 99.0$.

166

Table 6-1

Minute	Counts	Δ_i	Δ_i^2
1	89	-10	100
2	120	$+21$	441
3	94	-5	25
4	110	$+11$	121
5	105	$+6$	36
6	108	$+9$	81
7	85	-14	196
8	83	-16	256
9	101	$+2$	4
10	95	-4	16
Totals	990	0	1276

This average value is the best estimate that we can make of the "true" average, $\bar{x}_t$ which is the average we would find for an infinite number of observations.

Standard Deviation.   The distribution of the observed results about $\bar{x}_t$ is a measure of the precision of the data and can be described by giving all of the "moments" of the distribution; that is, the quantities

$$\frac{1}{N_0} \sum_{i=1}^{N_0} (x_i - \bar{x}_t)^n \tag{6-2}$$

for all values of n.   The first moment ($n = 1$) will always vanish because of the definition of $\bar{x}_t$; the other odd moments [expression (6-2) with "n" an odd number] will vanish only if the distribution is symmetrical about $\bar{x}_t$, and $\bar{x}_t$ is then the most probable value of x.   Usually just the second moment [expression (6-2) with $n = 2$, called the variance and denoted by σ_x^2] is given in practice.   The square root of the variance is called the standard deviation σ_x. This quantity is particularly significant because of the form of the so-called normal distribution law which is expected to describe the distribution of experimental results with random errors:

$$P(x)\, dx = \frac{1}{\sqrt{2\pi\sigma_x^2}} \exp \left[\frac{-(x - \bar{x}_t)^2}{2\sigma_x^2} \right] dx, \tag{6-3}$$

where $P(x)\, dx$ is the probability of observing a value of x in the interval $x \rightarrow x + dx$.

In our example, which contains a finite number of observations, we do not know $\bar{x}_t$; we have only an estimate of it: $\bar{x}$.   Under these circumstances the best possible estimate of the variance is

$$\sigma_x{}^2 = \frac{1}{N_0 - 1} \sum_{i=1}^{N_0} (x_i - \bar{x})^2. \tag{6-4}$$

For the data in table 6-1 we compute $\sigma_x{}^2 = 1276/9 = 141.8$; $\sigma_x = 11.9$. The difference between equations (6-2) and (6-4) is noteworthy. The division by $(N_0 - 1)$ in (6-4) instead of by N_0 is a consequence of estimating the unknown quantity $\bar{x}_t$ from N_0 observations; this estimation uses up one of the observations and leaves only $(N_0 - 1)$ independent quantities for the estimation of the variance. The validity of this reasoning becomes clear when we consider the extreme case of only a single observation. Evidently, from a single observation we can have no idea of the precision of the measurement, unless special assumptions are made. This problem is a fundamental one in statistical analysis and is discussed in standard texts on the subject (cf., for example, B1 and F1).

Precision of Average Value. In the preceding discussion we have been concerned with estimating, from N_0 observations, the results that would be obtained from a very large number of observations. It is now necessary to discuss the precision of our estimation which is not to be confused with the precision of the data, although the two quantities are related. We are here concerned with two things:

1. The distribution of the values of $\bar{x}$ given by (6-1) from many sets of experiments, each with a finite N_0.
2. The distribution of the quantities $\sigma_x{}^2$ obtained from the same sets of observations by (6-4).

The formal statistical analysis of these two problems, as discussed in standard texts (F1), is contained in the χ^2-test of the randomness of the data, the t-test of the reliability of $\bar{x}$ as an estimate of $\bar{x}_t$, and the F-test of the reliability of $\sigma_x{}^2$ as an estimate of the true variance of the sample.

Our main interest is in the first question, the reliability of $\bar{x}$; a measure of this reliability is the *variance of a mean* which is estimated by the variance of the set of observations divided by N_0:

$$\sigma_{\bar{x}}{}^2 = \frac{\sigma_x{}^2}{N_0} = \frac{1}{N_0(N_0 - 1)} \sum_{i=1}^{N_0} (x_i - \bar{x})^2. \tag{6-5}$$

The quantity $\sigma_{\bar{x}}{}^2$ is our best estimate of the second moment of the distribution of average values that would be found from an infinite number of *sets* of experiments, each containing N_0 observations of which table 6-1 is an example. The value of $\sigma_{\bar{x}}$ from table 6-1 is $\sqrt{141.8/10} = 3.76$.

The significance of this quantity, for a normal distribution, is found in the statement that the probability of observing a value of $\bar{x}$ between $\bar{x}$ and $\bar{x} + d\bar{x}$ is

$$P(\bar{x})\, d\bar{x} = \frac{1}{\sqrt{2\pi\sigma_{\bar{x}}^2}} \exp \left[\frac{-(\bar{x} - \bar{x}_t)^2}{2\sigma_{\bar{x}}^2} \right] d\bar{x}.$$

Rejection of Data.   The question often arises whether a particular datum should be rejected because of its relatively large deviation from the mean.   In table 6-1 the observation of 120 counts during the second minute is suspect, as perhaps, though to a lesser degree, is the observation of 83 counts during the eighth minute.   This is not necessarily to say that these observations are wrong (that the error is systematic rather than random), but that deviations of this magnitude among a small number of observations may have an undue influence on the mean value that is computed.   Thus the criteria for rejection should consider not only the magnitude of the deviation but also the number of observations made.   A criterion established by Chauvenet which includes both factors (the magnitude of the deviation and the number of observations) allows the rejection of an observation if deviations from the mean that are equal to or greater than the one in question have a probability of occurrence that is less than $1/(2N_0)$.   In our example the counting rate during the second minute may be rejected only if the probability of observing counting rates that deviate by at least 21 counts from the mean of 99 counts is less than 0.05. We compute this probability by using (6-3) to obtain the probability P of observing a count between 78 and 120:

$$1 - P = \int_{78}^{120} \frac{1}{\sqrt{2\pi \cdot 141.8}} \exp \left[\frac{-(x - 99)^2}{2 \cdot 141.8} \right] dx.$$

The value of the integral, as found in the *Handbook of Chemistry and Physics*, is 0.92; thus $1 - P$ is 0.08 and the datum must be retained.   If N_0 had been six, or less, then the datum would have been rejected.   When a datum is rejected, a new $\bar{x}$ must be computed, and Chauvenet's criterion may be applied to the remaining suspect data, but with N_0 being decreased by one each time that an observation is excluded.

B. PROBABILITY AND THE COMPOUNDING OF PROBABILITIES

The ideas and definitions just presented may be applied, with varying degrees of usefulness, to any set of data, whether or not strictly random phenomena are involved.   Before proceeding, we must consider the concept of probability in greater detail.   As illustrations we shall investigate the answers to questions such as these:

1. What is the probability that a card drawn from a deck will be an ace?

2. If a coin is flipped twice, what is the probability that it will fall "heads up" both times?

3. Given a sample of a radioactive material, what is the probability that exactly 100 disintegrations will occur during the next minute?

We shall define probability in this way: given a set of N_0 objects (or events, or results, etc.) containing n_1 objects of the first kind, n_2 objects of the second kind, and n_i objects of the ith kind, the probability p_i that an object specified only as belonging to the set is of the ith kind is given by $p_i = n_i/N_0$. By applying this definition we find that the probability that one card drawn from a full deck will be an ace is just $\frac{4}{52}$.

We may now rewrite the definition of the average value $\bar{x}$ of a set of quantities x_i, taking into account the possibility that any particular value may appear several, say n_i, times. Then

$$\bar{x} = \frac{1}{N_0} \Sigma n_i x_i = \Sigma p_i x_i.$$

This may be generalized, and the expression for the average value of any function of x is

$$\overline{f(x)} = \Sigma p_i f(x_i). \tag{6-6}$$

In particular,

$$\sigma_x{}^2 = \Sigma p_i (x_i - \bar{x})^2 = \overline{x^2} - \bar{x}^2 \tag{6-7}$$

a result that will be useful to us later.

In experimental measurements we may make a large number K of observations and find the ith result k_i times. Now the ratio k_i/K is not the probability p_i of the ith result as we have defined it, but for our purposes we assume that k_i/K approaches arbitrarily closely to p_i as K becomes very large:

$$\lim_{K \to \infty} \frac{k_i}{K} = p_i.$$

This assumption is not subject to mathematical proof because a limit may not be evaluated for a series with no law of sequence of terms.

Addition Theorem. We turn now to the compounding of several probabilities and consider first the addition theorem. Given a set of N_0 objects (or events, or results, etc.) containing n_i objects of the kind a_i and given that the kinds $a_1, a_2, \ldots a_i$ have no members in common, the probability that one of the N_0 objects belongs to a combined group $a_1 + a_2 + \cdots a_j$ is just $\sum_{i=1}^{i=j} p_i$. Thus for two mutually exclusive events with probabilities p_1 and p_2 the probability of one or the other occurring is just $p_1 + p_2$. When one card is drawn from a full deck, the chance of its being either a five or a ten is $\frac{4}{52} +$

$\frac{4}{52} = \frac{2}{13}$. (When one draws one card while already holding, say, four cards, none of which is a five or ten, the probability then of getting either a five or a ten is slightly greater, $\frac{4}{48} + \frac{4}{48} = \frac{1}{6}$, provided that there is available no information regarding the identity of other cards that may already have been withdrawn.) When a coin is tossed, the probability of either "heads" or "tails" is $\frac{1}{2} + \frac{1}{2} = 1$.

Multiplication Theorem. Another type of compounding of probabilities is described by the multiplication theorem. If the probability of an event i is p_i and if after i has happened the probability of another event j is p_j, then the probability that first i and then j will happen is $p_i \times p_j$. If a coin is tossed twice, the probability of getting "heads" twice is $\frac{1}{2} \times \frac{1}{2} = \frac{1}{4}$. If two cards are drawn from an initially full deck, the probability of two aces is $\frac{4}{52} \times \frac{3}{51}$. The probability of four aces in four cards drawn is $\frac{4}{52} \times \frac{3}{51} \times \frac{2}{50} \times \frac{1}{49}$. (The probability of drawing five aces in five cards is $\frac{4}{52} \times \frac{3}{51} \times \frac{2}{50} \times \frac{1}{49} \times \frac{0}{48} = 0$.)

Binomial Distribution. The binomial distribution law treats one fairly general case of compounding probabilities and can be derived by the application of the addition and multiplication theorems. Given a very large set of objects in which the probability of occurrence of an object of a particular kind w is p, then, if n objects are withdrawn from the set, the probability $W(r)$ that exactly r of the objects are of the kind w is given by

$$W(r) = \frac{n!}{(n-r)!\,r!}\, p^r (1-p)^{n-r}. \qquad (6\text{-}8)$$

To see how this combination of terms actually represents the probability in question, think for a moment of just r of the n objects. That the first of these is of the kind w has the probability p; that the first and second are of the kind w has the probability p^2, etc., and the probability that all r objects are of the kind w is p^r. But, if exactly r of the n objects are to be of this kind, the remaining $n - r$ objects must be of some other kind; this probability is $(1 - p)^{n-r}$. Thus we see that for a particular choice of r objects out of the n objects the probability of exactly r of kind w is $p^r(1 - p)^{n-r}$; this particular choice is not the only one. The first of the r objects might be chosen (from the n objects) in n different ways, the second in $n - 1$ ways, the third in $n - 2$ ways, and the rth in $n - r + 1$ ways. The product of these terms, $n(n - 1)(n - 2) \cdots (n - r + 1)$, is $n!/(n - r)!$, and this coefficient must be used to multiply the probability just found. But this coefficient is actually too large in that it not only gives the total number of possible arrangements of the objects in the way required but also includes the number of arrangements which differ only in the order of selection of the r objects. So we must divide by the number of permutations of r objects which is $r!$. Thus the final coefficient is $n!/(n - r)!\,r!$, which is that in (6-8). The law (6-8) is known as the binomial distribution law because this coefficient is just the coefficient of $x^r y^{n-r}$ in the binomial expansion of $(x + y)^n$. Since in (6-8),

$$x + y = p + (1 - p),$$

we have

$$\sum_{r=0}^{n} W(r) = 1,$$

and the binomial distribution is seen to be normalized.

C. RADIOACTIVITY AS A STATISTICAL PHENOMENON

Binomial Distribution for Radioactive Disintegrations. We may apply the binomial distribution law to find the probability $W(m)$ of obtaining just m disintegrations in time t from N_0 original radioactive atoms. We think of N_0 as the number n of objects chosen for observation (in our derivation of the binomial law), and we think of m as the number r that is to have a certain property (namely, that of disintegrating in time t), so that for this case the binomial law becomes

$$W(m) = \frac{N_0!}{(N_0 - m)!m!} p^m (1 - p)^{N_0 - m}. \qquad (6\text{-}9)$$

Now the probability of an atom not decaying in time t, $1 - p$ in (6-9), is given by the ratio of the number N that survive the time interval t to the initial number N_0,

$$\frac{N}{N_0} = e^{-\lambda t};$$

p is then $1 - e^{-\lambda t}$. We now have

$$W(m) = \frac{N_0!}{(N_0 - m)!m!} (1 - e^{-\lambda t})^m (e^{-\lambda t})^{N_0 - m}. \qquad (6\text{-}10)$$

Time Intervals between Disintegrations. Since the time of Schweidler's derivation of the exponential decay law from probability considerations, the applicability of these statistical laws to the phenomena of radioactivity has been tested in a number of experiments. As an example of the positive evidence obtained, we consider the distribution of time intervals between disintegrations. The probability of this time interval having a value between t and $t + dt$, which we write as $P(t)\,dt$, is given by the product of the probability of no disintegration between 0 and t and the probability of a disintegration between t and $t + dt$. The first of these two probabilities is given by (6-10) with $m = 0$:

$$W(0) = \frac{N_0!}{N_0!0!} (1 - e^{-\lambda t})^0 (e^{-\lambda t})^{N_0} = e^{-N_0 \lambda t}.$$

(Notice that $0! = 1$.) The probability of any one of the N_0 atoms disintegrating in the time dt is clearly, from the addition theorem, $N_0 \lambda \, dt$. [See chapter 1, p. 5, or obtain this result as $W(1)$ from equation 6-10 with $m = 1$, t replaced by dt, and all terms in $(dt)^2$ and higher powers of dt neglected.] Then

$$P(t) \, dt = N_0 \lambda e^{-N_0 \lambda t} \, dt. \tag{6-11}$$

Experiments designed to test this result usually measure a large number s of time intervals between disintegrations and classify them into intervals differing by the short but finite length Δt; then the probability for intervals between t and $t + \Delta t$ should be $N_0 \lambda e^{-N_0 \lambda t} \Delta t$, and the number of measured intervals between t and $t + \Delta t$ should be $s N_0 \lambda e^{-N_0 \lambda t} \Delta t$. For example, Feather found experimentally that the logarithm of the number of intervals between t and $t + \Delta t$ is proportional to t, as required by this formula.

Average Disintegration Rate. Another application of the binomial law to radioactive disintegrations may be seen if we calculate the average value of a set of numbers obeying the binomial distribution law. For the moment we shall revert to the notation of (6-8) and for further convenience represent $1 - p$ by q:

$$W(r) = \frac{n!}{(n - r)!r!} p^r q^{n-r}. \tag{6-12}$$

The average value to be expected for r is obtained from (6-6):

$$\bar{r} = \sum_{r=0}^{r=n} rW(r) = \sum_{r=0}^{r=n} r \frac{n!}{(n - r)!r!} p^r q^{n-r}.$$

To evaluate this awkward-appearing summation, consider the binomial expansion of $(px + q)^n$:

$$(px + q)^n = \sum_{r=0}^{r=n} \frac{n!}{(n - r)!r!} p^r x^r q^{n-r} = \sum_{r=0}^{r=n} x^r W(r).$$

Differentiating with respect to x, we obtain

$$np(px + q)^{n-1} = \sum_{r=0}^{r=n} rx^{r-1} W(r). \tag{6-13}$$

Now letting $x = 1$ and using $q = 1 - p$, we have the desired expression

$$np = \sum_{r=0}^{r=n} r\, W(r) = \bar{r}.$$

This result should not be surprising; it means that the average number $\bar{r}$ of the n objects which are of the kind w is just n times the probability for any given one of the objects to be of the kind w.

The foregoing result may be interpreted for radioactive disintegration if n is set equal to N_0 and $p = 1 - e^{-\lambda t}$, as before. Then the average number M of atoms disintegrating in the time t is $M = N_0(1 - e^{-\lambda t})$. For small values of λt, that is, for times of observation short compared to the half-life, we may use the approximation $e^{-\lambda t} = 1 - \lambda t$ and then $M = N_0 \lambda t$. The disintegration rate $\mathbf{R}$ to be expected is $\mathbf{R} = M/t = N_0 \lambda$. (This corresponds to the familiar equation $-dN/dt = \lambda N$.)

Expected Standard Deviation. What may we expect for the standard deviation of a binomial distribution? If we differentiate (6-13) again with respect to x, we obtain

$$n(n - 1)p^2(px + q)^{n-2} = \sum_{r=0}^{r=n} r(r - 1)x^{r-2}\, W(r).$$

Again letting $x = 1$ and using $p + q = 1$, we have

$$n(n - 1)p^2 = \sum_{r=0}^{r=n} r(r - 1)\, W(r) = \sum_{r=0}^{r=n} r^2\, W(r) - \sum_{r=0}^{r=n} r\, W(r),$$

$$n(n - 1)p^2 = \overline{r^2} - \bar{r}.$$

Recall from (6-7) that the variance $\sigma_r{}^2$ is given by

$$\sigma_r{}^2 = \overline{r^2} - \bar{r}^2.$$

Now, combining, we have

$$\sigma_r{}^2 = n(n - 1)p^2 + \bar{r} - \bar{r}^2,$$

and with $\bar{r} = np$

$$\sigma_r{}^2 = n^2 p^2 - np^2 + np - n^2 p^2 = np(1 - p) = npq,$$

$$\sigma_r = \sqrt{npq}.$$

For radioactive disintegration this becomes

$$\sigma = \sqrt{N_0(1 - e^{-\lambda t})e^{-\lambda t}} = \sqrt{Me^{-\lambda t}}. \tag{6-14}$$

In counting practice λt is usually small; that is, the observation time t is short compared to the half-life, and when this is so,

$$\sigma = \sqrt{M}. \qquad (6\text{-}14a)$$

We see here a particular example of a very important property of the binomial distribution which, as presently shown, is true for the Poisson distribution also; that is, there is a simple relationship between the true mean and the true variance of the distribution. As a consequence, a single observation from a distribution that is expected to be binomial, as is true of radioactive disintegration rates, gives both an estimate of the mean and an estimate of the variance of the distribution. Further, for a single observation, the estimate of the variance of the distribution is also an estimate of the variance of the mean. It must be immediately emphasized that these remarks are not true in general; the variance of a thermometer reading, of a length measured by a meter stick, or of the reading of a voltmeter cannot be estimated from a single observation and is not in general expected to be equal to the value observed.

If a reasonably large number m of counts has been obtained, that number m may be used in the place of M for the purpose of evaluating σ. Thus, if 100 counts are recorded in 1 minute, the expected standard deviation is $\sigma \cong \sqrt{100} = 10$, and the counting rate might be written 100 ± 10 counts per minute. If 1000 counts are recorded in 10 minutes, the standard deviation of this number is $\sigma \cong \sqrt{1000} = 32$; the counting rate is $(1000 \pm 32)/10 = 100 \pm 3.2$ counts per minute. Thus we see that for a given counting rate $\mathbf{R}$ the σ for the rate is inversely proportional to the square root of the time of measurement:

$$\mathbf{R} = \frac{m}{t};$$

$$\sigma_{\mathbf{R}} = \frac{\sqrt{m}}{t} = \frac{\sqrt{\mathbf{R}t}}{t} = \sqrt{\frac{\mathbf{R}}{t}}. \qquad (6\text{-}15)$$

What is the result in an experiment in which the counting time is long compared to the half-life? As $\lambda t \to \infty$, $e^{-\lambda t} \to 0$, and, in this limit, $\sigma = \sqrt{Me^{-\lambda t}} = 0$. The explanation is clear; if we start with N_0 atoms and wait for all to disintegrate, then the number of disintegrations is exactly N_0. However, in actual practice we observe not the number of disintegrations but that number times a coefficient c which denotes the probability of a disintegration resulting in an observed count. Taking this into account, we see that in this limiting case the proper representation of $\sigma = \sqrt{npq}$ is $\sigma = \sqrt{N_0 c(1 - c)}$. If $c \ll 1$, then $\sigma = \sqrt{N_0 c} = \sqrt{\text{number of counts}}$ as before. When $\lambda t \cong 1$ and c is neither unity nor very small, a more exact analysis based on $\sigma = \sqrt{npq}$ should be made, with the result that $\sigma = \sqrt{Mc(1 - c + ce^{-\lambda t})}$.

The introduction of the detection coefficient c in the preceding paragraph
may raise the question why it is not necessary to take account of this coefficient
in the more familiar case with λt small, where we have written $\sigma = \sqrt{m}$.   If
we do consider c in this case, we have for the probability of one atom producing a count in time t, $p = (1 - e^{-\lambda t})c$ and $q = 1 - p = 1 - c + ce^{-\lambda t}$.
Then

$$\sigma = \sqrt{N_0(1 - e^{-\lambda t})c(1 - c + ce^{-\lambda t})},$$

and for λt small and the same approximations as before

$$\sigma = \sqrt{N_0 \lambda t c} = \sqrt{Mc} = \sqrt{\text{number of counts recorded}}.$$

This is just the conclusion we had reached without bothering about the detection efficiency.   It should be emphasized, however, that actual counts and not
scaled counts from a scaling circuit must be used in these equations.

D. POISSON AND GAUSSIAN DISTRIBUTIONS

Poisson Distribution.   The binomial distribution law (6-10) can be put into
a more convenient form if we impose the restrictions $\lambda t \ll 1$, $N_0 \gg 1$, $m \ll N_0$,
that is, if we consider a large number of active atoms observed for a time short
compared to their half-lives.   The derivation of this more convenient form
requires the well-known mathematical approximation:

$$\ln (1 + x) \cong x - \frac{x^2}{2} \cdots \qquad \text{if } x \ll 1. \qquad (6\text{-}16)$$

Let us first define the average value of the distribution (6-10):

$$M = N_0(1 - e^{-\lambda t}).$$

The binomial distribution may then be written as

$$W(m) = \frac{N_0!}{(N_0 - m)!m!} \left(\frac{M}{N_0}\right)^m \left(1 - \frac{M}{N_0}\right)^{N_0} \left(1 - \frac{M}{N_0}\right)^{-m}.$$

Consider the term

$$\frac{N_0!}{(N_0 - m)!} = N_0(N_0 - 1) \cdots (N_0 - m + 1)$$

$$= N_0{}^m \left(1 - \frac{1}{N_0}\right) \cdots \left(1 - \frac{m - 1}{N_0}\right).$$

For $m \ll N_0$ this term may be estimated by taking its logarithm and using the first term of the approximation (6-16).   The result is

$$\frac{N_0!}{(N_0 - m)!} \cong N_0{}^m \exp\left[-\frac{m(m-1)}{2N_0}\right]. \tag{6-17a}$$

The term $[1 - (M/N_0)]^{N_0}$ may also be estimated by the use of (6-16), since $M/N_0 \ll 1$, a condition that is equivalent to $\lambda t \ll 1$:

$$\ln\left(1 - \frac{M}{N_0}\right)^{N_0} = N_0 \ln\left(1 - \frac{M}{N_0}\right)$$

$$\cong -M - \frac{M^2}{2N_0},$$

$$\therefore \left(1 - \frac{M}{N_0}\right)^{N_0} \cong e^{-M}e^{-M^2/2N_0}. \tag{6-17b}$$

Note that this time we use two terms of the expansion, since $M^2/2N_0$ is not necessarily small, even for $M/N_0 \ll 1$.

Again, for $M/N_0 \ll 1$, we immediately have from (6-16)

$$\ln\left(1 - \frac{M}{N_0}\right)^{-m} \cong \frac{mM}{N_0}$$

$$\left(1 - \frac{M}{N_0}\right)^{-m} \cong e^{mM/N_0}. \tag{6-17c}$$

When the three approximate results (6-17a, b, c) are put into the binomial distribution, the result is

$$W(m) \cong \frac{M^m e^{-M}}{m!}\left[e^{-(M-m)^2/2N_0}e^{m/2N_0}\right], \tag{6-18}$$

where $W(m)$ is the probability of obtaining the particular number of counts m when M is the average number to be expected.   The term outside the brackets in (6-18):

$$W(m) \cong \frac{M^m e^{-M}}{m!} \tag{6-19}$$

is the famous Poisson distribution; the term within the brackets may be considered as a correction factor and is a measure of how well the binomial distribution is approximated by the Poisson.   It is to be emphasized that the validity of (6-19) as an approximation to (6-10) requires not only that a large number of atoms be observed for a time short compared to their half-lives, but also that the absolute value of $(M - m)$ be substantially smaller than $\sqrt{N_0}$.   For example, if $N_0 = 100$ and $M = 1$, both the Poisson and binomial distributions give $W(0) = 0.37$; but the binomial distribution gives $W(10) =$

0.7×10^{-7}, whereas the Poisson distribution (6-19) gives $W(10) = 1.0 \times 10^{-7}$. The corrected Poisson distribution (6-18) gives $W(10) = 0.7 \times 10^{-7}$.

Two features of the Poisson distribution (6-19) might be noticed in particular. The probability of obtaining $m = M - 1$ is equal to the probability of obtaining $m = M$, or $W(M) = W(M - 1)$. For large M the distribution is very nearly symmetrical about $m = M$ if values of m very far from M be excluded.

Gaussian Distribution. A further approximation of the distribution law may be made for large m (say > 100) and for $|M - m| \ll M$. With these additional restrictions, with the approximate expansion,

$$\ln \left(1 + \frac{M - m}{m} \right) = \frac{M - m}{m} - \frac{(M - m)^2}{2m^2},$$

neglecting subsequent terms, and with the use of Stirling's approximation

$$x! = \sqrt{2\pi x}\, x^x e^{-x},$$

we may modify the Poisson distribution to obtain the Gaussian distribution:

$$W(m) = \frac{1}{\sqrt{2\pi M}}\, e^{-(M-m)^2/2M}. \tag{6-20}$$

It will be noticed that this distribution is symmetrical about $m = M$. For both the Poisson and Gaussian distributions[1] we may derive $\sigma = \sqrt{M}$, or, for large m, $\sigma \cong \sqrt{m}$.

E. STATISTICAL INFERENCE AND BAYES' THEOREM

As we mentioned at the outset of this chapter, the primary problem of statistical inference is to estimate, from information available after only a finite number of observations, the average value that would be obtained after an infinite number of experimental observations of a given physical quantity. In terms of (6-19), what we really wish to know, for example, is the proba-

---

[1] The functional dependence $\sigma = \sqrt{M}$ is a necessary condition of the Poisson but not of the Gaussian distribution. The general form of the Gaussian is

$$W(m) = \frac{1}{\sqrt{2\pi\sigma^2}} \exp \left[\frac{(M - m)^2}{2\sigma^2} \right],$$

where there is as a rule no relationship between M and σ. The relationship between σ and M for the Gaussian distribution of *counting rates* is a consequence of the particular source of random error: the fluctuation in the decay rate consistent with a decay probability per unit time which is independent of time.

bility that the number of detected disintegrations of a radioactive sample (counts) is characterized by a mean value M when we have observed a value m [we may denote this probability by $W'(M|m)$]. Equation 6-19 gives us the inverse of what we wish to know: the probability of observing m counts when the sample is characterized by a mean value M [this inverse probability we may denote as $W(m|M)$]. These two conditional probabilities are related to each other:

$$P'(m)W'(M|m) = P(M)W(m|M); \qquad (6\text{-}21)$$

where $P'(m)$ is the *prior probability* that the sample will give m counts before any observations have been made on the sample and $P(M)$ is the *prior probability* that the sample is characterized by a *mean number of counts* M before any observations have been made on the sample. The reader will readily perceive that these so-called prior probabilities are troublesome quantities. The two sides of (6-21) are equal to each other because each of them is equal to the joint probability that a sample will be characterized by a mean of M counts *and* will exhibit experimentally m counts. The quantity of interest $W'(M|m)$ may be readily obtained from (6-21):

$$W'(M|m) = \frac{P(M)W(m|M)}{P'(m)}, \qquad (6\text{-}22)$$

an expression which was first discussed by the Reverend Bayes some two centuries ago.[2] The *prior probabilities* $P'(m)$ and $P(M)$ are related:

$$P'(m) = \sum_{M=0}^{\infty} P(M)W(m|M),$$

which states that if, in some manner, $P(M)$ is known, then through a combination of the addition theorem and the multiplication theorem the *prior probability* $P'(m)$ must also be known. The final expression, then, is

$$W'(M|m) = \frac{P(M)W(m|M)}{\displaystyle\sum_{M=0}^{\infty} P(M)W(m|M)}. \qquad (6\text{-}23)$$

It is of interest to note the implications of (6-23) for a sample that complies with the restrictions required by the Poisson distribution, that is, a sample containing a large number of atoms which is observed for a time short compared to their half-life. Taking (6-19) for $W(m|M)$, we obtain from (6-23)

$$W'(M|m) = \frac{P(M)(M^m e^{-M}/m!)}{\displaystyle\sum_{M=0}^{\infty} P(M)(M^m e^{-M}/m!)}. \qquad (6\text{-}24)$$

[2] For a discussion of conditional probability, see chapter 5 of reference F1.

It is impossible to proceed without an explicit expression for $P(M)$; it is at this point that the analysis can become metaphysical. We shall proceed by taking all values of M as being equally probable:

$$P(M) \, dM = K \, dM. \qquad (6\text{-}25)$$

With this assumption, the summation in the denominator of (6-24) becomes an integral and we obtain

$$W'(M|m) \, dM = \frac{KM^m e^{-M}/m!}{\displaystyle\int_0^\infty K(M^m e^{-M}/m!) \, dM} \, dM = \frac{M^m e^{-M}}{m!} \, dM, \qquad (6\text{-}26)$$

since

$$\int_0^\infty M^m e^{-M} \, dM = m!. \qquad (6\text{-}27)$$

It is to be carefully noted that although the right side of (6-26) is similar to that of (6-19), it has a different meaning.   Equation 6-26 gives the probability, under our choice of $P(M)$, that the sample has a *mean* between M and $M + dM$ counts when m counts have been observed.   From (6-26) it is easily found that the most probable value of M is m; through the use of (6-6), (6-7) and (6-27), it is found that the average value of M is $m + 1$ and that the standard deviation of the distribution law (6-26) is $\sqrt{m + 1}$.   The difference between the average and the most probable value of M is unimportant for values of m that are not too small; for small values of m, for example $m = 0$, there is the question whether to estimate M by the average or by the most probable value.

To answer this question we must be clear about the meaning of the average value of M.   It is the value that would be obtained in the following experiment: take a very large collection of samples, each of which had given m counts in a given time interval.   Then the mean number of counts expected from *each* sample is determined from the average of a very large number of observations on *each* of the very large number of samples.   It is then the average value of this very large number of mean values that is given by $m + 1$; and m is the mean value that is most frequently observed.

Now, the observation of m counts was made on one of this large number of samples; the question is, which one?   The best answer is the most probable one; that is, the sample for which $M = m$.   This answer becomes more familiar if we consider the estimate of the mean counts expected from a sample after n observations which gave results $m_1, m_2, \ldots m_n$ have been made upon it.   An expression for $W(M|m_1, m_2, \ldots m_n)$, the probability that the sample is characterized by a mean value M when n observations give the results $m_1, m_2, \ldots, m_n$, can be derived in the same way as (6-26):

$$W(M|m_1, m_2 \cdots m_n) = \frac{n(nM)^{m_1 + m_2 + \cdots m_n} e^{-nM}}{(m_1 + m_2 + \cdots m_n)!}. \qquad (6\text{-}28)$$

The maximum of this distribution function occurs for

$$M = \frac{m_1 + m_2 + \cdots m_n}{n}, \qquad (6\text{-}29)$$

which is the average value of the set of observations just as is expected from (6-1).

Information on the precision of the estimate for M is contained in the expressions for the distribution function: (6-26) or (6-28). The precision of the estimate of M may be characterized by the variance of its distribution function: $m + 1$ for a single observation and $(m_1 + m_2 \cdots m_n + 1)/n^2$ for n observations.

Variance, as computed above, may be used in the normal distribution law (6-3). For small values of m, though, it is probably best to discuss the data directly in terms of the distribution function (6-28). For example, if there is a single observation that gives $m = 0$, (6-28) says that there is a probability of 0.99 that M will be less than 4.6. If the value of zero is obtained in 10 independent observations, then there is a probability of 0.99 that M will be less than 0.46.

In summary, then, for an observed number of counts in excess of about 100, the best statement that can be made is the customary one (6-14a) that the mean value is $m \pm \sqrt{m}$ (taking $m + 1 \cong m$); for a small number of counts the statement would be that the mean value is m and the confidence in the statement can be obtained from (6-28).

F. EXPERIMENTAL APPLICATIONS

Propagation of Errors. Whenever experimental data are used in the computation of a derived quantity, there is the question of the relationship between the precision of the computed values and the precision of the input information. For example, a background counting rate is to be subtracted from an observed counting rate; or the ratio of the counting rates of two samples is used as a measure of the relative numbers of atoms in the samples. The errors in the computed values may be more readily estimated from those of the input data if the error attached to each input datum is independent of that attached to any other.

Consider the independent measurements of two quantities x and y, which lead to the result that the probability of observing a value of x between x and $x + dx$ is $X(x)\,dx$, and similarly for y; then the independence of the measurements means that the probability of having a result with x between x and $x + dx$ while y is between y and $y + dy$ is

$$P(x, y)\,dx\,dy = X(x)\,Y(y)\,dx\,dy.$$

We now ask what is our best estimate of some quantity, f, which is a function, $f(x, y)$, of the variables x and y and what is the precision of our estimate of f? The answer to this question is suggested by (6-6). Our best estimate of $f(x, y)$ is its average value:[3]

$$\overline{f(x, y)} = \iint X(x) \ Y(y) \ f(x, y) \ dx \ dy. \qquad (6\text{-}30)$$

Since the quantity that is sought is $f(\bar{x}_t, \bar{y}_t)$, it is instructive to examine the properties of (6-30) by making a Taylor expansion of $f(x, y)$ about the point $\bar{x}, \bar{y}$ which is our best estimate of $\bar{x}_t, \bar{y}_t$:

$$\overline{f(x, y)} = \iint X(x) \ Y(y) \Bigg[f(\bar{x}, \bar{y}) + (x - \bar{x}) \ f_x(\bar{x}, \bar{y}) + (y - \bar{y}) \ f_y(\bar{x}, \bar{y})$$

$$+ \frac{(x - \bar{x})^2}{2} f_{xx}(\bar{x}, \bar{y}) + \frac{(y - \bar{y})^2}{2} f_{yy}(\bar{x}, \bar{y})$$

$$+ (x - \bar{x})(y - \bar{y}) \ f_{xy}(\bar{x}, \bar{y}) + \cdots \Bigg] dx \ dy, \quad (6\text{-}31)$$

where $f_x(\bar{x}, \bar{y})$, $f_{xx}(\bar{x}, \bar{y})$, $f_{xy}(\bar{x}, \bar{y})$ etc., mean the partial derivatives $\partial f/\partial x$, $\partial^2 f/\partial x^2$, $\partial^2 f/\partial x \ \partial y$, etc., evaluated at the point $\bar{x}, \bar{y}$.

If $f(x, y)$ is a sufficiently slowly varying function in the region of $\bar{x}, \bar{y}$ so that the higher derivatives are negligible, then

$$\overline{f(x, y)} \cong f(\bar{x}, \bar{y}), \qquad (6\text{-}32)$$

since

$$\iint X(x) Y(y)(x - \bar{x}) = \bar{x} - \bar{x} = 0$$

and

$$\iint X(x) Y(y)(y - \bar{y}) = \bar{y} - \bar{y} = 0.$$

For the three elementary arithmetic operations, addition, subtraction, and multiplication, the Taylor series terminates after a finite number of terms, and the exact results

$$\overline{x + y} = \bar{x} + \bar{y}, \qquad (6\text{-}33a)$$

$$\overline{x - y} = \bar{x} - \bar{y}, \qquad (6\text{-}33b)$$

$$\overline{xy} = \bar{x}\bar{y} \qquad (6\text{-}33c)$$

are obtained. This is not the result, however, for the elementary operation of division.[4]

---

[3] See discussion on p. 54 of reference B1 and p. 51 of reference B2.

[4] The quantity $\overline{x/y}$ as evaluated by (6-30) will be infinite unless $Y(y)$ approaches zero more rapidly than does y ($\lim_{y \to 0} [Y(y)/y] \neq \infty$). This infinity catastrophe is usually avoided by restricting the values of y to those that have a relatively large likelihood—that is, close to $\bar{y}$. When this is done, (6-32) gives the estimate of f.

The estimate of the variance is given by

$$\sigma_{\bar{f}}^2 = \overline{[f(x, y) - \overline{f(x, y)}]^2} = \iint X(x)\, Y(y)[f(x, y) - \overline{f(x, y)}]^2\, dx\, dy. \quad (6\text{-}34)$$

If again a Taylor expansion is used and the higher order terms are neglected, then

$$\sigma_{\bar{f}}^2 = f_x{}^2(\bar{x}, \bar{y})\sigma_{\bar{x}}{}^2 + f_y{}^2(\bar{x}, \bar{y})\sigma_{\bar{y}}{}^2 \cdots \quad (6\text{-}35)$$

Exact expressions again result for the variance of three of the elementary arithmetic operations:

$$\sigma^2_{\overline{x+y}} = \sigma_{\bar{x}}{}^2 + \sigma_{\bar{y}}{}^2 \quad (6\text{-}36a)$$

$$\sigma^2_{\overline{x-y}} = \sigma_{\bar{x}}{}^2 + \sigma_{\bar{y}}{}^2 \quad (6\text{-}36b)$$

$$\frac{\sigma^2_{\overline{xy}}}{\bar{x}^2\bar{y}^2} = \frac{\sigma_{\bar{x}}{}^2}{\bar{x}^2} + \frac{\sigma_{\bar{y}}{}^2}{\bar{y}^2} + \frac{\sigma_{\bar{x}}{}^2\sigma_{\bar{y}}{}^2}{\bar{x}^2\bar{y}^2} \quad (6\text{-}36c)$$

The third term in expression (6-36c) is usually small compared to the first two and may be neglected. Similarly, the first two terms of (6-35) are usually a good approximation for the variance of other functions of x and y.

As an example, suppose that the background counting rate of a counter is measured and 600 counts are recorded in 15 minutes. Then with a sample in place the total counting rate is measured, and 1000 counts are recorded in 10 minutes. We wish to know the net counting rate due to the sample and the standard deviation of this net rate. First the background rate $\mathbf{R}_b$ is

$$\mathbf{R}_b = \frac{600 \pm \sqrt{600}}{15} = \frac{600 \pm 24}{15} = 40 \pm 1.6 \text{ counts per minute.}$$

The total rate $\mathbf{R}_t$ is

$$\mathbf{R}_t = \frac{1000 \pm \sqrt{1000}}{10} = \frac{1000 \pm 32}{10} = 100 \pm 3.2 \text{ counts per minute.}$$

The net rate $\mathbf{R}_n = 100 - 40 = 60$ counts per minute, its standard deviation is $\sigma_n = \sqrt{1.6^2 + 3.2^2} = 3.6$, and $\mathbf{R}_n = 60 \pm 3.6$ counts per minute.

Gaussian Error Curve. Knowledge of the distribution law permits a quantitative evaluation of the probability of a given deviation of a measured result m from the proper average M to be expected. With the absolute error $|M - m| = \epsilon$, and with the assumption that the integral numbers are so large that the distribution may be treated as continuous, the probability $W(\epsilon)\, d\epsilon$ of an error between ϵ and $\epsilon + d\epsilon$ for the normal distribution is given by

$$W(\epsilon)\, d\epsilon = \frac{2}{\sqrt{2\pi M}}\, e^{-\epsilon^2/2M}\, d\epsilon. \quad (6\text{-}37a)$$

The factor 2 arises from the existence of positive and negative errors with equal probability within the limits of validity of this approximation.   Recalling that $\sigma = \sqrt{M}$, we have

$$W(\epsilon)\,d\epsilon = \frac{1}{\sigma}\sqrt{\frac{2}{\pi}}\,e^{-\epsilon^2/2\sigma^2}\,d\epsilon. \tag{6-37b}$$

The probability of an error greater than $k\sigma$ is obtained by integration from $\epsilon = k\sigma$ to $\epsilon = \infty$.   Numerical values of this integral as a function of k may be found in handbooks.   For example, we have taken for table 6-2 some representative values from the table, "Probability of Occurrence of Deviations" in the Chemical Rubber Publishing Company's *Handbook of Chemistry and Physics.*

Table 6-2

k	0	0.674	1	2	3	4
Probability of $\epsilon > k\sigma$	1.00	0.50	0.32	0.046	0.0027	0.00006

Notice that errors greater than and smaller than 0.674σ are equally probable; 0.674σ is called the "probable error" and is sometimes given rather than the standard deviation when counting data are reported.   In plots of experimental curves it can be convenient to indicate the probable error of each point (by a mark of the proper length); then on the average the smooth curve drawn should be expected to pass through about as many "points" as it misses. It is unfortunately not strictly correct to use (6-37b) with (6-35) in the estimation of the probability of an error of a function of random variables.   For example, the distribution of the differences of two random variables which have Gaussian distributions is not itself Gaussian.   Nevertheless, if the function does not vary too rapidly in the vicinity of its average, the distribution of values about the average is essentially Gaussian with a variance as given in (6-35).

Comparison with Experiment.   We now return to a consideration of the typical counting data in table 6-1.   We have already found from the deviations among the 10 measurements $\sigma = \sqrt{(N_0 - 1)^{-1}\Sigma(x_i - \bar{x})^2} = 11.9$.   If the counting rate measured there represents a random phenomenon, as we expect it should, we may evaluate the expected σ for the result in any minute as the square root of the number of counts.   For a typical minute, the ninth, we find $\sigma = \sqrt{101} = 10$, and for other minutes other values not much different.   Because these values agree reasonably with the 11.9 there is evidence for the random nature of the observed counting rate.   This test should occasionally be made on the data from a counting instrument.

In addition to estimating the σ for each entry in table 6-1, we may also estimate the $\sigma_{\bar{x}}$ for the average of the 10 observations.   This estimate can be performed in three different ways, and it is instructive to compare them:

1. Since the 10 data are observations of a radioactive decay, we expect from (6-14a) that each datum has a standard deviation given by the square root of the number of counts. The mean is calculated by summing the data and dividing by the number of observations (10). The standard deviation of the mean, then, can be obtained from (6-35) for the propagation of fluctuations for a function of random variables (the number 10 has zero standard deviation). The result is

$$\sigma_{\bar{x}} = 99 \sqrt{\frac{990}{990^2}} = \frac{1}{10} \sqrt{990} = 3.1.$$

2. The individual counting rates can be summed, which is equivalent to an observation of 990 counts in 10 minutes. Again, since we are dealing with radioactive decay, the standard deviation of the mean is given by (6-14a):

$$\sigma_{\bar{x}} = \frac{1}{10} \sqrt{990} = 3.1.$$

3. If the fact that these data are from radioactive decay is ignored and no special relation such as (6-14a) is assumed to exist between each observation and its standard deviation, then the standard deviation of the mean is computed from (6-5):

$$\sigma_{\bar{x}} = \sqrt{\frac{1276}{9 \times 10}} = 3.8.$$

It is important to note that methods 1 and 2 give the same answer, as they must; it is not possible to gain more information about the standard deviation of the mean by breaking a 10-minute observation into 10 one-minute observations. The $\sigma_{\bar{x}} = 3.1$ given by methods 1 and 2 is the correct answer. It is also of interest to see that relinquishing the information contained in (6-15), as in method 3, diminishes the precision of the estimate of the mean.

The average counting rate with its standard deviation is $\bar{x} = (990 \pm \sqrt{990})/10 = 99.0 \pm 3.1$ counts per minute. This means that the probability that the true average is between 95.9 and 102.1 is, from table 6-2, just $1 - 0.32 = 0.68$. Actually, when the counting data given in table 6-1 were obtained, the average rate was measured much more accurately in a 100-minute count, and the result was $(10,042 \pm \sqrt{10,042})/100 = 100.4 \pm 1.0$ counts per minute.

Counter Efficiencies. As another application of the methods of this chapter to counting techniques, we may estimate the efficiency of a Geiger counter for rays of a given ionizing power, with the assumptions that any ray that produces at least one ion pair in the counter gas is counted and that effects at the counter walls are negligible. Knowledge of the nature of the radiation and the information given in chapter 4 permit an estimate of the average number of ion pairs a to be expected within the path length of the radiation in the counter filling gas.   The problem then is to find the probability that a ray will pass through the counter, leaving no ion pairs, and thus will not be counted. We think of the path of the ray in the counter as divided into n segments of equal length; if n is very large, each segment will be so small that we may neglect the possibility of having two ion pairs in any segment.   Then just a of the n segments will contain ion pairs, and by definition the probability of having an ion pair in a given segment is $p = a/n$.   Now by (6-8) for the binomial distribution we have the probability of no ion pairs in n segments; that is, of $r = 0$

$$W(0) = \frac{n!}{n!0!} p^0 (1 - p)^n = (1 - p)^n = \left(1 - \frac{a}{n}\right)^n.$$

Since the probability[5] is evaluated correctly only as n becomes very large,

$$W(0) = \lim_{n \to \infty} \left(1 - \frac{a}{n}\right)^n = e^{-a}.$$

The probability of counting the ray, which is the efficiency to be determined, is then $1 - W(0) = 1 - e^{-a}$.   As a particular example, consider a fast β particle with the relatively low specific ionization of 5 ion pairs per millimeter in air and a path length of 10 mm in a counter gas which is almost pure argon at 7.6 cm pressure.   We estimate a from these assumptions, correcting for the relative densities of air and the argon:

$$a = 5 \times 10 \times \frac{7.6}{76} \times \frac{40}{29} = 7.$$

The corresponding estimated counter efficiency is $1 - e^{-7} = 99.9$ per cent. It should not be expected that an efficiency calculated in this way is very precise.   Wall effects may be important, and the assumption of random distribution of ion pairs along the β-ray path is not entirely consistent with the mechanism of energy loss by ionization presented in chapter 4.

Coincidence Correction.   If a counter has a recovery time (or dead time or resolving time) τ after each recorded count during which it is completely insensitive, the total insensitive time per unit time is $\mathbf{R}\tau$, where $\mathbf{R}$ is the observed counting rate.   If $\mathbf{R}^*$ is the rate that would be recorded if there were no coincidence losses, the number of lost counts per unit time is $\mathbf{R}^* - \mathbf{R}$ and is given by the product of the rate $\mathbf{R}^*$ and the fraction of insensitive time $\mathbf{R}\tau$:

$$\mathbf{R}^* - \mathbf{R} = \mathbf{R}^*\mathbf{R}\tau,$$

$$\mathbf{R}^* = \frac{\mathbf{R}}{1 - \mathbf{R}\tau}. \tag{6-38}$$

[5] We might have evaluated this probability more easily from the Poisson distribution expression: $W(0) = a^0 e^{-a}/0! = e^{-a}$.

A number of variants of this formula are also in use. One expression (the Schiff formula) is $R^* = Re^{R^*\tau}$; this is derived from a calculation of the probability $W(0)$ of having had no event during the time τ immediately preceding any event. An event, whether recorded or not, is here considered to prevent the recording of a second event occurring within the time τ.[6] Another approximate expression is derived from the first two terms in the binomial expansion of $(1 - R\tau)^{-1}$ appearing in (6-38):

$$R^* = R(1 + R\tau) = R + R^2\tau.$$

This form is especially convenient for the interpretation of an experiment designed to measure τ by measuring the rates R_1 and R_2 produced by two separate sources and the rate R_t produced by the two sources together, each of these rates including the background effect R_b. Obviously,

$$R_1^* + R_2^* = R_t^* + R_b,$$

where we have neglected the coincidence loss in the measurement of the low background rate. Replacing by $R_1^* = R_1 + R_1^2\tau$, etc., and rearranging, we have

$$\tau = \frac{R_1 + R_2 - R_t - R_b}{R_t^2 - R_1^2 - R_2^2}.$$

Statistics of Pulse Height Distributions. When a monoenergetic source of radiation is measured with a proportional- or scintillation-counter spectrometer, the observed pulse heights have a Gaussian distribution around the most probable value. The energy resolution of such an instrument is usually expressed in terms of the full width at half maximum of the pulse height distribution curve, stated as a fraction or percentage of the most probable pulse height H. The pulse height $h_{1/2}$ at the half maximum of the distribution curve may be obtained from the ratio of probabilities

$$\frac{W(h_{1/2})}{W(H)} = \exp\left[\frac{-(H - h_{1/2})^2}{2\sigma_h^2}\right] = 0.5.$$

Then $(H - h_{1/2})^2/\sigma_h^2 = \ln 2$, and the full width at half maximum is

$$\frac{2|H - h_{1/2}|}{H} = 2\sqrt{2\ln 2}\,\frac{\sigma_h}{H} = \frac{2.36\sigma_h}{H},$$

where σ_h is the standard deviation of the pulse height distribution.

In a proportional counter the spread in pulse heights for monoenergetic rays absorbed in the counter volume arises from statistical fluctuations in the num-

---

[6] It may be noticed that the Schiff formula might be expected to correspond more closely to the conditions of coincidence loss in a mechanical register, in which a new pulse within a dead time could initiate a new dead-time period, although it would not be recorded. There exists also the opportunity for coincidence losses in the electric circuits.

ber of ion pairs formed and statistical fluctuations in the gas amplification factor. The pulse height is proportional to the product of the gas amplification and the number of ion pairs, and therefore the fractional standard deviation of the pulse height equals the square root of the sum of the squares of the fractional standard deviations of these two quantities. As an example, consider the pulse height spectrum produced by the absorption of manganese K X rays in a proportional counter filled with 90 per cent argon and 10 per cent methane and operating with a gas gain of 1000. From table 4-1 the energy per ion pair is about 27 eV, and therefore the number of ion pairs formed by a 5.95-keV X ray is $5950/27 = 220 \pm \sqrt{220}$. If the numbers of ions collected per initial ion pair have a Poisson distribution, the fractional standard deviation in the gas gain is $\sqrt{1000}/1000$. Thus

$$\frac{\sigma_h}{H} = \sqrt{220/220^2 + 1000/1000^2} = \sqrt{0.00455 + 0.00100} = 0.0745,$$

and the full width at half maximum is $2.36 \times 0.0745 = 0.176$ or 17.6 per cent. If the gas gain is made sufficiently large, the fluctuations in the number of ion pairs determine the resolution, and in this case the resolution of a given counter is seen to be inversely proportional to the square root of the energy of the ionizing radiation absorbed.

In a scintillation counter the statistical fluctuations in output pulse heights arise from several sources (B3). The conversion of energy of ionizing radiation into photons in the scintillator, the electron emission at the photocathode, and the electron multiplication at each dynode are all subject to statistical variations. Although the photocathode emission has been shown to have somewhat larger fluctuations than correspond to the Poisson law, the observed pulse height distributions are for most practical purposes in sufficiently close agreement with those calculated on the assumption of Poisson distributions for all the statistical processes involved. With this assumption the standard deviation of a pulse height distribution for a single energy of ionizing radiation absorbed in the phosphor turns out to be approximately

$$\sigma_h \cong H \sqrt{\bar{n}/E\bar{q}\bar{f}\bar{p}(\bar{n} - 1)}, \tag{6-39}$$

where H is the most probable pulse height for an incident energy E keV, $\bar{q}$ is the mean value of the phosphor efficiency (number of light quanta emitted per 1000 eV of incident energy), $\bar{f}$ is the mean value of the light collection efficiency at the photocathode, $\bar{p}$ is the mean value of the photocathode efficiency (number of photoelectrons arriving at the first dynode for each photon incident on the photocathode), and $\bar{n}$ is the average electron multiplication per dynode.

In practice $\bar{f}$ can be made almost unity, $\bar{p}$ is of the order of 0.1, $\bar{n}$ is usually about 3 to 5, and $\bar{q}$ is approximately 30 for NaI(Tl), 15 for anthracene, and 7 for stilbene and for the best liquid scintillators. As an example we estimate the resolution attainable for the 662-keV photopeak of the Cs[137] γ rays with a sodium iodide scintillation counter.   Taking $\bar{f}\bar{p} = 0.1$ and $\bar{n} = 4$, we obtain

$$\frac{\sigma_h}{H} = \sqrt{\frac{4}{662 \times 30 \times 0.1 \times 3}} = 0.026.$$

The corresponding full width at half maximum is $2.36\,\sigma_h/H = 0.061$ or 6.1 per cent, which is indeed not far from the best resolution obtained experimentally. (See the experimental pulse height distribution with 8.5 per cent width at half maximum shown in figure 5-8.)

REFERENCES

B1 K. A. Brownlee, *Statistical Theory and Methodology in Science and Engineering*, Wiley, New York, 1960.

B2 C. A. Bennett and N. L. Franklin, *Statistical Analysis in Chemistry and Chemical Industry*, Wiley, New York, 1954.

B3 E. Breitenberger, "Scintillation-Spectrometer Statistics," *Progress in Nuclear Physics*, Vol. 4 (O. R. Frisch, Editor), Pergamon, London, 1955, pp. 56–94.

*E1 R. D. Evans, *The Atomic Nucleus*, McGraw Hill, New York, 1955, chapters 26–28.

F1 W. Feller, *Probability Theory and its Applications*, Wiley, New York, 1950.

EXERCISES

1. Mr. Jones's automobile license carries a six-digit number.   What is the probability that it has (a) exactly one 4, (b) at least one 4?   Make the assumption that the numbers 0 to 9 inclusive are equally probable for each of the six digits.

Answer: (b) 0.46856.

2. Consider the following set of observations:

Minute	Counts
1	203
2	194
3	201
4	217
5	195
6	189
7	210
8	207
9	230
10	188

(a) Calculate the average value.   (b) What is the standard deviation of the set? (c) What is the standard deviation of the mean?   (d) What is the probability that an eleventh observation would have a value greater than 230?   (e) What is the probability that a subsequent set of 10 one-minute observations will have an

average value that is greater than 212?   (f) Should any of the data be rejected? If so, what is the new average value?   *Answers:* (c) 4.16.
                                                                              (e) 0.019.

3. Given an atom of a radioactive substance with decay constant λ, what is (a) the probability of its decaying between 0 and dt, (b) the probability of its decaying between 0 and t?

4. A sample contains 4 atoms of Lw.   What is the probability that exactly 2 of the atoms will have decayed in (a) one half-life, (b) two half-lives?

5. A given proportional counter has a measured background rate of 900 counts in 30 minutes.   With a sample of a long-lived activity in place, the total measured rate was 1100 counts in 20 minutes.   What is the net sample counting rate and its standard deviation?   *Answer:* 25.0 $\pm$ 1.9 counts per minute.

6. Denote by $\mathbf{R}_t$ and $\mathbf{R}_b$ the total and background counting rates for a long-lived sample and calculate the optimum division of available counting time between sample and background for minimum σ on the net counting rate.   *Answer:* $\dfrac{t_t}{t_b} = \sqrt{\dfrac{\mathbf{R}_t}{\mathbf{R}_b}}$.

7. (a) Sample A, sample B, and background alone were each counted for 10 minutes; the observed total rates were 110, 205, and 44 counts per minute, respectively. Find the ratio of the activity of sample A to that of sample B and the standard deviation of this ratio.   (b) Sample C was counted on the same counter for 2 minutes and the observed total rate was 155 counts per minute.   Find the ratio, and its standard deviation, of the activity of C to that of A.
                                                                Answer: (a) 0.41 $\pm$ 0.027.

8. Derive (6-28) for the probability of a value M when given a set of observations m_1, $m_2, \ldots, m_n$.

9. The scintillation spectrometer of exercise 4, chapter 5, is to be used for the measurement of 120-keV conversion electrons.   What will be the full width at half maximum of the pulse height distribution?

10. Would the same spectrometer (exercise 9) completely resolve (i.e., give a dip between the pulse height peaks of) two conversion-electron groups of 44 and 52 keV, present in the abundance ratio 2:1?

7

Tracers in Chemical Applications

A. THE TRACER METHOD

Isotopic Tracers. Most of the ordinary chemical elements are composed of mixtures of isotopes, and each mixture remains essentially invariant in composition through the course of physical, chemical, and biological processes. That this is so is shown by the constant isotopic ratios found for elements from widely scattered sources[1] and by the fact that atomic weights reliable to many significant figures may be determined by chemical means. It is true that isotopic fractionation may be appreciable for the lightest elements in which the percentage mass difference between isotopes is greatest, and this effect must always be considered in the use of hydrogen tracer isotopes. However, apart from these isotopes and Be^7 which differs in mass from stable Be^9 by only about 25 per cent, the next heavier tracer is in carbon where already the specific isotope effect may be neglected in most tracer work of ordinary precision. In this section we shall assume that the fact that a given isotope may be radioactive does not in any way affect its chemical (or biological) properties until it actually undergoes the spontaneous radioactive change. The interesting and important divergences from this assumption are examined in section C.

Because the isotopic tracer atoms are detected by their radioactivity, they behave normally up to the moment of detection; after that moment they are not detected, and their fate is of no consequence. Of course, if the resulting atoms after the nuclear transformation are themselves radioactive and capable of a further nuclear change, the detection method must be arranged to give a response that measures the proper (in this case the first) radioactive species only. For example, if RaE (Bi^{210}) is used as a tracer for bismuth, the α particles from its daughter Po^{210} should not be allowed to enter the detection instrument but should be absorbed by a suitable absorber or by the counter wall. As a tracer for thorium, UX_1 is suitable in spite of the fact that most of the detectable radiation will be from its daughter UX_2; the reason is that the

---

[1] Some exceptions to the constancy of isotopic ratios were mentioned in chapter 2. p. 24.

short half-life, 1.18 minutes, of UX_2 ensures that it will be in transient equilibrium with the UX_1 by the time the sample is mounted and ready for measurement, so that the total activity will be proportional to the UX_1 content. Many multiple decays are found in the fission-product activities. If Ba^{140} ($t_{1/2}$ = 12.8 days) is to be used as a tracer for barium, the isolated samples either should be freed chemically of the daughter La^{140} ($t_{1/2}$ = 40 hours) or should be kept before measurement for a week or two until the transient equilibrium is achieved. The isomeric-transition activities present interesting cases; if the 4.5-hour Br^{80m} is chosen as a tracer for bromine, the 18-minute lower isomeric state of Br^{80} will always be present, and, because of chemical effects accompanying the isomeric transition (section E), the two isomers may be present in different chemical forms. However, the 4.5-hour Br^{80m} may be used with confidence, provided only that one measures isolated samples for the 4.5-hour period by analysis of the decay curves or simply by holding the prepared samples for a time long compared to 18 minutes before measurement. These special cases do not occur in the use of the great majority of popular tracer isotopes. Another source of interference with the tracer principle is a possible chemical (or biological) effect produced by the ionizing rays; this radiation chemistry effect is not often encountered at the usual tracer activity levels and may always be checked by experiments with a much higher or lower level of radioactivity.

A definite limitation of the radioactive tracer method is the absence of known active isotopes of suitable half-life for a few elements, especially oxygen and nitrogen. There are radioactive oxygen isotopes, O^{14}, O^{15}, and O^{19}, but these have half-lives of 72, 118, and 29 seconds, respectively. The 7-second N^{16} and 4-second N^{17} are useless as tracers, but some applications of the 10-minute N^{13} β^+ activity have been made. Also, there are no helium, lithium, and boron isotopes with half-lives longer than 1 second. The use of separated stable isotopes as tracers is a valuable technique. Enriched O^{18} and N^{15} are essential for many interesting and important purposes, and C^{13} offers significant advantages for some carbon tracer experiments. Deuterium (H^2) has found many applications as a hydrogen tracer; the use of tritium (H^3) is not entirely equivalent because its properties are even more different from those of protium (H^1).

Stable isotope tracers are most commonly assayed by mass spectrometers. (For the lightest elements, especially hydrogen, isotopic composition can be determined by measurement of physical properties, such as density or thermal conductivity, and for a number of separated isotopes assays could be made by measurement of a nuclear property, such as nuclear resonance absorption. Samples for introduction into a mass spectrometer are ordinarily put into gaseous form. Carbon-13 is commonly analyzed as CO_2. The same gas is often used for O^{18} measurements, and, since CO_2 and H_2O reach isotopic equilibrium by exchange in a day or so, a convenient analysis for O^{18} in H_2O is provided. It is interesting that the most precise determinations of the radioactive C^{14} have been made by mass spectrometry on samples of high specific activity.

The uses of isotopic tracers may be classified into two groups: (1) applications in which the tracer is necessary in principle and (2) applications in which a tracer, although not necessary in principle, may be a great practical convenience. Those applications which depend uniquely on the tracer principle—although the tracers may be either radioactive or separated stable isotopes—may be illustrated by studies of self-diffusion of an element or other substance into itself; no other investigative technique can give information on such matters. On the other hand, the studies on coprecipitation which have been made with radioactive tracers might have been done, at least at the higher concentrations, by careful application of conventional chemical methods or perhaps by spectroscopic or other means of analysis. In some of the more involved applications, particularly in biology, both aspects of tracer usefulness appear together.

Self-diffusion. To illustrate the unique tracer method we discuss first studies that have been made of self-diffusion. By the use of sensitive spectroscopic analyses the rates of diffusion of various metals (including gold, silver, bismuth, thallium, and tin) in solid lead at elevated temperatures have been investigated, but the first attempt (by G. Hevesy and his collaborators) to observe the diffusion of radioactive lead into ordinary lead failed, showing that the diffusion rate must be at least one hundred times smaller than that for gold in lead (which is the fastest of those just named, the others showing decreasing rates in the order listed). The method first used was a rather gross mechanical one, and the workers evolved a much more sensitive method based on the short range of the α particles from Bi^{212} in transient equilibrium with Pb^{212}. The lead, containing Pb^{212} isotopic tracer, was pressed into contact with a thin foil of inactive lead, which was chosen just thick enough to stop all the α rays, and then as diffusion progressed an α activity appeared and increased as measured through this foil. The diffusion coefficient $\mathbf{D}$ is obtained through a suitable integration of Fick's diffusion law, $\partial c/\partial t = \mathbf{D}(\partial^2 c/\partial x^2)$, where c is concentration of the diffusing tracer, t is time, and x is the coordinate along which the diffusion is measured; some typical values for $\mathbf{D}$ were 0.6×10^{-6} cm^2/day at 260°C, 2.5×10^{-6} at 300°C, and 47×10^{-6} at 320°C. A similar but even more sensitive technique than the α-range method was based on the much shorter ranges (a few millionths of a centimeter in lead) of the nuclei recoiling from α emission, with the radioactivity of the resulting Tl^{208} as an indicator of the emergence of recoil nuclei from the thin lead foils. At 200°C the diffusion of lead in lead is about 10 times slower than that of tin in lead and roughly 10^5 times slower than that of gold in lead.

The rates of diffusion of ions through aqueous salt solutions of uniform composition may be determined with radioactive tracers, and this information may be of special significance, since each rate is a property of the particular ion in that system, whereas salt-diffusion rates under a concentration gradient as ordinarily observed must necessarily depend on the diffusion tendencies of ions of both charges.   Determinations of the self-diffusion coefficients for a number of ions in solutions of their salts have been made chiefly by two techniques.   One is the diaphragm cell method, in which the solution with tracer is separated by a porous diaphragm from a chemically identical solution without added tracer.   The other technique is the open-ended capillary method.   Here a capillary tube about 0.06 cm in diameter and a few centimeters long is filled with the tracer solution and immersed open end up in an essentially infinite volume of a salt solution of the same concentration, but without tracer.

Self-diffusion coefficients have been obtained for a number of common ions, especially alkali metal and halide ions, in solutions over wide concentration ranges.   Unfortunately, the two experimental methods have so far given results that sometimes differ in details.   The coefficients for smaller "hydrated" cations such as Na^+ and K^+ exhibit a maximum at concentrations in the vicinity of one mole per liter.   The larger, "unhydrated" ions such as Rb^+, Cs^+, and I^- in their solutions show a more regular behavior; indeed for these ions the product of the self-diffusion coefficient and viscosity of the solution is a linear function of concentration at least up to concentrations well above one molar.   The slopes of such plots have been related to certain Debye-Hückel parameters.

Other Migration Problems.   Radioactive tracers are useful in the study of numerous migration problems other than self-diffusion, particularly when movements of very small amounts of material are involved.   In most applications of this kind the tracer is serving only as a sensitive and relatively convenient analytical tool.   Erosion and corrosion of surfaces may be measured with great sensitivity if the surface to be tested can be made intensely radioactive.   Transfer of minute amounts of bearing-surface materials during friction has been studied in this way.   Radioactive gases or vapors may be detected in small concentrations, and leakage, flow, and diffusion rates of gases may therefore be studied by the tracer method.

B. ISOTOPIC EXCHANGE AND OTHER TRACER REACTIONS

Qualitative Observations.   In an early exchange experiment in 1920 Hevesy demonstrated by the use of $ThB(Pb^{212})$ the rapid interchange of lead atoms between $Pb(NO_3)_2$ and $PbCl_2$ in water solution.   The experiment was performed by the addition of an active $Pb(NO_3)_2$ solution to an inactive $PbCl_2$ solution and by the subsequent crystallization of $PbCl_2$ from the mixture. The result is not at all surprising because the well-known process of ionization for these salts leads to chemically identical lead ions, Pb^{2+}.   This pioneering experiment opened an important field of chemical investigation, and many exchange systems, particularly since the advent of artificial radioactivity, have been examined since that time (S5).   A few qualitative examples are given in the following paragraphs.

For the majority of cases in which exchange is rapid at ordinary temperature there are known reversible reactions that would lead to interchange. For the other observed exchanges such reversible reactions also exist, which are possibly unknown, or the exchange occurs by a simple collision mechanism (which may amount to the same thing).

It was soon shown when artificially radioactive isotopes became available that aqueous Cl^- and Cl_2, Br^- and Br_2, and I^- and I_2 exchanged so quickly at room temperature that the rates could not be measured by ordinary methods. These exchanges are interpreted as occurring through the reactions illustrated by $I^- + I_2 \rightleftharpoons I_3^-$.   It has been found that Br_2 and HBr, either in the gas phase or in solution in dry carbon tetrachloride, exchange rapidly at room temperature, probably through reversible formation of a complex HBr_3, although the life of this intermediate may be very short and thus its concentration very low.

In dilute-acid solution at room temperature there is no rapid exchange of halogen atoms between Cl_2 and ClO_3^- or ClO_4^-, Br_2 and BrO_3^-, I_2 and IO_3^-, ClO_3^- and ClO_4^-, IO_3^- and IO_4^-; however, some of these pairs do exchange at measurable rates.   No exchange was found in alkaline solution between Cl^- and ClO_4^-, Br^- and BrO_3^-, I^- and IO_3^-.

Interesting exchange studies have been made with the tracer S^{35}.   Sulfur and sulfide ions exchange in polysulfide solution.   On the other hand, S^{2-} and SO_4^{2-}, SO_3^{2-} and SO_4^{2-}, H_2SO_3 and HSO_4^- do not exchange appreciably even at 100°C.   If active sulfur is reacted with inactive SO_3^{2-} to form $S_2O_3^{2-}$, and then the sulfur removed with acid, the H_2SO_3 is regenerated inactive; therefore, the two sulfur atoms in thiosulfate are not equivalent.   The ions $S_2O_3^{2-}$ and SO_3^{2-} exchange only very slowly at room temperature, but exchange one sulfur fairly rapidly at 100°C.   (Notice that this result can be found only by labeling the proper sulfur atom, the one attached directly to the oxygen atoms.)

Phosphoric and phosphorous acids, H_3PO_4 and $H_2(HPO_3)$, and also phosphoric and hypophosphorous acids, H_3PO_4 and $H(H_2PO_2)$, do not exchange phosphorus atoms even at 100°C, although the first of these exchanges might be expected to proceed (at some unknown rate) through the formation of hypophosphoric acid, $H_4P_2O_6$.   Arsenate and arsenite ions, and H_3AsO_4 and $HAsO_2$, do not exchange appreciably even at 100°C.

The availability of appreciably enriched stable isotope O^{18} has led to a number of observations on exchanges between oxygen-containing ions and water (T1).   In neutral or alkaline solution perchlorate, chlorate, nitrate, sulfate, and phosphate do not exchange appreciably with water; in acid solutions the exchange reactions are faster.   The hypothesis that these exchanges proceed through reversible formation of the acid anhydride has been advanced and receives support from the observation that the weaker acids such as H_2SO_3 and HIO_3 exchange more readily with water.   Moreover, the sulfate-water exchange is catalyzed by hydrogen ion quantitatively as if the reaction proceeded through undissociated H_2SO_4 molecules.   Exchange between water of hydration and solvent water also may be studied with O^{18}.   For most ions, including Al^{3+}, Fe^{3+}, Ga^{3+}, Th^{4+}, and Co^{2+}, this exchange is too rapid to have been observed by tracer techniques but has been studied through the observation of line-broadening in nuclear-magnetic-resonance spectra (S6).   For Cr^{3+}, actually $Cr(H_2O)_6{}^{3+}$, the exchange is about half complete in 40 hours.

The behavior of aluminum bromide in exchange reactions is remarkable, paralleling its extraordinary character as a catalyst.   It exchanges bromine atoms readily at room temperature with many alkyl bromides, with benzyl bromide, and with many aliphatic polybromides; it exchanges also, but more slowly, with aryl bromides.   Aluminum iodide appears to behave in a similar way.   Because these aluminum halides also exchange readily with gaseous halogen or hydrogen halide, a convenient synthesis of labeled organic halides is provided.   Obviously the presence of aluminum bromide will catalyze an exchange between two organic bromides.

It is difficult to overestimate the importance of the role played by tracers, usually C^{14}, in the detailed investigation of biological processes.   The demonstration of the citric acid cycle as the primary path for the oxidation of the products of lipid and amino acid metabolism is a classical example.   This example along with many others is reviewed in reference V1.

Quantitative Exchange Law.   Because exchange reactions occur at equilibrium with respect to the chemical species involved, although not with respect to the distribution of isotopes among the various chemical species, these reactions are particularly useful for the investigation of the theories of rates of reactions.   This is so because, strictly speaking, the existing theories for rates of reactions assume that equilibrium conditions prevail.   In this section we shall see how the rate of a chemical exchange reaction may be extracted from information on the rate at which a tracer atom is exchanged.

Consider a schematic exchange-producing reaction,

$$AX + BX^0 = AX^0 + BX,$$

where X^0 represents a radioactive atom of X.   The radioactive decay of this species will be neglected; in practice, if the decay is appreciable, correction of all measured activities to some common time must be used to avoid error from this condition.   The rate of the reaction between AX and BX in the dynamic equilibrium we call $\mathbf{R}$, in units of moles per liter per second; notice that $\mathbf{R}$ is quite independent of the concentration and even of the existence of the active tracer X^0, but is, in general, dependent on the total concentrations of the species $AX + AX^0$ and $BX + BX^0$.   We indicate mole-per-liter concentrations as follows: $(AX) + (AX^0) = a$, $(BX) + (BX^0) = b$, $(AX^0) = x$, and

$(BX^0) = y$.   The rate of increase (dx/dt) of (AX^0) is given by the rate of its formation minus the rate of its destruction.   The rate of formation of AX^0 is given by $\mathbf{R}$ times the factor y/b, which is the fraction of reactions that occur with an active molecule BX^0, and times the factor $(a - x)/a$, which is the fraction of reactions with the molecule AX initially inactive.   The rate of destruction of AX^0 is given by $\mathbf{R}$ times the factor x/a, which is the fraction of reactions in the reverse direction that occur with an active molecule AX^0, and times the factor $(b - y)/b$, which is the fraction of reverse reactions with the molecule BX initially inactive.   The differential equation is then

$$\frac{dx}{dt} = \mathbf{R}\frac{y}{b}\frac{(a - x)}{a} - \mathbf{R}\frac{x}{a}\frac{(b - y)}{b} = \mathbf{R}\left(\frac{y}{b} - \frac{x}{a}\right). \tag{7-1}$$

After a sufficiently long time, that is, at $t = \infty$, let $x = x_\infty$ and $y = y_\infty$. The conservation of radioactive atoms (after correction for any decay) demands that

$$x + y = x_\infty + y_\infty. \tag{7-2}$$

Further, at $t = \infty$ the exchange reaction is completed, which means that $dx/dt = 0$ and so, from (7-1)

$$\frac{x_\infty}{a} = \frac{y_\infty}{b}, \tag{7-3}$$

which constitutes an algebraic expression for the reasonable and well-known rule that when exchange is complete the specific activity (activity per mole or per gram of X) is the same in both chemical species.   By the use of (7-2) and (7-3), y may be eliminated from (7-1), resulting in

$$\frac{dx}{dt} = \mathbf{R}\frac{(a + b)}{ab}(x_\infty - x). \tag{7-4}$$

This differential equation with separable variables may be integrated to give

$$\ln\frac{(x_\infty - x_0)}{(x_\infty - x)} = \mathbf{R}\frac{(a + b)}{ab}t, \tag{7-5}$$

where x_0 is the value of x at $t = 0$.   Under the special, but common, condition that $x_0 = 0$, the more familiar forms emerge:

$$2.303 \log\left(1 - \frac{x}{x_\infty}\right) = -\frac{a + b}{ab}\mathbf{R}t, \tag{7-6a}$$

$$1 - \frac{x}{x_\infty} = \exp\left[-\frac{a + b}{ab}\mathbf{R}t\right]. \tag{7-6b}$$

The last result shows that $\mathbf{R}$ may be evaluated from the slope of a plot of $\log[1 - (x/x_\infty)]$ versus t.   Probably the most convenient procedure is to plot

$[1 - (x/x_\infty)]$ on semilog paper against t, read off the half-time $T_{\frac{1}{2}}$ at which the fraction exchanged, x/x_∞, is $\frac{1}{2}$, and find **R** from an equation derived immediately from (7-6b):

$$\mathbf{R} = \frac{ab}{a + b} \cdot \frac{0.693}{T_{\frac{1}{2}}}.$$

It is important to notice that if a or b or both should be varied the variation in half-time for the exchange would not directly reflect the variation in **R** because of the factor $ab/(a + b)$.

For a number of practical exchange studies the simple formulas AX and BX may not represent the reacting molecules; for example, AX_2 or BX_n might be involved. So long as the several atoms of X are entirely equivalent (or at least indistinguishable in exchange experiments) in each of these molecules, the equations just derived may be applied without modification, provided only that we redefine all the concentrations in gram atoms of X per liter rather than moles of AX or AX_2, etc., per liter. This is equivalent to considering (for this purpose only) one molecule of AX_2 as replaceable by two molecules of $A_{\frac{1}{2}}X$, etc., in the derivation. If in a molecule like AX_2 the two X atoms are not equivalent and if they exchange through two different reactions with rates $\mathbf{R}_1$ and $\mathbf{R}_2$, it may be seen that the resulting semilog plot will be not a straight line but a complex curve. The differential equations for the exchanges involving the several positions may be set up and solved simultaneously, so that the curve may, at least in principle, be resolved to give values for the several **R**'s; however, this becomes very difficult for more than about two rates. A simplification may be made if $a \ll b$, with the several nonequivalent positions in the molecule AX_n; here the value of y is very nearly a constant, and in this limit the complex semilog curve is resolvable in the same way as a radioactive decay curve into straight lines measuring $\mathbf{R}_1$, $\mathbf{R}_2$, etc. No example of a complex homogeneous exchange curve has been reported except the limiting case with $\mathbf{R}_1$ measurable, but $\mathbf{R}_2 = 0$ within experimental accuracy (like the sulfur exchange between $S_2O_3{}^{2-}$ and $SO_3{}^{2-}$). In this limiting case no unusual feature appears if x_∞ is used in an experimental sense, although after a much longer time x_∞ may be expected to reach a higher value.

It must also be noted that in the analysis leading to (7-6) it is assumed that there are no other chemical reactions involving AX and BX in the system. If there are other reactions, (7-6) will no longer be true in general; the problem will then usually entail the solution of a set of coupled differential equations.

Reaction Kinetics and Mechanisms. Radioactive tracers are finding an important place in the investigations of reaction kinetics and mechanisms. We shall discuss several examples to illustrate the kinds of information in this field that can be obtained with tracers but hardly in any other way. Consider the reversible reaction:

$$HAsO_2 + I_3{}^- + 2H_2O \rightleftharpoons H_3AsO_4 + 3I^- + 2H^+.$$

In the familiar theory of dynamic equilibrium, $K = k_f/k_r$, where K is the equilibrium constant and k_f and k_r are the rate constants of the forward and reverse reactions. Ordinarily K may be measured only at equilibrium and k_f or k_r far from equilibrium. Using radioactive arsenic to measure the rate of exchange between arsenious and arsenic acids induced by an iodine catalyst in accordance with the foregoing equilibrium reaction, J. N. Wilson and R. G. Dickinson were able to find the rate law and rate constant at equilibrium. For the reverse direction as written they found $\mathbf{R} = k_r (H_3AsO_4)(H^+)(I^-)$, with $k_r = 0.057$ liter^2 mol^{-2} min^{-1}, which is in satisfactory agreement with the information from ordinary rate studies made far from equilibrium, $\mathbf{R} = 0.071 (H_3AsO_4)(H^+)(I^-)$.

A theory of the Walden inversion calls for inversion at each substitution by the schematic mechanism:

$$I^- \rightarrow \quad \overset{R_1}{\underset{R_2 \ R_3}{C-I}} = \overset{R_1}{\underset{R_3 \ R_2}{I-C}} \quad + I^-.$$

As shown here the substitution is by a like group, and if the initial molecules are optically active the final product will be the racemic mixture. It has been shown that for sec-octyl iodide (or for α-phenyl ethyl bromide) the rate of exchange with radioactive iodide ion (or bromide ion) is identical with the rate of racemization, which is a verification of the mechanism.

A different type of racemization is that of chromioxalate ion, $Cr(C_2O_4)_3{}^{3-}$, which may be optically active through different linkings of the six octahedral bonds of the chromium with the carbon-oxygen groups. This racemization is fairly rapid in aqueous solution and apparently first order; it had been proposed that the mechanism involved an ionization as the rate-determining step:

$$Cr(C_2O_4)_3{}^{3-} = Cr(C_2O_4)_2{}^- + C_2O_4{}^{2-}.$$

Another theory favored an intramolecular rearrangement instead. The racemization has been allowed to proceed in a solution containing radioactive $C_2O_4{}^{2-}$. The activity did not enter the chromium compound; therefore the ionization mechanism is disproved, and the intramolecular rearrangement hypothesis is supported.

Some of the applications of tracers to reaction mechanism studies are essentially qualitative. For example, when HClO labeled with Cl^{38} oxidizes $ClO_2{}^-$, the product Cl^- contains the tracer, and the product $ClO_3{}^-$ is inactive. Also when Cl^- is oxidized by $ClO_3{}^-$ the product Cl_2 is formed from the Cl^-, and the product ClO_2 is formed from the $ClO_3{}^-$. When labeled I^- reduces $IO_4{}^-$ to $IO_3{}^-$ the tracer appears only in the I_2 product. Clearly any reaction intermediates containing the reactants must be unsymmetrical in that the two halogen atoms of initially different oxidation state are distinguished. This information at least rules out some of the conceivable reaction paths.

Another example may be taken from the many studies of organic reaction mechanisms made with radioactive carbon. Today almost all such studies use C^{14}, which is a very convenient long-lived tracer discovered in 1940 by S. Ruben and M. D. Kamen. But, because so little C^{14} was then available, these workers used the 20-min C^{11} to study the oxidation by alkaline permanganate of propionate to the products carbonate and oxalate (1 mole of each from 1 mole of propionate). It might have been a plausible guess that the CO_3^{2-} is formed from the carboxyl group; however, with the carboxyl carbon labeled they found that only about 25 per cent of the CO_3^{2-} was from that part of the molecule. In the oxidation of propionic acid by dichromate in acid they found that all the CO_2 did originate from the —COOH, demonstrating different mechanisms in the two instances. The oxidation of fumaric acid,[2] HOOC*CHCHC*OOH, by acid permanganate has been investigated; the product HCOOH (1 mole per mole of fumaric acid) is formed always from one of the secondary carbon atoms, and the CO_2 (3 moles per mole of fumaric acid) is from the carboxyl carbons and the other secondary carbon.

J. Halperin and H. Taube and other workers have used the stable isotope O^{18} as a tracer of oxygen atom transfers in oxidation reactions. When ClO_3^- is reduced to Cl^- by SO_3^{2-}, approximately 2.3 of the three oxygen atoms in ClO_3^- are found in the product SO_4^{2-}. Corresponding numbers are 1.5 for ClO_2^- and 0.36 for ClO^- as oxidizing agents. The interpretation is that the reduction occurs in steps $ClO_3^- \rightarrow ClO_2^- \rightarrow ClO^- \rightarrow Cl^-$ and that the reaction is an oxygen atom transfer from the halogen ion to the sulfite at least in the first two steps. (The reduction of ClO^- may also be predominantly by atom transfer but with a competing exchange of O^{18} between hypochlorite and water.) These studies show also that when sulfite is oxidized by BrO_3^-, by O_2, by MnO_2, and by MnO_4^- there is a transfer of 2, 2, 1, and 0.2 atoms of oxygen, respectively, per formula unit of the oxidizing agent. The most striking result is that labeled H_2O_2 transfers both oxygen atoms to one sulfite, giving a sulfate ion which contains two labeled oxygens. Incidentally, the O_2 evolved in the decomposition of H_2O_2 solutions under a wide variety of conditions is derived from the peroxide only.

Electron Exchange Reactions. In many oxidation-reduction reactions the net change appears to be a transfer of one or more electrons, as, for example,

---

[2] By common usage these asterisks indicate labeled positions (the two carboxyl carbons in this case) in the molecules; of course, because the active tracers are almost always very highly diluted with ordinary atoms it is very improbable for any given molecule actually to contain two radioactive atoms. Thus the asterisk denotes not a radioactive atom but an atom taken at random from a sample containing some active atoms; in other words, the position marked with the asterisk will in some of the molecules be labeled with a radioactive atom. Notice that on p. 196 we avoided the asterisk and chose a different superscript because there we wished to indicate an actual radioactive atom.

in the oxidation of ferrous ion by ceric ion:

$$Fe^{2+} + Ce^{4+} = Fe^{3+} + Ce^{3+}.$$

Some such reactions, including this one, are quite fast; others are much slower. An early observation was that the reactions generally are not fast unless the number of electrons lost by a mole of the reducing agent is the same as the number gained by a mole of the oxidizing agent. However, many reactions that meet this condition are slow, and the reasons are not known. Of course, two ions of like charge may so repel one another that close collisions are unlikely, but experiments would indicate that this factor is only a small part of the whole explanation.

Isotopic tracers make possible the study of a relatively simple class of electron transfer reactions, the exchange reactions between different oxidation states of the same element. For example, radioactive iron has been used by several investigators to study the rate of oxidation of ferrous ion by ferric ion,

$$Fe^{*2+} + Fe^{3+} = Fe^{*3+} + Fe^{2+}.$$

Such a reaction, of course, follows the quantitative exchange law already derived, and the rate is measured by the rate at which the tracer becomes randomly mixed between the two oxidation states. If the exchange is carried out in $6M$ HCl, the ferric iron is readily separated from the ferrous iron by ether extraction, but the rate is too rapid for measurement. In this system chloride complexes such as $FeCl_3$ are surely present, and the exchange observed may proceed through such species. In perchloric acid the exchange rate is fast, but measurable in dilute solutions at $0°C$. When fluoride ion is added, so that species such as FeF^{2+}, FeF_2^+, and FeF_3 are present, the variation of the rate with F^- concentration can be used to show that the reaction proceeds through all of these forms and that the exchange rate is fastest with FeF^{2+}.

Fast electron transfer exchange has been observed between $Fe(CN)_6^{4-}$ and $Fe(CN)_6^{3-}$, Ce^{3+} and Ce^{4+}, MnO_4^{2-} and MnO_4^-, Hg^{2+} and Hg_2^{2+}, NpO_2^+ and NpO_2^{2+}, Co^{2+} and Co^{3+}, and ClO_2^- and ClO_2, between the tris-(2,2'bipyridyl) complexes of Os(II) and Os(III), and between the tris-(5,6-dimethyl-1,10-phenanthroline) complexes of Fe(II) and Fe(III). In several of these cases the rate could not be followed quantitatively because the exchange is too fast either in the solution or at some stage in the separation process. (If the rate can be measured, a correction can be made for any exchange induced during separation.) Exchange half-times as short as 1 second at $0°C$ have been established in a few instances. The availability of high specific activities has sometimes made it possible to go to sufficiently low concentrations of reactants to bring exchange rates from an inaccessible into a measurable range. Very much slower exchanges have been reported between Eu^{2+} and Eu^{3+}, Sn(II) and Sn(IV) in HCl solutions, Pb(II) and Pb(IV) in glacial CH_3COOH, Sb(III) and Sb(V) in HCl, $Co(NH_3)_6^{2+}$ and $Co(NH_3)_6^{3+}$, and between Co(II) and Co(III) tris-ethylenediamine complexes.

Many of these exchange reactions have been the subject of detailed kinetic studies, including the determination of effects of reactant concentrations, pH, complexing agents, and temperature on the rates. Considerable detail on the mechanisms of electron exchange reactions has emerged from these studies (S5). For example, it was shown that the exchange between chromous and chromic ions in solutions containing chloride ion proceeds via the formation of a bridge and the transfer of a chlorine atom:

$$Cr^{*2+} + CrCl^{2+} = (Cr^*ClCr)^{4+} = Cr^*Cl^{2+} + Cr^{2+}$$

C. EFFECTS OF ISOTOPIC SUBSTITUTION UPON EQUILIBRIA AND RATES OF CHEMICAL REACTIONS

In sections A and B we have assumed that labeled species are physically distinguishable but chemically indistinguishable from their unlabeled counterparts. In this section we shall investigate the divergences from that assumption as they appear in equilibrium constants and rate constants. For example, (7-3) is not strictly true; there is some isotopic fractionation, although it is usually minute when compared to experimental error. Furthermore, the rate R in (7-1) is not exactly the same for all the isotopic species. But, again, the variation is usually too small to be detected experimentally.

Effect of Molecular Symmetry on Equilibrium Constants. The first matter that should be considered is the value of the equilibrium constant when there is *no* isotopic fractionation and (7-3) is obeyed. We may ask, for example, what the numerical value of K is for the following reaction when the two isotopes of hydrogen are randomly distributed:

$$H_2 + D_2 = 2HD.$$

It is tempting to conclude that a random distribution of isotopes implies that $K = 1$; but this would be wrong as can rather easily be seen: consider a sample containing N hydrogen atoms of which a fraction f_H is protium and a fraction f_D is deuterium ($f_H + f_D = 1$). A random distribution of isotopes means that any particular atom in the diatomic hydrogen molecule has a probability f_H of being a protium and f_D of being a deuterium, regardless of the nature of the other atom in the molecule. This means that the numbers of H_2 molecules and D_2 molecules are $f_H^2(N/2)$ and $f_D^2(N/2)$, respectively. The number of HD + DH molecules is $2f_Df_H(N/2)$; the factor of 2 arises because of the two ways to have the same molecule: HD and DH. It is to be noted that the total number of molecules is given by

$$\frac{N}{2}(f_H^2 + 2f_Hf_D + f_D^2) = \frac{N}{2}(f_H + f_D)^2 = \frac{N}{2}.$$

If the system is contained in a volume V, the equilibrium expression is

$$K = \frac{\left[\dfrac{2f_D f_H (N/2)}{V}\right]^2}{\left[\dfrac{f_H^2 (N/2)}{V}\right]\left[\dfrac{f_D^2 (N/2)}{V}\right]} = 4. \tag{7-7}$$

Thus because there are two ways of having the HD molecule the equilibrium constant is 4 instead of 1. The situation would be different if the two ways of having the molecule were distinguishable. For example, for the reaction

$$HCOOH + DCOOD = HCOOD + DCOOH$$

the equilibrium constant is

$$K = \frac{[HCOOD][DCOOH]}{[HCOOH][DCOOD]} = \frac{\left[\dfrac{f_H f_D (N/2)}{V}\right]\left[\dfrac{f_D f_H (N/2)}{V}\right]}{\left[\dfrac{f_H^2 (N/2)}{V}\right]\left[\dfrac{f_D^2 (N/2)}{V}\right]} = 1.$$

Here the constant is 1 because HCOOD and DCOOH are distinguishable chemical species. It should be noted that if the analysis of the mixture were done in a mass spectrometer in which HCOOD and DCOOH are not necessarily distinguished from each other, the reaction would be written

$$H_2CO_2 + D_2CO_2 \rightarrow 2HDCO_2$$

and the observed equilibrium constant would be 4.

This apparent dependence of the equilibrium constant on the state of innocence of the observer points up that the effect under consideration is contained in the entropy change in the reaction. Indeed, for exchange reactions in which there is no isotopic fractionation (there is no energy change in the reaction) it is just the entropy increase attendant on randomizing the distribution of isotopes which provides the driving force for the reaction. The equilibrium constant for the hydrogen reaction may be obtained from entropy considerations by realizing that the entropy of a particular isotopic molecular species at a temperature high enough so that the spacing between rotational energy levels is small compared to thermal energy may be written as

$$S^0 = S'^0 - R \ln \sigma, \tag{7-8}$$

where S'^0 is the entropy when no attention is paid to isotopic composition, and σ, the symmetry number, may be defined for our purposes as the number of *indistinguishable* ways that a molecule may be oriented in space under the condition that the various isotopes of an atom are considered to be *distinguishable*.[3] For example, $\sigma_{H_2} = \sigma_{D_2} = 2$ and $\sigma_{HD} = 1$; the entropy change for the

[3] This represents a restricted and special use of the concept of symmetry number. For a more general discussion of the whole question, cf. reference D1.

hydrogen exchange reaction, then, is

$$\Delta S^0 = 2R \ln 2$$

and from the usual thermodynamic relationships, $K = 4$. The apparent contradiction with the formic acid exchange hinges on the knowledge of whether the two hydrogens are equivalent; they, of course, are not.

As another example, consider the exchange reaction

$$PCl_3^{35} + PCl_3^{37} \rightarrow PCl_2^{35}Cl^{37} + PCl^{35}Cl_2^{37}.$$

The pyramidal PCl_3 molecule has $\sigma = 3$ when all three isotopes are the same and $\sigma = 1$ when only two are the same. From (7-8) the entropy change is

$$\Delta S^0 = 2R \ln 3$$

and the equilibrium constant is 9. It is worthy of note that the value of 9 can also be obtained from a probability argument of the type illustrated in (7-7). To use this method, though, it must be realized that the number of distinguishable ways of picking N objects when m are of one type and $N - m$ of another is

$$\frac{N!}{(N - m)!m!}.^{4}$$

Isotope Effect on Equilibrium Constants. The preceding discussion, based solely on entropy effects, neglected any energy effects that may accompany isotopic substitution and so represents the high-temperature limit in which equilibria are largely governed by entropy changes. The data in Table 7-1 for the hydrogen exchange reaction show the approach to 4 at high temperatures and also illustrate the significant divergence from random isotopic distribution at lower temperatures, which occurs because the energy change in the reaction is different from zero.

Table 7-1   Variation with temperature
of equilibrium constant for reaction
$H_2 + D_2 \rightleftharpoons 2HD$

T (°K)	K
100	2.3
300	3.3
500	3.6

The source of the energy change in exchange reactions does not lie, as it does in ordinary chemical reactions, in the change in the potential-energy field in which the atoms of the molecule find themselves. The potential-energy curve

[4] Refer to discussion following (6-8).

that defines the motion of two protium atoms in a hydrogen molecule, for example, is not significantly different from that for the two deuterium atoms in the isotopic molecule. What does change are the energies of the molecular translational, rotational, and vibrational quantum states. These changes arise directly from the mass differences of the isotopic molecules. The investigation of the effect of these changes on equilibrium constants (B1) has shown that the main effect stems from changes in the zero-point vibrational energies and from the spacing of vibrational states. It will be recalled that the vibrational energy states of the diatomic molecule AB are given by

$$E_{\text{vib}} = h\nu_{AB}(n + \tfrac{1}{2}), \qquad n = 0, 1, 2, \ldots,$$

where ν_{AB} is the fundamental vibration frequency of the AB molecule. This fundamental frequency depends upon the masses m_A and m_B of the atoms through the relation

$$\nu_{AB} = \frac{1}{2\pi}\left(\frac{f}{\mu_{AB}}\right)^{1/2}$$

where f is the force constant for the A—B chemical bond and

$$\mu_{AB} = \frac{m_A m_B}{m_A + m_B}$$

is the reduced mass of the system. The force constant f undergoes no significant change with isotopic substitution but obviously μ_{AB}, and therefore ν_{AB}, do. The consequences of the change in the fundamental frequencies are most easily seen for the dissociation constants for two isotopic molecules AB and AB'. The ratio of the two dissociation constants is (B1)

$$\frac{K_{AB'}}{K_{AB}} = \frac{(1 - e^{-U'})Ue^{-U/2}}{(1 - e^{-U})U'e^{-U'/2}},$$

where $U \equiv h\nu_{AB}/kT$ and $U' = h\nu'_{AB}/kT$. It will be noted that at very high temperatures (where U and U' approach 0) the ratio approaches unity and the isotope effect vanishes as is expected. At very low temperatures (where U and U' are very large) the ratio approaches $(U/U') \exp\left[-(U - U')/2\right]$ and is governed by the difference in zero point energies.[5] If $m_B < m_{B'}$, $\mu_{AB} < \mu_{AB'}$ and $\nu_{AB} > \nu_{AB'}$ and the AB molecule is less stable with respect to dissociation than is the AB' molecule $[(K_{AB'}/K_{AB}) < 1]$. This implies that in an exchange reaction such as

$$AB + CB' \to AB' + CB$$

---

[5] This approximation is valid only if the spacing of rotational energy states is small compared to thermal energies. This condition is not fulfilled for the various isotopic hydrogen molecules because of their small moments of inertia.

the light isotope will tend to concentrate in the compound with the smaller bond energy (smaller value of f). The isotope effect will be largest for the dissociation reaction in which there is no binding in the final state, and so the full difference in the vibrational energies of the isotopic molecules will appear.

The fact that equilibrium constants for exchange reactions differ from unity may be utilized for the separation of isotopes (J1). As an example, the exchange reaction between NO and HNO_3 (S1)

$$N^{15}O + HN^{14}O_3(aq) \rightleftharpoons N^{14}O + HN^{15}O_3(aq), \ K = 1.05 \text{ at } 25°C$$

has been used in a counter current apparatus (NO gas bubbles up and a nitric acid solution flows down) to enrich N^{15} to an abundance of more than 90 per cent from the normal 0.37 per cent.

The magnitude of the isotope effect, as seen from the preceding paragraphs, depends on the difference in the reduced masses of the two isotopic molecules and thus on the fractional difference in the masses of the two isotopes. As a consequence, the larger the atomic weight, the smaller the isotope effect.

Isotope Effect on Rate Constants. Although we often ignore quantitative differences in the rates of reactions of molecules containing different isotopes, these differences usually are measurable, especially for isotopes of the light elements. For example, it has been shown that the rate of the electron exchange reaction between Fe^{2+} and Fe^{3+} ions was diminished by a factor of 2 when the solvent was changed from H_2O to D_2O. This large effect demonstrated the important role played by the solvent molecules in the mechanism of the exchange reaction when they enter into the transition-state complex (S4).

These effects, when of significant magnitude, can be inconvenient in tracer studies, since they invalidate the straightforward interpretation outlined in section B. On the other hand, the magnitude of the isotope effect on reaction rates should depend on the details of the mechanism and would thereby afford an opportunity for new information if an adequate theory for these effects were available.

The most useful attempt to construct a theory for reaction rates of isotopic molecules is based on the transition-state approach to the problem (B2). These calculations can become very complicated (B3, M1) but, since the desired quantity is usually the ratio of rate constants for isotopic molecules rather than the actual rate constants, they probably represent the most successful application of transition-state theory.

As the theory is one in which equilibrium is assumed to exist between the reactants and the transition-state complex, the essence of the problem is similar to that encountered in the calculation of the ratios of equilibrium constants for reactions of isotopic molecules. Since one of the vibrations of the transition-state complex results in dissociation into the products of the reaction, the corresponding frequency is imaginary, and it turns out that the ratio of the rate constants depends on the ratio of these frequencies as well as the ratio of equilibrium constants for the various isotopic species.   The lack of information about the quantum states of the transition-state complex presents a difficulty which can often be overcome by reasonable approximations because again only the *ratio* of partition functions for isotopically substituted transition states is required.   At worst, the maximum effect may be estimated by assuming no isotopic bonding in the transition state.   These theoretical maximum effects occur when the isotopic substitution is at the bond being broken (primary isotope effect) and can be as large as 60 when the rate of H^1 is compared to that of H^3 ranging through 1.5 for C^{12} versus C^{14} to 1.02 for I^{127} versus I^{131}.

Intramolecular isotopic reactions, with both isotopes present initially in the same molecule, are a special case most easily treated theoretically.   For example, the decarboxylation of malonic acid containing a —$C^{13}OOH$ group may occur in either of two ways:

$$
\begin{array}{c}
C^{13}OOH \qquad {}_a\nearrow C^{12}O_2 + CH_3C^{13}OOH \\
\diagdown \qquad\qquad \diagup \\
CH_2 \\
\diagdown \qquad\qquad {}_b\diagdown \\
C^{12}OOH \qquad C^{13}O_2 + CH_3C^{12}OOH
\end{array}
$$

Theory predicts that the difference between the two rate constants a and b is largely caused by the different imaginary frequencies of the transition state rather than by differences in activation energies.   A calculation of just the frequency effect gives for the ratio a/b a value of 1.020.   The temperature dependence due to a difference in the activation energies is difficult to calculate, but should be small.   The best experimental results give a value of 1.029 for the ratio, independent of temperature within the experimental error. Another feature of the theoretical results, also independent of specific properties of the transition state, is that for small isotope effects expressed in per cent the magnitude of the effect is proportional to the differences in isotopic masses. For example, any C^{14} effect should be very closely twice the corresponding C^{13} effect, both compared to ordinary C^{12}.   Although some earlier experiments gave ratios of C^{14} to C^{13} effects in excess of two, the best experimental data on decarboxylation reactions now appear to agree substantially with the theoretical prediction.   It should be noted that the difference in activation energy is the primary source of the isotope effect when rates of compounds containing protium, deuterium, and tritium are compared.

D. ANALYTICAL APPLICATIONS

Test of Separations.   Radioactive tracers can be very conveniently used to follow the progress and test the completeness of chemical separation procedures.   If one component of a mixture is radioactive, it can often be followed satisfactorily through successive operations if beakers containing filtrates, funnels with precipitates, and so on, are merely held near a counter or ionization chamber. We have seen good chemical isolations made by these methods in the almost complete absence of knowledge of specific chemical properties. The crude qualitative procedure may be refined as far as desired, and valuable tests of analytical separation methods have been made with tracers. Further, it is possible to follow simultaneously the behavior of several radioactive tracers with characteristic γ-ray spectra by the use of a scintillation counter in conjunction with a multichannel analyzer.

Analysis by Isotope Dilution. It may frequently occur that quantitative analysis for a component of a mixture is wanted where no quantitative isolation procedure is known. Particularly in the case of complex organic mixtures, it may be possible to isolate the desired compound with satisfactory purity but only in low and uncertain yield. In such a case the analysis may be made by the technique of isotope dilution. To the unknown mixture is added a known weight of the compound to be determined containing a known amount (activity) of radioactively tagged molecules. Then the specific activity of that pure compound isolated from the mixture is determined and compared with that of the added material; the extent of dilution of the tracer shows the amount of inactive compound present in the original unknown. (One may think of the tracer as serving to measure the chemical yield of the isolation procedure. Obviously exchange reactions which would reduce the specific activity of the compound must be absent.) This powerful method may also be used with stable isotopes if mass-spectrometric analysis is employed. Accuracy of a few per cent may be obtained for some elements that are present in concentrations as low as parts per billion to parts per trillion (I1, H1).

Analysis by Activation. Throughout most of the tracer work discussed radioactive isotopes are assayed by measurement of their activities. This is actually an analytical procedure, but we have not emphasized that aspect because the samples are subject to analysis only if the tracer was provided earlier in the experiment. Of course, the naturally radioactive elements, including uranium, thorium, radium, potassium, and rubidium, may be assayed by radioactive measurement; a practical although not very sensitive procedure for potassium assay by means of its radioactivity has been reported.

A somewhat different technique can be useful, in which an unknown sample is subjected to activation by neutrons for appropriately chosen lengths of time and chemical elements are identified and assayed by measurement of characteristic radionuclides formed. In general, the irradiation must be followed by chemical isolation of the desired radionuclides, carried out in the usual manner after the addition of appropriate carriers. (For a discussion of carriers in radiochemistry see chapter 12, section D.) Standardization is provided by irradiation, along with the unknown sample, of a standard sample containing a known amount of the element to be analyzed. It is sometimes desirable to use a standard of similar composition to the unknown, or else to use small samples, to avoid errors due to strong neutron absorption by other constituents.   In the event that the chemical isolation of the desired radionuclides is not required, activation analysis can be performed without physical alteration of the sample.   This possibility has proved to be of great use in the analysis of valuable archeological objects (S3).

The specificity of activation analysis is usually excellent since the purity of the radionuclide measured may be checked by half-life and energy determinations.   Sensitivity depends on the flux of bombarding particles, nv, the cross section for the reaction involved, σ, the decay constant of the nuclide measured, λ, the duration of the irradiation, t, and the efficiency of the detector, ϵ. The counting rate at the end of the irradiation of a sample that contains m grams of the isotope of atomic weight M to be assayed, if the total sample is thin enough so that attenuation of the neutron flux in the sample may be ignored, is given by

$$\epsilon \left(nv \, \frac{m}{M} \times 6.023 \times 10^{23} \times \sigma \right) (1 - e^{-\lambda t}).$$

With a neutron flux of 10^{12} cm^{-2} sec^{-1}, sensitivities for most elements are in the range of 10^{-16} to 10^{-11} g.   Thus analysis by neutron activation is practical for impurities present in parts per million or even parts per billion. A guide to the voluminous literature on activation analysis may be found in reference S3.

Although slow-neutron irradiation is by far the most widely used technique in activation analysis, some applications of charged-particle activation have also been reported.   By deuteron irradiation small impurities of gallium in iron, of copper in nickel, and of iron in cobalt oxide have been found.   A simple quantitative method for the determination of carbon in steels by means of the reaction $C^{12}(d, n)N^{13}$ has been described.

Either neutron or charged-particle irradiation can also be used for the determination of the isotopic composition of a sample.   For example, the amount of Fe^{56} present in enriched Fe^{54} was estimated through the (p, n) reaction which produced radioactive Co^{56}.

E. HOT-ATOM CHEMISTRY

Szilard-Chalmers Process.   In 1934 L. Szilard and T. A. Chalmers showed that after the neutron irradiation of ethyl iodide most of the iodine activity formed could be extracted from the ethyl iodide with water; they used a small amount of iodine carrier, reduced it to I$^-$, and finally precipitated AgI. Evidently the iodine-carbon bond was broken when an I^{127} nucleus was transformed by neutron capture to I^{128}.   This type of process has since been used to concentrate the products of a number of n, γ reactions and of some $\gamma, n, \ n, 2n,$ and d, p reactions. It is referred to as the Szilard-Chalmers process. Three conditions have to be fulfilled to make a Szilard-Chalmers separation possible. The radioactive atom in the process of its formation must be broken loose from its molecule, it must neither recombine with the molecular fragment from which it separated nor rapidly interchange with inactive atoms in other target molecules, and a chemical method for the separation of the target compound from the radioactive material in its new chemical form must be available.

Most chemical bond energies are in the range of 1 to 5 eV (20,000 to 100,000 cal per mole). In any nuclear reaction involving nucleons or heavier particles entering or leaving the nucleus with energies in excess of 10 keV the kinetic energy imparted to the residual nucleus far exceeds the magnitude of bond energies.[6] In thermal-neutron capture, in which the Szilard-Chalmers method has its most important applications, the incident neutron does not impart nearly enough energy to the nucleus to cause any bond rupture. But neutron capture is almost always followed by γ-ray emission, and the nucleus receives some recoil energy in this process. A γ ray of energy E_γ has a momentum $p_\gamma = E_\gamma/c$. To conserve momentum, the recoiling atom must have an identical momentum, and therefore the recoil energy $R = p_\gamma{}^2/2M = E_\gamma{}^2/2Mc^2$, where M is the mass of the atom. For M in atomic mass units and E_γ in millions of electron volts we have

$$R = \frac{537E_\gamma{}^2}{M} \text{ eV.} \tag{7-9}$$

Table 7-2 shows values of R for a few values of E_γ and M. Neutron capture

Table 7-2   Recoil energies in electron volts imparted to nuclei by gamma rays of various energies

M	$E_\gamma = 2$ MeV	$E_\gamma = 4$ MeV	$E_\gamma = 6$ MeV
20	107	430	967
50	43	172	387
100	21	86	193
150	14	57	129
200	11	43	97

usually excites a nucleus to about 6 or 8 MeV, and a large fraction of this excitation energy is dissipated by the emission of one or more γ rays. Unless all the successive γ rays emitted in a given capture process have low energies

[6] For reactions other than n, γ, particularly for d, p reactions, the Szilard-Chalmers technique is less useful because the energy dissipated by the incident radiation in the target is so great that many inactive molecules are also disrupted.

(say below 1 or 2 MeV), which is a relatively rare occurrence, the recoiling nucleus receives more than sufficient energy for the rupture of one or more bonds. Of course, it is not the entire recoil energy but something more like its component in the direction of a bond that should be compared with the bond energy; furthermore, the momenta of several γ rays emitted in cascade and in different directions may partially cancel one another. There is no evidence that two capture γ rays in cascade are preferentially emitted in opposite directions, and momentum cancelation is therefore hardly expected to reduce the probability of bond rupture by a very large factor. In most n, γ processes the probability of rupture is certainly very high.

The second condition for the operation of the Szilard-Chalmers method requires at least that *thermal* exchange be slow between the radioactive atoms in their new chemical state and the inactive atoms in the target compound. The energetic recoil atoms may undergo exchange more readily than atoms of ordinary thermal energies. These exchange reactions and other reactions of the high-energy recoil atoms, called "hot atoms," determine to a large extent the separation efficiencies obtainable in Szilard-Chalmers processes. Hot-atom reactions are considered further after a discussion of some examples of Szilard-Chalmers separations.

A large amount of work in the field of Szilard-Chalmers separations has been done on halogen compounds. Many different organic halides (including CCl_4, $C_2H_4Cl_2$, C_2H_5Br, $C_2H_2Br_2$, C_6H_5Br, CH_3I) have been irradiated, and the products of neutron capture reactions (Cl^{38}, Br^{80}, Br^{82}, I^{128}) removed by various techniques. Extraction with water, either with or without added halogen or halide carrier, results in rather efficient separations. Yields are often improved, especially in the case of iodine, by extraction with an aqueous solution of a reducing agent such as HSO_3^-. In an interesting study, the yields of extractable iodine in d,p, $n,2n$, γ,n, and n,γ activations of ethyl iodide were shown to be the same when other conditions were kept the same. Similar results were found with methyl iodide. This demonstrates that the chemical effects which determine the eventual fate of the active atom are rather independent of the initial recoil energy.

Szilard-Chalmers separations of halogens with 70 to 100 per cent yields have been obtained in neutron irradiations of solid or dissolved chlorates, bromates, iodates, perchlorates, and periodates; from these the active halogen can be removed as silver halide after addition of halide ion carrier. Szilard-Chalmers separations based on differences in oxidation state before and after the neutron capture have been successful for a number of other elements. For example, about half the P^{32} activity formed in neutron irradiation of phosphates (solid or in solution) is found in $+3$ phosphorus.

Collection of charged fragments on electrodes has been used successfully for a number of Szilard-Chalmers separations. Arsenic activity has been separated by this method from arsine gas with yields as high as 34 per cent. Deposition occurs on both positive and negative electrodes in this case.

The bombardment of metal-organic compounds and complex salts is often useful for Szilard-Chalmers separations if the free metal ion does not exchange with the compound and if the two are separable. Some of the compounds used successfully are cacodylic acid, $(CH_3)_2AsOOH$, from which As^{76} can be separated as silver arsenite in 95 per cent yield; copper salicylaldehyde o-phenylene diamine, from which as much as 97 per cent of the Cu^{64} activity can be removed as Cu^{2+} ion; and uranyl benzoylacetonate, $UO_2(C_6H_5COCH-COCH_3)_2$, from which U^{239} activity has been extracted in about 10 per cent yield.

Many more examples of these various methods for isotopic enrichment by the Szilard-Chalmers process can be found in review articles (H2, M2, W1, W2).

Chemistry of Recoil Atoms. Aside from its importance as a method for isotopic enrichment, the Szilard-Chalmers process provides an opportunity for the study of the chemical reactions of the energetic recoil atoms. These atoms, as already mentioned, are often referred to as "hot atoms" and the field of study as "hot-atom chemistry." That there are such reactions is immediately seen from the observation made in the preceding discussion to the effect that only a fraction of the radioactive atoms is in a chemical form different from that of the parent compound. The fraction of the active atoms that is in the same chemical form as the parent compound is often called the retention. The observation, for example, that the retention of pure CCl_4 is about 43 per cent and diminishes to about 5 per cent on the addition of 50 mole per cent C_6H_{12} indicates that neither lack of bond rupture nor recombination with the fragment from which the "hot" atom had broken away is of much importance.

It has been shown that in reacting with a molecule a "hot" atom may replace another atom or group. For example, after the slow-neutron irradiation of CH_3I, 11 per cent of the I^{128} activity was found in the form of CH_2I_2; furthermore, this result was shown to be temperature-independent between -195 and $15°C$, which proves that the substitution is not an ordinary thermal reaction. The formation of labeled CH_2Br_2 in the irradiation of $CH_2BrCOOH$, of labeled CH_3I and C_2H_5I in the irradiation of iodine dissolved in ethyl alcohol, and of labeled C_6H_5Br in the irradiation of aniline hydrobromide shows that the excited halogen atoms can replace such groups as $-COOH$, $-OH$, $-CH_2OH$, $-NH_2$, and probably many others. The yield of active atoms in one of these substitution products is usually less than about 10 per cent. Reactions of this type may be used to synthesize labeled compounds of high specific activity. The formation of C^{14}-labeled compounds in high specific activity by slow-neutron irradiation of nitrogenous materials has been demonstrated. For example, the formation of labeled anthracene in the neutron irradiation of acridine has been reported. The interpretation is that a small fraction of the "hot" C^{14} atoms formed by $N^{14}(n, p)$ reactions end up, after losing most of their kinetic energy in other collisions, by replacing nitrogen atoms in acridine to form anthracene. Many such replacement reactions induced by hot C^{14} atoms and also by hot C^{11} [e.g. from the reactions $C^{12}(n, 2n)$, $C^{12}(p, pn)$, $C^{12}(\gamma, n)$] have been studied and found useful in the interpretation of organic reaction mechanisms (W3).

It has also proved possible to label organic compounds with tritium by means of hot-atom reactions resulting from the $He^3(n, p)H^3$ and the $Li^6(n, \alpha)H^3$ reactions. The procedure with the He^3 simply involves irradiating a gaseous mixture of He^3 and the organic compound with neutrons from a reactor; whereas for the Li^6 reaction an intimate mixture of Li_2CO_3 and the organic compound is irradiated. More details are given in references E1 and W3.

Theoretical attempts to understand "hot-atom" chemistry in liquids and vapors have divided the events into two classes: those that occur before and those that occur after the "hot atom" has been reduced to thermal equilibrium. Analysis of the "hot" processes requires an expression for the energy spectrum of the recoil species while it is being cooled by collisions and an expression for the probabilities of the various possible reactions in each collision as a function of the energy of the recoil atom. The energy spectrum may be obtained easily under the assumption that the collisions are elastic atom-atom collisions (M3), an assumption that is probably not justified in the energy region just above thermal, where, unfortunately, most of the "hot" reactions are expected to occur. The energy dependence of the probabilities for reactions in each collision, except for a simple model proposed by Libby (L1), has yet to be treated theoretically and is left to be experimentally determined (E1, M4).

The reactions of the thermalized recoil atom are the usual ones expected at thermal energies, but there is, in addition, the possibility of recombination with the fragments created by the recoil atom while it was being slowed down. This process should be particularly important in liquids and in solids (H2, W2).

Considerable effort has been expended on the study of hot-atom reactions in solid inorganic compounds. The chemical fate of the radioactive atom has so far been determined only by separation of the various chemical species after dissolution of the irradiated solid in an appropriate solvent, usually water. Retentions determined in this fashion can be as high as 90 per cent ($K_2Cr_2O_7$) and as low as 19 per cent ($NaClO_3$).

Much of the recent interest in these studies has been focused upon the post-recoil annealing effects in which the increase in retention is investigated as a function of the time and temperature at which the irradiated crystal is stored before being dissolved for analysis. These data are of interest in connection with problems in solid-state chemistry and radiation damage (H2).

Chemical Effects of Radioactive Decay. "Hot" atoms may result not only from nuclear reactions but also from radioactive decay processes. The chemistry of hot atoms formed as a result of β-decay processes has been studied in a number of cases. For example, reactions such as

$$TeO_3{}^{2-} \rightarrow IO_3{}^- + \beta^- \quad \text{and} \quad MnO_4{}^- \rightarrow CrO_4{}^{2-} + \beta^+$$

can occur in addition to molecular disruption leading to other forms.   Of course, for these studies the nucleus resulting from the β decay must itself be radioactive if its fate is to be investigated.

Hot-atom reactions are so complex in solutions that it is difficult to obtain quantitative information on the primary bond ruptures accompanying decay processes.   This difficulty was avoided by R. Wolfgang, R. C. Anderson, and R. W. Dodson in a gas-phase study of the chemical effects of C^{14} decay in ethane.   Ethane was synthesized from C^{14} of high specific activity, so that an appreciable fraction of the molecules were doubly labeled.   Decay of one C^{14} in a doubly labeled ethane molecule should lead to C^{14}-labeled methylamine if the C-C bond remains intact or, rather, becomes a C-N bond.   The experiment showed that indeed about 50 per cent of the bonds remained unbroken, and other possible explanations for the labeled methylamine were carefully eliminated.

The chemistry of recoil atoms following isomeric transitions has been studied more than the hot-atom chemistry of other radioactive decay processes.   It is perhaps not immediately clear why isomeric transitions may lead to bond rupture.   The γ-ray energies in isomeric transitions are much lower than in neutron-capture processes, often below 100 keV and rarely above 500 keV. According to (7-9), a 100-keV γ ray would give a nucleus of mass 100 a recoil energy of only about 0.05 eV, which is not nearly enough to break a chemical bond.   Although internal-conversion electron emission gives rise to roughly 10 times greater recoil energy than γ emission at the same energy,[7] even this is not sufficient for bond rupture in most cases.   However, the vacancy left in an inner electron shell by the internal conversion leads to electronic rearrangements and emission of Auger electrons.   The atom is therefore in a highly excited state (and positively charged), and molecular dissociation may take place if the atom is bound in a molecule.[8]

Separations of nuclear isomers analogous to Szilard-Chalmers separations have been performed in a number of cases in which the isomeric transition proceeds largely by conversion-electron emission.   The 18-minute Br^{80} has been separated from its parent, the 4.5-hour Br^{80m}, by a number of different methods analogous to the Szilard-Chalmers methods used for bromine.   The lower states of Te^{121} (17 days), Te^{127} (9.3 hours), Te^{129} (72 minutes), and Te^{131} (25 minutes) have been separated as tellurite in good yield from tellurate solutions containing the corresponding upper isomeric states.   Isomer separations are sometimes useful for the assignment of isomer activities and for

---

[7] In nonrelativistic approximation an atom of mass M receives a recoil energy $R = E_e(m/M)$ from a conversion electron of energy E_e.

[8] An experimental measurement of the charges of Xe^{131} atoms following isomeric transition of Xe^{131m} shows that on the average 8.5 electrons are lost.   Similar measurements give $+3.4$ for the average charge of Cl^{37} atoms following electron capture in Ar^{37}.   In both isomeric transitions and electron captures the high charges are believed to result largely from Auger processes (P1).   In β^- decay atomic charges in the neighborhood of $+1$ have been reported.

the elucidation of genetic relationships.   That the possibility of obtaining isomer separations depends on internal conversion was shown in experiments using the gaseous compounds $Te(C_2H_5)_2$ containing 105-day Te^{127m} and 33-day Te^{129m}, and $Zn(C_2H_5)_2$ containing the 13.8-hr isomer Zn^{69m}.   The lower isomeric states of the tellurium isomers could be separated on the walls of the vessels or on charged plates, but under identical conditions no separation of the zinc isomers was obtained; the isomeric transition in Zn^{69} proceeds by an unconverted 435-keV γ ray, whereas the tellurium transitions have energies of only about 100 keV but are almost completely converted.

F. ARTIFICIALLY PRODUCED ELEMENTS

More than half a century ago the methods of chemistry conventional at that time had already reached a limit in the search for new and missing elements; discoveries since that time have depended on the introduction of new physical methods.   Through studies of optical spectra the elements rubidium, cesium, indium, helium, and gallium were found.   The first evidence for hafnium and rhenium came from X-ray spectra.   Early investigations of the natural radioelements revealed the existence (often in extremely small amounts) of polonium (84), radon (86), radium (88), actinium (89), and protactinium (91), and more recently the missing element number 87 has been found in the same way. Through studies of nuclear reactions and artificially induced radioactivities, the elements 43, 61, and 85 have been identified, and elements 93 to 103 have been added to the periodic chart.

Technetium and Astatine.   The first missing elements to be synthesized by nuclear reactions were technetium (P2) and astatine (C1).   It appears quite unlikely that technetium exists on the earth in either stable or very long-lived form, since all mass numbers between 94 and 102 are occupied by stable isotopes of one or both of the neighboring elements, molybdenum and ruthenium, and all the radioactive isotopes are apparently accounted for.   The instability of astatine is expected, since all nuclides above Bi^{209} are either alpha or beta unstable.

The chemical properties of the two elements are, of course, those that are expected from their positions in the periodic table: Tc lying between Mn and Re and At being the heaviest member of the halogens.   A summary of the chemical properties of these two elements has been given by Anders (A1).

Promethium.   The fission of uranium produces several radioactive isotopes of element 61, which have been investigated and definitely characterized (M5) by concentration of tracer activities with the ion-exchange resin adsorption and elution technique.   The name promethium suggested by the discoverers has been officially adopted for element 61.   Many milligrams of Pm^{147}, a

β^- emitter of 2.6 years half-life, have been isolated, and visible amounts of pink promethium salts have been exhibited. An even longer-lived isotope, 18-year Pm^{145}, has been reported.

Transuranium Elements. When Fermi and his group in Rome first exposed uranium to slow neutrons, they observed a number of activities, and in the following few years many more active species were found to be produced, most of which were at that time assigned to transuranium elements. The assignments were made because the substances were transformed by successive β^- emissions which led to higher Z values and because they could be shown by chemical tests to be different from all the known elements in the neighborhood of uranium in the periodic chart. This situation was resolved in the discovery by O. Hahn and F. Strassman that these activities could be identified with known elements much lighter than uranium and that therefore the neutrons produce fission of the uranium nuclei. Further investigation of the fission process and products led to the proof by E. M. McMillan and P. Abelson (M6) that one of the activities, the one with 2.3 days half-life, could not be a product of fission and was actually the daughter of the 23-minute β-particle-emitting U^{239} which resulted from $U^{238}(n, \gamma)U^{239}$. Also, they devised a procedure for separating chemically the element 93 tracer from all known elements through an oxidation-reduction cycle, with bromate as the oxidizing agent in acid solution and with a rare-earth fluoride precipitate as carrier for the reduced state. They gave the name neptunium, symbol Np, to the new element, taking the name from Neptune, the planet next beyond Uranus in the solar system.

Both Np^{239} discovered by McMillan and Abelson, and Np^{238} from $\alpha, p3n$ or $d, 2n$ reactions on U^{238} emit β particles and lead to known isotopes of element 94, very naturally named plutonium after Pluto (a planet beyond Neptune), with symbol Pu. These isotopes, Pu^{238} and Pu^{239}, are moderately long-lived α emitters first studied by McMillan, Seaborg, Segrè, Wahl, and Kennedy; Pu^{239} is distinguished for its practical usefulness in slow- and fast-neutron fission.

Transuranium elements with atomic numbers up to 103 have been synthesized since that time, largely by Seaborg and co-workers (S7), through nuclear reactions of various types with lighter transuranium elements. The chemical properties of each newly discovered element were first investigated on a tracer level; elements up to 98 have since been produced in weighable quantities. The first synthesis (G1) of mendelevium ($_{101}Mv$) provides an example of the remarkable tracer-level chemical manipulations which were developed for the preparation and investigation of the transuranium elements:

Element 101 was first prepared by the α-particle bombardment of a target which contained approximately 10^9 atoms of $_{99}Es^{253}$ (half-life 20 days) covering an area of about 0.05 cm^2 on a gold foil. Those atoms of $_{99}Es^{253}$ which reacted with α particles were ejected from the target and were caught on another gold foil adjacent to the target; the atoms which had been transmuted were thereby removed from the bulk of the target. The gold catcher foil was dissolved and the gold was removed from the transmutation products by adsorption on an anion-exchange column from 6M HCl. The transuranium elements in the solution were then separated from one another by elution with α-hydroxy-isobutyrate through a cation-exchange column. The fraction eluted just before the one identified as containing $_{100}Fm^{256}$ should contain any element 101 that was produced; this fraction was found to contain a spontaneous-fission activity which was ascribed to element 101 or to one of its decay products. The production of element 101 was demonstrated by the observation of a total of 17 spontaneous-fission events in several separate experiments. Thus the various steps in the chemical separation were performed on less than 100 atoms of element 101.

Actinide Series. The transuranium elements (at least through californium) and uranium and thorium all have similar precipitation properties when in the same oxidation state; they differ principally in the ease of formation and in the existence of the various oxidation states. Seaborg (S2) has advanced the hypothesis, and with considerable evidence, that a new rare-earth series begins with actinium (number 89), with the 5f electron orbitals being filled in subsequent elements. This would be analogous to the lanthanide rare-earth series beginning with lanthanum (57), with the 4f orbitals filling in the next 14 elements. Some of the evidence for this actinide series may be seen in these facts: (1) actinium is chemically similar to lanthanum; (2) thorium is similar to cerium in the +4 state; (3) the ease of removal of more than three electrons decreases from uranium to curium. There is additional evidence for the second rare-earth series from spectroscopic and crystal-structure data, from magnetic susceptibilities, and from ion-exchange elution sequences.

It does seem evident that this new series differs from the familiar rare-earth series in that the resemblance of successive elements is less than for the lanthanide series. The lanthanide earths are for the most part separable only by multiple fractionation processes, or better by adsorption and elution from ion-exchange resins. The elements from 89 to 95 are separable by oxidation-reduction processes, but the separation of 95 to 103 is best done with an ion-exchange column as indicated in chapter 12, section D. On the actinide hypothesis, curium, by analogy to gadolinium, is expected to resist oxidation or reduction in the +3 state, because the $5f^7$ and $4f^7$ structures, with one electron in each of the seven f orbitals, are particularly stable. Actually no state of curium other than +3 has been observed in solution. Americium, by analogy to europium, should be reducible to a +2 state. Berkelium, with the configuration $5f^8$, might be oxidized by powerful oxidizing agents from the ordinary Bk^{3+} to Bk^{4+}; the potential of this couple is about -1.6 V.

Some of the difficulties in work with substances like Cm^{242} may be mentioned here, difficulties in addition to those naturally associated with work on the ultramicrochemical scale. The heavy short-lived α emitters are extremely dangerous as radioactive poisons, and amounts of the order of a few micrograms taken into the body may produce harmful effects.   Also, the high level of α radiation in concentrated samples can be expected to have some effect on chemical reactions; a Cm^{242} preparation glows in the dark.   In fact, the rate of energy release is so great that if cooling effects are neglected it may be estimated that a 0.1 molar Cm^{242} solution would begin to boil in about 15 seconds and reach dryness in about 2 minutes.   The more recently discovered longer-lived isotopes will make it possible to minimize these difficulties.

REFERENCES

A1   E. Anders, "Technetium and Astatine Chemistry," *Ann. Rev. Nuclear Sci.* **9**, 203 (1959).

B1   J. Bigeleisen and M. G. Mayer, "Calculation of Equilibrium Constants for Isotopic Exchange Reactions," *J. Chem. Phys.* **15**, 261 (1947).

B2   J. Bigeleisen, "The Relative Reaction Velocities of Isotopic Molecules," *J. Chem. Phys.* **17**, 675 (1949).

B3   J. Bigeleisen and M. Wolfsberg, "Isotope Effects in Chemical Kinetics," *Advances in Chemical Physics*, Vol. 1 (I. Prigogine, Editor), Interscience, New York, 1958.

C1   D. R. Corson, K. R. MacKenzie, and E. Segrè, "Possible Production of Radioactive Isotopes of Element 85," *Phys. Rev.* **57**, 459 (1940).

D1   N. Davidson, *Statistical Mechanics*, Chapter 9, McGraw-Hill, New York, 1962.

E1   P. J. Estrup and R. Wolfgang, "Kinetic Theory of Hot Atom Reactions: Application to the System $H + CH_4$," *J. Am. Chem. Soc.* **82**, 2665 (1960).

G1   A. Ghiorso, B. G. Harvey, G. R. Choppin, S. G. Thompson, and G. T. Seaborg, "New Element Mendelevium, Atomic Number 101," *Phys. Rev.* **98**, 1518 (1955).

H1   H. Hintenberger, "High-Sensitivity Mass Spectroscopy in Nuclear Studies," *Ann. Rev. Nuclear Sci.* **12**, 435 (1962).

H2   G. Harbottle and N. Sutin, "The Szilard-Chalmers Reaction in Solids," *Advances in Inorganic Chemistry and Radiochemistry*, Vol. 1 (H. J. Emeleus and A. G. Sharpe, Editors) Academic, New York, 1959, p. 267.

I1   M. Inghram, "Stable Isotope Dilution as an Analytical Tool," *Ann. Rev. Nuclear Sci.* **4**, 81 (1954).

J1   J. F. Johns, "Isotope Separation by Multistage Methods," *Progress in Nuclear Physics*, Vol. 6 (O. R. Frisch, editor) Pergamon, New York, 1957.

L1   W. F. Libby, "Chemistry of Energetic Atoms Produced by Nuclear Reactions," *J. Am. Chem. Soc.* **69**, 2523 (1947).

*M1   L. Melander, "Isotope Effects on Reaction Rates," Ronald, New York, 1960.

*M2   H. A. C. McKay, "The Szilard-Chalmers Process," *Progress in Nuclear Physics*, Vol. I (O. R. Frisch, Editor) Pergamon, London, 1950.

M3   J. M. Miller, J. W. Gryder, and R. W. Dodson, "Reactions of Recoil Atoms in Liquids," *J. Chem. Phys.* **18**, 579 (1950).

M4   J. M. Miller and R. W. Dodson, "Chemical Reactions of Energetic Chlorine Atoms Produced by Neutron Capture in Liquid Systems," *J. Chem. Phys.* **18**, 865 (1950).

M5   J. A. Marinsky, L. E. Glendenin, and C. D. Coryell, "The Chemical Identification of Radioisotopes of Neodymium and of Element 61," *J. Am. Chem. Soc.* **69**, 278 (1947).

M6   E. M. McMillan and P. Abelson, "Radioactive Element 93," *Phys. Rev.* **57**, 1185 (1940).

P1   M. L. Perlman and J. A. Miskel, "Average Charge on the Daughter Atom Produced in the Decay of Ar^{37} and Xe^{131m}," *Phys. Rev.* **91**, 899 (1953).

P2   C. Perrier and E. Segrè, "Some Chemical Properties of Element 43," *J. Chem. Phys.* **5**, 715 (1937), and **7**, 155 (1939).

S1   W. Spindel and T. I. Taylor, "Preparation of 99.8% Nitrogen-15 by Chemical Exchange," *J. Chem. Phys.* **24**, 626 (1956).

*S2   G. T. Seaborg and J. J. Katz (Editors), "The Actinide Elements," National Nuclear Energy Series, Div. IV, Vol. 14A, McGraw-Hill, New York, 1954.

*S3   E. Sayre, "Methods and Applications of Activation Analysis," *Ann. Rev. Nuclear Sci.* **13**, 145–162 (1963).

S4   N. Sutin, J. K. Rowley, and R. W. Dodson, "Chloride Complexes of Iron (III) Ions and the Kinetics of the Chloride-catalyzed Exchange Reactions between Iron (II) and Iron (III) in Light and Heavy Water," *J. Phys. Chem.* **65**, 1248 (1961).

*S5   N. Sutin, "Electron Exchange Reactions," *Ann. Rev. Nuclear Sci.* **12**, 285 (1962).

S6   T. Swift and R. Connick, "NMR-Relaxation Mechanism of O^{17} in Aqueous Solutions of Paramagnetic Cations and the Lifetime of Water Molecules in the First Coordination Sphere," *J. Chem. Phys.* **37**, 307 (1962).

*S7   G. T. Seaborg, *Man-Made Transuranium Elements*, Prentice-Hall, Englewood Cliffs, New Jersey, 1963.

T1   H. Taube, "Application of Oxygen Isotopes in Chemical Studies," *Ann. Rev. Nuclear Sci.* **6**, 277 (1956).

V1   C. A. Villee, "Radioisotopes in Biochemical and Medical Research," *Ann. Rev. Nucl. Sci.* **1**, 325 (1952).

*W1   A. C. Wahl and N. A. Bonner (Editors), *Radioactivity Applied to Chemistry*, Wiley, New York, 1951.

W2   J. E. Willard, "Chemical Effects of Nuclear Transformations," *Ann. Rev. Nucl. Sci.* **3**, 193 (1953).

*W3   A. P. Wolf, "Labeling of Organic Compounds by Recoil Methods," *Ann. Rev. Nucl. Sci.* **10**, 259 (1960).

EXERCISES

1. Would you expect almost complete exchange, and by what mechanism, in 1 hr at room temperature, between the members of each of the following pairs?
   (a) Hg^{2+}, very finely divided Hg   (b) $Cr^*O_4{}^{2-}$, $CrO_2{}^-$
   (c) $Cr^*O_4{}^{2-}$, $Cr_2O_7{}^{2-}$ (d) CH_3I^*, C_6H_5I
   (e) $Hg^*(CH_3)_2$, $Hg(C_2H_5)_2$ (f) H_2O^*, H_2SeO_3

2. Refer to the information on fumaric acid oxidation on p. 200. The authors of that report (M. B. Allen and S. Ruben) in the experiment measured the specific activities of the original fumaric acid, the CO_2 evolved, and the formic acid remaining. If the C^{11} activity of the original acid had been 20,000 counts min^{-1} mg^{-1} of carbon at 6:00 P.M., what would they have found for the specific activity of the CO_2 if measured at 7:22 P.M. and of the HCOOH if measured at 8:30 P.M.? What must have been the activity per milligram of carbon of the KC^*N used to synthesize the fumaric acid from ethylene dichloride, corrected to 6:00 P.M.?
   Answer to last part: 40,000 counts min^{-1} mg^{-1}.

3. A mixture is to be assayed for penicillin. You add 10.0 mg of penicillin of specific activity 0.405 μc mg^{-1} (possibly prepared by biosynthesis). From this mixture you are able to isolate only 0.35 mg of pure crystalline penicillin, and you determine its specific activity to be 0.035 μc mg^{-1}. What was the penicillin content of the original sample?

4. The exchange between I_2^* and $IO_3{}^-$ has been studied under these conditions: $(I_2) = 0.00050M$, $(HIO_3) = 0.00100M$, $(HClO_4) = 1.00M$, at 50°C. At specified

times samples were taken and measured for total (I_2 plus IO_3^-) radioactivity by γ counting. These counting rates corrected to the time $t = 0$ on the basis of the 8.0-day half-life of I^{131} are given below in the column "Corrected Total Activity." The I_2 fractions were removed by extraction with CCl_4 and the residual (IO_3^-) radioactivity measured and corrected in the same way; these rates are in the column "Corrected IO_3^- Activity."

Time (hours)	Corrected Total Activity	Corrected IO^{-3} Activity
0.9	1680	9.9 $\pm$ 3.0
19.1	1672	107 $\pm$ 4.1
47.3	1620	246 $\pm$ 6.6
92.8	1653	438 $\pm$ 9.4
169.2	1683	610 $\pm$ 13
"∞"	1640	819 $\pm$ 9.8

Find the half-time $T_{1/2}$ for the exchange and the rate $\mathbf{R}$ of the exchange reaction.
        Answer: $T_{1/2}$ = 90.6 hr; $\mathbf{R}$ = 3.83 $\times$ 10^{-6} mole liter^{-1} hr^{-1}.

5. The experiment described in exercise 4 was repeated but with this difference, (I_2) = $0.00100M$. The results are tabulated as before.

Time (hours)	Corrected Total Activity	Corrected IO_3^- Activity
0.9	1717	7.6 $\pm$ 2.7
19.1	1483	70.1 $\pm$ 3.6
47.3	1548	178 $\pm$ 5.2
92.8	1612	305 $\pm$ 7.3
169.2	1587	413 $\pm$ 9.8
"∞"	1592	534 $\pm$ 5.7

For these conditions find $T_{1/2}$ and $\mathbf{R}$. What is the apparent order of the exchange reaction with respect to I_2? *Note:* Do not be surprised if the order is not an integer; according to O. E. Myers and J. W. Kennedy, *J. Am. Chem. Soc.* **72,** 897 (1950), the order is consistent with this rate law for the exchange-producing reaction: $\mathbf{R}$ = $k(I^-)(H^+)^3(IO_3^-)^2$.

6. Estimate the sensitivity of the neutron activation method for the detection of (a) arsenic, (b) dysprosium with a thermal-neutron flux of 1 $\times$ 10^{11} cm^{-2} sec^{-1}.

7. In an experiment (by T. C. Hoering) bromate ion synthesized to contain 1.13 per cent O^{18} in its oxygen was reacted with excess sulfurous acid in ordinary water. The product sulfate was isolated, and its oxygen was found to contain 0.314 per cent O^{18}. What average number of the three oxygen atoms in BrO_3^- appeared in the SO_4^{2-}?
        Answer: 1.4.

8. Neglecting isotopic fractionation, what is the equilibrium constant for the reaction
$$CCl_4^{35} + CCl_4^{37} \rightarrow 2CCl_3^{35} Cl_2^{37}?$$
        Answer: 36.

9. What is the recoil energy imparted to a Te^{129} atom by the emission of a 74-keV conversion electron? (Use the relativistic expression for the electron energy.)
        Answer: 0.34 eV.

10. Suggest hopeful easily prepared compounds for use in Szilard-Chalmers processes of (a) iron, (b) mercury, (c) technetium.

8

Radioactive Decay Processes

In the discussion of binding energies in chapter 2 we indicated the condition for the stability of any nuclide toward spontaneous (radioactive) decay: a nuclide will be energetically stable toward decay by some specified mode (such as α emission, β emission, or spontaneous fission) if its atomic mass is smaller than the sum of the masses of the products that would be formed in that decay mode. We also noted that in this sense all nuclei with A $\gtrsim$ 100 are unstable toward fission into two approximately equal fragments and all nuclei with A $\gtrsim$ 140 are unstable towards α emission. These trends as well as the energetics of β-decay processes were discussed in terms of the properties of the nuclear-energy surface, which, in turn, can be understood as resulting from the interplay of the various terms in the binding-energy equation (2-3): volume energy, surface term, Coulomb energy, symmetry energy, and pairing interaction. In nuclear, as in chemical, systems a statement about thermodynamic stability tells us only part of the story. For any system that is energetically *not* stable, we are usually interested in the rates of the possible processes, since a thermodynamically unstable system may, for all practical purposes, behave as if it were stable. The so-called stable nuclei with $A > 140$ are an example. In considering the various forms of radioactive decay we thus always inquire about the decay *rates* or half-lives, and in this chapter we shall be chiefly concerned with the various factors that affect these decay rates. In other words, we shall attempt to outline the theoretical framework within which each decay mode is described, explore the predictions the theory makes about the dependence of the decay rate on such factors as the energy change ΔE, the spin change ΔI, and the parity change $\Delta \Pi$ involved in the transition, and when possible, to compare these predictions with experimental data. In addition to striving for a basic understanding of the decay processes themselves, we shall be interested in the information that can be obtained about the properties of nuclear energy levels (energy spacings, spins, and parities[1]) via the study of decay

---

[1] We shall use the notation $I+$ (or $I-$) to denote a state of spin I and even (or odd) parity.

processes.   Such knowledge of nuclear spectroscopy is vital for any systematic understanding of nuclei and forms the basis of the various nuclear models discussed in chapter 9.   Needless to say, the development of each of these models has in turn stimulated a great deal of work in nuclear spectroscopy designed to test model predictions.

A. ALPHA DECAY

Alpha-Particle Spectra.   Much empirical information on decay was accumulated in the three decades following the discovery of radioactivity.   The three naturally occurring radioactive series provided a sizable number of α emitters spanning a very large range of half-lives.   As we saw in chapter 1, the identity of α particles as He^{2+} ions was established as early as 1903 and the monoenergetic nature of α rays was also soon recognized.   Until 1929 it was in fact thought that each α-emitting species had only one α-particle energy associated with it; only then was the so-called fine structure of α spectra discovered.   This phenomenon had escaped attention for so long because of the strong dependence of transition probability on decay energy, which is discussed later and which usually leads to a marked preference for transitions to the product ground state, with smaller transition probabilities to the lowest lying excited states; the latter had not been resolved from the dominant ground-state branches until magnetic spectrometry was used.

Since the 1930's the increasing sophistication of measuring and analyzing instruments and the augmentation of the list of about two dozen naturally occurring α emitters known at that time by almost 200 artificially produced α-emitting nuclides have led to an enormous increase in experimental information on α decay.   From data on the energies of the different α groups emitted by a given nuclide an energy-level diagram of the daughter product can be constructed; the observation of γ rays with energies corresponding to the energy differences between the disintegration energies associated with pairs of observed α-particle groups[2] serves as confirmation of the level scheme, and the α- and γ-ray intensities must, of course, bear a definite relation to one another.   To illustrate these points, the energy-level diagram of Ra^{224} as derived from the α decay of Th^{228} is shown in figure 8-1.   We shall not concern ourselves here with the relative intensities of γ transitions depopulating a given level; the relevant selection rules for γ transition are discussed in section D.[3]

[2] Recall that the total disintegration energy in α emission is the sum of the kinetic energies of α particle and recoil nucleus (chapter 2, p. 49).

[3] An interesting and somewhat different situation occurs in the ThC and RaC β-decay branches.   Here some of the excited states of $ThC'(Po^{212})$ and $RaC'(Po^{214})$ are so unstable with respect to α emission that α decay from these states can compete with γ emission. This is the origin of the so-called long-range α particles of ThC' and RaC' with energies up

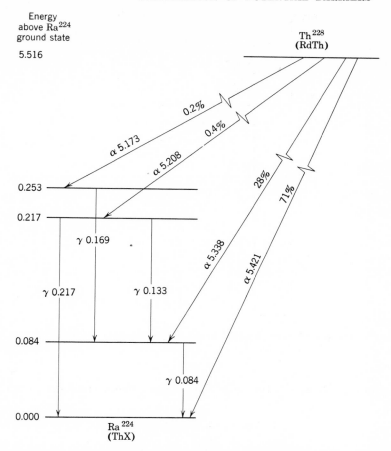

Fig. 8-1   Energy-level diagram for $_{88}Ra^{224}$ as obtained from the observed α and γ radiations of $_{90}Th^{228}$.   All energies are in MeV.   The α-particle energies are the kinetic energies, not total disintegration energies.

Penetration of Potential Barriers.   The 11.7-MeV α particles of Po^{212m} are the most energetic α particles known from radioactive sources.   Near the other extreme are the 2.0-MeV α particles of samarium.   Most of the naturally occurring α-active species emit α particles with energies between 4 and 8 MeV.   This relatively small range in energies is associated with an enormous range in half-lives (from $< 10^{-6}$ second to $> 10^{10}$ years).   A qualitative (inverse) correlation between energy release and half-life was recognized by Rutherford in 1906, and the first quantitative expression of this relationship came in 1911

---

to 10.5 MeV.   The α- to γ-branching ratio and a knowledge of the partial half-lives for α decay have been used to obtain partial γ-half-lives for the ThC′ and RaC′ states involved.

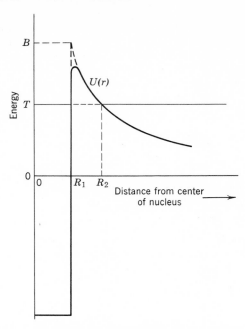

Fig. 8-2   Potential energy in the neighborhood of a nucleus.

with the famous Geiger-Nuttall rule (G1) relating decay constant λ with range in air r:

$$\log \lambda = a + b \log r, \tag{8-1}$$

where b is a constant and a takes on a different value for each of the three radio-active series.   Because of the functional connection between range r, energy E, and velocity v, (8-1) can be replaced by expressions of the same form (but with different numerical values of the coefficients) relating λ with E or v.   The Geiger-Nuttall rule fits experimental data for the natural series[4] remarkably well over factors of $\sim 10^{15}$ in λ (see, e.g., H1 p. 94).   However, until 1928 no theoretical interpretation existed for the very strong dependence of decay constant on energy release (in terms of the Geiger-Nuttall rule, λ varies about as E^{90}).   In fact, before the advent of quantum mechanics it was very puzzling that α particles could be emitted at all with energies that were much lower than the heights of the potential barriers known from other evidence to surround the α-emitting nuclei.   For example, scattering experiments with 8.78-MeV α particles (from ThC$'$) on uranium show that around a uranium nucleus no

---

[4] The fit breaks down when the large body of data on artificially produced nuclides beyond the natural series is included.

deviation from the Coulomb law exists at least up to 8.78 MeV; yet U^{238} emits 4.2-MeV α particles. How do these α particles get outside the potential barrier, which is at least 8.78 MeV high (and actually quite a bit higher)?

This question was answered by the wave-mechanical treatment of the problem which was first presented in 1928 independently by G. Gamow (G2) and by R. W. Gurney and E. U. Condon (G3). In this treatment the Schrödinger wave equation for an α particle of energy E inside the nuclear potential well is set up and solved. The wave function representing the α particle does not go abruptly to zero at the wall of the potential barrier (R_1 in figure 8-2) and has finite, although small, values outside the radial distance R_1. By applying the boundary conditions that the wave function and its first derivative must be continuous at R_1 and R_2, one can solve the wave equation for the region between R_1 and R_2, that is, inside the barrier where the potential energy $U(r)$ is greater than the total kinetic energy T (sum of kinetic energies of α particle and recoil nucleus). The probability P for the α particle of mass M_α to penetrate that region, the so-called barrier penetrability factor, is given by the square of the wave function and turns out to be

$$P = \exp\left(-\frac{4\pi}{h} \sqrt{2M} \int_{R_1}^{R_2} \sqrt{U(r) - T}\, dr\right) \qquad (8\text{-}2)$$

where

$$M = \frac{M_\alpha M_R}{M_\alpha + M_R}$$

is the reduced mass of α particle and recoil nucleus. [For a derivation of (8-2), see, for example, E1, pp. 45–74.] It is clear from (8-2) that the probability for barrier penetration decreases with increasing value of the integral in the exponent, that is, with increasing barrier height and width. (The higher the barrier, the larger the difference $U(r) - T$; and the wider the barrier, the greater the range of the integration.)

The decay constant λ may be considered as the product of P and the frequency f with which an α particle strikes the potential barrier; the order of magnitude of f may be estimated as follows. The de Broglie wavelength h/Mv of the α particle of velocity v and momentum Mv inside the nucleus is taken[5] comparable to R_1, thus

$$\frac{h}{Mv} \cong R_1, \quad \text{or} \quad v \cong \frac{h}{MR_1}.$$

If the α particle is considered as bouncing back and forth between the potential walls,

$$f = \frac{v}{2R_1}, \quad \text{or} \quad f \cong \frac{h}{2MR_1{}^2}.$$

---

[5] Note that we are making an order-of-magnitude estimate only.

Therefore the decay constant

$$\lambda \cong \frac{h}{2MR_1{}^2} \exp\left[-\frac{4\pi}{h} \sqrt{2M} \int_{R_1}^{R_2} \sqrt{U(r) - T}\, dr \right]. \tag{8-3}$$

By a more elaborate treatment more accurate expressions for f and λ are obtained.

For certain simple forms of the potential energy $U(r)$, the integral in the exponential of (8-2) and (8-3) can be solved in closed form, and explicit expressions for λ in terms of R_1, Z, and E can thus be obtained. For more realistic shapes of the nuclear potential, (8-2) has to be solved by numerical integration.

For a square-well nuclear potential of radius R_1 and a Coulomb potential[6] $U(r) = Zze^2/r$ for $r > R_1$ (the dotted line in figure 8-2), the integral in (8-2) and (8-3) becomes

$$I = \int_{R_1}^{R_2} (Zze^2 - Tr)^{\frac{1}{2}} \frac{dr}{r^{\frac{1}{2}}}, \tag{8-4}$$

which, by the substitutions $x = r^{\frac{1}{2}}$ and $a^2 = Zze^2/T$, turns into the readily integrable form $2\sqrt{T} \int_{\sqrt{R_1}}^{\sqrt{R_2}} \sqrt{a^2 - x^2}\, dx$, with the solution

$$I = \sqrt{T}\left[x(a^2 - x^2)^{\frac{1}{2}} + a^2 \sin^{-1}\frac{x}{a} \right]_{\sqrt{R_1}}^{\sqrt{R_2}}. \tag{8-4a}$$

Values of the radii R_1 and R_2 are obtained from the expressions for the total kinetic energy T and the barrier height B (see figure 8-2):

$$T = \frac{Zze^2}{R_2} \quad \text{and} \quad B = \frac{Zze^2}{R_1}.$$

After substitution of the integration limits and some algebraic manipulations we obtain

$$I = \frac{Zze^2}{\sqrt{T}}\left[\cos^{-1}\left(\frac{T}{B}\right)^{\frac{1}{2}} - \left(\frac{T}{B}\right)^{\frac{1}{2}}\left(1 - \frac{T}{B}\right)^{\frac{1}{2}} \right]. \tag{8-5}$$

Finally, remembering that $T = \frac{1}{2}Mv^2$, substitution of (8-5) in (8-3) gives

$$\lambda \cong \frac{h}{2MR_1{}^2} \exp\left\{ -\frac{8\pi Zze^2}{hv}\left[\cos^{-1}\left(\frac{T}{B}\right)^{\frac{1}{2}} - \left(\frac{T}{B}\right)^{\frac{1}{2}}\left(1 - \frac{T}{B}\right)^{\frac{1}{2}} \right] \right\}. \tag{8-6}$$

---

[6] Note that Z is the atomic number of the daughter nucleus. However, R_1 cannot be directly interpreted as the radius of the daughter nucleus. Rather it is an "effective" radius corresponding to the maximum in the potential energy $U(r)$ and containing a contribution ρ_α from the finite size of the α particle. The value of ρ_α has been estimated (B1, p. 357) as 1.2×10^{-13} cm on the basis of excitation functions for α-induced reactions; these involve barrier penetration by α particles also and can be investigated over a wider range of Z (and thus R) than can α radioactivity and therefore give more definite information on ρ_α.

Table 8-1   Comparison of decay constants calculated from (8-6) with experimental data

Alpha Emitter	T (MeV)	$R_1 \times 10^{13}$ cm*	λ_{calc} (sec^{-1})	λ_{exp}# (sec^{-1})
Th^{232}	4.077	9.142	0.69×10^{-18}	1.20×10^{-18}
Th^{230}	4.765	9.118	1.6×10^{-13}	2.10×10^{-13}
Th^{228}	5.518	9.095	7.4×10^{-9}	8.20×10^{-9}
Th^{226}	6.444	9.072	2.4×10^{-4}	2.96×10^{-4}
Fm^{254}	7.315	9.390	1.3×10^{-4}	5.05×10^{-5}
Po^{214}	7.826	8.927	4.5×10^{3}	4.23×10^{3}
Po^{210}	5.402	8.878	9.6×10^{-7}	5.80×10^{-8}
Sm^{146}	2.62	7.982	19×10^{-16}	4.4×10^{-16}

* Radii were calculated from the formula $R_1 = (1.30 A^{1/3} + 1.20) \times 10^{-13}$ cm.
# The λ values listed are partial decay constants for the ground-state α transitions only.

As an illustration of the remarkable success of the Gamow-Gurney-Condon approach, we show in table 8-1 a few values calculated with (8-6) and the corresponding experimental values. The calculations were done with $R_1 = (1.30 \times A^{1/3} + 1.20) \times 10^{-13}$ cm and with no other adjustable parameters. The agreement between calculated and measured values is seen to be within a factor of 4 in all cases except Po^{210} (a nucleus with 126 neutrons which decays to a nucleus with 82 protons—both tightly bound closed-shell nuclei with abnormally small radii), even though the λ values extend over a range of more than 10^{21}. The absolute values of λ_{calc} depend sensitively on the nuclear radii assumed, each increase by 0.03×10^{-13} cm in the nuclear radius parameter r_0 giving rise to an approximate doubling of all λ values. The fact that (8-6) gives good agreement with experimental data when r_0 is taken as 1.30×10^{-13} cm (1.30F) should not be considered significant, since (1) we used a square-well potential rather than a more realistic potential shape with tapering sides, (2) the effective α-particle radius $\rho_\alpha = 1.20 \times 10^{-13}$ cm was chosen somewhat arbitrarily, (3) the pre-exponential factor was derived on the rather naïve assumption that α particles pre-exist and oscillate in nuclei ("one-body model"). The important point is that none of these assumptions strongly affects the spectacular dependence of λ on T, which is given principally by the exponential in (8-6), but they have a significant effect on the relation between λ and R_1, the effective nuclear radius. Many attempts have been made to use more sophisticated models (see, e.g., H1 and E1 for brief reviews of such calculations) and thus to obtain reliable values of nuclear radii from α-decay data. The results obtained by different authors differ, depending on their model assumptions, but they can usually be expressed either in the form

$$R_1 = r_0' A^{1/3}, \tag{8-7}$$

with r_0' values between 1.45 and 1.57F, or in the form

$$R_1 = r_0 A^{1/3} + \rho_\alpha, \tag{8-8}$$

with r_0 values between 1.25 and 1.4F if ρ_α is taken as 1.20F. We have already mentioned that apparent anomalies in the half-life-versus-energy relationships at shell crossings ($N = 126$ and $Z = 82$) can be interpreted by postulating that the radii of closed-shell nuclei are a few per cent smaller than the smooth A-dependence of (8-7) or (8-8) would predict.

Hindered Alpha Decay. The α emitters listed in table 8-1 all have even Z and even N, and the dependence of half-life on decay energy predicted by the barrier-penetration theory in its simple form applies to even-even α emitters only. In fact, good agreement with experimentally determined partial half-lives is generally found only for transitions to ground states and first excited states of even-even nuclei. All other α transitions (those to higher excited states in even-even nuclei and most transitions in other nuclear types) tend to be slower (by factors up to about 10^4) than would be predicted by the simple theory. This is indicated by a few points for odd-A nuclei entered on figure 8-3, which otherwise shows only the regular behavior of log t_α (partial half-life) versus T for ground- and first-excited-state transitions in even-even nuclei. The smooth curves correspond to (8-6) with $R_1 = 1.51 \times 10^{-13} A^{1/3}$ cm.

Since the ground states of even-even nuclei have 0 spin and even parity (first excited states have spin 2 and even parity) and since (8-6) was derived without regard to angular-momentum effects (that is, for emission of s-wave α-particles only) we might at first sight be tempted to ascribe the relative slowness of other α transitions to angular-momentum changes, or, in other words, to the existence of a centrifugal barrier in addition to the Coulomb barrier. However, the so-called hindrance factors—the ratios of calculated (for $\Delta L = 0$) to observed transition probabilities—are larger than can be accounted for by inclusion of angular-momentum effects in the theory.[7]

The other important simplifying assumption implicit in the discussion so far —spherical shape for all nuclei—is known to be incorrect over most of the region of α emitters, and much effort has gone into the development of theoretical methods for treating the α-decay probabilities of spheroidal and other deformed nuclei (H1). In spite of these attempts, no complete theory of hindered α transitions appears to be available, but some empirical observations

---

[7] For an α particle carrying away L units of angular momentum, the barrier penetration factor (8-2) has to be modified by substituting $\{Zze^2/r + [L(L+1)/2Mr^2]\hbar^2 - T\}^{1/2}$ for $[(Zze^2/r) - T]^{1/2}$. The added term is small and the integral can be solved after suitable expansion of the square root (H1). In addition to this effect of L on barrier penetration which goes in the direction of decreasing penetrability with increasing L, angular momentum is also expected to have an effect on the pre-exponential "striking frequency" term; however, not only the magnitude, but even the direction of this effect varies with the shape and depth of the potential well assumed (H1).

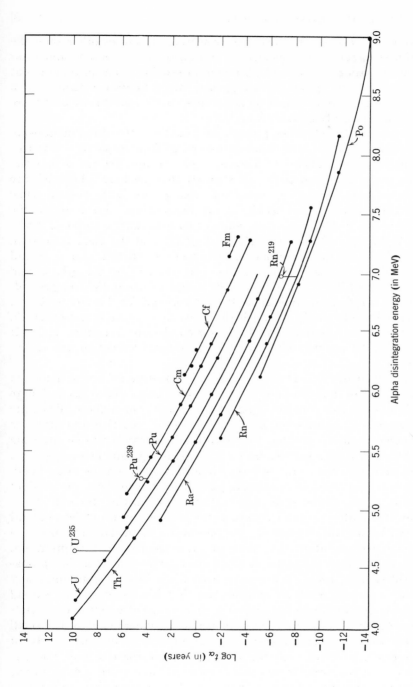

Fig. 8-3 Relation between partial α half-life and α-disintegration energy for even-even α emitters. [After I. Perlman, A. Ghiorso, and G. T. Seaborg, *Phys. Rev.* **77**, 26 (1950).] Experimental points for each element are connected by a curve. A few even-odd species are shown also.

229

may be made.   In even-even nuclei the transitions to states other than the ground and first excited states (mostly $4+$, $6+$, $8+$, and $1-$ states, and a few other odd-parity states) have hindrance factors ranging from unity to about 12,000.   There is some general trend toward larger hindrance factors with increasing ΔI, and for some transition types ($0+ \rightarrow 4+$ and $0+ \rightarrow 1-$) there is a fairly regular progression with Z or N.

Among the even-odd, odd-even, and odd-odd nuclides, the situation appears to be even more complex.   Hindrance factors range from unity to $\sim 3 \times 10^4$, with systematic trends difficult to discern.   One striking feature is that the ground-state transitions, especially for strongly deformed odd-A nuclei, are highly hindered even when there is no spin change involved, whereas some transition to an excited state is usually almost unhindered.   For example, the α transitions of Am^{241} ($5/2-$) to the ground and first excited states of Np^{237} ($5/2+$ and $7/2+$) have hindrance factors of $\cong 500$, and the main transition, almost unhindered, is to the second excited state at 60 keV above ground ($5/2-$).   Similarly, the transitions of the even-odd U^{235} to the ground and first excited states of Th^{231} are hindered by factors of about 10^3 (the ground-state transition is indicated on figure 8-3); if it were not for this circumstance, this important nuclide would be much too short-lived to be found on earth. The possibility of large hindrance factors for ground-state transitions makes it sometimes difficult to decide whether the ground-state transition has in fact been found.

A qualitative explanation of the phenomena just mentioned has been offered by Perlman, Ghiorso, and Seaborg (P1).   They suggest that a ground-state transition from a nucleus containing an odd nucleon in the highest filled state can take place only if that nucleon becomes part of the α particle and therefore if another nucleon pair is broken; this is certainly a less favorable situation than the formation of an α particle from already existing pairs in an even-even nucleus and may give rise to the observed hindrance.   If, on the other hand, the α particle *is* assembled from existing pairs in such a nucleus, the product nucleus will be in an excited state, and this may explain the "favored" transitions to excited states.   Detailed data obtained from other evidence about the spins, parities, and other quantum numbers of the particular states between which these favored transitions take place appear to confirm this interpretation.

Although, as we have indicated, a complete theory of α-transition probabilities is still lacking, great strides have been made in understanding many of the phenomena.   It must be remembered that hindrance factors extend over four orders of magnitude, whereas the whole range of α half-lives extends over a factor of 10^{24}.   Thus in a sense we are now talking about rather small perturbations to the general theory.   Furthermore, the detailed study of α-particle spectra, together with the associated γ-ray spectroscopy, and $\alpha\gamma$-coincidence and angular-correlation measurements, have aided greatly in the mapping out and characterization of energy levels and have given impetus to the development of the collective and unified models discussed in chapter 9.

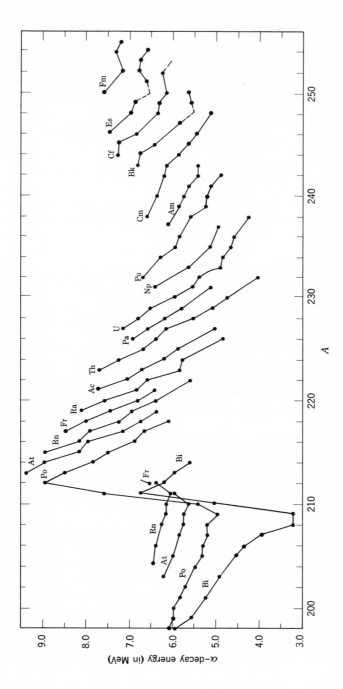

Fig. 8-4 Plot of decay energies versus mass numbers for α emitters from bismuth to fermium. [Based on a graph by I. Perlman, A. Ghiorso, and G. T. Seaborg, *Phys. Rev.* **77**, 26 (1950), but with addition of more recent data.]

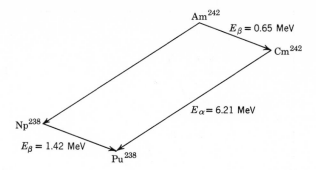

Fig. 8-5   Closed decay cycle for determination of Am^{242} α-decay energy. From the measured decay energies shown the α-decay energy of Am^{242} is computed to be 0.65 + 6.21 − 1.42 = 5.44 MeV.

Alpha-Decay Energies.   In addition to the regularities in lifetimes, α-particle emitters exhibit some interesting systematic trends in their decay energies. Most of these features can be derived from the general properties of the nuclear-energy surface discussed in chapter 2.   By obtaining, with the aid of (2-3), a general expression for the energy difference between ground states of two nuclei with $\Delta A = 4$ and $\Delta Z = 2$ and examining the properties of the first and second partial derivatives of this expression with respect to A and Z we arrive at the following conclusions.   For the isotopes of any element in the region of the α emitters the α-decay energies are expected to decrease monotonically with increasing A, and for a series of isobars they will increase with increasing Z.

These predictions of the liquid-drop model of nuclei are largely borne out by experimental data as shown in figure 8-4, in which α-decay energies for ground-state transitions are plotted against the mass number of the α emitter.   Points belonging to one element are joined.   A few of the points were obtained not by direct measurements but by the method of closed decay cycles which is illustrated for Am^{242} in figure 8-5.

In a diagram such as figure 8-4 the abrupt interruption of the predicted regularities in the neighborhood of $A = 210$ is as striking as the regular trends themselves.   Had we plotted neutron number rather than mass number as the abscissa,[8] it would have been more immediately evident that the sharp drop in decay energy occurs for each element between the α emitter with 128 and that with 126 neutrons.   This indicates exceptionally strong binding of the last few neutrons below neutron number 126 and is one of the strongest pieces of evidence for a closed shell at $N = 126$.   Similarly, the sharp decrease in E_α in going from the heavy polonium to the heavy bismuth isotopes is a consequence

[8] This was not done because the data for different elements would be less well separated on such a plot.

of the closed proton shell at $Z = 82$. Evidently this shell effect is responsible for the absence of observable α decay in lead and thallium isotopes. On the other hand, the absence of a sharp drop in decay energy between $Po^{200-204}$ and $Bi^{199-203}$ is puzzling and might conceivably indicate that these light bismuth isotopes do not contain a filled 82-proton shell. The slight breaks in the curves of figure 8-4 below Cf^{252}, Es^{253}, and Fm^{254} have been interpreted as evidence of a closed neutron subshell at $N = 152$.

In the rare-earth region, just above neutron number 82, an "island" of α emitters has been found that includes the naturally occurring $_{62}Sm^{147}$ and $_{60}Nd^{144}$ and about 20 artificially produced nuclides, most of which are neutron deficient relative to the stable isotopes. The highest decay energies occur, as expected, for emitters with $N = 84$.

B. OTHER HEAVY-PARTICLE DECAY MODES

Proton Radioactivity. The question is frequently asked why proton decay with measurable lifetimes does not appear to occur from nuclear ground states, whereas α radioactivity is such a common phenomenon. The reasons can be readily given in the light of our discussions of α decay. In the vicinity of the β-stable region proton emission is, in contrast to α emission, energetically impossible over the entire mass range. The difference lies simply in the large binding energy of the four nucleons in the free α particle; they are more tightly bound than the most loosely bound nucleons in any heavy nucleus. On the other hand, a proton to be emitted must be supplied with an energy of several million electron volts—its own binding energy to the residual nucleus.

Far on the proton-excess side of the β-stability valley nuclei do indeed become unstable with respect to proton emission, largely as a result of the increasing importance of the Coulomb and symmetry terms in the binding-energy equation (2-3). Just beyond the limit of proton stability we thus might expect proton emission with measurable lifetimes. However, for nuclei so far removed from β stability (e.g., Mn^{45} or Se^{60}) the β-decay energies will be extremely high, hence, as we shall see in section C, the half-lives for positron decay or electron capture very short; proton emission can therefore be observable only if it, too, has comparably short half-lives (say <1 second). The half-life for proton emission can be estimated by use of (8-6), and it turns out that a half-life range of 1 second to 10^{-10} second corresponds to a decay-energy range of only about 30 to 80 keV at $Z = 10$ or 0.2 to 0.5 MeV at $Z = 30$. It would thus be a very fortuitous circumstance to find a nuclide with a proton-decay energy in the required narrow interval and with a sufficiently favorable ratio of proton- to positron-emission probability for observation. There may be a few such cases among light nuclei (G4). At higher Z values where the possibly observable half-life region corresponds to larger decay-energy intervals the proton-unstable nuclei lie so far from the region of β stability that they are not likely to be produced in any nuclear process now known.

Delayed proton emission of a different sort *has* been observed in products of high-energy nuclear reactions. In this process, which is quite analogous to the well-known delayed neutron emission first observed among fission fragments (cf. p. 315), a β^+ decay leads to a proton-unstable product which instantaneously (in $<10^{-12}$ second) emits a proton. The observed half-life for proton emission is just the half-life for the preceding β^+ decay. The β^+ emitters Ne^{17} ($t_{1/2} = 0.7$ second) and Mg^{21} ($t_{1/2} = 0.13$ second) have been identified as delayed-proton precursors.

Emission of Other Heavy Particles. Binding-energy considerations alone would certainly permit spontaneous emission of tightly bound nuclear entities other than α particles. Deuterons are not sufficiently tightly bound for their emission from nuclei anywhere near the β-stability valley. But many α-unstable nuclei could, on purely energetic grounds, emit nuclei such as C^{12}, O^{16}, or Ne^{20} also. However, the barrier heights for the emission of such nuclei are so high that the expected half-lives are far beyond presently detectable limits.

Two-Proton Radioactivity. The possibility of an interesting additional class of radioactive decay has been pointed out by V. I. Goldanskii (G4, G5): the simultaneous emission of two protons. This process can presumably occur among light ($Z \lesssim 50$) nuclides of even Z which lie near the proton-stability limit. Such nuclei, though slightly stable toward proton emission, may be unstable with respect to the emission of two protons, for as a result of the pairing energy [last term in (2-3)] the last (even) proton in the nucleus Z^A may be positively bound, whereas the last (odd) proton in the nucleus $(Z-1)^{A-1}$ may not be. Furthermore, if the absolute value of the positive binding energy of the Zth proton is less than that of the negative binding energy of the $(Z-1)$th proton, the energetic condition for two-proton radioactivity is fulfilled. Proton pairing energies are of the order of 1 to 3 MeV, and the decay energy for two-proton emission must always be less than that pairing energy.

On the basis of detailed binding-energy considerations[9] Goldanskii (G5) has predicted the nuclides that can be expected to undergo two-proton decay and their decay energies. With the usual barrier penetration formulas, we can then estimate half-lives and find that they are expected to span an appreciably wider range than in single-proton emission. Furthermore, the total barrier penetration probability for a pair of protons turns out to be sharply peaked when they are emitted with equal energies. The emission of two protons of nearly equal energies should be a fairly characteristic event observable in emulsions or cloud chambers, even when β^+ emission is the predominant decay

---

[9] Our equation (2-3) is not sufficiently reliable at large distances from β stability to be useful for this purpose.

mode. Among promising candidates as two-proton-radioactive species are Ne^{16}, Ti^{38}, Kr^{67}.

Spontaneous Fission (H4). The process of nuclear fission was first observed to take place under neutron bombardment (O. Hahn and F. Strassmann, 1939, H2). However, only about a year later K. A. Petrzhak and G. N. Flerov reported that uranium (U^{238}) nuclei undergo fission spontaneously, although with a half-life of about 10^{16} years, which is long compared to the α-decay half-life. Since then the spontaneous fission rates of some 35 nuclides (all with $Z \geq 90$) have been measured, and spontaneous fission is thus firmly established as another mode of radioactive decay. The observed rates extend from about 3×10^{-4} fissions g^{-1} sec^{-1} for U^{235} (an upper limit of $\sim 6 \times 10^{-8}$ fissions g^{-1} sec^{-1} has been reported for Th^{232}) to about 10^{17} fissions g^{-1} sec^{-1} for Fm^{256}. The latter nuclide appears to decay almost exclusively by spontaneous fission, with a three-hour half-life.

We have already seen in chapter 2 that the break-up of any heavy nucleus ($A \gtrsim 100$) into two nuclei of approximately equal size is exoergic. The question is then why spontaneous fission has been observed only for nuclei with $A \geq 230$ and what factors govern the half-lives for the process. The answers to these questions are to be found in considerations somewhat analogous to those used in the discussion of α decay. Clearly, the separation of a heavy nucleus into two positively charged fragments is hindered by a Coulomb barrier, and fission can therefore be treated as a barrier penetration problem. The height of the barrier can be roughly estimated [from (2-10)] as the Coulomb energy between the two fragments when they are just touching. With a radius parameter $r_0 = 1.57 \times 10^{-13}$ cm (consistent with α-decay data, cf. p. 228), this simple approach gives for the Coulomb barrier for a split into two equal fragments 197 MeV in the case of U^{238} and 158 MeV in the case of Hg^{200}.[10] On the other hand, the energy release, calculated with the aid of semiempirical mass data, is about 180 MeV for the symmetrical fission of U^{238} and about 120 MeV for that of Hg^{200}. In other words, the barrier height increases more slowly with increasing nuclear size than does the decay energy for fission. In view of the steep dependence of barrier penetrability on the ratio of available energy to barrier height it is thus not surprising that spontaneous fission is observed only among the very heaviest elements and that spontaneous-fission half-lives in general decrease rapidly with increasing Z.

It is instructive to pursue the energetics of fission a little further with the aid of the semiempirical mass equation (2-5). The energy release Q_f in the fission of nucleus Z^A into two equal fragments $(Z/2)^{A/2}$ is given by

$$Q_f = M_{Z,A} - 2M_{Z/2,A/2},$$

[10] These are presumably high estimates because the fragments are surely not spherical at the moment of separation. Also, break-up into *equal* fragments does not necessarily give the largest energy release, but for the rather crude estimates given here this is not important.

and, from (2-5), we obtain

$$Q_f = 0.216Z^2A^{-\frac{1}{3}} + 0.152ZA^{-\frac{1}{3}} - 3.41A^{\frac{2}{3}}.$$

The height of the Coulomb barrier between two spherical nuclei of mass number $A/2$ and atomic number $Z/2$ is, according to (2-10), and with $r_0 = 1.57 \times 10^{-13}$ cm, given by $V_c = 0.144Z^2A^{-\frac{1}{3}}$. A nucleus will be expected to break up into two fragments within a few nuclear vibrations (no barrier penetration required) if $Q_f \geq V_c$, that is, if

$$0.216Z^2A^{-\frac{1}{3}} + 0.152ZA^{-\frac{1}{3}} - 3.41A^{\frac{2}{3}} \geq 0.144Z^2A^{-\frac{1}{3}}$$

or

$$\frac{Z^2}{A} \geq 47.4 - 2.1\frac{Z}{A}.$$

Since Z/A is always $\cong 0.4$ for heavy nuclei, we can write the condition for a measurable lifetime for spontaneous fission as

$$\frac{Z^2}{A} \lesssim 46.6. \tag{8-9}$$

A somewhat better value for this critical value of Z^2/A can be obtained by an approach first used in the classic 1939 paper of Bohr and Wheeler (B2). If a nucleus is considered as a uniformly charged, spherical drop of incompressible liquid, any small distortion of the sphere will be associated with changes in the surface tension (nuclear binding) and Coulomb energies. The surface area and therefore the surface energy will increase, thus decreasing the total binding, whereas the accompanying increase in the average separation of protons decreases the Coulomb energy and thus tends to increase binding. Furthermore, it can be shown that for small, axially symmetric distortions of a sphere the changes in Coulomb and surface energies are given by

$$\Delta E_c = -\tfrac{1}{5}\alpha_2{}^2E_c{}^0 \quad \text{and} \quad \Delta E_s = \tfrac{2}{5}\alpha_2{}^2E_s{}^0,$$

respectively, where $E_c{}^0$ and $E_s{}^0$ are the Coulomb and surface energies of the undistorted spheres, and α_2 is a parameter characterizing the amount of distortion. It is clear that the drop will be stable against distortions of this type only if $|\Delta E_c| < \Delta E_s$, that is, if $(E_c{}^0/2E_s{}^0) < 1$. Unless this condition is met, there will be no net restoring force to return the drop to its spherical shape, and even the slightest distortion will therefore lead to fission. By substituting the surface and Coulomb terms of the semiempirical binding-energy equation (2-3) for $E_s{}^0$ and $E_c{}^0$ [replacing $Z(Z - 1)$ by Z^2] we have for the limiting condition

$$\frac{0.585Z^2A^{-\frac{1}{3}}}{2 \times 13.1A^{\frac{2}{3}}} < 1 \quad \text{or} \quad \frac{Z^2}{A} < 44.8. \tag{8-9a}$$

The significance of this result [which is not very different from that in (8-9)] is (1) that fissionability may be expected to correlate with the parameter Z^2/A and (2) that all nuclei with Z^2/A values exceeding some critical value $(Z^2/A)_{\text{crit}}$

are expected to undergo fission instantaneously, that is, in a time comparable to a nuclear vibration period, $\cong 10^{-20}$ second. Thus, a limit is set to the synthesis of transuranium nuclei at $Z \cong 110$–120. The exact value of $(Z^2/A)_{\text{crit}}$ depends, of course, on the particular coefficients in the mass formula used.

If the logarithm of spontaneous-fission half-lives is plotted against Z^2/A for even-even nuclides (figure 8-6), as first suggested by Seaborg (S1), the straight-line relationship suggested by the simplest liquid-drop considerations is not observed. Rather, for each element the partial spontaneous-fission half-lives of even-even nuclides go through a maximum. There is a general trend with Z^2/A, but a similar set of curves would be obtained if some other parameter such as A were used as the abscissa (H3). Even the envelope of the curves in figure 8-6 is not a straight line, and it appears to extrapolate to a $(Z^2/A)_{\text{crit}}$ value higher than that predicted by the simple liquid-drop model. Correlations such as those shown in figure 8-6, even though at present largely empirical,

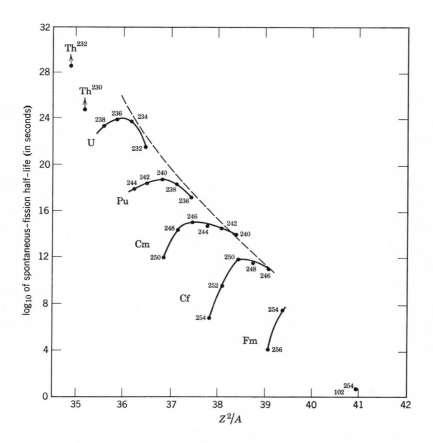

Fig. 8-6   Partial half-lives for spontaneous fission of even-even nuclides, plotted against Z^2/A.

Table 8-2   Partial spontaneous-fission half-lives for odd-A and odd-odd nuclides

Nuclide	Partial Spontaneous-Fission Half-life (years)	Principal Mode of Decay	Half-life
$_{91}\text{Pa}^{231}$	10^{16}	α	3.5×10^4 y
$_{92}\text{U}^{233}$	$>3 \times 10^{17}$	α	1.6×10^5 y
$_{92}\text{U}^{235}$	1.9×10^{17}	α	7.1×10^8 y
$_{93}\text{Np}^{237}$	$\geq 10^{18}$	α	2.2×10^6 y
$_{93}\text{Np}^{239}$	$>5 \times 10^{12}$	β^-	2.35 d
$_{94}\text{Pu}^{239}$	5.5×10^{15}	α	2.44×10^4 y
$_{95}\text{Am}^{241}$	2×10^{14}	α	458 y
$_{97}\text{Bk}^{249}$	$>1.4 \times 10^9$	β^-	314 d
$_{98}\text{Cf}^{249}$	1.5×10^9	α	360 y
$_{99}\text{Es}^{253}$	3×10^5	α	20 d
$_{99}\text{Es}^{254}$	1.5×10^5	α	480 d
$_{100}\text{Fm}^{255}$	1×10^4	α	20 h

are, of course, quite useful in predicting the spontaneous-fission half-lives of still undiscovered nuclides.

The partial spontaneous-fission half-lives for odd-nucleon nuclides are significantly longer than would correspond to interpolations from figure 8-6. The available data for odd-A nuclides and for the odd-odd Es^{254} are given in table 8-2. The abnormally long spontaneous-fission half-lives of these nuclides compared to the even-even species cannot be accounted for in the framework of the liquid-drop model but are qualitatively explained by the effect of nuclear deformation on the energy states of odd-nucleon nuclides (see the discussion of "Nilsson states" in chapter 9, section E).

We have considered here only the lifetime for spontaneous fission and the factors influencing it. Other characteristics of spontaneous fission, such as the division of mass and charge between the two fragments, their kinetic energies, and the emission of neutrons accompanying fission, are essentially similar to what is observed in induced fission reactions. Discussion of these subjects is therefore postponed until chapter 10.

C. BETA DECAY

Phenomenology.   Much of the phenomenology of β-decay processes has been discussed in chapter 2. Here we wish to recapitulate only a few essential points. We concluded from the parabolic cross section of the nuclear-mass surface at constant A (2-8) that for each odd A there could be only one and for each even A at most three β-stable nuclides. The decay energies of β-unstable nuclei vary rather systematically with distance from the so-called β-stability line, as predicted by these mass parabolas, except for shell-edge perturbations. On the other hand, although there is an obvious qualitative connection between half-life and decay energy (large decay energies being generally associated with short lifetimes), the quantitative relations are not nearly so simple as in the case of α decay.

The most striking aspect of β decay and one which has occasioned much of the experimental and theoretical studies in this field is the continuous energy spectrum with which β particles are emitted even though they correspond to transitions between two discrete energy states. We have already discussed (chapter 2, section E3) how this puzzling phenomenon and the equally disturbing apparent lack of conservation of angular momentum and statistics in β decay were qualitatively accounted for by the introduction of the neutrino. It now remains for us to review the simplest aspects of the quantitative theory of β decay which was developed in 1934 by Fermi (F1) and is still the cornerstone in our understanding of β decay.

Fermi Theory. As already discussed in chapter 2, electrons cannot exist in nuclei, and the observed emission of electrons in β decay therefore had to be explained in terms of the creation of an electron (and a neutrino) at the moment of emission. Negatron decay (which was the only mode of β decay known when Fermi originally formulated his theory) can thus be represented by the equation

$$n \rightarrow p + \beta^- + \bar{\nu}, \qquad (8\text{-}10)$$

where $\bar{\nu}$ stands for antineutrino. Similarly, positron decay is represented by

$$p \rightarrow n + \beta^+ + \nu, \qquad (8\text{-}11)$$

where ν is the symbol for the neutrino. We should note that in (8-10) and (8-11) n and p are to be considered not as free particles but as bound in the nucleus. In the free state the neutron has a mass that exceeds the mass of a hydrogen atom by 0.782 MeV. Free neutrons therefore decay by process (8-10) (with a half-life of about 12.8 minutes), whereas free protons are stable. Inside a nucleus, process (8-11), too, is possible without violation of energy conservation because the nucleus as a whole can supply the energy necessary to drive the reaction.

Fermi's theory of β emission from nuclei was in many ways patterned after the theory of light emission from atoms (light quanta, too, are "created" at the moment of emission). The well-known electromagnetic interaction, characterized by the electronic charge e, has to be replaced by a new type of interaction characterized by a new universal constant, the Fermi constant g whose magnitude must be determined by experiment.[11]   The probability $P(p_e)\,dp_e$ that an electron of momentum between p_e and $p_e + dp_e$ is emitted per unit time may be written as

$$P(p_e)\,dp_e = \frac{4\pi^2}{h}\left|\psi_e(0)\right|^2\left|\psi_\nu(0)\right|^2\left|M_{if}\right|^2 g^2 \frac{dn}{dE_0}. \qquad (8\text{-}12)$$

Here ψ_e and ψ_ν are the electron and neutrino wave functions (plane waves in Fermi's theory), and $\left|\psi_e(0)\right|^2$ and $\left|\psi_\nu(0)\right|^2$ are the probabilities of finding electron and neutrino, respectively, at the nucleus. M_{if} represents the matrix element characterizing the transition from the initial to the final nuclear state; the square of its magnitude $\left|M_{if}\right|^2$ is a measure of the amount of overlap between the wave functions of initial and final nuclear states.   The so-called statistical factor dn/dE_0 is the density of final states (number of states of the final system per unit decay energy) with the electron in the specified momentum interval $p_e \rightarrow p_e + dp_e$.[12]

Energy Spectrum.   Following Fermi, we shall first derive the shape of the β-energy (or -momentum) spectra predicted by (8-12).   Integration over all electron momenta from zero to the maximum possible momentum should then give us transition probabilities or lifetimes.

Let us consider so-called allowed transitions, that is, transitions in which both electron and neutrino are emitted with zero orbital angular momentum or, what is classically equivalent, with zero impact parameter.   The magnitudes of $\left|\psi_e\right|^2$ and $\left|\psi_\nu\right|^2$ at the position of the nucleus will certainly be much larger for these s-wave neutrinos and electrons than for electrons and neutrinos emitted with larger orbital angular momenta.   Therefore the largest transition probabilities will be associated with s-wave electron and neutrino emission. The treatment of these allowed transitions is relatively simple.   The magnitudes of $\left|\psi_\nu(0)\right|$ and $\left|M_{if}\right|$ are independent of the division of energy between electron and neutrino, and the spectrum shape is thus determined entirely by $\left|\psi_e(0)\right|$ and by dn/dE_0.   The first of these factors enters only through the Coulomb interaction between nucleus and emitted electron, and we begin by neglecting this effect (a good approximation at low Z) and evaluating the statistical factor alone.

---

[11] This same interaction constant g which turns out to be $\cong 10^{-49}$ erg cm^3 (see p. 247) is now believed to govern all processes involving interactions between four particles of spin $\frac{1}{2}$ (fermions): nucleons, electrons ($+$ and $-$), μ mesons ($+$ and $-$), neutrinos and antineutrinos, and some hyperons.   Such interactions are called weak interactions to distinguish them from the much stronger interactions governed by nuclear forces and from the electromagnetic interactions of intermediate strength.   The fourth type of fundamental interaction known, the gravitational interaction, is still weaker.

[12] Despite its appearance, the statistical factor dn/dE_0 is an infinitesimal quantity, as is required for (8-12) to be correct as written (with dp_e on the left side).   This can be understood when we remember that dn is the number of states with *both* electron and neutrino in specified momentum intervals [see (8-15)].   Perhaps it would be preferable to write the statistical factor as $(dn)^2/dE_0$ and the number of states in (8-15) as $(dn)^2$.

The density of final states of the system, dn/dE_0, with the electron in the momentum interval $p_e \rightarrow p_e + dp_e$, can be found as follows. Consider an infinitesimal interval dE_0 of the total (electron plus neutrino) kinetic energy E_0. An electron with kinetic energy E_e has associated with it a neutrino with kinetic energy $E_\nu = E_0 - E_e$ (if we neglect the minute amount of recoil energy given to the nucleus), and the range of E_ν is dE_0. For the neutrino of zero rest mass the relation between momentum and kinetic energy is

$$p_\nu = \frac{E_\nu}{c} = \frac{(E_0 - E_e)}{c}.$$ (8-13)

Therefore for a given electron energy E_e we have

$$dp_\nu = \frac{dE_0}{c}.$$ (8-14)

The number of neutrino states with neutrino momentum between p_ν and $p_\nu + dp_\nu$ is[13]

$$\frac{4\pi p_\nu{}^2 \, dp_\nu}{h^3}.$$

This, we should emphasize, is the number of neutrino momentum states associated with a *given* electron momentum. However, the number of electron states in the momentum interval $p_e \rightarrow p_e + dp_e$ is $4\pi p_e{}^2 \, dp_e/h^3$, and with each of these electron states the number of neutrino states given above can be associated. Therefore the total number of states of the system in the interval

---

[13] The number of translational states of a particle in a certain momentum interval is derived from the quantum-mechanical treatment of the particle in a box. The form of the expression can be made plausible by recourse to the uncertainty principle: for a particle whose position is specified by the Cartesian coordinates x, y, z, and whose momentum is given by the momentum components p_x, p_y, p_z, the product of the uncertainty in a space coordinate and the uncertainty in the corresponding momentum component is of the order of Planck's constant. Thus

$$\Delta x \, \Delta p_x \cong h; \qquad \Delta y \, \Delta p_y \cong h; \qquad \Delta z \, \Delta p_z \cong h.$$

In the six-dimensional "phase space" characterized by three space coordinates and three momentum coordinates a particle can therefore be specified only as being in a volume $\Delta x \, \Delta y \, \Delta z \, \Delta p_x \, \Delta p_y \, \Delta p_z \cong h^3$, called the unit cell in phase space. The number of translational states available to a particle in a certain volume and in a certain momentum interval is taken as the number of unit cells in the corresponding volume of phase space. For further discussions of these concepts, the reader is referred to standard works on statistical mechanics.

In (8-15) we have taken the volume in coordinate space as unity. Any arbitrary volume could have been used; but the wave functions ψ_e and ψ_ν in (8-12) would then have to be normalized over the same volume, and the volume used would subsequently cancel out of the equations.

The volume in (three-dimensional) momentum space corresponding to momentum between p and $p + dp$ is given by the spherical shell with inner radius p and outer radius $p + dp$; this is equal to $4\pi p^2 \, dp$.

dE_0 and with electron momentum in the range $p_e \rightarrow p_e + dp_e$ is

$$dn = \frac{4\pi p_\nu^2\, dp_\nu}{h^3} \cdot \frac{4\pi p_e^2\, dp_e}{h^3}. \tag{8-15}$$

We can now substitute (8-13) and (8-14) into (8-15) and obtain

$$dn = \frac{16\pi^2}{h^6 c^3}\, p_e^2 (E_0 - E_e)^2\, dp_e\, dE_0. \tag{8-16}$$

It is customary to express momentum in units of $m_0 c$ and *total* energy W (kinetic plus rest energy) in units of $m_0 c^2$, that is, to set $p_e/m_0 c = \eta$, $(E_e/m_0 c^2) + 1 = W$, and $(E_0/m_0 c^2) + 1 = W_0$. We can then rewrite (8-16) as

$$\frac{dn}{dE_0} = \frac{16\pi^2 m_0^5 c^4}{h^6}\, \eta^2 (W_0 - W)^2\, d\eta, \tag{8-17}$$

or, making use of the relativistic relation (appendix B) $\eta^2 = W^2 - 1$ and therefore $\eta\, d\eta = W\, dW$,

$$\frac{dn}{dE_0} = \frac{16\pi^2 m_0^5 c^4}{h^6}\, W(W^2 - 1)^{\frac{1}{2}} (W_0 - W)^2\, dW. \tag{8-18}$$

We readily see from (8-18) that dn/dE_0 goes to zero both at $W = 1$ and at $W = W_0$. The characteristic bell shape of β spectra (figure 2-9) is thus at least qualitatively reproduced by the statistical factor (8-18). For β emitters of low Z the agreement with experimental spectrum shapes is almost quantitative.

Coulomb Correction. So far we have neglected the Coulomb interaction between the nucleus and the emitted electron. The effect of this interaction is to decelerate negatrons and to accelerate positrons, so that negatron spectra may be expected to contain more, positron spectra fewer, low-energy particles than predicted by the purely statistical considerations of the preceding paragraphs. This indeed corresponds to experimental observations, as shown, for example, by the measured shapes of the negatron and positron spectra of Cu^{64} which happen to have similar endpoint energies (0.57 and 0.65 MeV). They are displayed in figure 8-7. Formally, the Coulomb interaction may be treated as a perturbation on the electron wave function $\psi_e(0)$; the entire spectrum (8-18) then has to be multiplied by a Coulomb correction factor $F(Z, W)$ which is defined as the ratio of $|\psi_e(0)|^2_{\text{Coul}}$ to $|\psi_e(0)|^2_{\text{free}}$. The nonrelativistic result for $F(Z, W)$ is

$$F(Z, W) = \frac{2\pi x}{1 - \exp(-2\pi x)}, \tag{8-19}$$

where $x = \pm Ze^2/\hbar v$, with the $+$ sign applicable to negatrons, the $-$ sign to positrons, v the velocity of the β particle far from the nucleus, and Z the atomic number of the *product* nucleus.

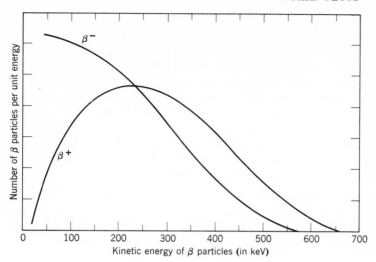

Fig. 8-7   Energy spectra of the positrons and negatrons emitted by Cu⁶⁴.   The pronounced difference between the two spectral shapes results largely from the Coulomb effect.   [Data from J. R. Reitz, *Phys. Rev.* **77**, 10 (1950).]

Since the Coulomb effect is most important for the lowest-energy electrons emitted, the nonrelativistic Coulomb correction (8-19) is a fairly useful approximation in many cases.   For precise computations, however, the much more complex relativistic form of $F(Z, W)$ given by Fermi (F1) must be used. Values of this relativistic Coulomb correction factor for a wide range of Z and W are available in tabular form (U1).   Additional correction terms for the screening effects of extranuclear electrons have also been calculated (see, for example, E1, p. 552) and may be important, especially in enhancing the emission of very-low-energy positrons.

Kurie Plots.   Much effort has been expended by experimenters in checking the theoretical predictions about spectrum shapes.   In most β-ray spectrometers β particles are analyzed according to their momenta, and the quantity measured is the (relative) number of β particles per unit momentum.   Therefore, for comparisons between theory and experiment, the spectrum as given by (8-17), modified by the Coulomb function $F(Z, W)$, is convenient.   We then have for the probability of electron momentum between η and $\eta + d\eta$ (with η in units of $m_0 c$)

$$P(\eta)\, d\eta \propto F(Z, W)\eta^2(W_0 - W)^2\, d\eta, \qquad (8\text{-}20)$$

provided that the transition matrix element M_{if} is independent of the energy partition between electron and neutrino.   As pointed out by Kurie et al. (K1), it follows from (8-20) that a plot of $[P(\eta)/\eta^2 F(Z, W)]^{1/2}$ against W should be a straight line for allowed transitions, with the intercept on the energy axis at W_0. Such plots, with $P(\eta)$ from spectrometer data, are known as Kurie plots, Fermi plots, or F-K plots, and have proved exceedingly useful for the analysis of β spectra. Extrapolation of Kurie plots to the energy axis is the only reliable means for the determination of β-ray endpoint energies.[14] The theoretical predictions of allowed spectrum shapes have now been extremely well verified, a major triumph for the Fermi theory. However, for many years much confusion and misinterpretation resulted from the effects of electron scattering in β-spectrometer sources and their mountings, which can cause sizable shifts of β spectra to lower energies. Extremely thin sources and source backings are required for careful measurements of spectral shapes.

Comparative Half-Lives. Returning now to (8-12) and inserting the spectrum shape from (8-18) and the Coulomb correction $F(Z, W)$, we obtain for the probability per unit time that an electron in the energy interval $W \rightarrow W + dW$ is emitted:

$$P(W) \, dW = \frac{64\pi^4 m_0^5 c^4 g^2}{h^7} \left| M_{if} \right|^2 F(Z, W) \, W(W^2 - 1)^{\frac{1}{2}}(W_0 - W)^2 \, dW. \quad (8\text{-}21)$$

[Note that the electron and neutrino wave functions $\psi_e(0)$ and $\psi_\nu(0)$ no longer appear because of the normalization used; see footnote 13.]

Integrating (8-21) over all values of W from 1 to W_0, we obtain the total probability per unit time that a β particle is emitted, which is just the decay constant λ:

$$\lambda = \frac{64\pi^4 m_0^5 c^4 g^2}{h^7} \cdot \left| M_{if} \right|^2 \cdot \int_1^{W_0} F(Z, W) \, W(W^2 - 1)^{\frac{1}{2}}(W_0 - W)^2 \, dW. \quad (8\text{-}22)$$

We have here again assumed that the nuclear matrix element is independent of β energy, an assumption valid for allowed transitions only. If we denote the integral in (8-22) as f_0 and combine the natural constants as $K = 64\pi^4 m_0^5 c^4 g^2/h^7$, we can write

$$\lambda = \frac{\ln 2}{t_{\frac{1}{2}}} = K \left| M_{if} \right|^2 f_0. \quad (8\text{-}23)$$

Thus the product $f_0 t_{\frac{1}{2}}$, usually denoted as the $f_0 t$ (or, more loosely, ft) value and called the comparative half-life of a transition should be approximately the same for all transitions with similar matrix elements. The ft value may be thought of as the half-life corrected for differences in Z and W_0.

---

[14] It should be mentioned that Kurie plots, even for allowed transitions, will be straight all the way to the energy axis only if the neutrino rest mass m_ν is zero. For finite values of m_ν the spectrum shape will be modified [through the necessary modification of (8-13)] to make the Kurie plot near the endpoint turn down toward the axis. Careful measurements of the shape of the β spectrum of H^3 near the end point have set an upper limit of 0.25 keV for m_ν.

The integral f_0 can be evaluated in closed form only when $F(Z, W) = 1$, that is, for very small Z. Then we find that

$$\lambda \propto (W_0^2 - 1)^{\frac{1}{2}} \left(\frac{W_0^4}{30} - \frac{3W_0^2}{20} - \frac{2}{15} \right) + \frac{W_0}{4} \ln \left[W_0 + (W_0^2 - 1)^{\frac{1}{2}} \right]. \quad (8\text{-}24)$$

For sufficiently high values of W_0, (8-24) approaches $\lambda \propto W_0^5$, a result that accounts for the approximately linear plot of $\log \lambda$ versus $\log E_{\max}$ (although with slope corresponding to somewhat less than the fifth power) obtained by B. W. Sargent before the development of the Fermi theory. Sargent, in fact, found that the data for the naturally occurring β emitters grouped themselves around two lines on such a plot, separated by $\log \lambda \cong 2.5$, and the terms "allowed" and "forbidden" β transitions originated from this classification. With the much greater number of data points available since the discovery of artificial radioactivity, the scatter on a Sargent plot has become so great that it obscures any groupings, and this method of data presentation has been abandoned.

To obtain $f_0 t$ values from measured decay energies and half-lives, in the general case in which $F(Z, W)$ cannot be neglected, it is convenient to use either the extensive graphs of $\log f_0$ versus W_0 and Z published by Feenberg and Trigg (F2) or the nomograph of $f_0 t$, half-life, and energy given by Moszkowski (M1) and reproduced in a number of standard works (e.g., D1, p. 542). Sufficient accuracy for most purposes (errors less than 0.3 in $\log f$ for $0 < Z < 100$ and for 0.1 MeV $< E_0 < 10$ MeV) can also be obtained with the following approximate expressions:

$$\log f_{\beta^-} = 4.0 \log E_0 + 0.78 + 0.02Z - 0.005(Z - 1) \log E_0, \quad (8\text{-}25)$$

$$\log f_{\beta^+} = 4.0 \log E_0 + 0.79 - 0.007Z - 0.009(Z + 1) \left(\log \frac{E_0}{3} \right)^2. \quad (8\text{-}26)$$

Note that in these equations Z, as before, is the atomic number of the *product* nuclide and E_0 is the kinetic energy, in MeV, of the upper limit of the spectrum.

Electron Capture. Before proceeding with the discussion of ft values and their significance, we note that we have talked so far entirely about β^- and β^+ emission and have ignored the third type of β transition, orbital electron capture (EC). In electron capture an electron bound with energy E_B MeV is captured and a neutrino of energy E_0 MeV is emitted.[15] The calculation of the statistical factor thus becomes much simpler, since only the neutrino phase space needs to be considered and no integration over energy is involved. Again values of f_{EC} can be obtained from the sources quoted or they can be approximated in the spirit of (8-25) and (8-26) (but with appreciable errors for $E_0 <$

---

[15] Note that in EC decay the decay energy (i.e., the difference between the atomic masses of parent and daughter) equals $E_0(\nu) + E_B$, whereas in β^+ decay it is $E_0(\beta^+) + 2m_0c^2$. The binding energy E_B is that of the electron *before* capture, i.e., in the parent atom.

0.5 MeV at high Z):

$$\log f_{\rm EC} = 2.0 \log E_0 - 5.6 + 3.5 \log (Z + 1). \tag{8-27}$$

It is to be noted that whenever β^+ emission is energetically possible it competes with electron capture. Since initial and final nuclear states are the same for the two modes of decay, the ratio $\lambda_{\rm EC}/\lambda_{\beta^+}$ is, at least for allowed transitions, expected to be completely independent of the nuclear matrix element and just equal to $f_{\rm EC}/f_{\beta^+}$. Measurements of electron-capture-to-positron branching ratios thus constitute an important test for β-decay theory. In general, these ratios increase with decreasing decay energy (going to infinity when β^+ emission becomes impossible at decay energies $\leq 2m_0c^2$) and with increasing Z. The latter trend comes about through the increase in the expectation value for finding orbital (especially K) electrons at the nucleus and through the increasing suppressive effect of the Coulomb factor $F(Z, W)$ on β^+ emission (8-19).

Whenever energetically possible, capture of K electrons predominates over capture of electrons with higher principal quantum numbers because, of all the electron wave functions, those of the K electrons have the largest amplitudes at the nucleus. However, at decay energies below the binding energy of the K electrons, only L-, M-, etc., electron capture is possible. The ratio of $L_{\rm I}$ capture[16] to K capture as a function of decay energy has been calculated (R4) for allowed transitions. The results for $Z \gtrsim 14$ can be represented by the approximate formula

$$\frac{L_{\rm I}}{K} = (0.06 + 0.0011Z) \left[\frac{E_0^{\,L}(\nu)}{E_0^{\,K}(\nu)} \right]^2, \tag{8-28}$$

where $E_0^{\,L}(\nu)$ and $E_0^{\,K}(\nu)$ are the neutrino energies accompanying the two processes; $E_0^{\,L}(\nu)$ exceeds $E_0^{\,K}(\nu)$ by the difference between the binding energies of the two shells. At decay energies not too far in excess of the K-binding energy, $E_0^{\,L}(\nu)/E_0^{\,K}(\nu)$ differs appreciably from unity, and a measurement of the L- to K-capture ratio then permits an estimation of the decay energy by use of (8-28).

Selection Rules. We remarked (on p. 240) that transitions in which electron and neutrino carry away no orbital angular momentum are expected to have the largest transition probabilities (for a given energy release) and we have called these transitions "allowed." If electron and neutrino do not carry off angular momentum, the spins of initial and final nucleus cannot differ by more than one unit of $\hbar$ and their parities must be the same. In fact, if electron and neutrino are emitted with their intrinsic spins antiparallel (singlet state), the nuclear spin change ΔI must be strictly zero; if electron and neutrino spins are parallel (triplet state), ΔI may be $+1$, 0, or -1 (but $0 \to 0$ transitions are forbidden). The former selection rule was the one originally proposed by

---

[16] In allowed transitions the capture probability is small for $L_{\rm II}(p_{1/2})$ electrons and zero for $L_{\rm III}(p_{3/2})$ electrons. The contribution of M capture can usually be neglected too.

Fermi; the latter was subsequently suggested by Gamow and Teller.   Which of these selection rules applies depends on the exact form assumed for the interaction between nucleons and the electron-neutrino field, but the precise form of this interaction is still not completely known.   Here we merely state that five different types of interaction known as scalar (S), vector (V), tensor (T), axial vector (A), and pseudoscalar (P), as well as their linear combinations, are possible.   For S and V interactions Fermi selection rules apply; for T and A interactions Gamow-Teller rules are appropriate.   Much effort has been expended in learning about the interaction types from the ft values of observed transitions.

The information on allowed transitions between states whose spins are known from other data indicates that neither pure Fermi nor pure Gamow-Teller selection rules can account for all the observations.   The best present evidence suggests a mixture of V and A interactions with approximately equal coupling strengths (D1, K3).

From what has been said so far we might expect that all allowed β transitions, that is, all transitions between states of $\Delta I = 0$ or 1 with no parity change, should have (1) the allowed spectrum shape, (2) closely similar $f_0 t$ values. Whereas the first expectation is borne out by all experiments to date, the second is not.   The values of $f_0 t$ extend from $\sim 10^3$ (for example $n \to H^1$) to $\sim 10^9$ (e.g., $C^{14} \to N^{14}$).   There is, however, a strong clustering of $\log f_0 t$ values around 3 to 3.5 and another broader peak with $\log f_0 t$ between 4 and 7. Transitions characterized by the very low $f_0 t$ values in the first of these groups are called "favored" or "superallowed."   They are found mainly among β emitters of low Z and particularly between so-called mirror nuclei.   Two nuclei constitute a mirror pair if one contains n neutrons and $n + 1$ protons, the other $n + 1$ neutrons and n protons; examples are $_1H^3$ and $_2He^3$, $_{12}Mg^{23}$ and $_{11}Na^{23}$.   Provided neutron-neutron and proton-proton forces are the same except for a Coulomb interaction, the wave functions characterizing two mirror nuclei are certainly expected to be very nearly the same, and therefore the square of the nuclear matrix element $|M_{if}|^2$ for a mirror transition should be $\cong 1$.   It is from the decay rates of these superallowed transitions (the simplest one being the decay of the free neutron with $f_0 t \cong 1200$ sec) that the magnitude of the β-decay coupling constant $g \cong 10^{-49}$ erg cm^3 has been estimated.   Once the value of g is known, ft values of other β transitions can be used to obtain information about nuclear matrix elements.

The rather wide range of $f_0 t$ values found for allowed transitions (other than the superallowed ones) indicates that our assumption of approximately equal $|M_{if}|^2$ values for all transitions with $\Delta I = 0, \pm 1$ without parity change was too naïve.   The nuclear matrix elements are evidently sensitive to other factors.   As an extreme illustration we mention the so-called l-forbidden transitions, of which $P^{32} \to S^{32} + \beta^- + \bar{\nu}$ is an example.   Here the spins of P^{32} and S^{32} have been measured as 1 and 0, respectively, and both parities are unquestionably even.   Yet the $\log f_0 t$ value is 7.9, and this large value apparently comes about (as will be made clearer in the discussion of shell-model states in chapter 9, section D) because a $d_{3/2}$ neutron is transformed into an $s_{1/2}$ proton, so that $\Delta l = 2$ even though $\Delta I = 1$.

Forbidden Transitions. The discussion so far has been confined to allowed transitions. Let us now consider briefly what happens when the transition from initial to final nucleus cannot take place by the emission of s-wave electron and neutrino. That electron and neutrino emission with orbital angular momenta other than zero is possible at all comes about because of the finite size of nuclei. The wave functions $\psi_e(0)$ and $\psi_\nu(0)$ "at the nucleus" which appear in (8-12) thus have to be evaluated over the entire nuclear volume and therefore do not vanish for p-wave, d-wave, etc., emission. However, the magnitudes of these electron and neutrino wave functions over the nuclear volume decrease rapidly with increasing orbital angular momentum. Hence, for each unit of angular momentum l_l carried off by the two light particles together, the β-transition probability decreases by several orders of magnitude, and β transitions with $l_l = 1, 2, 3$, etc., are classified as first, second, third, etc., forbidden transitions. The various transition orders, the ranges of $\log f_0 t$ corresponding to them, and some examples are listed in table 8-3.

The selection rules for the various orders of forbiddenness are readily derived. If l_l is odd, initial and final nucleus must have opposite parities ($\Delta\Pi$ yes); for even l_l values the parities must be the same ($\Delta\Pi$ no). Furthermore, as in allowed transitions, the emission of electron and neutrino in the

Table 8-3   Selection rules for beta decay

Type	l_l	ΔI	$\Delta\Pi$	Log ft	Log $[(W_0{}^2 - 1)^{\Delta I-1} ft]$	Examples
Allowed (favored)	0	0 or 1	no	3		H^3, Mg^{23}
Allowed (normal)	0	0 or 1	no	4 to 7		S^{35}, Zn^{69}
Allowed (l-forbidden)	0	1	no	6 to 9		C^{14}, P^{32}
First forbidden	1	0 or 1	yes	6 to 10		Ag^{111}, Ce^{143}
First forbidden (unique)	1	2	yes	($\sim$9)	$\sim$10	Cl^{38}, Sr^{90}
Second forbidden	2	2	no	10 to 14		Cl^{36}, Cs^{135}
Second forbidden (unique)	2	3	no	($\sim$14)	$\sim$15	Be^{10}, Na^{22}
Third forbidden	3	3	yes	17–19		Rb^{87}
Third forbidden (unique)	3	4	yes	($\sim$18)	$\sim$21	K^{40}
Fourth forbidden	4	4	no	$\sim$23		In^{115}
Fourth forbidden (unique)	4	5	no		$\sim$28	

singlet state (Fermi selection rules) requires $\Delta I \leq l_l$, whereas triplet-state emission (Gamow-Teller selection rules) allows $\Delta I \leq l_l + 1$. Assuming again a mixture of Fermi- and Gamow-Teller-type interactions, the selection rules listed in table 8-3 result. Note that values of $\Delta I < l_l$ appear only in first forbidden transition ($l_l = 1$), because in all other cases transitions with such spin changes ΔI are also possible with lower degrees of forbiddenness ($l_l - 2$, etc.). The ranges of $\log f_0 t$ values show a fair amount of overlap, especially between allowed and first forbidden transitions. Thus the determination of $\log f_0 t$ alone can rarely give unambiguous information on ΔI and $\Delta \Pi$; but with other data, and particularly in conjunction with the predictions of nuclear models (see chapter 9), $\log f_0 t$ values are an important aid in making spin and parity assignments.

An illustration of these concepts may be seen in figure 8-8, which shows the decay scheme of $_{11}\text{Na}^{24}$. Almost all the β^- decays are to the second excited state of $_{12}\text{Mg}^{24}$ at 4.14 MeV. Thus, for this transition, $t = 15 \times 60 \times 60 = 5.4 \times 10^4$ sec. From (8-25) $\log f = 1.6$, and $\log ft = 6.3$, which agrees with the "normal allowed" classification for $4 + \rightarrow 4+$, $\Delta I = 0$, no. For the rare 4.15-MeV β^- transition we compute $t = 5.4 \times 10^4 \times 100/0.003 = 1.8 \times 10^9$ sec, $\log ft = 12.7$; again the value is consistent with the level assignments, $4+ \rightarrow 2+$, the transition being second forbidden, $\Delta I = 2$, no. The fourth

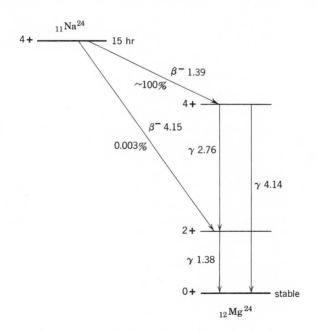

Fig. 8-8   Decay scheme of Na^{24}.   The transition energies are in MeV.   To the left of each level is shown its spin and parity.

forbidden β^- transition directly to the ground state is not observed. For it we compute $\log f = 4.0$, and, taking $\log ft \cong 23$, obtain as an estimate $\log t \cong 19$. Thus the half-life of Na^{24} if it decayed only to the Mg^{24} ground state might be $\sim 3 \times 10^{11}$ years. The expected branching ratio for this mode of decay is $\sim 5 \times 10^{-15}$, which is quite unobservable. Other applications of β-decay selection rules to decay-scheme determinations are discussed in chapter 12, section G.

Additional identification of transition types sometimes comes from spectrum shapes. As noted earlier, the assumption that M_{if} is independent of the energy partition between electron and neutrino applies in general to allowed transitions only. For other transition types (8-20) is usually not valid, and Kurie plots therefore do not give straight lines. However, for each of the various forms of basic β-decay interaction (p. 247) and for each order of forbiddenness it is possible to calculate the additional energy dependence and (in good approximation) to factor out of the matrix element an energy-dependent term (B1, p. 726ff.; K2). By multiplying the right-hand side of (8-20) by the appropriate one of these shape correction factors, we again obtain a function which, if plotted against the β energy W, gives a straight line. The correction factors take on a particularly simple form for the transitions with $\Delta I = l_l + 1$, which are forbidden by the Fermi selection rules and therefore cannot involve interactions other than T and A. This restriction makes the predictions of the theory much less ambiguous for these than for any other transitions— hence they are called "unique" (see table 8-3). The shape correction factor for a unique transition of order l_l can be written

$$\frac{(p_e + p_\nu)^{2l_l+2} - (p_e - p_\nu)^{2l_l+2}}{4p_e p_\nu},$$

which, for first forbidden unique transitions ($l_l = 1$) reduces to $2(p_e{}^2 + p_\nu{}^2)$, in the literature often called the "a_1 correction factor." As illustrated in figure 8-9 for the Y^{91} β spectrum, the use of this correction term indeed linearizes the Kurie plots of β spectra emitted in decays between states characterized by $\Delta I = 2$, yes. The same is true for the unique transitions of higher order when the appropriate correction factors are used. It is now well established that spectra for almost all nonunique first forbidden transitions ($\Delta I = 0$ or 1, yes) have the "allowed" shape (8-20). For higher forbidden nonunique transitions the situation can become quite complex.

If the β spectrum does not have the allowed shape, (8-22) and (8-23) for the decay constant are no longer strictly correct because M_{if} is then not independent of W. Thus we should not use $f_0 t$ as the "comparative half-life," but a corrected ft value. However, this is not customary, and the tabulated ft values are almost always $f_0 t$ values. For the unique forbidden transitions the corrected f value can be approximated by $(W_0{}^2 - 1)^{\Delta I - 1} f_0$. In table 8-3, therefore, we list $\log [(W_0{}^2 - 1)^{\Delta I - 1} f_0 t]$ for these transitions; this quantity is much more nearly constant for a given order of unique transition than is

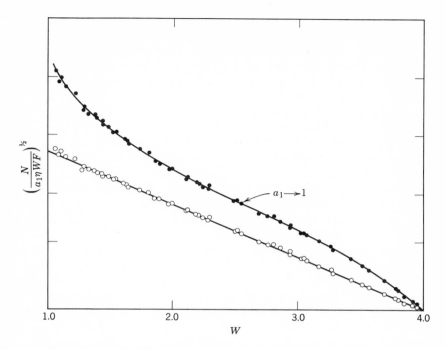

Fig. 8-9   Kurie plot of the Y^{91} β^- spectrum (from reference K2).   The open circles are the data points corrected by the shape factor a_1; they are seen to fall on a straight line.   The closed circles represent the same data but treated as if the transition were allowed ($a_1 = 1$).

log $f_0 t$.   Note, however, that this method of correcting ft values of "unique" transitions, based as it is on considerations of spectrum shapes, should not be applied to EC transitions.

Nonconservation of Parity.   We mentioned in passing (chapter 2, p. 39) that the conservation of parity, long accepted as one of the universal conservation laws, does not hold for weak interactions.   This possibility was suggested in 1956 by T. D. Lee and C. N. Yang (L1) to explain what appeared to be two different decay modes of a single type of particle, a K meson—one to an even-parity (two-pion), one to an odd-parity (three-pion) final state.   Lee and Yang pointed out that no then-existing experimental data proved parity conservation in weak interactions (whereas it was well established for strong and electromagnetic interactions) and suggested some experimental tests. The first experimental verification of nonconservation of parity in weak interactions came in the historic experiment by Wu et al. (W1) in which the emission of β particles from Co^{60} nuclei whose spins were aligned by a magnetic field at very low temperatures (to suppress thermal agitation) was found to be preferentially along the direction opposite to the Co^{60} spin vector.

To understand the implications of this experiment, we must consider the properties of different quantities under space inversion, that is, reflection through a point, or change of sign of space coordinates. So-called polar vectors, such as linear momentum, velocity, or electric field, change sign under this operation, whereas so-called axial vectors, such as angular momentum or magnetic field (which are characterized not only by direction but also by a screw sense), do not change sign. Any observed quantity that is the (scalar) product of two polar vectors or of two axial vectors will be invariant under space inversion; such quantities are called scalars. A number that is the scalar product of one polar vector and one axial vector changes sign under space inversion; the occurrence of such quantities, called pseudoscalars, is prohibited by the requirement of parity conservation.

The important point made by Lee and Yang in their 1956 paper was that none of the experimental data on weak interactions then available could throw any light on the question of parity conservation because the observed quantities were always scalars. The experiment with aligned Co^{60} nuclei was specifically designed to look for a pseudoscalar quantity, namely a component of β-particle intensity proportional to the product of the nuclear spin (an axial vector) and the electron velocity (a polar vector). The asymmetry found established the existence of this pseudoscalar component and thus proved that parity was not conserved in β decay. Since then many other experiments have corroborated nonconservation of parity in all weak interactions. Thus we now know that nature, in this class of processes, distinguishes left from right. In fact, an ingenious experiment by Goldhaber, Grodzins, and Sunyar (K3) has shown that the neutrinos accompanying electron capture (and presumably those accompanying β^+ emission) are "left-handed," that is, they have their spins antiparallel to their direction of motion. Positrons then must be "right-handed," negatrons "left-handed," and the antineutrinos accompanying β^- decay "right-handed."

It is worthwhile to emphasize once more that everything that has been said in this section about spectrum shapes and lifetimes in β decay is unaffected by the overthrow of parity conservation because only scalar quantities are involved. Thus Fermi's basic theory is largely unaffected, except for the need for the inclusion of some additional parity-nonconserving coupling constants in the interaction. On the other hand, the discovery of parity nonconservation has stimulated whole new classes of experiments involving observations of (1) asymmetry of emission from aligned nuclei, (2) polarization of β particles, and (3) correlations between β particles and polarized γ rays. These experiments, which can shed some light on the form of the β-decay interaction are discussed in D1 and K3.

Neutrinos and Antineutrinos. Partly as a result of the discovery of parity nonconservation and partly through difficult experimental work on the very rare neutrino interactions, there has been an important clarification and simplification of our ideas about neutrinos (see, for example, K3 and R1).   As we mentioned on p. 55, Reines and Cowan (R1) have experimentally established the capture of antineutrinos (from a nuclear reactor) by protons and have measured a cross section of about 10^{-43} cm^2 for this process, in rough agreement with theoretical expectations.   On the other hand, R. Davis obtained a null result in attempts to measure the capture of reactor antineutrinos ($\bar{\nu}$) in Cl^{37} to form Ar^{37} (a reaction that presumably requires neutrinos, since it is the inverse of the electron-capture reaction Ar$^{37} + e^- \rightarrow$ Cl$^{37} + \nu$); he could set an upper limit to the cross section which was about $\frac{1}{10}$ of that expected for neutrinos.   Thus neutrinos and antineutrinos are evidently different particles.   This conclusion is in accord with the expectations from the new β-decay theory without parity conservation.   All present evidence is consistent with the view that in β-decay processes there are two and only two types of neutral massless particles: left-handed neutrinos and right-handed antineutrinos.[17]

The distinction between ν and $\bar{\nu}$ removes an ambiguity that has existed with respect to the expected half-lives for double β decay.   If neutrino and antineutrino were identical, this process could take place through a virtual intermediate state, with a "neutrino" produced in the first step and absorbed in the second, each step producing one β^-.   With $\nu \neq \bar{\nu}$, this type of process is excluded, since the first step produces $\bar{\nu}$ and the second would require the absorption of ν.   Instead, the production of $2\beta^- + 2\bar{\nu}$ is required, and the expected lifetime for this process is several orders of magnitude greater than that of the neutrinoless one.   Several experimental lower limits for double-β-decay half-lives (e.g., 10^{18}y for Pd^{110}, 6×10^{18}y for U^{238}, 3×10^{17}y for Sn^{124}) are consistent with the theoretically expected $2\bar{\nu}$ emission.

D. GAMMA DECAY

In chapter 2 we defined as a γ transition any de-excitation of an excited nuclear state to a state of lower excitation but with the same Z and A.   Excited states appear as the result of α- and β-decay processes, of nuclear reactions, of direct excitation from the ground state (by electromagnetic radiation or Coulomb interaction with charged particles), and of γ transitions from higher excited states.   The de-excitation may proceed by emission of a quantum of electromagnetic radiation (γ ray), of an internal conversion electron, of an e^+e^- pair, or of two quanta simultaneously.[18]   Different modes of de-excitation

---

[17] Recent experiments indicate, however, that the neutrinos associated with μ mesons are not identical with those emitted in β decay, so that there appear to be four neutrinos: ν_e, $\bar{\nu}_e$, ν_μ, and $\bar{\nu}_\mu$ (L2).

[18] The last-named process has not been experimentally observed; see p. 257.

from the same level may occur; in fact γ emission and internal conversion generally compete.

An excited state may also emit particles in competition with its γ decay. When the threshold for neutron emission is exceeded by $\gtrsim 1$ MeV, neutron emission usually predominates over gamma decay; however, the gamma decay may not always be negligible, for, although lifetimes for neutron emission to a given level are generally very much shorter than those for γ emission to a given level, the large number of final states available for γ decay may partly compensate for this effect.   The interesting case of competing α and γ emission from the excited states of RaC' (Po^{214}) and ThC' (Po^{212}) fed by β decays has already been mentioned (p. 56).   Usually, lifetimes for β-decay processes are much longer than those for γ decay, and β decay from excited states is therefore observed only when γ decay is strongly forbidden by selection rules, that is, from isomeric states.

Lifetimes of Excited States.   The overwhelming majority of γ transitions take place on a time scale too short for direct measurement, that is, in less than about 3×10^{-11} second, which is about the limit of present electronic techniques for delayed-coincidence measurements (chapter 3, p. 83).   Some indirect methods have been used to measure γ-ray lifetimes down to $\cong 10^{-15}$ second (D2).   Among them is the measurement of the probability of the inverse reaction called Coulomb excitation; this is the excitation from the ground state to an excited state by purely electromagnetic interaction brought about by charged particles which move with insufficient velocity to overcome the Coulomb repulsion and to come within the range of nuclear forces.   Other experiments are designed to give information on the energy width ΔE (full width at half maximum) of the excited level of interest;[19] the mean life Δt is then deduced from application of the uncertainty principle: $\Delta E \cdot \Delta t = h/2\pi$.

From the measurements just sketched as well as from the relative probabilities of nucleon and photon emission already mentioned it is clear that most γ transitions take place in 10^{-13} to 10^{-16} second.   This is in agreement with the order of magnitude of radiative lifetimes to be expected for a dipole of nuclear dimensions and unit electronic charge.   Whether or not one wishes to consider γ transitions of such short lifetimes as radioactive decay processes is a matter of taste and convention.   Since they so generally occur with, or rather

[19] The processes used are resonance scattering and resonance absorption of γ rays.   To excite a nucleus of mass M to a state that lies at an energy E above the ground state, an incident γ ray must have an energy $E + E^2/2Mc^2$ because of the requirement of momentum conservation.   Furthermore, if the γ ray from the de-excitation of the same level is to be used, an additional energy increment $E^2/2Mc^2$ is lost to the recoil of the source.   Resonance absorption or scattering of this γ ray becomes possible only if this energy deficit E^2/Mc^2 is somehow supplied (for an exception, see chapter 13, section A).   This can be brought about by mechanical motion of the source, by heat (thermal agitation of the source atoms), or by taking advantage of the recoil imparted to the source nuclei by some preceding radiation such as β particles.   By varying the extra energy supplied, one can measure the width of the resonance and thus the level width ΔE.

following, α- and β-decay processes, they are certainly of vital importance in all types of radioactivity measurements and in the establishment of nuclear level schemes, even when their half-lives cannot be measured. However, in this section we shall be mostly concerned with the occurrence of γ transitions of much longer lifetimes and with the factors that make possible the existence of the metastable or isomeric nuclear states from which such γ transitions occur.

As we remarked in chapter 2 (p. 57), the definition of a nuclear isomer in terms of a "measurable half-life" has become quite vague in view of the newer direct and indirect measuring techniques. Lifetimes of excited states evidently range all the way from the "instantaneous" region ($< 10^{-13}$ second) to many years. We shall now discuss the connection between transition probability, decay energy, and spins and parities of initial and final states.

Multipole Radiation and Selection Rules. Gamma radiation arises from purely electromagnetic effects which may be thought of as changes in the charge and current distributions in nuclei. Since charge distributions give rise to electric moments and current distributions to magnetic moments, γ-ray transitions are correspondingly classified as electric (E) and magnetic (M). In addition, it is convenient, as in β decay, to characterize transitions according to the angular momentum l (in units of $\hbar$), which the γ ray carries off. We shall see that, as in β decay, transition probabilities fall off rapidly with increasing angular-momentum changes. The accepted nomenclature[20] is to refer to radiations carrying off $l = 1, 2, 3, 4, 5$ units of $\hbar$ as dipole, quadrupole, octupole, 2^4-pole, and 2^5-pole radiations. The shorthand notation for electric (or magnetic 2^l-pole radiation is El (or Ml); thus $E2$ means electric quadrupole, $M4$, magnetic 2^4 pole, etc. The electric and magnetic multipole radiations differ in their parity properties. If we denote even and odd parity of the radiation by $+1$ and -1, then electric 2^l-pole radiation has parity $(-1)^l$, whereas magnetic 2^l-pole radiation has parity $(-1)^{l+1}$.

We can now formulate the selection rules for γ transition between an initial state of spin I_i and a final state of spin I_f and with either equal or opposite parities. It follows immediately from what was said above about angular momenta associated with 2^l-pole radiation that $l \geq |I_i - I_f|$. However, consideration of the vector addition of the angular momenta involved leads to the

---

[20] The multipole nomenclature developed because the radiation field around a system of oscillating charges can always be expressed as an expansion in spherical harmonics of orders $1, 2, 3 \ldots$; for a pure dipole radiator the first nonvanishing term in this series is the first term, for a quadrupole radiator the second term, etc. Furthermore, the successive terms in this multipole expansion correspond to the photon carrying off $1, 2, 3$, etc., units of angular momentum. The lth term in the expansion of the field is proportional to $(R/\lambda)^l$, where R is the dimension of the radiator ($\cong$ nuclear radius) and λ the wavelength of the emitted radiation divided by 2π. For γ rays λ is always large compared with nuclear dimensions (for a 1-MeV photon, $\lambda \cong 2 \times 10^{-11}$ cm), so that the series converges rapidly and only the first nonvanishing term usually needs to be considered.

further restriction that l cannot exceed $I_i + I_f$. Thus we have, for both electric and magnetic radiations,

$$I_i + I_f \geq l \geq |I_i - I_f|. \tag{8-29}$$

If initial and final state have the same parity, electric multipoles of even l and magnetic multipoles of odd l are allowed. If initial and final state have opposite parities, electric multipoles of odd l and magnetic multipoles of even l are allowed. As an example, if the transition is between a $4+$ and a $2+$ state, multipole orders l can range from 2 to 6, but because of the parity rules $E2$, $M3$, $E4$, $M5$, and $E6$ are the only radiations possible.

The actual situation is simpler because as a rule only the lowest multipole order (sometimes the lowest two) allowed by the selection rules contributes appreciably to the intensity. This comes about as follows: the transition probability is proportional to the square of the matrix element for the interaction; the contribution of each term in the power-series expansion of the field (see footnote 20) to the transition probability is therefore proportional to $(R/\lambda)^{2l}$. Since R/λ is always a small number ($\cong 10^{-2} - 10^{-3}$), only the lowest allowed multipole order will generally contribute. Exceptions to this rule occur commonly when the lowest allowed radiation is magnetic dipole ($M1$); here electric quadrupole ($E2$) transitions evidently can often compete favorably. This can be understood if we remember that the current densities in nuclei (which give rise to the magnetic multipoles) are smaller than the charge densities (which produce the electric multipoles) by $\sim v/c$, where v represents the speed of the charges (protons) in the nucleus; for a given multipole order the magnetic transitions therefore can be expected to be weaker than the electric ones by a factor of the order of $(v/c)^2 \cong 10^{-2}$. (This neglects the contributions of the intrinsic magnetic moments of nucleons.) Thus we might expect $E(l + 1)$ radiation to compete with Ml; this expectation, as we remarked, is often borne out experimentally for $l = 1$ but has not been clearly established for higher-order transitions.

Table 8-4   Selection rules for gamma transitions

ΔI	0^a	0^a	1	1	2	2	3	3	4	4	etc.
$\Delta \Pi$	No	Yes	No	Yes	No	Yes	No	Yes	No	Yes	
Transition type	$M1$	$E1$	$M1$	$E1$	$E2$	$M2$	$M3$	$E3$	$E4$	$M4$	
	$E2$		$E2^b$			$(E3^b)$	$(E4^b)$			$(E5^b)$	

[a] The more complete selection rule (8-29) excludes *any* single-photon transitions when $I_i = I_f = 0$ (photons have intrinsic spin 1). For alternative de-excitation processes in this situation see text. For transitions between two $I = \frac{1}{2}$ states of equal parity $E2$ is forbidden by (8-29), but $M1$ is allowed.

[b] When one of the states involved in a transition has spin 0 and the allowed transition of lowest order is a magnetic multipole, the next higher electric multipole is strictly forbidden by (8-29).

The selection rules discussed are summarized in table 8-4.   Some significant special cases not yet mentioned are noted in the table.   They all stem from the restriction $l \leq I_i + I_f$.   In particular, $0 \rightarrow 0$ transitions ($I_i = 0$, $I_f = 0$) cannot take place by photon emission, essentially because a photon has spin 1 and therefore must (vectorially) remove at least one unit of angular momentum (this condition can always be fulfilled for other $\Delta I = 0$ transitions by proper orientation of the vectors $\mathbf{I}_i$ and $\mathbf{I}_f$).   If there is no change in parity in a $0 \rightarrow 0$ transition, de-excitation may occur by emission of an internal-conversion electron or, if $\Delta E > 1.02$ MeV, by simultaneous emission of an electron-positron pair.   Examples of the former mode are known for transitions to the ground states (0+) from 0+ states in Ge^{72} ($\Delta E = 0.68$ MeV; $t_{1/2} = 0.3$ μsec) and in Po^{214} ($\Delta E = 1.42$ MeV).   Pair emission occurs, for example, from a 6.05-MeV state in O^{16} ($t_{1/2} = 7 \times 10^{-11}$ second).   Transition between two $I = 0$ states of opposite parity cannot take place by any first-order process; it would require simultaneous emission of two γ quanta or two conversion electrons.   No such transition has been established.

Isomeric Transitions.   Having stated the selection rules, we are now ready to return to a more quantitative discussion of actual lifetimes for γ transitions, with the eventual aim of comparing theoretical predictions with the experimental observations on isomeric transitions.   We have already stated that the transition probability for emission of 2^l-pole radiation of wavelength $2\pi\lambda$ from a nucleus of radius R should be roughly proportional to $(R/\lambda)^{2l}$.   Since $R \propto A^{1/3}$ and transition energy $E \propto 1/\lambda$, we get for the transition probability or partial decay constant for γ emission

$$\lambda_\gamma \propto E^{2l} A^{2l/3}. \tag{8-30}$$

Thus for a given spin change half-lives will decrease rapidly with increasing A and even more rapidly with increasing E (a more detailed analysis actually gives an E^{2l+1} dependence), and both the A and E dependence become steeper with increasing multipole order.

To go beyond these qualitative statements and to calculate absolute transition probabilities or half-lives, it becomes necessary to make more specific assumptions about the charge and current distributions in nuclei, that is, to choose a nuclear model.   The simplest model for this purpose is the extreme single-particle model (see chapter 9, section D).   The assumption is that a γ transition can be described as the transition of a single nucleon from one angular-momentum state to another, the rest of the nucleus being represented as a potential well.   On this basis Weisskopf has derived expressions for decay constants for electric and magnetic 2^l-pole transitions (B1, p. 583).   These somewhat unwieldy general formulas (see, for example, W2 and D3) reduce to the simple expressions given in table 8-5 for the first few multipole orders. The numerical values given for the illustrative case of $A = 125$, $E = 0.1$ MeV indicate the enormous effect of multipole order on half-life.   Measured half-

Table 8-5   Partial half-lives for gamma transitions calculated on the single-particle model[a]

Transition Type	Partial Half-Life t_γ (sec)		Illustrative t_γ Values (sec) for $A = 125$, $E = 0.1$ MeV
$E1$	5.7×10^{-15}	$E^{-3}A^{-\frac{2}{3}}$	2×10^{-13}
$E2$	6.7×10^{-9}	$E^{-5}A^{-\frac{4}{3}}$	1×10^{-6}
$E3$	1.2×10^{-2}	$E^{-7}A^{-2}$	8
$E4$	3.4×10^{4}	$E^{-9}A^{-\frac{8}{3}}$	9×10^{7}
$E5$	1.3×10^{11}	$E^{-11}A^{-\frac{10}{3}}$	1×10^{15}
$M1$	2.2×10^{-14}	E^{-3}	2×10^{-11}
$M2$	2.6×10^{-8}	$E^{-5}A^{-\frac{2}{3}}$	1×10^{-4}
$M3$	4.9×10^{-2}	$E^{-7}A^{-\frac{4}{3}}$	8×10^{2}
$M4$	1.3×10^{5}	$E^{-9}A^{-2}$	8×10^{9}
$M5$	5.0×10^{11}	$E^{-11}A^{-\frac{8}{3}}$	1×10^{17}

[a] The energies E are expressed in MeV. The nuclear radius parameter r_0 has been taken as 1.3F. Note that t_γ is the partial half-life for γ emission only; the occurrence of internal conversion will always shorten the measured half-life.

lives for γ decay[21] are usually sufficiently close to the values predicted by this single-particle theory to allow determination of the spin change. Particularly for the $M4$ transitions, which are very common among isomers, the agreement is good (usually within a factor of 2 or 3), and for other transitions with $\Delta I > 2$ the calculated values are rarely wrong by more than a factor of 100.

The success of the independent-particle calculations of isomer lifetimes was a strong argument in favor of the shell model (which is a particular form of single-particle model—see chapter 9) and gave great impetus to its further development in the early 1950's (G6, G7). As outlined in chapter 9 section D, the shell model predicts, for a given nucleus, low-lying states of widely differing spins in certain regions of neutron and proton numbers, namely just preceding the shell closures at N or Z values of 50, 82, and 126. These regions coincide exactly with the so-called "islands of isomerism" empirically found, and the shell model indeed accounts remarkably well for the properties of these isomers, including their lifetimes.

The Weisskopf formula (table 8-5) is thought to give lower limits for γ half-lives in the sense that transitions between states whose spins and parities cannot be ascribed completely to the properties of individual nucleons, but come about through interaction between several nucleons outside a closed shell,

[21] Note that the expressions in table 8-5 give the partial half-life t_γ for γ emission only. If the internal-conversion coefficient is α the half-life for de-excitation (which is the measured half-life if there is no other competing decay mode) is $t_{1/2} = t_\gamma/(1 + \alpha)$.

(see chapter 9, p. 283) are expected to be slower.   Many deviations from the Weisskopf lifetimes are in this direction and are ascribed to various forms of multiple-particle configurations, although still in the spirit of the single-particle model.

On the other hand there is a large group of $E2$ transitions in heavy nuclei which are of the order of 100 times faster than the single-particle model would predict.   This suggests some type of collective motion involving not one but many protons.   The properties of these fast $E2$ transitions which occur mainly between the low-lying states of nuclei with neutron numbers between 90 and 120 and above $\sim$140 are indeed best accounted for by the collective model (see chapter 9, section E) which predicts bands of "rotational" states for these spheroidally deformed nuclei; for the even-even nuclei the spins of the successive rotational states are 0, 2, 4, 6 . . . .   Interestingly enough, in the same general region of nuclei some extremely slow (10^3 to 10^8 times single-particle lifetimes) $E1$ transitions between states of opposite parity are observed. On the basis of the collective model this phenomenon is ascribed to the occurrence of less symmetric forms of deformation (such as pear shapes) and the operation of special selection rules for transitions between such states and the normal spheroidal or ellipsoidal states (W2).   The same phenomenon can be described in terms of single-particle states (see chapter 9, section G).

Internal Conversion.   We have noted repeatedly that emission of internal-conversion electrons is an alternative de-excitation process that frequently competes with γ-ray emission.   Total and partial internal-conversion coefficients (designated as α, α_K, α_L, etc.) were defined on p. 57.   The calculation of internal-conversion coefficients is a problem in atomic physics.   It involves the computation of the amplitudes of electron wave functions at the nucleus and can be carried out without regard to nuclear forces.   Calculations have been done in various approximations.   The most complete tabulation is the one by Rose (R2 reprinted in part in S2).   For a wide range of Z and E_γ and for all multipoles up to 2^5 he calculated conversion coefficients for the K shell and the three L subshells (L_I, L_{II}, L_{III}),[22] as well as some M-shell and M-subshell values.   The K and L coefficients were calculated with inclusion of the effects of electron screening and finite nuclear size; the M-shell coefficients are for point nuclei and were obtained without screening corrections.   A small sample of the K and L values is shown in table 8-6.   For accurate values at other energies and in other elements the more complete tables (R2), should, of course, be consulted.   However, the data in table 8-6 can be used to obtain approximate results at other values of E and Z.   Interpolation to other energies is best made on plots of log α versus log E for each multipole order.   In general, the coefficients increase with decreasing energy, increasing Z, and increasing ΔI.

---

[22] The L_I, L_{II}, L_{III} subshells are those containing $2s_{1/2}$, $2p_{1/2}$, and $2p_{3/2}$ electrons, respectively.

Table 8-6   *K- and L-shell conversion coefficients*[a,b]

| | | $E_\gamma =$ | | | | | | | |
| | | 0.15 | | 0.40 | | 1.00 | | 2.00 $m_e c^2$ | |
Z	Transition Type	α_K	α_L	α_K	α_L	α_K	α_L	α_K	α_L
25	E1	7.76(−2)	6.81(−3)	3.86(−3)	3.34(−4)	2.98(−4)	2.57(−5)	6.48(−5)	5.56(−6)
	M1	5.52(−2)	4.98(−3)	4.31(−3)	3.81(−4)	5.04(−4)	4.38(−5)	1.19(−4)	1.02(−5)
	E2	1.11(0)	1.12(−1)	2.56(−2)	2.33(−3)	1.01(−3)	8.87(−5)	1.47(−4)	1.28(−5)
	M2	7.18(−1)	7.45(−2)	2.50(−2)	2.33(−3)	1.58(−3)	1.39(−4)	2.61(−4)	2.28(−5)
	E3	1.31(1)	1.88(0)	1.45(−1)	1.45(−2)	3.04(−3)	2.76(−4)	3.05(−4)	2.68(−5)
	M3	8.90(0)	1.15(0)	1.40(−1)	1.42(−2)	4.66(−3)	4.26(−4)	5.37(−4)	4.75(−5)
	E4	1.48(2)	3.41(1)	7.73(−1)	9.12(−2)	8.75(−3)	8.29(−4)	6.11(−4)	5.46(−5)
	M4	1.11(2)	1.86(1)	7.84(−1)	8.82(−2)	1.36(−2)	1.29(−3)	1.08(−3)	9.68(−5)
	E5	1.66(3)	6.30(2)	4.09(0)	5.91(−1)	2.48(−2)	2.50(−3)	1.21(−3)	1.10(−4)
	M5	1.40(3)	3.08(2)	4.40(0)	5.61(−1)	3.95(−2)	3.92(−3)	2.14(−3)	1.96(−4)
45	E1	2.73(−1)	3.31(−2)	1.65(−2)	1.91(−3)	1.49(−3)	1.70(−4)	3.41(−4)	3.83(−5)
	M1	6.42(−1)	8.07(−2)	4.31(−2)	5.25(−3)	4.29(−3)	5.12(−4)	8.78(−4)	1.03(−4)
	E2	2.57(0)	7.41(−1)	8.84(−2)	1.34(−2)	4.61(−3)	5.72(−4)	7.89(−4)	9.15(−5)
	M2	7.86(0)	1.45(0)	2.42(−1)	3.45(−2)	1.37(−2)	1.72(−3)	2.09(−3)	2.49(−4)
	E3	1.94(1)	2.13(1)	4.10(−1)	1.05(−1)	1.27(−2)	1.87(−3)	1.62(−3)	2.00(−4)
	M3	7.44(1)	2.70(1)	1.22(0)	2.26(−1)	3.89(−2)	5.41(−3)	4.31(−3)	5.35(−4)
	E4	1.40(2)	5.02(2)	1.77(0)	8.66(−1)	3.37(−2)	6.24(−3)	3.20(−3)	4.28(−4)
	M4	6.77(2)	5.02(2)	6.24(0)	1.54(0)	1.08(−1)	1.70(−2)	8.51(−3)	1.12(−3)
	E5	1.01(3)	9.89(3)	7.67(0)	6.93(0)	8.85(−2)	2.14(−2)	6.17(−3)	9.12(−4)
	M5	6.09(3)	9.36(3)	3.03(1)	1.08(1)	3.00(−1)	5.47(−2)	1.65(−2)	2.33(−3)
70	E1	5.54(−1)	9.35(−2)	4.47(−2)	6.44(−3)	4.91(−3)	6.94(−4)	1.23(−3)	1.67(−4)
	M1	6.32(0)	8.75(−1)	3.89(−1)	5.37(−2)	3.37(−2)	4.66(−3)	5.94(−3)	8.24(−4)
	E2	1.52(0)	5.81(0)	1.55(−1)	7.23(−2)	1.33(−2)	2.79(−3)	2.98(−3)	4.71(−4)
	M2	5.99(1)	2.17(1)	1.86(0)	4.18(−1)	1.00(−1)	1.81(−2)	1.44(−2)	2.35(−3)
	E3	2.53(0)	2.11(2)	4.90(−1)	8.85(−1)	3.38(−2)	1.23(−2)	6.23(−3)	1.22(−3)
	M3	2.02(2)	4.94(2)	7.09(0)	2.97(0)	2.52(−1)	5.78(−2)	2.80(−2)	5.07(−3)
	E4	3.50(0)	4.45(3)	1.55(0)	8.32(0)	9.37(−2)	5.13(−2)	1.22(−2)	3.01(−3)
	M4	4.65(2)	9.99(3)	2.61(1)	2.15(1)	6.22(−1)	1.87(−1)	5.23(−2)	1.08(−2)
	E5	5.16(0)	7.49(4)	5.12(0)	6.64(1)	2.04(−1)	2.00(−1)	2.32(−2)	7.18(−3)
	M5	8.73(2)	1.78(5)	9.64(1)	1.56(2)	1.52(0)	6.24(−1)	9.52(−2)	2.29(−2)
95	E1		1.80(−1)	7.79(−2)	1.62(−2)	1.15(−2)	2.32(−3)	3.34(−3)	6.63(−4)
	M1		1.25(1)	3.25(0)	7.63(−1)	2.62(−1)	6.52(−2)	4.12(−2)	1.13(−2)
	E2		4.60(1)	1.43(−1)	5.31(−1)	3.14(−2)	1.97(−2)	9.52(−3)	3.14(−3)
	M2		2.93(2)	1.05(1)	5.14(0)	6.08(−1)	2.04(−1)	8.79(−2)	2.56(−2)
	E3		1.42(3)	2.86(−1)	6.89(0)	7.85(−2)	1.09(−1)	2.11(−2)	9.72(−3)
	M3		6.47(3)	2.07(1)	3.22(1)	1.18(0)	5.80(−1)	1.47(−1)	5.00(−2)
	E4		2.19(4)	5.77(−1)	5.67(1)	1.80(−1)	4.64(−1)	4.06(−2)	2.63(−2)
	M4		1.21(5)	3.69(1)	2.18(2)	2.24(0)	1.76(0)	2.37(−1)	9.98(−2)
	E5		2.62(5)	1.17(0)	3.88(2)	3.99(−1)	1.73(0)	7.33(−2)	6.47(−2)
	M5		1.84(6)	6.48(1)	1.48(3)	4.32(0)	5.54(0)	3.79(−1)	2.04(−1)

[a] From M. E. Rose (R2).
[b] Following each number in parentheses is the power of 10 by which the number is to be multiplied; thus 4.06(−2) means 4.06×10^{-2}. Note that the transition energies are given in units of $m_e c^2$ (= 511 keV).

Experimental determination of absolute conversion coefficients is difficult, since it entails the measurement of conversion-electron and γ-ray intensities with known detection efficiencies. In practice, it is much easier to determine, in an electron spectrograph, the *relative* intensities of two or more conversion-electron lines belonging to the same transition and to compare these ratios with theoretical values. As can be seen from table 8-6, the α_K/α_L ratios, although they do not vary over ranges as wide as the individual coefficients, can be used to great advantage to characterize multipole order and thus ΔI and $\Delta\Pi$, especially at relatively high Z and low energy. As pointed out by Goldhaber and Sunyar (G7), α_K/α_L ratios for a given multipole type vary approximately as Z^2/E.

Still further help in identification can be obtained from α_L/α_M ratios and especially from the L-subshell ratios $\alpha_{L_I}/\alpha_{L_{II}}$, $\alpha_{L_I}/\alpha_{L_{III}}$, etc. These subshell ratios usually differ strongly for electric and magnetic multipoles of the same order. For example, at $Z = 70$ and $E = 0.40m_ec^2 = 204$ keV the α_K/α_L ratios for $E1$ and $M1$ transitions are nearly the same (6.94 and 7.25 from table 8-6); yet the L_{III}/L_I ratio is 0.17 for $E1$ and 0.011 for $M1$. In heavy elements transition energies may often be below the K-binding energy, and then the L-subshell-, L/M, etc., ratios are particularly important tools for identification. They are also helpful for determination of mixing ratios in mixed $M1$-$E2$ transitions. The resolution of electron lines originating from the different L subshells and from higher shells requires very thin sources as well as electron spectrometers of high resolution. A useful compilation of electron-binding energies in the various shells of all the elements is available (H5, S2).

The atomic rearrangement processes which follow internal conversion and the attendant emission of X rays and Auger electrons have already been discussed in chapter 2 (p. 52).

Angular Correlations (B3, F3). In discussing the various techniques for identification of multipole character of γ transitions—half-lives and conversion coefficients—we have made the tacit assumption that once removed from the decaying nuclei the γ rays themselves bear no recognizable mark of the multipole interaction that gave birth to them. This is indeed correct under ordinary circumstances. However, it is true that different multipole fields give rise to different angular distributions of the emitted radiation with respect to the nuclear-spin direction of the emitting nucleus. We ordinarily deal with samples of radioactive material which contain randomly oriented nuclei and therefore the observed angular distribution of γ rays is isotropic.

If it were possible to align the nuclear spins in a γ-emitting sample in one direction, the angular distribution of emitted γ-ray intensity would depend in a definite and theoretically calculable way on the initial nuclear spin and the multipole character of the radiation. One method for obtaining alignment of nuclear spins is the application of strong external electric or magnetic fields at temperatures near 0°K. This technique, which requires costly specialized equipment, has found limited but important applications; we shall not discuss it here but refer the reader to some recent reviews (see, for example, A1, R3).

A second and more widely applicable method for obtaining partially "oriented" nuclei is to observe a γ ray in coincidence with a preceding radiation (α, β, or γ). By selecting a particular direction of emission for this first radiation, we then in effect select a preferred direction for the spin orientation of the intermediate nucleus; provided the lifetime of this intermediate state is short enough for the spin orientation to be preserved until the γ ray is emitted, the direction of the γ-ray emission will be correlated with the direction of emission of the preceding radiation.   If a coincidence experiment is done, in which the angle θ between the two sample-detector axes is varied, the coincidence rate will in general vary as a function of θ.

Theoretical correlation functions $W(\theta)\, d\Omega$ have been calculated for a great variety of situations (see, for example, D3).   Here $W(\theta)\, d\Omega$ denotes the relative probability that the second radiation will be emitted into solid angle $d\Omega$ if the angle between the two directions of emission is θ.   Usually the correlation function is normalized so that $\int W(\theta)\, d\Omega = 1$.   The correlation function can then always be written in the form

$$W(\theta) = 1 + a_2 \cos^2 \theta + a_4 \cos^4 \theta + \cdots , \tag{8-31}$$

where only even powers of $\cos \theta$ appear.   If the angular momenta carried away by the first and second radiation are denoted by l_1 and l_2 and the spin of the intermediate state by I, the highest power of $\cos \theta$ that occurs is less than or equal to twice the smallest of these three numbers (l_1, l_2, I).[23]   If something is known about two or at least one of these quantities, angular-correlation experiments can then give information on the other(s).   For example, the angular correlations involving a $\Delta I = 1$ transition without parity change are different, depending on whether the transition is $M1$ or $E2$.   Comparison of measured and calculated correlation functions can thus give the $M1/E2$ mixing ratio for the transition.   The coefficients a_2, a_4, etc., have been tabulated for many types of cascades (D3, B3).   Directional correlations in themselves cannot distinguish electric and magnetic transitions of the same multipole order; the parity change in a transition *can* be inferred if, in addition to the directional correlation, the polarization of the γ rays is measured.

Angular correlation experiments can usually be performed only when transitions with low multipole orders are involved; for $l_2 > 2$, the lifetime of the intermediate state is usually so long that the correlation is destroyed.   Thus the $\cos^4 \theta$ term in (8-31) is the highest term that appears in practical cases, and only two parameters need to be determined experimentally.   Angular-correlation measurements therefore do not usually require data at many angles.   A quantity often used to express experimental results is the anisotropy parameter

$$A = \frac{W(180°) - W(90°)}{W(90°)}, \tag{8-32}$$

which is, of course, related to a_2 and a_4 in (8-31): $A = a_2 + a_4$.

---

[23] Angular correlations thus cannot occur if the intermediate state has spin 0 or $\frac{1}{2}$.

The interesting effects which external fields and chemical environment can have on angular correlations are discussed in chapter 13.

REFERENCES

A1 E. Ambler, "Nuclear Orientation," in *Methods of Experimental Physics*, Vol. 5B, *Nuclear Physics* (L. C. L. Yuan and C. S. Wu, Editors), Academic, New York, 1963, pp. 162–214.

*B1 J. M. Blatt and V. F. Weisskopf, *Theoretical Nuclear Physics*, Wiley, New York, 1952.

B2 N. Bohr and J. A. Wheeler, "The Mechanism of Nuclear Fissions," *Phys. Rev.* **56,** 426 (1939).

B3 L. C. Biedenharn and M. E. Rose, "Theory of Angular Correlation of Nuclear Radiations," *Rev. Mod. Phys.*, **25,** 729 (1953).

*D1 M. Deutsch and O. Kofoed-Hansen, "Beta Rays," in *Experimental Nuclear Physics*, Vol. III (E. Segrè, Editor), Wiley, New York, 1959, pp. 427–638.

D2 S. Devons, "The Measurement of Very Short Lifetimes," in *Nuclear Spectroscopy*, Part A, (F. Ajzenberg-Selove, Editor), Academic, New York, 1960, pp. 512–547.

*D3 M. Deutsch and O. Kofoed-Hansen, "Gamma Rays," in *Experimental Nuclear Physics*, Vol. III, (E. Segrè, Editor), Wiley, New York, 1959, pp. 258–425.

*E1 R. D. Evans, *The Atomic Nucleus*, McGraw Hill, New York, 1955.

F1 E. Fermi, "Versuch einer Theorie der β-Strahlen," *Z. Physik* **88,** 161 (1934).

F2 E. Feenberg and G. Trigg, "The Interpretation of Comparative Half-lives in the Fermi Theory of Beta Decay," *Rev. Mod. Phys.* **22,** 399 (1950).

F3 H. Frauenfelder, "Angular Correlation," in *Methods of Experimental Physics*, Vol. 5B, *Nuclear Physics* (L. C. L. Yuan and C. S. Wu, Editors), Academic, New York, 1963, pp. 129–151.

G1 H. Geiger and J. M. Nuttall, "The Ranges of the α-Particles from Various Radioactive Substances and a Relation between Range and Period of Transformation," *Phil. Mag.* **22,** 613 (1911); **23,** 439 (1912).

G2 G. Gamow, "Zur Quantentheorie des Atomkernes," *Z. Physik* **51,** 204 (1928).

G3 R. W. Gurney and E. U. Condon, "Quantum Mechanics and Radioactive Disintegration," *Nature* **122,** 439 (1928); *Phys. Rev.* **33,** 127 (1929).

G4 V. I. Goldanskii, "On Neutron-Deficient Isotopes of Light Nuclei and the Phenomena of Proton and Two-Proton Radioactivity," *Nucl. Phys.* **19,** 482 (1960).

G5 V. I. Goldanskii, "Two-Proton Radioactivity," *Nucl. Phys.* **27,** 648 (1961).

G6 M. Goldhaber and R. D. Hill, "Nuclear Isomerism and Shell Structure," *Rev. Mod. Phys.* **24,** 179–239 (1952).

G7 M. Goldhaber and A. W. Sunyar, "Classification of Nuclear Isomers," *Phys. Rev.* **83,** 906 (1951).

*H1 G. C. Hanna, "Alpha Radioactivity," in *Experimental Nuclear Physics*, Vol. III (E. Segrè, Editor), Wiley, New York, 1959, pp. 54–257.

H2 O. Hahn and F. Strassmann, "Ueber den Nachweis und das Verhalten der bei der Bestrahlung des Urans mittels Neutronen entstehenden Erdalkalimetalle," *Naturwiss.* **27,** 11 (1939).

H3 J. R. Huizenga, "Spontaneous Fission Systematics," *Phys. Rev.* **94,** 158 (1954).

*H4 I. Halpern, "Nuclear Fission," *Ann. Rev. Nucl. Sci.* **9,** 245–342 (1959).

H5 R. D. Hill, E. Church, and J. Mihelich, "The Determination of Gamma-Ray Energies from Beta-Ray Spectroscopy and a Table of Critical X-Ray Absorption Energies," *Rev. Sci. Instr.* **23,** 523 (1952).

K1 F. N. D. Kurie, J. R. Richardson, and H. C. Paxton, "The Radiations Emitted from Artificially Produced Radioactive Substances. I. The Upper Limit and Shapes of the β-Ray Spectra from Several Elements," *Phys. Rev.* **49,** 368 (1936).

*K2 E. J. Konopinski and L. M. Langer, "The Experimental Clarification of the Theory of β Decay," *Ann. Rev. Nucl. Sci.* **2**, pp. 261–304 (1953).

*K3 E. J. Konopinski, "The Experimental Clarification of the Laws of β Radioactivity," *Ann. Rev. Nucl. Sci.* **9**, 99–158 (1959).

L1   T. D. Lee and C. N. Yang, "Question of Parity Conservation in Weak Interactions," *Phys. Rev.* **104**, 254 (1956).

L2   L. Lederman, "The Two-Neutrino Experiment," *Scientific American* **208**, No. 3, 80 (March 1963).

M1   S. A. Moszkowski, "Rapid Method of Calculating Log (ft) Values," *Phys. Rev.* **82**, 35 (1951).

*P1 I. Perlman, A. Ghiorso, and G. T. Seaborg, "Systematics of Alpha Radioactivity," *Phys. Rev.* **77**, 26 (1950).

R1   F. Reines, "Neutrino Interactions," *Ann. Rev. Nucl. Sci.* **10**, 1–26 (1960).

R2   M. E. Rose, *Internal Conversion Coefficients*, North Holland, Amsterdam, 1958.

R3   L. D. Roberts and J. W. T. Dabbs, "Nuclear Orientation," *Ann. Rev. Nucl. Sci.* **11**, 175–212 (1961).

R4   M. E. Rose and J. L. Jackson, "The Ratio of L_I to K Capture," *Phys. Rev.* **76**, 1540 (1949).

S1   G. T. Seaborg, "Some Comments on the Mechanism of Fission," *Phys. Rev.* **85**, 157 (1952).

*S2 K. Siegbahn (Editor), *Beta- and Gamma-Ray Spectroscopy*, North Holland, Amsterdam, 1955.

U1   U. S. National Bureau of Standards, "Table for Analysis of Beta Spectra," Appl. Math Series No. 13, 1952.

W1   C. S. Wu, E. Ambler, R. W. Hayward, D. D. Hoppes, and R. P. Hudson, "Experimental Test of Parity Conservation in Beta Decay," *Phys. Rev.* **105**, 1413 (1957).

W2   D. H. Wilkinson, "Analysis of Gamma Decay Data," in *Nuclear Spectroscopy*, Part B (F. Ajzenberg-Selove, Editor), Academic, New York, 1960, pp. 852–889.

EXERCISES

1. The nuclide Cf^{254} decays predominantly by spontaneous fission (SF). Estimate its α-decay energy from systematics and, from that estimate and from the known half-life (appendix E), derive a rough prediction of its α/SF branching ratio.

2. Pu^{244} decays by α emission to the ground state of U^{240}. From figure 8-3 and from data in appendix E estimate the mass of Pu^{244}. Can you predict whether Pu^{244} is β stable?

3. With the aid of the semiempirical mass equation (2-5), show how α-decay energies in the heavy-element region are expected to vary with Z for a given A. Compare your conclusion with experimental data (e.g., those plotted in figure 8-4).

4. (a) Use (8-6) to verify the value listed in table 8-1 for the calculated decay constant of the α transition of Th^{228} to the ground state of Ra^{224}. (b) With the aid of the same equation calculate the partial decay constant for the decay of Th^{228} to the first excited state of Ra^{224} at 84 keV above ground. (c) Compare the calculated branching ratio for the two α transitions which you have just obtained with the experimentally determined one shown in figure 8-1 and comment on the degree of agreement.

5. Cm^{243} decays by α emission with a 35-year half-life, mostly to an excited state of Pu^{239} at 286 keV above the ground state. The transitions to the ground and first excited states of Pu^{239} take place in only 1 and 5 per cent of the disintegrations with α-particle kinetic energies of 6.061 and 6.054 MeV, respectively. Estimate the hindrance factors for these two transitions.

6. From figures 8-3, 8-4, and 8-6, predict (in order-of-magnitude fashion) the partial half-lives for α decay and spontaneous fission of (a) Cf^{242}, (b) Pu^{246}.

7. What is the approximate rate of neutron emission from 1 curie of Cf^{252}? Compare this with the neutron emission rate of 1 curie of Po^{210} mixed with beryllium (see chapter 11, section C).

8. Using the semiempirical mass equation, calculate the energy release for spontaneous fission of Cf^{252} (a) into two equal fragments ($_{49}In^{126}$), (b) into $_{54}Xe^{140}$ and $_{44}Ru^{112}$. (c) Which of these two fission modes will have the lower Coulomb barrier?
Answers: (a) 193 MeV; (b) 194 MeV.

9. With the aid of data in appendix E, calculate approximate values of log $f_0 t$ for (a) the β^+ decay of K^{37}, (b) the β^- decay of Ca^{45}, (c) the EC decay of Ca^{41}. Note that no γ emission accompanies any of these decays. Identify the most likely transition type for each of the transitions.

10. (a) Derive an expression for the average total energy of electrons in an allowed β spectrum with maximum total energy W_0, neglecting the Coulomb correction and assuming that the nuclear matrix elements are independent of the electron energy. (b) Compute, under the assumptions of part (a), the ratios of average to maximum kinetic energies for β spectra with maximum kinetic energies of 0.51 and 1.53 MeV. (c) Qualitatively, how would the results of part (b) for β^- emission be affected by inclusion of the Coulomb correction?

Answers:

(a) $$\frac{W_0(W_0^2 - 1)^{1/2}(W_0^4 - 2W_0^2 + 12.25) - 7.5(W_0^2 + 0.5)\ln[W_0 + (W_0^2 - 1)^{1/2}]}{(W_0^2 - 1)^{1/2}(2W_0^4 - 9W_0^2 - 8) + 15\,W_0\ln[W_0 + (W_0^2 - 1)^{1/2}]}$$

(b) 0.38; 0.42.

11. Recalling that a particle of charge e emu and momentum p g cm sec^{-1} in a magnetic field of H gauss moves in a circle of radius $\rho = p/He$ cm, derive an expression for the magnetic rigidity $H\rho$ (in gauss cm) of an electron in terms of its kinetic energy T (in MeV). [*Note.* the required relativistic relations are given in appendix B.]
Answer: $H\rho = 3.3356 \times 10^3 (T^2 + 1.022T)^{1/2}$.

12. The following data were obtained in a measurement of the H^3 β spectrum with a magnetic spectrometer:

$H\rho$ (gauss cm)	150	200	250	300	350	400	425	450	460
Intensity (counts/min)	4995	6812	7645	7274	5336	2434	1040	150	85

Using the expression derived in exercise 11, construct a Kurie plot from these data, taking the Coulomb correction as constant throughout the spectrum. What endpoint energy do you find for the β spectrum? Comment on the probable causes for any deviation from a straight-line Kurie plot that you may find.

13. The ground state of Ni^{61} has spin $I = \frac{3}{2}$ and negative parity. Co^{61} decays with a half-life of 1.65 hours and with the emission of a β^- spectrum with end-point energy of 1.22 MeV to an excited state of Ni^{61} 0.068 MeV above the ground state. The 0.068-MeV transition has a K-conversion coefficient of about 0.12 and a K/L conversion ratio of about 10.3. The β^- transition of Co^{61} to the Ni^{61} ground state takes place in less than 10^{-6} of the disintegrations. What are the most likely spin and parity assignments for (a) Co^{61}, (b) the 0.068-MeV state of Ni^{61}? Give all your reasoning. *Answers:* (a) $7/2-$, (b) $5/2-$.

14. A sample of 79-hr Zr^{89} is found to emit positrons (of 0.905 MeV maximum energy) and yttrium K X rays in the intensity ratio 0.38:1. No γ-ray emission in coincidence with the positrons or X rays is observed. (a) What is the ratio of electron capture to positron decays? (Include the contribution of L-electron capture!) (b) Estimate the log ft values for the two branches and the degree of forbiddenness of the transition.

15. The nuclide Mn^{54} ($t_{1/2} = 280d$, $I = 3$) decays by electron capture to the first excited state of Cr^{54} (2+, 0.835 MeV above ground).   (a) From the masses in appendix E determine the energy of this EC transition.   (b) Calculate its approximate ft value.   (c) From the information you now have can you deduce the parity of Mn^{54}?   (d) What would be the end-point energy of the β^+ spectrum corresponding to the decay of Mn^{54} to the Cr^{54} ground state?   (e) Estimate an upper limit for the fraction of the Mn^{54} decays that might go by this ground-state transition.

16. The 104-day isomer Te^{123m_2} decays by an 88-keV isomeric transition to Te^{123m_1}, which in turn decays with a 1.9×10^{-10}-sec half-life to the Te^{123} ground state. For the 88-keV step the internal conversion coefficient is very large ($\gg 100$), and $\alpha_K/\alpha_L = 0.68$.   The second transition has an energy of 159 keV, $\alpha_K = 0.17$, and $\alpha_K/\alpha_L = 6.6$.   On the basis of these data, select the most likely transition types for the two steps.   Check the partial γ half-lives of the two transitions against the approximate formulas in table 8-5.

17. The first excited state of Pt^{194} is 329 keV above the ground state.   The 329-keV transition has a K-conversion coefficient of 0.045 and a K/L-conversion ratio of 2.2.   (a) Deduce the transition type, hence the spin and parity of the 329-keV state.   (b) The half-life of this state has been measured as 4.5×10^{-12} second. Compare this result with the half-life calculated from the single-particle model. (c) What is the natural width of the 0.329-keV state (in eV)?

18. Show how the special rules stated in the footnotes to table 8-4 follow from the general selection rule 8-29.

19. The 40-hour isomeric state of Hg^{195} decays 63 per cent by electron capture, 37 per cent by isomeric transition (a 123-keV transition followed by 16.2-keV and 37.1-keV transitions in cascade).   The 123-keV transition has $\alpha_K \cong 45$ and $\alpha_K/(\alpha_K + \alpha_L + \alpha_M) \cong 0.080$.   What is the partial half-life for γ emission for the Hg^{195} isomer?   *Answer:* $\cong 6.9$ years.

20. The nuclide Ta^{179} ($t_{1/2} = 1.6y$) decays by electron capture to the ground state of Hf^{179}.   From measurements of K and L X-ray intensities, a value of 1.4 has been deduced for the ratio of L_I capture to K capture.   Approximately what is the energy difference between the ground states of Ta^{179} and Hf^{179}?   The binding energies of K and L_I electrons in hafnium are 65.3 keV and 11.3 keV, respectively.
                                                                         Answer: 90 keV.

9

Nuclear Models

The central theoretical problem of nuclear physics is the derivation of the properties of nuclei from the laws that govern the interactions among nucleons. The central problem in theoretical chemistry is entirely analogous: the derivation of the properties of chemical compounds from the laws (electromagnetic and quantum-mechanical) that determine the interactions among electrons and nuclei. The chemical problem defies any solution other than an approximate one, except for the hydrogen atom, largely because we have no mathematical techniques, other than approximate ones, for analyzing the properties of systems that contain more than two particles. The nuclear problem also suffers from this difficulty, but in addition, it has three others.

1. The law that describes the force between two free nucleons is not completely known.
2. There is reason to believe that the force exerted by one nucleon on another when they are both also interacting with other nucleons is not identical to that which they exert on each other when they are free.
3. Even if the force law were known, it is not certain that the present scheme of quantum mechanics would provide the proper framework within which to analyze nuclear properties.

Under these circumstances there is no alternative but to make simplifying assumptions which provide approximate solutions of the fundamental problem. These assumptions lead to the various models employed; or, more usually, a model for a nucleus or an atom is suggested by experimental results, and subsequently the assumptions consistent with the model are worked out. Consequently, several different models may exist for the description of the same physical situation; each model is used to describe a different aspect of the problem. For example, the Fermi-Thomas model of the atom is particularly useful for calculating quantities such as atomic form factors, which depend mainly on the spatial distribution of electron charge within the atom, but is

267

less good than Hartree's self-consistent field approximation when questions of chemical binding are under analysis.

In the following sections we shall describe the models that have been found useful in codifying a large array of nuclear data; in particular, the energies, spins, and parities of nuclear states, as discussed in chapter 8, as well as nuclear magnetic and quadrupole moments. First we shall sketch what is known about nuclear forces and their implications for the properties of complex nuclei.

A. NUCLEAR FORCES

Information about the forces that exist between two free nucleons may be obtained most directly from observations on the scattering of one nucleon by another and from the properties of the deuteron. The quantity that is immediately useful for calculation is not the force between two nucleons, but rather the potential energy as a function of the coordinates (space, spin, and nucleon type) of the system. The quantity that we seek, therefore, plays a role similar to that of the Coulomb potential in the analysis of atomic and molecular properties and of the gravitational potential in the analysis of the motion of planets and satellites. The nuclear potential, though, seems to be considerably more complex than either the Coulomb or the gravitational potential. Although it is not yet possible to write down a unique expression for the nuclear potential, several of its properties are well known.

Characteristics of Nuclear Potential. The potential energy of two nucleons shows great similarity to the potential-energy function that describes the stretching of a chemical bond.

1. *It is not spherically symmetrical.* For the chemical system this is simply a statement of the directional character of the chemical bond, the direction being determined by the other atoms in the molecule. For the nuclear interaction the direction is determined by the angles between the spin axis of each nucleon and the vector that connects the two nucleons. The quadrupole moment of the deuteron gives unambiguous evidence that the ground state of the deuteron lacks spherical symmetry, hence the potential cannot be a purely central one. The spherically symmetric part of the potential is called a central potential; the asymmetric part is the tensor interaction.

2. *It has a finite range and becomes large and repulsive at small distances.* The potential energy involved in the stretching of a chemical bond is adequately described by the well-known Morse potential which is large and repulsive for the small distances at which electron clouds start to overlap, goes through a minimum several electron volts deep at distances of a few angstroms, and then essentially vanishes at distances of several angstroms. The nuclear potential behaves in much the same way, except that the distances are about

10^5 times smaller and the energies are about 5×10^6 times larger: the nuclear potential becomes repulsive at distances smaller than about 0.4×10^{-13} cm and becomes attractive with a depth of the order of 25 MeV out to about 2.4×10^{-13} cm, by which point it has essentially vanished (see figure 9-1).

The detailed knowledge of the potential energy of the chemical bond comes mainly from information about excited vibrational states and from the determination of bond lengths from either diffraction studies or rotational spectra. The range and depth of the nuclear potential are derived from the binding energy of the only bound state of the deuteron (there are no excited states of the deuteron that are stable with respect to decomposition) and from studies of the collisions between nucleons with energies up to several MeV. The existence of the repulsive core was discovered only when collisions at energies of several hundred MeV were studied.

The factor of 5×10^6 in the relative strengths of the nuclear and chemical forces is the source of the usual remark that nuclear forces are very strong; nevertheless, in view of their short range nuclear forces behave, in point of fact, as if they were *very weak*.   This apparently paradoxical statement can be easily understood when it is recalled that if two particles are to be confined within a distance R of each other they must have a de Broglie wavelength in the center-of-mass system ($\lambda = h/\mu v$, where $\mu = m_1 m_2/(m_1 + m_2)$ is the reduced mass of the two particles and h is Planck's constant) that is no larger than $2R$; the

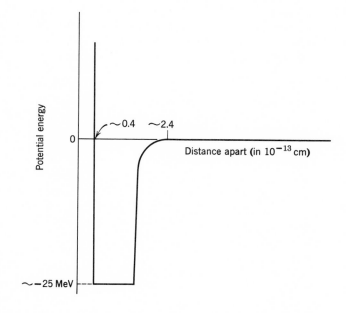

Fig. 9-1   Schematic diagram of nucleon-nucleon potential energy as a function of separation.

wave function that describes the relative motion of the two particles may then vanish at zero distance and at a distance R. If μv is the relative momentum of the two particles in the system, we may write

$$\mu v \geq \frac{h}{2R},$$

hence the kinetic energy of the particles is

$$\tfrac{1}{2}\mu v^2 \geq \frac{h^2}{8\mu R^2}.$$

The kinetic energy of two nucleons which are to remain within the range of nuclear forces must be at least

$$\frac{(6.6 \times 10^{-27})^2}{8 \times \tfrac{1}{2}(1.67 \times 10^{-24}) \times (2.4 \times 10^{-13})^2 \times 1.6 \times 10^{-6}} = 71 \text{ MeV},$$

which is greater than the depth of the potential well that is meant to hold them together. Thus the absence of excited states of the deuteron, its low binding energy ($\cong 2.2$ MeV), and its large size (the proton and neutron spend about one half the time outside the range of the nuclear force) result from the weakness of the nuclear force when viewed in the context of its small range.

The chemical bond, on the other hand, has a range of about 10^5 times that of a nuclear force, and so the kinetic energy requirement is 10^{10} times smaller, or only 10^{-2} eV, which is but a small fraction of the depth of the potential. This large difference between the "real" strengths of the interatomic and internucleon forces is of great importance to our understanding of the properties of nuclear matter.

3. *It depends on the quantum state of the system.* The potential-energy curve that describes the stretching of a chemical bond depends on the electronic state of the molecule. For example, the stable H_2 molecule is one in which the two electrons have opposed spin (singlet state); when the electrons have parallel spin (triplet state), the molecule is unstable with respect to dissociation into two atoms.

The stable state of the deuteron is the one in which neutron and proton have parallel spins (triplet state); the potential energy of the singlet state is sufficiently different from that of the triplet so that there are no bound states of the isolated system consisting of one neutron and one proton with opposed spins.

Scattering experiments suggest that the interaction of two nucleons depends also on their relative angular momentum. For example, a force called the Serber force which has been useful in interpreting scattering experiments is taken to be attractive for all states in which the angular momentum is an even number of $\hbar$ and to vanish when the angular momentum is an odd number of $\hbar$.

It has also been found that the analysis of certain high-energy scattering experiments requires a term in the potential energy that depends on the relative orientations of the vectors describing the intrinsic spins of the nucleons and the orbital angular momentum of the system.   In these experiments it was found that protons scattered out of an initially unpolarized beam of protons (random spin direction) by an unpolarized scatterer were partly polarized.   The interaction that is responsible for this polarization is known as spin-orbit coupling.

4. *It has exchange character.*   Our understanding of the chemical bond entails the exchange of electrons between the bonded atoms.   If, for example, a beam of hydrogen ions were incident on a target of hydrogen atoms and many hydrogen atoms were observed to be ejected in the same direction as the incident beam, any analysis of the problem would have to include the process in which a hydrogen atom in the target merely handed an electron over to a passing hydrogen ion.   The formal result would be that a hydrogen ion and a hydrogen atom would have exchanged coordinates.

It has been observed that the interaction between a beam of high-energy neutrons and a target of protons leads to many events (more than can be explained by "head-on" collisions), in which a high-energy proton is emitted in the direction of the incident neutron beam.   The analysis of the observation entails the idea that the neutron and proton, when within the range of nuclear forces, may exchange roles.   The observation is an excellent example of what is meant by the exchange character of the nuclear potential.   The exchange character of the potential in conjunction with the requirement that the wave function describing the two-nucleon system be antisymmetric can give rise to the type of force described in (3) (B1).

Charge Symmetry and Charge Independence.   So far we have not distinguished among neutron-neutron forces, proton-proton forces, and neutron-proton forces.   The first evident difference is the Coulomb repulsion that must exist between two protons.   At a distance of 10^{-13} cm the Coulomb potential is about $+1.5$ MeV (the zero of potential energy is chosen to be at infinite separation of the two particles), which is to be compared with about -25 MeV for the nuclear potential energy.   Second, since the neutron and proton have differing magnetic moments, there will be different potential energies because of the magnetic interaction; this effect is even smaller than the Coulomb repulsion.   From the observation that the difference in properties of a pair of mirror nuclei (nuclei in which the number of neutrons and the number of protons is interchanged, for example, $_{20}Ca^{41}$ and $_{21}Sc^{41}$) can be accounted for by the differing Coulomb interactions in the two nuclei, the purely nuclear part of the proton-proton interaction in a given quantum state has been taken to be identical to that of two neutrons in the same quantum state as the protons. This identity is known as "charge symmetry."   A more powerful generalization arises from the similarity between neutron-proton scattering and proton-proton scattering when the two systems are in the same quantum state.[1]   This

---

[1] This does not mean that the scattering of neutrons by protons is identical to that of protons by protons; the Pauli exclusion principle makes certain states inaccessible to the two protons

similarity leads to the assumption of "charge independence" which asserts that the interaction of two nucleons depends only on their quantum state and not at all on their type. It is not yet clear that "charge independence" is completely valid, but so far there is no important divergence between this assertion and experimental results.

Isotopic Spin. The charge independence of nuclear forces leads to the idea that the proton and neutron can be considered as two different quantum states of a single particle, the nucleon. Since only two states occur, the situation is analogous to that of the two spin states an electron may exhibit, and thus the whole quantum-mechanical formalism developed for a system of electron spins was taken over for the description of the charge state of a group of nucleons. The physical property that is operative is called the "isotopic spin" (T), and each nucleon has a total isotopic spin of $\frac{1}{2}$ just as the electron has a total spin of $\frac{1}{2}$. The z component of the isotopic spin (T_z) may be either $+\frac{1}{2}$ or $-\frac{1}{2}$; the $+\frac{1}{2}$ state corresponds to a proton and the $-\frac{1}{2}$ state corresponds to a neutron. The vector addition of isotopic spin occurs just as it does for intrinsic spin. Two nucleons, for example, may have a total isotopic spin of either 1 or 0. For $T = 1$, T_z may be $+1$ (2 protons), 0 (a proton and a neutron), or -1 (2 neutrons); for a total spin of 0 the z component can only be 0 (a proton and a neutron). Thus a system containing a proton and a neutron must have $T_z = 0$ but may have $T = 1$ or $T = 0$; a system of two neutrons or of two protons must have $T = 1$. The demands of the Pauli principle are satisfied within this formalism by requiring antisymmetry of the wave function describing the system, which is now, however, a function of three classes of variables: space, spin, and isotopic spin.

$$\psi \text{ (system)} = \psi \text{ (space)} \, \psi \text{ (spin)} \, \psi \text{ (isotopic spin)}.$$

In the ground state of the deuteron, for example, ψ (space) is symmetric (it is a mixture of an s state and a d state), ψ (spin) is symmetric (the two spins are parallel), so that the ψ (isotopic spin) must be antisymmetric and thus $T = 0$ (the two isotopic spins are oppositely oriented). The lowest state of the deuteron in which the two nucleon spins are opposed [ψ (spin) is then antisymmetric] is the lowest one in which $T = 1$. The concept of isotopic spin and its applications are described in detail in refs. B1 and B2.

Meson Theories of Nuclear Forces. The qualitative similarity between the properties of chemical forces and nuclear forces led early investigators, notably H. Yukawa (Y1), to explore the possibility that nuclear forces resulted from the exchange of a particle between two nucleons in a manner analogous to the chemical force (which depends on the exchange of an electron between

---

which may be quite important in the neutron-proton scattering. For example, in low-energy scattering which occurs in states that have no orbital angular momentum (s states) the two protons must have opposite spins ($^1s_0$), whereas the neutron and proton may have either opposite spins ($^1s_0$) or parallel spins ($^3s_1$).

two atoms).   This is not to say that the nucleon was now to be considered a composite particle, as is the atom, but rather that the particle to be exchanged, so to speak, was created at the instant of emission from one nucleon and vanished at the instant of absorption by the other nucleon.   Processes of this type in which "virtual" particles are exchanged are operative in all aspects of modern field theory which go beyond the classical idea of action-at-a-distance. For example, the Coulomb interaction between two charged particles in now analyzed in terms of the exchange of virtual photons between the two charges. The creation of the virtual particle immediately brings up the question of energy conservation.   It takes energy to create particles; where does this energy come from?   The answer is "nowhere," and that is why the particle is "virtual"; energy conservation is accounted for by making sure that the virtual particle does not live too long.   From the Heisenberg uncertainty principle we know that

$$\Delta E \, \Delta t \gtrsim \frac{h}{2\pi},$$

where Δt is the time available for measuring the energy of a system and ΔE is the accuracy within which the energy may be determined in the time Δt.   All that is required, then, for energy conservation is that the life Δt of the virtual particle be such that

$$\Delta t \gtrsim \frac{h}{2\pi \, \Delta E}.$$

Since the energy required to create a particle of mass m is given by the Einstein equation

$$\Delta E = mc^2,$$

$$\Delta t \gtrsim \frac{h}{2\pi mc^2}.$$

If the virtual particle moves with the velocity of light, then the range of the force is about

$$R \cong c \, \Delta t \gtrsim \frac{h}{2\pi mc}.$$

A range of about 2×10^{-13} cm requires a virtual particle with a mass about 200 times that of an electron.   Further, just as the quantum of the electromagnetic field (the virtual photon) may become a real particle in the physical world by absorbing some of the energy available in the collision between two charged particles, so the quantum of the nuclear field should become a physical particle in a collision between nucleons in which sufficient energy is available to supply the rest-mass energy of the quantum.   This process does indeed occur, and the pi-meson, a particle of 273 electron masses, is observed and is taken to be the quantum of the nuclear field.[2]   Unfortunately, as the available energy is increased other "strange" particles are also created whose role in the nuclear force field is not understood.   So far no complete field theory of nuclear forces in terms of meson exchange exists, but the approximate theories provide a valuable guide.

B. NUCLEAR MATTER

We shall first consider the properties of an infinite chunk of nuclear matter that contains essentially equal numbers of neutrons and protons.   This hypothetical infinite nucleus is probably a good description of the central region of heavy nuclei.   It is a good starting point in a discussion of nuclei because the complexities caused by boundary conditions at the surface of the nucleus may be ignored.

There are two immediately evident and important characteristics of nuclear matter exhibited by nuclei of mass number larger than about 20:

1. The binding energies per nucleon are essentially independent of mass number as reflected by the first term in the binding-energy formula (2-3). This means that all nucleons in a nucleus *do not* interact with all other nucleons (if they did, the binding energy *per nucleon* would be proportional to the mass number).

2. The densities are also essentially independent of mass number, which means that all nuclei *do not* simply collapse until the diameter is about equal to the range of nuclear forces so that all nucleons may be within one another's force field.   Thus these two general characteristics of nuclear matter are related and should have a common explanation.

Two different possible causes of these characteristics which have immediate analogies in the domain of chemical forces come to mind.

a. A drop of liquid argon, for example, has a density and a binding energy per atom independent of the size of the drop as long as it is not too small. These characteristics result from the Van der Waals forces which are attractive and large only for nearest neighbors and which sharply limit the number of nearest neighbors by becoming repulsive when the atoms "touch."   Thus, as an approximation, each argon atom interacts strongly with at most 12 other argon atoms.   The key to the situation here is the Van der Waals repulsion which sets in when the atoms touch.   The corresponding repulsion that exists

---

[2] The first particle with approximately the right mass that was discovered was the μ meson. The discovery, though, was quite a blow to the theory, since the μ meson interacted only very weakly with nuclei—hardly an acceptable behavior for the quantum of the nuclear field.   Several years later it was found that the μ meson was the decay product of another meson, the π meson, which does interact strongly with nuclei.

in nuclear forces at small distances would lead to the same effect. The observed density of nuclear matter, though, is much smaller than this effect by itself would give. Thus an additional factor must be operative.

b. A piece of diamond also exhibits a density and a binding energy per atom independent of size; but the reason is different from that for a drop of liquid argon. In diamond each carbon atom is covalently bonded to four other carbon atoms and thus interacts strongly with only these four. It pays little attention to a fifth that may be brought near to it because the chemical bond has *saturation properties* and the first four carbon atoms have saturated the valency of the central carbon atom. The saturation property of the chemical bond arises from the limited number of valence electrons available for exchange between bonded atoms. The exchange character of nuclear forces also causes the interaction between nucleons to be strong only if the nucleons are in the proper states of relative motion. For example, the Serber force, which has been mentioned, will give a strong attraction between two nucleons only if their relative angular momentum vanishes.

Unfortunately, it is not simple to show that the repulsive core, in conjunction with the exchange character of nuclear forces, results in the constancy of the density of nuclear matter and of the binding energy per nucleon. This result is difficult to obtain because it also involves the many-body aspects of the problem in an essential manner (G1).

Another interesting question about nuclear matter is concerned with the motion of the neutrons and protons; or, in quantum-mechanical language, what is the wave function (a function of the coordinates of all the constituent nucleons) that describes nuclear matter? The *effective* weakness of nuclear forces, discussed in section A, and the Pauli exclusion principle combine to give a surprising and simple approximate answer to these questions: the nucleons in nuclear matter move about much as free particles, with little perturbation of their motions by collisions with other nucleons; or, equivalently, a good first approximation to the wave function for nuclear matter is simply a properly antisymmetrized product of the free-particle wave functions for each of the nearly free nucleons in the nucleus. The collisions between nucleons are essentially quenched because for a collision to be effective the colliding particles must transfer some momentum to one another; but all states of lower momentum are already occupied by other nucleons, and the Pauli principle therefore forbids the occurrence of any momentum transfers. The effectiveness of the Pauli principle could be diminished if the internucleon forces were strong enough. For example, deuterium atoms, which must also obey the Pauli principle, will not move as essentially free particles at low temperatures but will couple together to form D_2 molecules; this is a manifestation of the *effective* strength of chemical forces compared to nuclear forces.

It is expected that nucleons would also form clusters (analogous to the D_2 molecules) when the nuclear density is so low that the average distance between them is fairly large compared to the range of the nucleon-nucleon force. The diffuse surface of finite nuclei must have such a region of very low density, and there is some evidence that alpha-particle clusters may have a transient existence in this diffuse edge.

To summarize, the characteristics of nucleon-nucleon forces in conjunction with the Pauli exclusion principle cause nuclear matter to exhibit apparently contradictory behavior: the macroscopic properties, such as density and binding energy, resemble those of a drop of liquid; the microscopic properties, such as nuclear wave functions and particle motions, resemble those of a weakly interacting gas. The resemblance to a drop of liquid has already been exploited in the development of the binding-energy equation in chapter 2 and appears again in the treatment of the collective model; the resemblance to a weakly interacting gas serves as the basis for the Fermi gas model and for the shell model.

C. FERMI GAS MODEL

The simplest nuclear model that emphasizes the "free-particle" character of the motion of nucleons within the nucleus is the Fermi gas model. In this model the nucleus is taken to be composed of a degenerate Fermi gas of neutrons and protons contained within a volume defined by the nuclear surface. The gas is considered degenerate because all the particles are crowded into the lowest possible states in a manner consistent with the requirements of the Pauli principle. The gas, for each type of particle, may be characterized by the kinetic energy of the highest filled state, the Fermi energy. The Fermi energy is easily determined through the condition that there be enough states up to and including the highest filled state to accommodate the particles in the nucleus. Recalling that there may be two identical nucleons with opposed spins in each quantum state, this condition gives, for neutrons [refer to footnote 13, chapter 8]

$$\frac{N}{2} = \frac{(4\pi/3)p_f{}^3 V}{h^3},\tag{9-1}$$

where N is the neutron number of the nucleus, V is the volume of the nucleus, and p_f is the momentum of the neutron in the highest filled state, the Fermi momentum. By rearrangement of (9-1) and substitution of the classical relation between kinetic energy ϵ and momentum, $p = (2M\epsilon)^{1/2}$, the expression for the Fermi energy of the neutron gas is

$$\epsilon_f = \frac{1}{8}\left(\frac{3}{\pi}\right)^{2/3}\left(\frac{N}{V}\right)^{2/3}\frac{h^2}{M},\tag{9-2}$$

where M is the mass of the neutron.   For the center of complex nuclei, where the density is about 2×10^{38} nucleons cm^{-3}, the Fermi energy for $N = A/2$ is about 43 MeV.   To give the proper value for the binding energy of the neutron, approximately 8 MeV, this value of the Fermi energy implies that the neutron gas is contained in a potential-energy well which, at the center of complex nuclei, is about 50 MeV deep.

The Fermi gas model is not useful for the prediction of the detailed properties of low-lying states of nuclei observed in the radioactive decay processes described in chapter 8.   It is useful, though, for the estimation of the momentum distribution of nucleons within the nucleus and for the approximate thermodynamic treatment of the properties of nuclei that are excited up into the continuum.   These two aspects of the model are of importance in the study of nuclear reactions, and we return to the Fermi gas model in chapter 10.

D. SHELL MODEL

The shell model of the nucleus is essentially the same as the Fermi gas model in that the interactions among the nucleons in the nucleus are again replaced by a potential-energy well within which each particle moves freely.   In the Fermi gas model, as we have just seen, the nucleus is characterized by the energy of the highest filled level, the Fermi energy.   In the shell model one is concerned with the detailed properties of the quantum states; these properties are determined by the shape of the potential-energy well.   Before discussing the shell model in detail, it would be well to review briefly the experimental evidence that forced the shell model into nuclear theory.   It did not develop from first principles; indeed, it appeared despite first principles, and much of the theoretical work sketched in section B was motivated by a desire to make the successes of the shell model respectable.

Experimental Evidence.   In addition to the evidence for "magic numbers" (closed shells) cited in chapter 2, information about energies, spins, parities, and magnetic moments of nuclear states, partly gathered by the methods discussed in chapter 8, gave decisive impetus to a serious consideration of the shell model.   Briefly, the observations of

1. ground-state spin of 0 for all nuclei with even neutron- and proton-number,
2. the systematics of the ground-state spins (half-integral) of odd-mass-number nuclei, and
3. the form of the dependence of magnetic moments of nuclei upon their spins

suggested that the properties of the ground states odd-mass-number nuclei could be considered to be those of the odd nucleon alone.   The important point is the implication that all the other nucleons play no role except that of providing a potential-energy field which determines the single-particle quantum states and of filling those quantum states up to the one in which the odd particle moves.   For details on the experimental evidence for the shell model, as well as the calculations of wave functions and energy levels, the reader is referred to Mayer and Jensen (M1).

Shell-Model States.   The important simplification in the shell model is to replace the nucleon-nucleon interactions inside the nucleus with an effective potential energy acting on each nucleon which may be a function of its coordinates but is not a function of the coordinates of the other nucleons.   Thus it is the nuclear counterpart of the Hartree method for the many-electron atom. The problem then reduces to solving the Schrödinger equation for a particle moving in the chosen potential-energy field.

Two potentials are usually discussed.   Both are taken to be spherically symmetric, but they differ in their radial dependence.   The first is the harmonic-oscillator potential

$$V(r) = -V_0\left[1 - \left(\frac{r}{R}\right)^2\right], \tag{9-3}$$

and the other is the square-well potential

$$\begin{aligned} V(r) &= -V_0 \qquad r < R \\ V(r) &= \infty \qquad\quad r > R, \end{aligned} \tag{9-4}$$

where $V(r)$ is the potential at a distance r from the center of the nucleus and R is the nuclear radius.   To simplify the mathematical solution of the problem, both potentials, unrealistically, go to infinity rather than to zero outside of the nucleus.   This further simplification has only a small effect on the energies of the states and on their relative stabilities.   Because both potentials, (9-3) and (9-4), are spherically symmetric, the resulting quantum states may be characterized by two quantum numbers: n, the principal quantum number, related to the number of radial nodes in the wave function, and l, a measure of the orbital angular momentum of the particle, which is a constant of the motion.   As in atomic spectroscopy, the states with $l = 0, 1, 2, 3, \ldots$ will be designated as s, p, d, f, . . . , respectively; thus states are identified as $3s$, $1d$, $2f$, etc.[3]

The solution of the Schrödinger equation for the isotropic harmonic oscillator gives for the energy levels

$$\epsilon = h\omega_0[2(n - 1) + l],$$

$$\omega_0 = \left(\frac{2V_0}{MR^2}\right)^{\frac{1}{2}},$$

---

[3] The existence of states such as $1d$ and $2f$, which apparently violates the terms in atomic spectroscopy (where $n \geq l + 1$) is merely a consequence of the definition of n.   For the usual hydrogenic wave functions n is defined so that each state has $n - l - 1$ radial nodes; the definition of n used above gives each state $n - 1$ radial nodes.

where M is the nucleon mass.   These energy levels (taking the lowest state as zero) are shown schematically on the left in figure 9-2.   Two important properties of these levels should be noted:

1. All states with the same value of $2n + l$ have the same energy and are therefore accidentally degenerate.

2. Since the energy goes as $2(n - 1) + l$, the states of a given energy must either all have even or all have odd values of l; hence all degenerate states have the same parity.

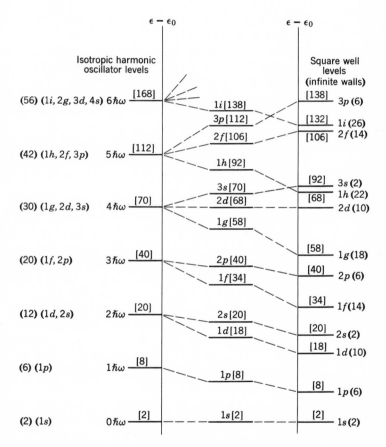

Fig. 9-2   Energy levels of the three-dimensional isotropic harmonic oscillator and of the square well with infinitely high walls.   The numbers in parentheses are the numbers of nucleons of one kind required to fill the various levels; the numbers in brackets are the numbers of nucleons of one kind that are required to fill all the levels up to and including a given level.   (From reference M1.)

The pattern of eigenstates for the square-well potential shown on the right of figure 9-2 is similar to that of the harmonic oscillator except that all of the accidential degeneracies are removed.   The change from a harmonic oscillator to a square well lowers the potential energy (a negative quantity) near the edge of the nucleus and thus enhances the stability of those states that concentrate particles near the edge of the nucleus; this means that states with largest angular momentum are most stabilized.   The sequence of levels in real nuclei might be expected to be someplace between these two extremes and is indicated by the levels in the center of figure 9-2.

As has been mentioned in chapter 2, the experimental evidence for the shell structure of nuclei points to the "magic numbers" of 2, 8, 20, 28, 50, 82, and 126 as the numbers of neutrons or protons which occur at closed shells and correspond to the atomic numbers of the rare gases in chemistry.   The sequence of levels in figure 9-2 show the possibility of predicting the first three of these numbers, but the others are certainly not evident.   The same situation occurs with the chemical elements; hydrogenic wave functions predict closed shells at atomic numbers 2, 10, 28, 60, 110, . . . , only the first two of which correspond to experimental fact.   The atomic problem is now thoroughly understood in terms of the removal of degeneracies by the interactions of the electrons with one another; this suggests the search for an interaction in nuclear matter which would split the levels in figure 9-2 even further and perhaps reveal the "magic numbers."

Spin-Orbit Interaction.   This important interaction which had not yet been included was pointed out independently by Mayer (M2) and by Haxel, Jensen, and Suess (H1); it was the interaction between the orbital angular momentum and the intrinsic angular momentum (spin) of a particle.   This interaction was already well known in the atomic problem, where, however, it played a relatively minor role.   It is also seen (see section A) in the polarization of scattered particles.

Consider, for example, a nucleon in a $1p$ state; it has an orbital angular momentum of $1\hbar$ and a spin of $\frac{1}{2}\hbar$.   By the rules of quantum mechanics the total angular-momentum of the particle may be either $J = \frac{3}{2}\hbar$ or $J = \frac{1}{2}\hbar$, states which we shall designate as $1p_{3/2}$ and $1p_{1/2}$, respectively.   Spin-orbit interaction means that the energies of the $1p_{3/2}$ and $1p_{1/2}$ states are not the same; the sixfold degenerate $1p$ state is split into the fourfold degenerate $1p_{3/2}$ and the twofold degenerate $1p_{1/2}$ states (the remaining degeneracy is simply that of the orientation of the total angular momentum vector, $\mathbf{J}$, in space).

If, in particular, the energy difference of states split by spin-orbit interaction is taken to be of the same order as the spacing between shell-model states and if the states with the higher j ($j = l + \frac{1}{2}$) are made more stable as those with the lower j ($j = l - \frac{1}{2}$) are made less stable, then the sequence of levels becomes something like that illustrated in figure 9-3.   There it is seen that the closed shells at 28, 50, 82, and 126 nucleons appear because of the splitting of the $1f$, $1g$, $1h$, and $1i$ levels, respectively.

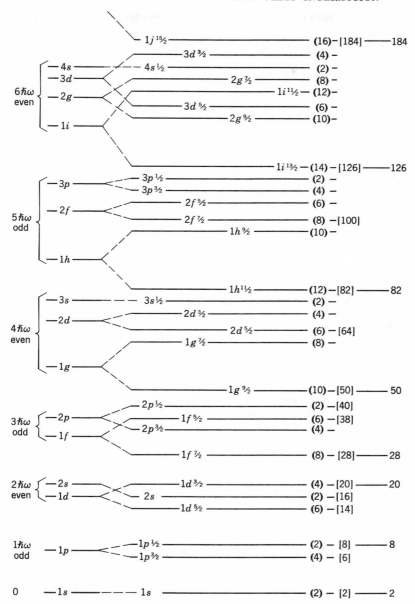

Fig. 9-3 Splitting of the energy levels of the three-dimensional isotropic harmonic oscillator by spin-orbit coupling. The numbers in parentheses and brackets have the same meaning as in Fig. 9-2; the "magic numbers" are given on the far right. (From reference M1.)

There are several important features of this energy level diagram. First, the level order given is to be applied independently to neutrons and to protons. Thus the nucleus $_2\text{He}^4$ contains two protons and two neutrons all in the $1s_{\frac{1}{2}}$ level; $_4\text{Be}^9$ contains four protons, two in the $1s_{\frac{1}{2}}$ and two in $2p_{\frac{3}{2}}$ (indicated more briefly by $1s_{\frac{1}{2}}{}^2 2p_{\frac{3}{2}}{}^2$), and five neutrons, $1s_{\frac{1}{2}}{}^2 2p_{\frac{3}{2}}{}^3$. On an absolute energy scale the proton levels are increasingly higher than neutron levels as Z increases. This is the familiar Coulomb repulsion effect, and in first approximation it does not change the order of the levels for a particular kind of nucleon. But there is a small tendency for the proton levels in nuclei of large Z to shift in relative stability, those levels with maximum orbital angular momentum ($1f$, $1g$, $1h$, $1i$) appearing at relatively lower energies, apparently because the proton suffers less from Coulombic repulsion when traveling in the outermost region of the nucleus.

Second, the order given within each shell is essentially schematic and may not represent the exact order of filling; indeed this order may differ slightly in different nuclides, depending on the number of nucleons in the outermost shell. (Similar level shifts are quite familiar in the atomic structure of the heavier elements.)

Ground States of Nuclei. If a nucleus contains 2, 8, 20, 28, 50, 82, or 126 neutrons, the level scheme just described permits a good prediction of the quantum states occupied by the neutrons. Thus $_{38}\text{Sr}^{88}$ has its 50 neutrons filling the five shells: $(1s^2)$, $(1p^6)$, $(1d^{10}2s^2)$, $(1f_{\frac{7}{2}}{}^8)$, $(1f_{\frac{5}{2}}{}^6 2p^6 1g_{\frac{9}{2}}{}^{10})$. Similarly, the proton structure is obvious for nuclides with magic atomic numbers: He, O, Ca, Ni, Sn, and Pb. It is a well-known theorem in atomic structure that filled shells are spherically symmetric and have no spin or orbital angular momentum and no magnetic moment. To this is added the postulate in the extreme single-particle nuclear model that not only filled nucleon shells but any even number of neutrons or protons have no net angular momentum and no magnetic moment in the ground state. Thus we expect $I = 0$ (and $\mu = 0$ and even parity) not only for $_2\text{He}^4$ and $_8\text{O}^{16}$ but also for $_{38}\text{Sr}^{88}$, $_{76}\text{Os}^{192}$, $_{92}\text{U}^{238}$, and all the other even-even nuclei.

In any nucleus of odd A all but one of the nucleons are considered to have their angular momenta paired off, forming an even-even "core"; the single odd nucleon is thought to move essentially independently in (or outside) this core, and the net angular momentum of the entire nucleus is determined by the quantum state of this nucleon. For example, consider the even-odd nucleus $_6\text{C}^{13}$. The six protons and six of the seven neutrons are paired up (in the configuration $1s^2 1p_{\frac{3}{2}}{}^4$); the odd neutron is in the $1p_{\frac{1}{2}}$ level, and the entire nucleus in its ground state is characterized by the $p_{\frac{1}{2}}$ designation. The nuclear spin of C^{13} has been measured, and the value $I = \frac{1}{2}$ corresponds to the resultant angular momentum indicated as the subscript in $p_{\frac{1}{2}}$. As a second example consider $_{23}\text{V}^{51}$. The odd nucleon in this case is the twenty-third proton; it belongs in the $1f_{\frac{7}{2}}$ level; the ground state of the nucleus is expected to be $f_{\frac{7}{2}}$.

The measured spin of this nucleus is $\frac{7}{2}$. Without going into details we may say that the measured magnetic moments for C^{13} and V^{51} lend some support to the spin evidence for the correct assignment of these ground states. For a given spin the magnitude of the magnetic moment of a nucleus depends on whether the spin and orbital angular momenta of the odd nucleon are parallel or antiparallel. An $s_{\frac{1}{2}}$ and a $p_{\frac{1}{2}}$ nucleus will, for example, have quite different magnetic moments, and the differences can be at least qualitatively predicted.

For nuclei in general the situation is not nearly so simple as is indicated in the examples just cited. The order of the levels within each shell may often be different from that in figure 9-3, especially for two or three adjacent levels which differ in n or l. In such a case we conclude from the single-particle model only that several particular states of the nucleus are close together in energy without knowing which is the lowest, or ground, state. (Sometimes this information alone is useful.) As an example, the odd nucleon in $_{56}Ba^{137}$ is the eighty-first neutron, and figure 9-3 may be said to indicate that the ground state is probably $h_{11\frac{1}{2}}$, $s_{\frac{1}{2}}$, or $d_{3\frac{1}{2}}$, depending on the order of filling of these three levels. For Ba^{135} the ground state is probably $h_{11\frac{1}{2}}$ or $d_{3\frac{1}{2}}$. For each of these nuclei the measured spin is $I = \frac{3}{2}$.

What can be said about the states of odd-odd nuclides? Most of these nuclides are radioactive (the stable ones known are $_1H^2$, $_3Li^6$, $_5B^{10}$, $_7N^{14}$), and there are fewer directly measured data on spins and magnetic moments. The single-particle model assumption of pairing leaves in every case one odd proton and one odd neutron, each producing an effect on the nuclear moments. No universal rule can be given to predict the resultant ground state; however, the following rules proposed by L. W. Nordheim are very helpful. If for the two odd nucleons $j_1 + j_2 + l_1 + l_2 = $ *an even number*, the resultant spin is $I = |j_1 - j_2|$. To this rule there is one known exception, K^{40}. If $j_1 + j_2 + l_1 + l_2 = $ *an odd number*, I is probably large, approaching $j_1 + j_2$. For example, in $_{23}V^{50}$, both the twenty-seventh neutron and the twenty-third proton are expected to be in $1f_{7\frac{1}{2}}$ levels (for the proton this is confirmed by the measured spin $\frac{7}{2}$ of $_{23}V^{51}$). Since $j_1 + j_2 + l_1 + l_2 = 13$, Nordheim's rule predicts a high resultant spin for V^{50}, the maximum possible being $\frac{7}{2} + \frac{7}{2} = 7$. The measured spin of V^{50} is 6. Even $(+)$ parity is predicted unambiguously for this nucleus in its ground state, since each of the two odd nucleons is in an f state and therefore has odd parity.

For many nuclides the single-particle model is certainly an oversimplification. As an illustration of one kind of evidence for this statement, $_{11}Na^{23}$ would be expected to have a $d_{5\frac{1}{2}}$ ground state, and the only other reasonable single-particle model possibility is $s_{\frac{1}{2}}$. The measured spin is $\frac{3}{2}$. This is not to be attributed to an odd proton in the $1d_{3\frac{1}{2}}$ level because $1d_{3\frac{1}{2}}$ certainly should lie higher than $1d_{5\frac{1}{2}}$; moreover, the magnetic moment is definitely in disagreement with the $d_{3\frac{1}{2}}$ interpretation. Another such case is $_{25}Mn^{55}$, in which the odd nucleon clearly should be $1f_{7\frac{1}{2}}$, yet the measured spin is $\frac{5}{2}$. Anomalies such as these can be caused by the interactions among all of the nucleons outside the closed shells that have not already been included in the effective potential that determines the shell-model states. Thus, evidently the interactions among the five $f_{7/2}$ protons and the two neutrons beyond the closed shell of twenty eight cause the ground state of Mn^{55} to have a spin of $\frac{5}{2}$ instead of $\frac{7}{2}$. It is interesting to note that Mn^{53} which does not have the two neutrons beyond the closed shell of twenty eight has the expected ground-state spin of $\frac{7}{2}$. We shall return to anomalous ground-state spins in section H and discuss them in terms of nuclear deformation.

Application of Single-Particle Model to Nuclear Isomerism. The concept of closed shells and the single-particle model have applications in the study of excited states of nuclei, particularly for low excitation energies. Generally, when the model predicts several possible low-lying configurations, all except the one that happens to be energetically favored are eligible for existence as excited states, particularly for odd-nucleon nuclides. We have seen in chapter 8 that γ transitions, especially where ΔI is large and E is small, can have appreciable lifetimes and are then known as isomeric transitions. An important aspect of nuclear shell structure has been the correlation of nuclear spins and isomer lifetimes.

E. Feenberg (F1) in 1949 called attention to the fact that there were abundant groupings of isomers with odd Z or odd N just below the magic number values 50, 82, and 126. This phenomenon is connected with the appearance, just before shell closure at these numbers, of a new level of very high spin ($1g_{9/2}$ before 50, $1h_{11/2}$ before 82, $1i_{13/2}$ before 126). As an illustration consider $_{48}Cd^{113}$; its odd nucleon — the sixty-fifth neutron — is assigned to the $3s_{1/2}$ state to accord with the measured ground-state spin $I = \frac{1}{2}$. Other possible unfilled states within the same shell, all probably low-lying, are $2d_{3/2}$ and $1h_{11/2}$. If $1h_{11/2}$ happens to be the first excited level, which is the case for this nuclide, then the γ transition to ground is $h_{11/2} \rightarrow s_{1/2}$, $\Delta I = 5$, yes (the parity changes), an $E5$ transition which should be very long-lived. The isomer actually observed, Cd^{113m}, has $t_{1/2} = 14$ years; it decays predominantly by β decay so that t_{γ} is greater than 200 years.

For Cd^{111}, to choose a related example, the situation is similar. For the ground state $I = \frac{1}{2}$ and the sixty-third neutron is $3s_{1/2}$. The $1h_{11/2}$ state is 0.396 MeV above the ground state, and its half-life (Cd^{111m_2}) is 49 minutes. The γ transition is not directly to the ground state but to a $2d_{5/2}$ state which happens to lie between at 0.247 MeV above ground; therefore $\Delta I = 3$, yes, and the classification is $E3$. The $2d_{5/2}$ state in turn decays to the ground state; for this transition $\Delta I = 2$, no, and the half-life of this $E2$ transition is measured as 8×10^{-8} second.

The large number of even-odd isomers from $_{48}Cd^{111}$ to $_{56}Ba^{137}$, with 63 to 81 neutrons, have similar explanations. The upper state in most of these pairs is $1h_{11/2}$, and the corresponding isomeric transition in most is $h_{11/2} \rightarrow d_{3/2}$, $\Delta I = 4$, yes, classification $M4$, often followed by the $M1$ transition $d_{3/2} \rightarrow s_{1/2}$. When the next neutron shell (82 to 126) is partly filled in the even-odd nuclei such as $_{78}Pt^{195}$, Pt^{197}, $_{80}Hg^{197}$, Hg^{199}, and $_{82}Pb^{207}$, the long-lived isomeric level is $1i_{13/2}$; the transition is generally $i_{13/2} \rightarrow f_{5/2}$, which is again $M4$. Another group of $M4$ isomeric transitions is between $g_{9/2}$ and $p_{1/2}$ states in the region of odd nucleon numbers just below 50.

There is a sizable number of odd-odd isomers, but because of the difficulties in assignment of configurations to the two-nucleon states these isomers are not easy to classify in any organized way. There are some very interesting even-even isomers. In one, $_{32}Ge^{72m}$, $t_{1/2} = 3 \times 10^{-7}$ sec, $E = 0.69$ MeV, the ground and the first excited state both have $I = 0+$; the transition is thus of the $0 \rightarrow 0$ type, and, as required by the selection rules (chapter 8, p. 257), takes place entirely by emission of internal conversion electrons, in spite of the rather large transition energy. Most even-even isomers have very short half-lives; one of the interesting exceptions, $_{72}Hf^{180m}$ (5.5 hr), is discussed in section G.

E. COLLECTIVE MOTION IN NUCLEI

We have already illustrated or implied several deficiencies of the single-particle model in regard to the spins of nuclear ground states; when nuclear electric quadrupole moments are considered,[4] the failure of the single-particle model is hardest to repair. Even in nuclei with just one nucleon more or just one nucleon less than a closed shell, where the single-particle model should be at its best, the measured quadrupole moments are several times larger than can reasonably be attributed to the odd nucleon. Related to this enhancement of static quadrupole moments is the observation that $E2$ transitions (electric quadrupole transitions) are often much faster than would be expected for a transition between single-particle states.

Physically, quadrupole moments and enhanced $E2$ transition rates mean that the nucleus has a spheroidal rather than a spherical charge distribution. The shell model assumes a spherical distribution for the nucleons in closed shells and ascribes the spheroidal deformation only to the nucleons outside closed shells. Evidently this assumption is not valid.

It was first suggested by Rainwater (R1) that these discrepancies might be overcome in odd-A nuclei, for example, by considering the polarization of the even-even core by the motion (not spherically symmetric) of the odd nucleon. In this manner all of the nucleons in the nucleus could contribute collectively to static quadrupole moments and to quadrupolar transition rates. The hypothesis that the even-even core may have a spheroidal rather than a spherical shape has other implications also, and these implications were investigated in an important series of papers mainly by Bohr and Mottelson (B3).

---

[4] Quadrupole moments are defined on p. 38 of chapter 2.

Before proceeding to a discussion of these investigations it should be pointed out that again nuclear theory was forced to make major revisions because of experimental facts.   This time it would be necessary to consider both aspects of nuclear matter simultaneously: the resemblance to a liquid and the resemblance to a gas.   This flexibility in the theory of nuclear structure is to be expected for the theory of any system that contains more than two particles; all such theories must always be approximate, and the most useful approximations will nearly always be suggested by the facts that the theory is trying to explain.

Collective Model.   The spheroidal deformation suggested for the even-even core of a nucleus of odd mass number means that the odd nucleon no longer moves in a spherically symmetric potential-energy field and so its orbital angular momentum is no longer conserved.   However, the total angular momentum of the system must be conserved, which leads to the conclusion that the spheroidal core must have angular momentum coupled to that of the odd nucleon.   Further, when another nucleon is added to the odd-A nucleus, the even-even nucleus which may be formed will still be spheroidal (unless a shell has been completed) and can have angular-momentum states that reflect the coherent motion of the nucleons in the nucleus as a whole rather than the motions of a few of the nucleons moving independently in shell-model states. It is the quantization of this coherent motion of the nucleons that forms the basis for the collective model.

The simplest way to quantize this motion, and one that is consistent with the properties of nuclear matter described in section B, is to treat the even-even core as a drop of incompressible liquid and to quantize the classical hydrodynamical equations that describe its oscillations.   Even this simplification, though, must be carried out in an approximate manner both because of the complexity of the equations and because enough of the shell model must somehow be left in the analysis to yield spheroidal equilibrium shapes for even-even nuclei with partly filled shells.

Since the treatment of the problem is both complicated and approximate, it is constantly being refined mathematically by the addition of higher-order terms and physically by the consideration of more subtle interactions.   We shall content ourselves here with listing some of the consequences of the model that are immediately relevant to understanding the energy spectra of even-even nuclei for which the model should be best, since there is no "odd nucleon" to which the collective motions must be coupled:

1. Rotational states of the well known form for symmetric-top molecules are expected:

$$E = \hbar^2 \frac{[I(I+1) - K^2]}{2\mathcal{I}_1} + \frac{\hbar^2 K^2}{2\mathcal{I}_3}. \tag{9-5}$$

In this expression $[I(I+1)]^{1/2}\hbar$ is the total angular momentum of the nucleus,

$K\hbar$ is the component of angular momentum along the symmetry axis of the nucleus, and $\mathcal{I}_1$ and $\mathcal{I}_3$ are the "moments of inertia" perpendicular to and parallel to the symmetry axis, respectively.   Classically, the rotational motion of the liquid drop is that of an irrotational flow and is best viewed as the motion of waves around the surface of the drop rather than rotation of the whole drop; the moments of inertia are thus fictitious ones which are not immediately related to the classical moment of inertia of the drop.

2. Both the model and experimental fact indicate that $\mathcal{I}_3$ is quite small; this leads to low-lying rotational states for which $K = 0$, and the energy expression then resembles that for a diatomic molecule:

$$E = \hbar^2 \frac{I(I+1)}{2\mathcal{I}}. \tag{9-6}$$

States with $K \neq 0$, other rotational bands, may occur if one or more of the nucleons are excited to a higher shell-model state so that the intrinsic angular momenta no longer cancel by pairs.

3. The symmetry of the nucleus dictates that these low-lying rotational levels ($K = 0$ band) be characterized by the sequence of states with $I = 0$, 2, 4, 6, . . . , all with even parity.   The accuracy of this prediction is well exemplified by the ground-state rotational band of Pu^{238} which is illustrated

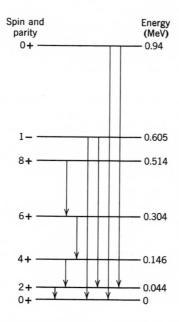

Fig. 9-4   Some of the energy levels of Pu^{238}.   Spins and parities are listed on the left, energies above the ground state on the right.   The first five levels are members of the ground-state rotational band.

in figure 9-4: the energy levels are, within a precision of a few per cent, given by (9-6) with $\hbar^2/2\mathscr{I} = 7.31$ keV which corresponds to a moment of inertia about half that of the rigid-body moment of inertia of the nucleus.

4. Low-lying excited states that are easily identifiable as members of a rotational band are expected for those nuclei with relatively large $\mathscr{I}$ and large permanent deformations and are therefore expected in the regions between closed shells (A between 150 and 190 and greater than 220).

5. At or near closed shells nuclei are essentially spherical and quite resistant to deformation. The resultant small value of $\mathscr{I}$ causes the spacing of rotational levels and single-particle levels to become similar, and the two types of excitation are no longer separable.

6. Excited vibrational states are expected in which the nucleus oscillates about its equilibrium deformation. These vibrations are of two types: the β vibrations that preserve the spheroidal symmetry of the nucleus and the γ vibrations that go through ellipsoidal symmetries. This subject is not settled and is under experimental and theoretical investigation.

To summarize, the treatment of the collective motion of nucleons is reasonably successful in interpreting the pattern of excited states of even-even nuclei and also the dependence of that pattern on mass number. For odd-A nuclei, though, the situation is complicated by the fact that the motion of the core and the motion of the odd nucleon must be coupled and thus the motion of the odd nucleon is no longer adequately described by the shell-model states which correspond to a static spherical core. The analysis of this problem gave rise to the unified model, one that attempts to fuse the independent motion and the correlated motion of nucleons in complex nuclei.

Unified Model. The first step toward the construction of a unified model is the investigation of the effect of nuclear deformation on shell-model states. Qualitatively, the main effects are the removal of part of the degeneracy that arose from the $(2j + 1)$ possible orientations of the angular-momentum vector in space and the loss of the angular momentum of the odd nucleon as a constant of the motion, although the component of the angular momentum along the axis of symmetry of the nuclear potential does remain a constant of the motion. The quantitative treatment of the problem for axially symmetric deformations was carried out by Nilsson (N1), and the resultant states are known as Nilsson states. The odd nucleon, then, is placed in the appropriate perturbed shell-model state (Nilsson state) and its motion is coupled to the collective states of the residual even-even core.

A full analysis of this approximate treatment of odd-mass nuclei has been given by Mottelson and Nilsson (M3); a summary of some of the more important results follows.

1. The quantum state is no longer given simply by that of the odd nucleon; it is now defined by the quantum numbers I which measures the total (even-

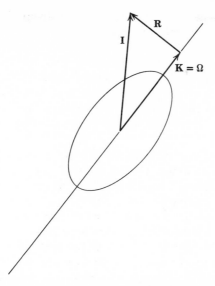

Fig. 9-5 Vector diagram of the total angular momentum of a deformed nucleus in the unified model.

even core plus odd nucleon) angular momentum of the state, K which is the projection of I on the symmetry axis, and Ω which is the projection of the angular momentum (orbital plus spin) of the odd nucleon on the symmetry axis. These quantities are illustrated in figure 9-5 in which **R** is the collective angular momentum of the even-even core. For nuclei with cylindrical symmetry **K** $= \Omega$, and **R** is perpendicular to the axis of symmetry.

2. The ground state will have $I = K = \Omega$ (unless $\Omega = \frac{1}{2}$); Ω can be determined by filling the Nilsson states with nucleons just as the shell-model states were filled in section D.

3. The new single-particle states, which are at most doubly degenerate, are now closer together and there will be low-energy states (100–200 keV separation) that correspond to excitation of the odd nucleon to higher single-particle states.

4. For each assignment of the odd nucleon to a single-particle state there is a band of rotational-energy states for which $K = \Omega$ and for which the energies are given by (9-5), with I taking on the values $K, K + 1, K + 2, \ldots$.

F. PAIR CORRELATIONS

The importance of paired nucleons within nuclei has already been seen in the pairing-energy term of the binding-energy formula [cf. (2-3)] and in the pairing of neutrons or protons to zero net angular momentum in the ground states of both spherical and spheroidal even-even nuclei.   Both observations lead to the conclusion that the residual interactions among the nucleons in a nucleus, the interactions not included in the average nuclear potential which a nucleon experiences, are strongest when the two nucleons are in particular "conjugate" states.   Further, these two conjugate states are evidently characterized by the same sets of quantum numbers except that they have opposite signs for the projection of the angular momentum of the state on the symmetry axis of the nucleus.

As mentioned earlier, any residual interaction among the nucleons will invalidate the extreme single-particle model, which simply fills the shell-model states from the bottom up, and will introduce configuration mixing.   Interactions by pairs will produce a particular configuration mixing analogous to that found for conduction electrons in metals and which has been invoked to explain the phenomenon of superconductivity.   This interaction will affect the energies of the ground state and of the low-lying excited states but will, in general, have no effect on spin and parity assignments based on shell-model considerations. The requirements of the Pauli exclusion principle cause even the approximate mathematical analysis of this problem to become complex, and so we shall give only a brief qualitative description of a few of the results that are significant for nuclei.

The main result of the interactions by conjugate pairs of nucleons is to change the energy spectrum of low-lying single-particle states for nuclei as given in section D to one in which the energies of the single-particle states, ϵ_j, are replaced by the modified "single-particle states" of energies, with respect to the Fermi energy (cf. section C),

$$E_j = \sqrt{(\epsilon_j - \lambda)^2 + \Delta^2}. \tag{9-7}$$

The quantity λ is very nearly equal to the Fermi energy; the quantity 2Δ is called the "energy gap" and is approximately the pairing-energy defined in (2-3).   These modified states are states for the whole system and so transition among them no longer corresponds to a change of a single nucleon from one state to another.   Nuclear excitations within this context correspond to the creation of quasi-particles with excitation energy as given by (9-7) just as vibrational excitation of the normal modes of crystals are described in terms of the creation of phonons; (9-7) has been useful in explaining the fact that even-even nuclei always have a much higher energy for the first-excited intrinsic state ($\cong 2\Delta$) than is expected from that for odd-mass nuclei and in understanding the values of the moments of inertia in (9-5) (N2).   The pairing effect is most important for the low-lying states of even nuclei; at higher excitations the effect vanishes ($\Delta \to 0$) just as the phenomenon of superconductivity vanishes above a critical temperature.

G. TRANSITION PROBABILITIES AND NUCLEAR MODELS

It has already been mentioned in section E of this chapter that enhanced $E2$ transition rates were an important impetus to the development of the collective model. This means that a quantitative evaluation of the way transition rates depend on changes in spin and parity (discussed in chapter 8) is affected by the model assumed for the nucleus. The same situation occurs in molecular spectroscopy in which a spin change of 1 could be a transition either from a triplet to a singlet electronic state or from one rotational state to another. These two types of transition, after correction for the energy changes, have very different half-lives. The general analysis of this dependence is quite complicated, and so we shall illustrate the results only by a particular example.

There is an excited state of Hf^{180} which has a negative parity and a spin of 9; this state decays by an $E1$ transition to another excited state of Hf^{180} with positive parity and spin 8. The half-life of the upper excited state is 5.5 hours, a factor of 10^{15} or 10^{16} longer than would be expected for an $E1$ transition on the basis of the single-particle model. The enormous retardation of the transition can be understood when it is realized that the 5.5-hour state is the lowest state of a $K = 9$ rotational band, and the state to which it decays is the fourth excited state of a $K = 0$ rotational band. The quantum number K for even-even nuclei, as mentioned in section E, is essentially a measure of the component of single-particle momenta on the symmetry axis of the nucleus. Thus, although the transition is characterized by a spin change of 1, it actually corresponds to a transition with a spin change in the single-particle state (intrinsic state) of 9. The half-life, then, would be that for an $E9$ transition except for the interaction between the single-particle and collective modes of motion.

H. SUMMARY AND COMPARISONS OF NUCLEAR MODELS

The various nuclear models we have discussed result in a spectrum of nuclear states that bears a strong resemblance to that for a polyatomic molecule: there are intrinsic states (single-particle for nuclei and electronic for molecules), rotational states, and vibrational states. It must be immediately stated, though, that this resemblance is much more a consequence of the interaction of scientists with the many-body problem than it is of any resemblance between the interactions in molecules and those in nuclei.

The fundamental model is the independent-particle model. The difficulty lies in the nucleon-nucleon interactions that are not included in the effective potentials exemplified in (9-3) and (9-4); the so-called residual interactions.

The residual interactions cause any description of the nucleus with a particular assignment of nucleons to single-particle states (the configuration) to be inaccurate; rather the nucleus must be described by a superposition of many different configurations (configuration mixing).

The extreme single-particle model which was discussed in section D uses the residual interactions to cause an even number of identical nucleons with the same n, l, and j quantum numbers to couple to a net angular momentum of zero. Configuration mixing is neglected. This extreme assumption is found to work rather well at or near closed shells, a fact that suggests that configuration mixing is important primarily for the nucleons outside closed shells and that the mixed configurations include mainly the single-particle states within a given shell.

The pairing of nucleons in the extreme single-particle model roughly takes account of the short-range correlations in nucleon motions expected from the residual interactions; the collective model and the unified model attempt to include the long-range correlations also. They accomplish this by replacing the configuration mixing by a spheroidal deformation which represents a time average of the spatial distribution expected for the appropriate mixture of single-particle configurations. It is assumed that the oscillations of the deformed nucleus about its equilibrium shape are slow compared to single-particle motions, and thus the single-particle states and the collective states may be treated separately. This approximation is roughly equivalent to the Born-Oppenheimer approximation in the theory of molecular structure.

The range of applicability and the successes of these various approximations to configuration mixing is most easily seen in the particular examples that follow.

Intrinsic States. The outstanding success of the extreme single-particle model lies in its ability to predict the ground-state spins and parities of nearly all odd-mass nuclei. Where it fails, such as in Na^{23}, as mentioned on p. 283, the failure may usually be remedied by taking account of the spheroidal deformation in the region between closed shells which splits the single-particle states. The $1d_{5/2}$ single-particle state, for example, is split into three doubly degenerate states with spins and parities $1/2+$, $3/2+$, and $5/2+$. In Na^{23} the ninth and tenth protons are coupled to zero spin in the $1/2+$ state and the ground-state properties come from the eleventh proton in the $3/2+$ state. Similarly the $7/2+$ isomeric states of Ag^{107m} and Ag^{109m} apparently arise from the splitting of the $1g_{9/2}$ single-particle proton state by the spheroidal deformation; the $1/2+$, $3/2+$, and $5/2+$ states so produced are filled and the forty-seventh proton is in the $7/2+$ state.

The extreme single-particle model should be useful in the characterization of excited states for nuclei that are very near closed shells. The low-lying states of Pb^{207} (filled shell of 82 protons, a single hole in the 126-neutron shell) provide an excellent example. In figure 9-6 it is seen that the first four excited

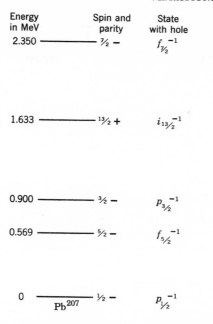

Fig. 9-6   Energy levels of Pb^{207} with energies given on the left side and spins and parities on the right.   The superscript -1 on any spectroscopic term indicates a "hole" in that state.

states in Pb^{207} correspond to transitions of the neutron hole among the various available single-particle states: different intrinsic states.   It should be noted that the relative stabilities of these states emphasize that the order given in figure 9-3 for states within a given shell is not to be taken seriously.

Vibrational States.   As the number of particles or holes beyond a closed shell increases, the residual interactions make it impossible to describe the excited states by a particular configuration, as is done for Pb^{207}.   The configuration mixing that results for nuclei not too far from closed shells is described by vibrations about the spherical shape and the low-lying excited states are described as vibrational states.   The first two excited states of Fe^{56} (two neutrons and two proton holes beyond the closed shell of 28), shown in figure 9-7, are characterized as being the first two excited vibrational states or, in terms of single-particle configuration, as being different mixtures of single-particle configurations rather than different unique configurations or intrinsic states.   The main justification for this assignment comes from the rates for the $E2$ transitions between these states which are much faster than expected for single-particle transitions.   It would be desirable to trace this vibrational series of states up to higher energies; this is made difficult by the mixing in of other intrinsic states at higher excitation energies.

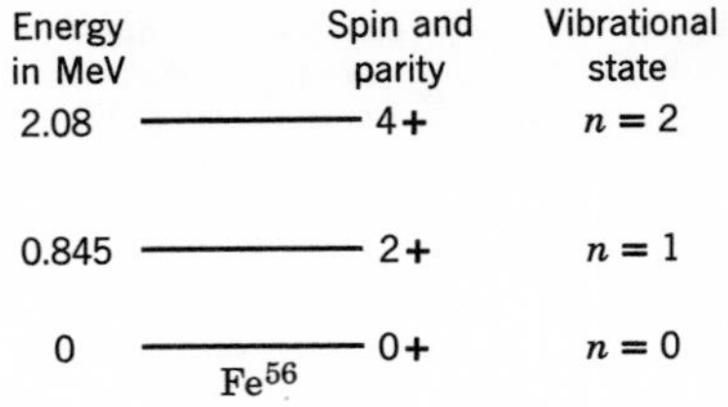

Energy in MeV	Spin and parity	Vibrational state
2.08	4+	$n = 2$
0.845	2+	$n = 1$
0	0+	$n = 0$

Fe^{56}

Fig. 9-7   The first three states of Fe^{56}, with energies given on the left and spins and parities on the right; n is the vibrational quantum number for each state.

Rotational States.   As still more nucleons or holes are added beyond the closed shells, the configuration mixing that results from the residual interactions causes a permanent spheroidal deformation of the nucleus and the excited states are better described as rotational states.   This disagreeable metamorphosis also occurs in molecular spectroscopy: CO_2, because it is linear,

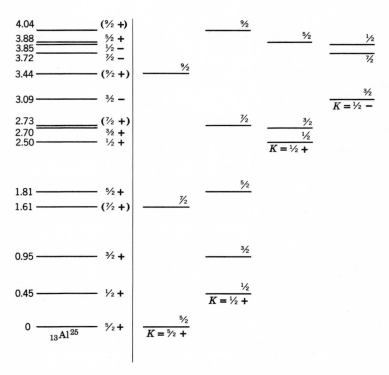

Fig. 9-8   Energy level spectrum of Al^{25}.   On the left are drawn all of the levels of Al^{25} that have been observed up to 4 MeV; the measured spins and parities are also shown.   On the right these levels have been classified in terms of the rotational bands associated with the different intrinsic states.   (From reference B4.)

has four degrees of vibrational freedom and two degrees of rotational freedom; H_2O, because it is nonlinear, has three degrees of vibrational freedom and three degrees of rotational freedom.   Thus straightening out the molecule turns a rotation into a vibration.   An example of a rotational band built on the ground intrinsic state of Pu^{238} has been shown in figure 9-4, in which the energy levels conform to (9-6).   Another illuminating example comes from the excited states of Al^{25}.   Figure 9-8, taken from Bohr and Mottelson (B4) shows that the observed spectrum in Al^{25} may be interpreted as rotational states built on the first four intrinsic states.   It is to be recalled that (9-5) does not hold for $K = \frac{1}{2}$; a much more complicated expression which can even cause the crossing of rotational states must be used.

The specific examples that have been given for intrinsic, vibrational, and rotational states illustrate the usefulness of the appropriate nuclear models. It must also be stated, though, that the unambiguous identification of the character of nuclear states is still the exception rather than the rule because of the mixing of the three kinds of states for the majority of the nuclides which are neither near enough to nor far enough from closed shells.

REFERENCES

*B1   H. A. Bethe and P. Morrison, *Elementary Nuclear Theory*, Wiley, New York, 1956.
 B2   W. E. Burcham, "Isotopic Spin and Nuclear Reactions," *Progress in Nuclear Physics*, Vol. 4 (O. R. Frisch, Editor), Pergamon, London, 1955.
 B3   A. Bohr and B. R. Mottelson, "Collective and Individual Particle Aspects of Nuclear Structure," *Dan. Mat.-Fys. Medd.* **27,** No. 16 (1953); and "Collective Nuclear Motion and the Unified Model," *Beta and Gamma Ray Spectroscopy* (K. Siegbahn, Editor), North Holland, Amsterdam, 1955.
*B4   A. Bohr and B. R. Mottelson, "Collective Motion and Nuclear Spectra," *Nuclear Spectroscopy* Part B (Fay Ajzenberg-Selove, Editor) Academic, New York, 1960.
 F1   E. Feenberg, "Nuclear Shell Structure and Isomerism," *Phys. Rev.* **75,** 320 (1949).
 G1   L. C. Gomes, J. D. Walecka, and V. F. Weisskopf, "Properties of Nuclear Matter," *Ann. Phys.* **3,** 241 (1958).
 H1   O. Haxel, J. H. D. Jensen, and H. E. Suess, "Modellmässige Deutung der ausgezeichneten Nukleonen-Zahlen im Kernbau," *Z. Physik* **128,** 295 (1950).
*M1   M. G. Mayer and J. H. D. Jensen, *Elementary Theory of Nuclear Shell Structure*, Wiley, New York, 1955.
 M2   M. G. Mayer, "Nuclear Configurations in the Spin Orbit Coupling Model.   I. Emperical Evidence," *Phys. Rev.* **78,** 16 (1950).
 M3   B. R. Mottelson and S. G. Nilsson, "Classification of Nucleonic States in Deformed Nuclei," *Phys. Rev.* **99,** 1615 (1955).
*M4   S. A. Moszowski, "Models of Nuclear Structure," *Encyclopedia of Physics*, Vol. 39 (S. Flügge, Editor), Springer-Verlag, Berlin, 1957.
 N1   S. G. Nilsson, "Binding States of Individual Nucleons in Strongly Deformed Nuclei," *Dan. Mat.-Fys. Medd.*, **29,** No. 16 (1955).
 N2   S. G. Nilsson and O. Prior, "The Effect of Pair Correlation on the Moment of Inertia and the Collective Gyromagnetic Ratio of Deformed Nuclei," *Dan. Mat.-Fys. Medd.* **32,** No. 16, (1961).

R1   L. J. Rainwater, "Nuclear Energy Level Argument for a Spheroidal Nuclear Model," *Phys. Rev.* **79**, 432 (1950).

Y1   H. Yukawa, "On the Interaction of Elementary Particles. I.," *Proc. Physico-Math. Soc. Japan* **17**, 48 (1935).

EXERCISES

1. Estimate the radii of the nuclei Sc^{41} and Ca^{41} from the observation that the maximum energy of the β^+ spectrum emitted in the decay of Sc^{41} to the ground state of Ca^{41} is 5.6 MeV.   Approximate both nuclei as uniformly charged spheres for which the electrostatic energy is $(3/5)[Z(Z-1)e^2]/R$ where Ze and R are the charge and the radius of the sphere, respectively.   *Answer:* 5.2×10^{-13} cm.

2. (a) Estimate the Fermi energies of neutrons and protons in the center of U^{238}. Assume the density of nuclear matter in the center of the U^{238} nucleus to be 2×10^{38} nucleons cm^{-3}.   *Answer:* 36 MeV for protons.
   (b) Compare the difference in Fermi energies of neutrons and protons in U^{238} with the Coulomb repulsion experienced by a proton approaching Pa^{237}.   Take the radius constant to be 1.5×10^{-13} cm.

3. The β^+ and EC decay of Zr^{89} (see exercise 8-14) lead to the 16-second Y^{89m} rather than to the stable Y^{89} of spin $I = \frac{1}{2}$.   The isomer is de-excited by a 913 keV transition with $\alpha_K \cong 0.01$ and $\alpha_K/\alpha_L = 7.0$.   (a) Using these data and shell structure considerations, assign spins and parities to Y^{89m} and to the 79h Zr^{89}. (b) Estimate the partial half-life for direct decay of Zr^{89} to the ground state of Y^{89}. (c) The 1.463-MeV β^- spectrum of 51d Sr^{89} is not accompanied by γ radiation and has the unique first-forbidden shape.   What is the log ft value for this transition and the spin and parity of Sr^{89}?   (d) Estimate the fraction of Sr^{89} decays that might lead to Y^{89m}.   *Answer:* (d) $\sim 10^{-5}$.

4. What do you expect the ground state spins and parities to be for (a) Ar^{39}; (b) Pt^{196}; (c) Zr^{89} (d); Co^{55}; (e) N^{16}?   *Answer:* (a) $7/2-$; (b) $0+$; (e) $2-$.

5. Estimate the smallest distance within which a neutron and proton can move so as just to give a bound deuteron.   (The average separation of neutron and proton is about half this distance and is the "size" of the deuteron.)   *Answer:* $\cong 4.0 \times 10^{-13}$ cm.

6. (a) What is the smallest possible value for the isotopic spin of $_{27}Co^{59}$?
   (b) The isotopic spin of a π-meson is 1 with z components $+1$, 0, and -1 corresponding to π^+, π°, and π^-, respectively.   What total isotopic spins are possible in the interaction of a π^+ + proton, π° + proton, π^- + proton?   *Answer:* (a) $\frac{5}{2}$.

7. The first excited state of W^{182} is $2+$ and is 0.100 MeV above the ground state. Estimate the energies of the lowest lying $4+$ and $6+$ states of W^{182}.   (The observed values are 0.329 and 0.680 MeV.)
   (b) The half-life of the $2+$ state is 1.5×10^{-9} second.   Compare this value with that calculated from Table 8-5.   Explain the relative values.

10

Nuclear Reactions

The energy balance in nuclear reactions as well as their rates in terms of the relevant measurable quantity, the cross section, were discussed in chapter 2. In this chapter we first (in sections A to C) give a survey of the various processes that may occur during a collision between a nucleus and another particle and then (in sections E to H) give a sketch of the theoretical ideas that have been useful in the interpretation of nuclear reactions.

Nuclear reactions have been produced by a wide variety of particles incident on complex nuclei. A limitation on the nature of the bombarding particle has been that of its acceleration to an energy sufficient to overcome the Coulomb repulsion between the reacting nuclei. As discussed in chapter 11, accelerator technology has now reached the point at which nuclei up to $_{18}Ar^{40}$ have been used as bombarding particles and energies up to 30,000 MeV (for protons) have been attained. Most studies of nuclear reactions are made with neutrons, protons, deuterons, and alpha particles.

The Coulomb repulsion problem does not, of course, exist for neutrons; but they are not found in chemical catalogs and must be produced from other nuclear reactions before they can be used as bombarding particles. A similar remark holds for photons, except that they are usually produced by electromagnetic rather than nuclear interactions.

A. TYPES OF EXPERIMENTS IN INVESTIGATIONS OF NUCLEAR REACTIONS

There are three main types of experiment which are made in the study of nuclear reactions and which provide complementary information about them. The three types include measurements of excitation functions, particle spectra, and recoil spectra. These experiments and the information obtained from each of them are sketched in the following paragraphs.

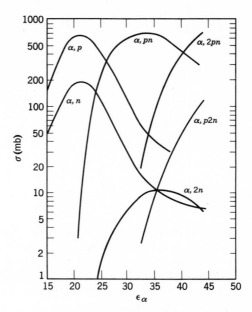

Fig. 10-1 Excitation functions for various reactions between alpha particles and Fe^{54} nuclei. The ordinate is the cross section in units of 10^{-27} cm^2 (millibarns) and the abscissa is the kinetic energy of the alpha particle in the laboratory system. (Data from reference M1.)

Excitation Functions. The determination of an excitation function involves the measurement of the cross section for the formation of a particular product, usually radioactive, of a nuclear reaction as a function of the energy of the bombarding particle. The curve of cross section versus energy is known as the excitation function. An example of excitation functions in the medium-energy region is shown in figure 10-1.

Excitation functions provide some information about the probabilities for the emission of various kinds of particles and combinations of particles in nuclear reactions because the formation of a given product implies what particles were ejected from the target nuclide. For example, figure 10-1 shows that in the reactions of alpha particles with Fe^{54} the emission of a single proton is about three times as likely as that of a single neutron and the emission of a proton-neutron pair is about 50 times more probable than that of two neutrons. It is not possible from these data alone, though, to know whether the proton-neutron pair was emitted as a deuteron; in this instance it very likely was not.

It is possible to get some information about the kinetic energies of the emitted particles both from the energies at which the various excitation functions reach their maxima and from the slopes of the excitation functions, but these are at best rather crude estimates.   Excitation functions will not yield any information about the angular distribution of the emitted particles.

Particle Spectra.   In contrast to excitation functions the second type of experiment focuses attention on the energy and angular distributions of the emitted light particles.   In its simplest form this information may be collected experimentally by detection of the emitted particles in an energy-sensitive detector placed at various angles θ with respect to the incident beam.   The quantity that is usually reported is $\partial^2\sigma/\partial\epsilon\,\partial\Omega$, which is a function of the kinetic energy of the emitted particle ϵ, and of the angle of emission θ.   This quantity is the differential cross section for the emission of the particle with kinetic energy between ϵ and $\epsilon + d\epsilon$ into an element of solid angle $d\Omega$ at an angle θ with respect to the incident beam.   In the laboratory system of coordinates the solid angle $d\Omega$ may be roughly approximated by dividing the area of the detector normal to the emission direction of the particle by the square of the distance between the target and the detector.   Bearing in mind that the differential cross section is a function of ϵ and θ, the total cross section for the emission of the particle is obtained by integrating over all angles and energies:

$$\sigma = 2\pi \int_0^\infty \int_0^\pi \frac{\partial^2\sigma}{\partial\epsilon\,\partial\Omega} \sin\theta\,d\theta\,d\epsilon.$$

Some examples of energy and angular distributions are shown in figures 10-2 and 10-3.

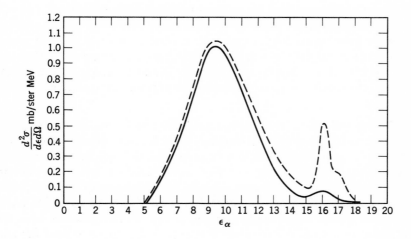

Fig. 10-2   Spectrum of alpha particles from Ni(p,α)Co for a nickel target of normal isotopic composition (68% Ni^{58} and 26% Ni^{60}) bombarded with 17.6-MeV protons.   The dashed curve is the spectrum at 30° with respect to the incident beam, the solid curve is that at 120°.   (Reproduced from reference S4.)

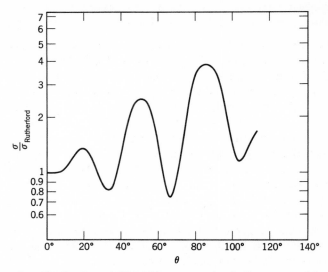

Fig. 10-3 Angular distribution of 22-MeV protons elastically scattered by nickel. The ordinate is the observed scattering cross section divided by the scattering expected from a purely Coulombic interaction (Rutherford scattering) and the abscissa is the angle of scattering in the center-of-mass system.   (Data from reference C1.)

An obvious limitation on the information that measurements of particle spectra provide for theoretical analysis lies in the lack of knowledge about the other particles that may be emitted in the same event with the one being detected.   This difficulty may be circumvented either by using an energy so low that the probability for the emission of more than one particle is negligible or by having several detectors and demanding coincidences among them before an event is recorded.   The latter technique is fine in principle, but the rate of gathering information with it diminishes rapidly as the number of detectors exceeds two.

Recoil Spectra (H1).   The measurement of the kinetic-energy and angular distribution of the products of a nuclear reaction represents an effort to use the advantages of both product cross section and particle spectrum measurements simultaneously.   The identity of the recoil product uniquely defines the nucleons emitted from the target nucleus in the reaction; however, whether they are emitted singly or in clusters, such as deuterons or alpha particles, remains ambiguous.   The angle of recoil of the product nucleus with respect to the beam direction and its kinetic energy will define the *total* momentum of all of the emitted particles but will give no information about the division of momentum among them.

The experimental techniques that are employed for the measurements vary in complexity.   In the simplest experiment, catcher foils placed before and behind the target determine the fraction of a given product that recoils out of the target in the forward and in the backward directions. This measurement is sensitive to the relative amounts of momentum carried away by the particles emitted in the forward and in the backward direction. In the more elegant experiment, stacks of very thin (thin compared to the range of the recoil product) catcher foils are placed at various angles with respect to the incident beam, and a direct measurement of the angular and kinetic-energy distribution of the products is made.

B. MODELS FOR NUCLEAR REACTIONS

The variety of processes that can occur in a nuclear reaction makes it difficult to give a coherent survey of the field. To assist in the organization of the survey given in the following section, we first anticipate some of the later sections of this chapter and give a brief description of three of the models which have proved to be of value in codifying the vast amount of nuclear-reaction data that have been collected. They are the optical model, the compound-nucleus model, and the direct-interaction model.

Optical Model. The optical model for nuclear reactions takes the same view of the nucleus as the shell model discussed in chapter 9. The interactions of the incident particle with the nucleons in the nucleus are replaced by the interaction of the incident particle with a potential-energy well, as illustrated in figure 10-4; the classical analogy of this model is that of a beam of light passing through a transparent glass ball. With this model the only effect of

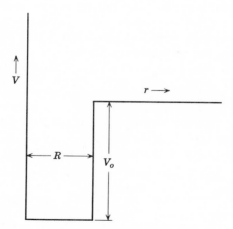

Fig. 10-4   Schematic diagram of square-well optical-model potential energy.

the nuclear collision would be to deflect the incident particle away from its original direction. To allow for the possibility of more complicated events, the particle is also given a mean free path for being absorbed when it is in the potential-energy well representing the target nucleus. In terms of the optical analogy, the ball of glass is made somewhat absorbent, rather than transparent, and is known as the "cloudy crystal ball."

The optical model allows the calculation of the following:

1. The cross section for deflection of the incident particle.
2. The angular distribution of this deflection.
3. The cross section for absorption of the incident particle.

It is silent about what happens after the particle is absorbed. Further details about the optical model are given in section F.

Compound-Nucleus Model. The compound-nucleus model is one of the two models that attempt to predict what happens in those events in which the incident particle is absorbed by the target nucleus. In the compound-nucleus model it is assumed that the energy carried in by the incident particle is randomly distributed among all the nucleons in the resultant compound nucleus so that none of them has enough energy to escape immediately, and thus the compound nucleus has a lifetime which is long (10^{-14}–10^{-18} second) compared to the time for a nucleon to traverse a nucleus (10^{-20}–10^{-23} second). There is a finite lifetime because there can always be a statistical fluctuation in the energy distribution which concentrates enough energy on a nucleon (or a cluster of nucleons) to allow it to escape. The most probable fluctuations are those that concentrate only a part of the excitation energy on the escaping particle, and so we expect that its kinetic energy will be less than the maximum possible and that the residual nucleus will still be in an excited state. Thus, if the original excitation energy of the compound nucleus is great enough, there may be the sequential emission of several particles from the excited compound nucleus, each with a relatively low kinetic energy. The similarity of this model to that for the escape of molecules from a drop of hot liquid has caused the emission of particles from excited nuclei to be called "evaporation."

In the compound-nucleus model, then, a nuclear reaction is divided into two distinct and independent steps.

1. Capture of the incident particle with a random sharing of the energy among the nucleons in the compound nucleus.
2. The evaporation of particles from the excited compound nucleus.

Further details on the compound-nucleus model are given in section G.

Direct Interaction. The direct-interaction model is the second one that attempts to predict the consequences of the absorption of the incident particle. It differs from the compound-nucleus model in that it does *not* assume that the energy of the incident particle is randomly distributed among all the nucleons in the target nucleus. Rather, in one aspect of the direct-interaction model, it is assumed that the incident particle collides with only one, or at most a few, of the nucleons in the target nucleus, some of which may thereby be directly ejected. It is also possible for the incident particle to leave the target nucleus after losing but a part of its energy in these few collisions. Thus the reaction does not proceed through the formation of an intermediate excited nucleus, and it is expected that the kinetic energies of the emitted particles will usually be higher than those of particles that are evaporated from an excited compound nucleus.

In addition, direct interactions include those events in which only a part of an incident complex particle, such as a deuteron, strikes the target nucleus. The part that does not collide with the target nucleus will then continue on after being deflected. This kind of reaction is particularly important with deuterons and is known as a stripping reaction. Another direct reaction that is frequently observed is the so-called pickup process (which may be considered the inverse of stripping); it involves the formation of a complex particle, such as a deuteron or He^3, by interaction of an incident particle (e.g., a proton) with a nucleon or group of nucleons in the nucleus.

Further analysis of direct interactions is given in section H.

C. TYPES OF NUCLEAR REACTIONS

In the following paragraphs we give a survey of the types of nuclear reactions which have been observed. The variety of information is correlated, whenever possible, with the models for nuclear reactions described in section B.

1. Elastic Scattering

The simplest consequence of a nuclear collision, and one that occurs at all energies and with all particles, is elastic scattering. An event is called an elastic scattering if the particles do not change their identity during the process and if the sum of their kinetic energies (ignoring molecular and atomic excitations and bremsstrahlung) remains constant. Elastic scattering of charged particles with energies below the Coulomb barrier of the target nucleus is the Rutherford scattering described in chapter 2. As the energy of the bombarding particle is increased, the particle may penetrate the Coulomb barrier to the surface of the target nucleus, and the elastic scattering will then also have a contribution from the nuclear forces. For neutrons, of course, elastic scattering is caused by nuclear forces at all energies.

Except for neutrons of very low energy, elastic scattering may be considered to arise largely from the optical-model potential energy described in the preceding section.   The compound nucleus formed by the capture of a neutron of very low energy has an appreciable probability of evaporating a neutron with all its original energy, and so there is also a compound-nucleus contribution to the elastic scattering of low-energy neutrons.

The angular distribution in elastic scattering, as exemplified in figure 10-3, may be analyzed in terms of the optical model to give information about the shape of the potential-energy function, which is shown only schematically in figure 10-4.   Indeed, it may be said that we "see" nuclei as much by their scattering of particles as we "see" the words in this book by its scattering of photons.   It is because our eyes are such gross objects that we need instruments and theories with which to "see" nuclei.

2. Slow-Neutron Reactions

Since there is no Coulomb barrier to be surmounted, it should be possible for neutrons of even very low kinetic energies to enter into target nuclei and cause nuclear reactions.   This was found to be true, and indeed slow neutrons can induce nuclear reactions with a cross section far exceeding that of any other particle for reasons discussed in section E.

Slow-neutron reactions proceed through the formation of a compound nucleus; actually the compound-nucleus model was first evolved just for the interpretation of these reactions.   Since the excitation energy of the compound nucleus formed in the capture of a slow neutron by a target nucleus is only a little greater than the binding energy of the captured neutron, it takes a relatively long time for a fluctuation which concentrates the escape energy back on a neutron to occur, and there is a greater probability that the excitation energy will be emitted as gamma rays.   Thus the main reaction with slow neutrons is the (n, γ) reaction.

Excitation functions for (n, γ) reactions, as exemplified in figure 10-5, are characterized by narrow peaks called resonances.   They occur because the quantum states of the excited compound nuclei are relatively widely spaced at these low excitation energies and the cross section for the (n, γ) reaction can become very large when the energy of the incident neutron is just right to form one of these excited states.   The situation is similar to that observed in the absorption of light by chemical systems.

Since the positions and heights of the resonances depend on the properties of nuclear excited states, they vary in an unsystematic way from target nucleus to target nucleus.   A table of cross sections for reactions with 0.025-eV neutrons is given in appendix C.

Although the (n, γ) reaction is the most important one for slow neutrons, it is also occasionally possible to have (n, p) and (n, α) reactions, particularly with targets of low Z, for which the Coulomb barrier is not so high as to hinder the emission of charged particles decisively.   A well-known and important

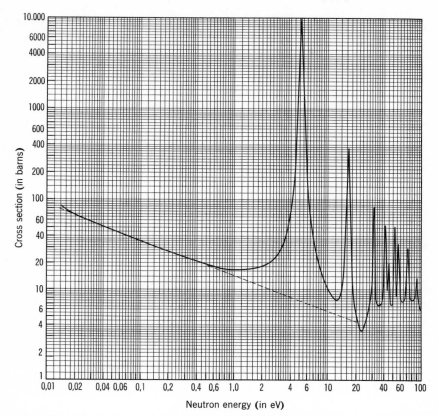

Fig. 10-5   Neutron cross section of silver as a function of energy in the region of 0.01 to 100 eV.   (From Atomic Energy Commission Document AECU-2040 and its supplements.) The dashed line indicates the $1/v$ slope.   Note that the data as plotted are for silver of normal isotopic composition; however, each individual resonance peak can be assigned to one or the other silver isotope.   For example, the first resonance at 5.12 eV belongs to Ag^{109}.

(n, p) reaction with thermal neutrons is that with N^{14} which produces the valuable tracer, C^{14}.   The (n, α) reaction with B^{10} is of great value in the detection of neutrons because of its large cross section and because of the ease with which the alpha particles produced can be detected.

3. Medium-Energy Reactions

The Coulomb barrier makes it impossible, except with the very lightest nuclei, to study nuclear reactions with low-energy charged particles.   For charged-particle-induced reactions we are therefore concerned with the medium-energy region that extends from a few MeV up to about 50 MeV. Nuclear reactions induced either by charged particles or by neutrons in this energy region differ in two important ways from those with slow neutrons:

1. The isolated resonances are no longer observable because their spacing becomes small compared to their width.

2. The increased energy that is available permits a wide variety of reactions to occur.

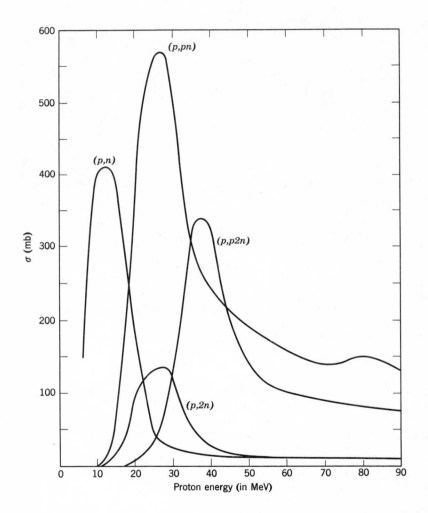

Fig. 10-6   Excitation functions for proton-induced reactions on Cu^{63} (taken from reference M4).

The approximate upper limit for this energy region is chosen to reflect the observation that new kinds of processes appear to become important at energies between 50 and 100 MeV which have not, unfortunately, received much study.

Excitation Function. Characteristic excitation functions for reactions induced by α particles in this energy region have been given in figure 10-1; others for protons are shown in figure 10-6.

In contrast to that for slow neutrons, the reaction cross section in the medium-energy region does not exhibit resonances. For charged particles the reaction cross section rises from essentially zero at energies just a little below the Coulomb barrier and asymptotically approaches πR^2 where R is the distance between centers of incident and target nuclei when they "feel" one another's nuclear forces (interaction radius). The asymptotic value of the cross section is of the order of 10^{-24} cm^2 (1 barn). For neutrons the reaction cross section descends from the very high values (hundreds or thousands of barns) in the electron-volt energy region and also approaches πR^2.

It can be seen from figures 10-1 and 10-6 that there are many reactions possible in the medium-energy region and that there is competition among them. The consequences of the competition lead to the conclusion that a major part of the reactions in this energy region proceed through the formation of a compound nucleus. For example, the observation that (α, n) and (α, p) excitation functions, as shown in figure 10-1, go through a maximum at the energy at which the $(\alpha, 2n)$ and (α, pn) excitation functions are starting to rise and that these in turn go through a maximum at the energy at which the $(\alpha, 2pn)$ and $(\alpha, p2n)$ reactions set in is just what is expected from the compound-nucleus model, in which each emitted particle carries away only a fraction of the available excitation energy. The reactions observed may be considered as proceeding in the following manner:

$$_2\text{He}^4 + {}_{26}\text{Fe}^{54} \longrightarrow {}_{28}\text{Ni}^{58}$$

The relative probabilities of the various paths depend on the excitation energy of the $_{28}\text{Ni}^{58}$ compound nucleus and may be calculated in a manner outlined in section G. A similar description can be given for the proton excitation functions exhibited in figure 10-6, at least for energies up to about 40 MeV.

Types of Emitted Particles. The particles that are usually emitted in medium-energy reactions include n, p, α, H^3, He^3, d, and fission fragments. The relative probabilities for the emission of various combinations of these particles and the number of particles emitted depends, as might be expected, on the identity of the target nucleus and the energy and type of the incident particle. In compound-nucleus reactions the probability for the emission of a particular particle essentially depends on the sum of its binding energy and the height of the Coulomb barrier that the particle must surmount on its way out; the larger this quantity, the smaller the probability for the emission of the particle. This statement suggests that the emission of neutrons would, in general, be favored. But, on the other hand, the symmetry term in the mass equation (cf. p. 41) tells us that the binding energy of a neutron increases as we move out on the neutron-deficient side of stability, and thus charged-particle emission can also be of importance. This effect is particularly important for compound nuclei with atomic numbers below about 30 where the Coulomb barrier is not very high and where the neutron binding energy increases rather rapidly with increasing neutron deficiency. Thus for these lighter targets we expect to find most reactions leading to products that are just a little over on the neutron-deficient side of stability; as the atomic number of the target increases, the main products will be found moving farther over on the neutron-deficient side. Further, it is of interest to see in figure 10-1 that proton emission is more probable than neutron emission from the Ni^{58} compound nucleus despite the fact that the binding energy of the neutron is probably a little less than the sum of the binding energy and the Coulomb barrier of the proton. This result has been ascribed to special properties of the residual nuclei (Ni^{57} and Co^{57} in this example) which, as is discussed in section G, depend on their neutron and proton numbers. To get some orientation on the reactions that are to be expected, let us consider the bombardment of targets in the vicinity of copper and in the vicinity of bismuth with protons of various energies.

The predominant reactions to be expected in the bombardment of copper with protons of about 5 to 15 MeV are the (p, n) and (p, p') reactions. It is also possible for a significant amount of (p, α) reaction to occur, particularly with targets of atomic number lower than that of copper. As the energy of the proton is increased into the 15–25 MeV interval, reactions such as $(p, 2n)$, (p, pn), $(p, 2p)$, $(p, \alpha n)$ will become dominant at the expense of the reactions in the lower energy interval. In this energy region reactions such as (p, He^3), (p, H^3), and (p, d) are also observed, although usually with smaller cross sections. These reactions often show characteristics of a pickup rather than a compound-nucleus mechanism. As the energy of the proton is further increased, the compound-nucleus reactions in which three or more particles are emitted begin to dominate. These reactions will include, for example, $(p, 3n)$, $(p, p2n)$, $(p, \alpha pn)$, and $(p, \alpha 2n)$. The pickup reactions will continue to occur, still with rather smaller cross sections. In the vicinity of 40 to 50 MeV, reactions with four or more emitted particles will predominate.

If the incident particle is an α particle instead of a proton, the pattern of reactions will be much the same, except that there may be some enhancement of α-particle emission because of direct reactions that may occur. Again there will be some emission of He^3 and H^3, but now the direct contribution will be due to stripping rather than to pickup.

The pattern of reactions will again be similar if the incident particle is a deuteron; however, it will also include the important stripping reactions of the deuteron which are discussed in section C-4 of this chapter.

When the atomic number of the target is increased, the increasing Coulomb barrier progressively suppresses the emission of charged particles until at bismuth the main processes are (p, xn) reactions where x, the number of neutrons emitted, increases with bombarding energy, reaching a value of 4 or 5 around 50 MeV. But even here there will still be some proton emission, partly from compound-nucleus reactions but largely from direct reactions. Alpha particles will still be seen (although with low cross section); the binding energy of the α particle becomes negative in the heavy elements and thus can partly compensate for the increase in Coulomb barrier. Again, a similar pattern holds for α-particle irradiation and again there can be some enhancement of α-particle emission because of direct reactions.

The emission of d, H^3, and He^3, probably by direct processes, is still found with cross sections not very different from what they are with the lighter targets.

It should be mentioned that fission is the most probable reaction in the irradiation of targets with atomic numbers in excess of about 90.

Energy Spectra. The energy spectrum of the α particles emitted in the Ni (p, α) reaction shown in figure 10-2 again substantiates the expectation of the compound-nucleus model: the energy spectrum of the α particles has a maximum in the vicinity of the Coulomb barrier (about 10 MeV), and thus most of the α particles are emitted with the minimum possible energy. It is also to be noted that the energy spectra for the emitted α particles are very similar both at 30° and at 120° with respect to the incident beam. This is again what would be expected from the random motion of the nucleons within the excited compound nucleus. But there is an important divergence from this conclusion which is also seen in figure 10-2. There is a second peak in the energy spectrum taken at 30°; this peak corresponds to the emission of α particles with up to all of the available energy and leads to formation of the ground state and first few excited states of the residual nucleus. The second peak is contrary to the expectation of the compound-nucleus model and suggests that there are also some direct interactions, particularly for high-energy particles emitted in the forward direction.

Further evidence of direct interaction can be seen in the (p, pn) and $(p, p2n)$ excitation functions in figure 10-6. Although both of these excitation functions exhibit a characteristic maximum, they start to level off above 50 MeV rather than continuing to drop as more complex reactions become more important.   The flat portion of the (p, n) excitation functions above 40 MeV also suggests the direct ejection of relatively high-energy neutrons rather than their evaporation from a compound nucleus.

In summary, the excitation function for a medium-energy reaction rises to a maximum and then diminishes because of competition from other reactions which become energetically possible.   The energy spectra of most of the emitted particles have peaks in the vicinity of the lowest energy which allows the particle to escape from the nucleus; however, in addition, some high-energy particles are usually emitted preferentially in the forward direction.

4. Deuteron Reactions

It was found early in the study of nuclear reactions that (d, p) reactions occur at deuteron energies well below the Coulomb barrier of the target nucleus and that the cross sections are considerably larger than those for the corresponding (d, n) reactions, particularly for heavy nuclei.   These two observations are completely at odds with what would be expected from the compound-nucleus model: there should be essentially no reactions at energies below the Coulomb barrier, and neutron emission should predominate over proton emission from the few compound nuclei that may be formed, particularly with elements of high atomic number.   This apparent anomaly has been explained by Oppenheimer and Phillips (O1) as being the result of polarization of the deuteron by the Coulomb field of the nucleus.   As the deuteron approaches the nucleus, its "neutron end" is thought to be turned toward the nucleus, the "proton end" being repelled by the Coulomb force. Because of the relatively large neutron-proton distance in the deuteron (several times 10^{-13} cm) the neutron reaches the surface of the nucleus while the proton is still outside most of the Coulomb barrier.   Since the binding energy of the deuteron is only 2.23 MeV, the action of the nuclear forces on the neutron tends to break up the deuteron, leaving the proton outside the potential barrier. The process just described is generally called an Oppenheimer-Phillips (or O-P) process.   An analogous mechanism appears to be responsible for the low-energy (He^3, p) reaction.   An interesting feature of the O-P process is that the emergent protons have a spread of energies which includes values in excess of the incident deuteron energy, so that in a fraction of the events the excitation of the compound nucleus is that which would result from the capture of a neutron of negative kinetic energy.

With increasing deuteron energy, reactions other than the O-P process become possible; but because of the large size and low binding energy of the deuteron there seems to be no energy range in which deuteron-induced reactions can be described completely, or even largely, by the simple compound-nucleus picture.   At high energies (> 100 MeV) the dominant process becomes deuteron stripping, in which either the proton or neutron is stripped off by collision with a nucleus and the other nucleon continues essentially in the original direction of the deuteron with its share of the deuteron momentum (S2). The inverse reaction, the pickup process, is also observed: a fast proton picks up a neutron while making a glancing collision with a nucleus and leaves as a high-energy deuteron in the forward direction.

In addition to stripping, there are also compound-nucleus reactions with deuterons in the medium-energy range; the consequences are similar to those already discussed for medium-energy reactions with protons and α particles. The excitation functions, though, will often look quite different from those with protons and alphas because of the stripping reactions that are also occurring.

5. Heavy-Ion Reactions

Nuclear reactions induced by heavy ions (mass and charge greater than that of an α particle) exhibit the characteristics both of the compound-nucleus and of the stripping and pickup mechanisms. The compound-nucleus reactions usually involve the evaporation of several particles because the heavy ion must have considerable kinetic energy before it can surmount the Coulomb barrier to form the compound nucleus. For targets with atomic numbers greater than about 35 or 40, most of the cross section for the compound-nuclear reactions goes into the evaporation of neutrons; with targets of lower Z, charged-particle emission may compete significantly with neutron emission. The excitation functions for these reactions exhibit the maxima that are expected from the evaporation process.

Reactions are also observed in which the heavy ion evidently makes a grazing collision with the target nucleus and one or more nucleons are transferred either to or from the heavy ion. As is expected for this kind of direct interaction, the residual heavy ion continues on in the forward direction, as was observed, for example, in the reaction

$$N^{14} + Pb^{207} \rightarrow \dot{N}^{13} + Pb^{208}.$$

The cross section for these transfer reactions diminishes as the number of nucleons transferred increases. The major part of the nuclear-reaction cross section for heavy ions at energies below that of the Coulomb barrier is found in transfer reactions.

6. Fission

The possibility that an excited nucleus may split into two roughly equal fragments is suggested by the discussion of spontaneous fission in section B of chapter 8; fission is indeed one other possible mode of de-excitation of an excited compound nucleus and, in the region of high atomic numbers, competes with the evaporation of nucleons and small clusters of nucleons.

Nature of the Process. The fission process is usually accompanied by the emission of neutrons and much more rarely by the emission of α particles and possibly other light fragments. Fission has been produced in some nuclides (notably U^{235}, U^{238}, and Th^{232}) by neutrons, protons, deuterons, helium ions, and γ and X rays of moderate energies; and with higher bombarding energies (50 to 450 MeV) lighter elements such as bismuth, lead, gold, tantalum, and some rare earths have been shown to undergo fission. By far the most important of these reactions is neutron-produced fission. The species U^{232}, U^{233}, U^{235}, Pu^{239}, Am^{241}, and Am^{242} undergo fission either with thermal or fast neutrons, whereas fission of Th^{232}, Pa^{231}, and U^{238} requires fast neutrons.

The analogy between a nucleus and a liquid drop which Bohr used when he proposed the idea of the compound nucleus can be extended to explain fission, at least in a qualitative way. As is discussed on p. 236, there is a certain critical size for nuclei, depending on Z^2/A, above which the force of electrostatic repulsion will be greater than the surface forces holding the nucleus together. This critical size has been calculated to occur for Z somewhere near 110–120, and it is therefore reasonable that for a nucleus only slightly below this limit of stability a small excitation should be sufficient to induce breakup into two fragments. Bohr and Wheeler (B5) calculated the energetic conditions for fission of various heavy nuclear species on the basis of this model, and their theory is in fair agreement with the facts. They were able to predict the fission of Pa^{231} and to estimate its threshold energy before the reaction had been discovered.

Because nuclei of medium weight have higher binding energies per nucleon than those of the heaviest elements, the fission process is accompanied by a large energy release, close to 200 MeV. The unique importance of the fission reaction is due to this energy release and especially to the fact that in each neutron-produced fission process more than one neutron is emitted, which makes a divergent chain reaction possible. The number ν of neutrons produced per thermal-neutron fission is 2.5 for U^{235} and 2.9 for Pu^{239}. It is to be noted, however, that not only fission but also radiative capture occurs with the heaviest elements. Therefore, to obtain the number η of neutrons produced per thermal neutron absorbed, the ν value has to be multiplied by the ratio of the thermal fission cross section to the total thermal-neutron absorption cross section. The resulting values of η are 2.1 for U^{235}, 1.3 for normal uranium, and 2.1 for Pu^{239}.

Fission Product Chains (K2). The fission process may occur in many different modes, and a very large number of fission products, ranging from $Z = 30$ (zinc) to $Z = 65$ (terbium) and from $A = 72$ to $A = 161$ in the thermal-neutron fission of U^{235}, are known. Fission into two equal fragments is by no means the most probable mode in thermal-neutron fission. Quite asymmetric modes are much more favored, the maximum fission product yields occurring at $A = 95$ and $A = 138$. The asymmetry appears to become less pronounced with increasing bombarding energy.

An enormous amount of radiochemical work was required to arrive at our present state of knowledge about fission products. It was necessary to develop chemical separation procedures, to analyze radioactive decay and growth patterns, to determine β- and γ-ray energies, to establish mass assignments of many previously unknown nuclides, and to measure the fission yields. The fission yield of a nuclide is the fraction or the percentage of the total number of fissions which leads directly or indirectly to that nuclide.

As would be expected from the different neutron-proton ratios for U^{235} and for the stable elements in the fission product region, the primary products of fission are generally on the neutron-excess side of stability. Each such product decays by successive β^- processes to a stable isobar. Chains with as many as six β decays have been established, and undoubtedly some fission products still further removed from stability (higher up on the parabolic slope of the stability valley) have escaped detection because of their very short half-lives. No neutron-deficient nuclides have been found among the products of thermal-neutron fission; however, a few so-called shielded nuclides occur among the fission products. A shielded nuclide is one that has a stable isobar one unit lower in Z so that it is not formed as a daughter product in a β-decay chain. The fission yield of such a nuclide is presumably due entirely to its direct formation as a primary product.

Charge Distribution. Direct information on the distribution of yields along any given chain is confined to the relatively few good measurements of independent fission yields of individual chain members and of the shielded nuclides. A hypothesis which appears to account fairly well for most of the data is the postulate of "equal charge displacement." To state this, we make use of the quantity Z_A, the value of Z corresponding to the highest binding energy for a given A (chapter 2, section D). Furthermore, we define Z_p as the most probable charge for a primary fission fragment of mass number A. The postulate of equal charge displacement is that the two complementary fragments in a given fission event always have equal $Z_A - Z_p$ values and furthermore that the probability distribution around Z_p is the same for all values of A. Remembering that on the average 2.5 neutrons are emitted per U^{235} fission, we can write $Z_A - Z_p = Z_{233.5-A} - (92 - Z_p)$, or $Z_p = 46 + \frac{1}{2}(Z_A - Z_{233.5-A})$. From this formula and (2-7), Z_p can be calculated for any A. The calculated values for $Z_A - Z_p$ range from about 3.0 for the lightest and heaviest fission products to about 4.0 near $A = 115$. The measured independent yields indicate a probability distribution such that 50 per cent of the total chain yield occurs for $Z = Z_p$, about 25 per cent each for $Z = Z_p \pm 1$, about 2 per cent each for $Z = Z_p \pm 2$, and much less for other Z values. The isobaric yield distribution around Z_p appears to be Gaussian, and the postulate of a universal

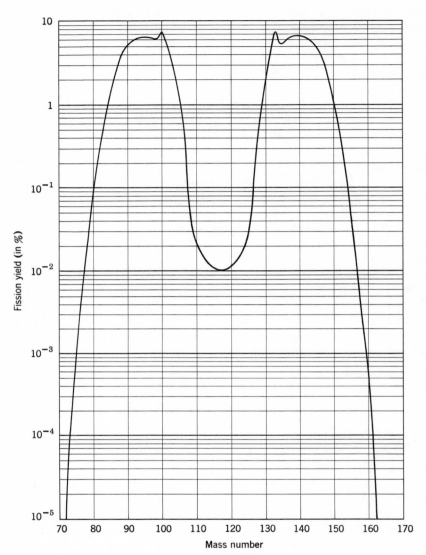

Fig. 10-7   Yields of fission-product chains as a function of mass number for the slow-neutron fission of U^{235}.

distribution in Z at all values of A is borne out by experiment, but not understood theoretically.   A more detailed analysis of charge distribution in fission is given in (W2).

The measured total yield of a particular fission product represents the sum of its independent yield and of the independent yields of all its precursors. Wherever fission yields have been measured for several isobars, the results are in fair agreement with the postulate of equal charge displacement. We may therefore assume that the measured total fission yield of a nuclide which is only one or two units of Z from stability represents most of the total yield for its mass number.

Mass Distribution.   When the total fission yield at each mass number is plotted against mass number, the curve shown in figure 10-7 results. The curve is essentially symmetrical about the minimum at $A = 233.5/2$ and has two rather broad maxima around mass numbers 95 and 138. The yields in each of the two peaks add up to approximately 100 per cent.

Sufficient information is available on the fission yields in the thermal-neutron fission of Pu^{239} to draw a mass-yield curve for this case. The general shape is similar to the U^{235} curve, but there are certain significant differences. The yield at the minimum is not so low as for U^{235}; it is about 0.04 per cent fission yield at $A = 119$. The heavy peak appears not to be appreciably displaced, compared with U^{235} fission, but the light peak has its maximum at about $A = 99$.

Shell Effects.   Considerable attention has been given to some of the details of the fission yield curves. Among the most interesting result of these investigations was the discovery of the two "spikes" near the peaks of the mass-yield curve (figure 10-7). These spikes have been studied both by mass-spectrographic and by radiochemical techniques. The effects are explained in terms of nuclear shell structure and appear to arise from three different causes.

In the first place there are among the fission products six delayed-neutron emitters, with half-lives of 55.6, 22.0, 5.6, 2.7, 0.6, and 0.18 seconds. These are actually β^- emitters which decay to nuclides that are unstable with respect to instantaneous neutron emission. The first two have been radiochemically identified as $_{35}Br^{87}$ and $_{53}I^{137}$. The daughters in these two cases, $_{36}Kr^{87}$ and $_{54}Xe^{137}$, have 51 and 83 neutrons, respectively, each just one neutron more than a "magic" number. Because of these delayed-neutron emitters the yields of mass numbers 87 and 137 are expected to fall slightly below the smooth yield curve, with the yields at masses 86 and 136 correspondingly raised. Similar effects are expected from the other, still unassigned, neutron emitters. But from the yields of delayed neutrons it is clear that altogether these effects cannot account for the magnitude of the observed spikes.

A second possible cause for the observed phenomena might be the prompt boil-off of a neutron from primary fragments containing 51 or 83 neutrons. An analysis of this hypothesis shows that it would account reasonably well for the anomalies in the mass range 133–135, but it could not predict the almost complementary spike near $A = 100$.

The complementarity of the two spikes suggests that they are connected with the fission act itself. The assumption that in the primary process fragments with closed-shell configurations ($N = 82$ and $N = 50$) are formed preferentially leads to rather good agreement with the observations, not only for thermal-neutron fission of U^{235}, but also for several other fission reactions which yield "spikes" in somewhat different positions.   However, there is still evidence for some contribution of neutron boil-off after fission.

A number of attempts have been made to account theoretically for the observed fission-yield distributions as well as the angular distribution and the energy spectra of the emitted particles.   These calculations include a hydro-dynamical approach which emphasizes the modes of vibration of a liquid drop, a statistical approach which resembles the evaporation theory of section G, and a detailed examination of the effect of nuclear deformation on single-particle states as discussed in section E of chapter 9.   None of these efforts, which are reviewed along with the relevant experimental information in ref. H4, has yet succeeded in unifying a substantial portion of the information on fission.

7. Reactions at High Energies (M3)

In the discussion of the excitation functions and particle spectra found in medium-energy nuclear reactions it was pointed out that the compound-nucleus model does not adequately describe all of the reactions that occur with protons up to 40 MeV and is surely in trouble between 40 and 100 MeV.   In the high-energy region above 100 MeV the compound-nucleus model is no longer useful; the nuclear reactions seem to proceed nearly completely by direct interactions.

One reason for this remark may be seen in the contrast among the nuclear reactions induced by 40-, 400-, and 4000-MeV (4-GeV) protons with bismuth (Bi^{209}), as illustrated schematically in figure 10-8, where the cross section for the formation of a given mass-number product is plotted against the mass number.   At 40 MeV, where most of the reactions proceed through the formation of a compound nucleus with a given excitation energy, we find that nearly all of the reactions, as expected, lead to products with mass numbers ranging from 206 to 208.   At 400 MeV, on the other hand, there is a wide distribution of products which roughly divide themselves into two groups: those down to mass number 150, which are called spallation products, and those between mass numbers 60 and 140, which we shall designate as fission products.   At 4 GeV, there is a continuous distribution of products with no evident division between fission and spallation.   There is evidence that there is not much further change in the mass-yield curve as the energy is increased to 30 GeV.   It is evident from figure 10-8 that at the higher bombarding energies we do not observe the relatively few spallation products expected from the formation of a compound nucleus at a given excitation energy but rather a large array of products corresponding to various amounts of excitation energy from zero up to the maximum possible.   Measurements of the energy and angular distribution of the emitted particles show some of them to be of high energy, approaching that of the incident particle, and to be preferentially emitted in the forward direction.

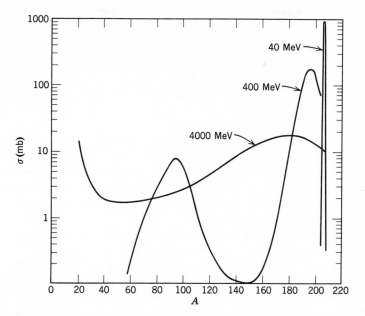

Fig. 10-8 Comparison of the approximate mass distributions of the products of the reactions of 40-, 400-, and 4000-MeV protons with Bi^{209}.

Intranuclear Cascades. In explanation of these phenomena R. Serber (S3) pointed out that, if the energy of the incident proton is significantly larger than the interaction energy between the nucleons in the nucleus, and its wavelength is less than the average distance between nucleons, then the incident proton will collide with one nucleon at a time within the nucleus. Further, the cross section for each collision and the angular distribution will be very nearly the same as if the collision occurred in free space rather than in the interior of a complex nucleus. These assumptions are an extreme form of what is known as the "impulse approximation." Under these circumstances the incident proton may be considered to have a mean free path in nuclear matter

$$\Lambda = \frac{1}{\rho\bar{\sigma}},$$

where ρ is the density of nucleons in the nucleus and $\bar{\sigma}$ is the effective nucleon-nucleon interaction cross section.

Taking the density of nuclear matter as about 10^{38} nucleons cm^{-3} and the effective interaction cross section for incident protons of a few hundred MeV as about 30×10^{-27} cm^2, the mean free path of the proton in nuclear matter is about 3×10^{-13} cm, a distance that is of the same order of magnitude as nuclear radii. From this result it is no longer surprising that a high-energy proton may make only a few collisions while traversing a complex nucleus, leaving behind only a fraction of its energy and sometimes directly ejecting a nucleon with which it collides.   The struck nucleons also often have considerable kinetic energy, and their passage through the nucleus can be considered in the same manner as that for the incident proton; in this fashion an intra-nuclear knock-on cascade of fast nucleons is generated; at energies in excess of about 350 MeV the cascade must also include the pi-mesons which can be created in nucleon-nucleon collisions.

This model is represented schematically in figure 10-9 for a proton incident on a complex nucleus at an impact parameter b.   A cascade nucleon may either immediately escape from the nucleus, as is shown in figure 10-9 for a neutron and a proton, or it may be reduced to (or formed with) an energy so low that it is considered captured by the nucleus and gives up its energy to excitation of the whole nucleus.

It must be mentioned at this point that the other nucleons in the nucleus are not totally without effect on a collision; they occupy quantum states and so, because of the Pauli exclusion principle, make those states unavailable as final states to the two colliding nucleons.   The result is a lowering of the effective collision cross section primarily through the decreased probability of very small or very large energy transfers.   Collisions forbidden by the Pauli principle are shown as open circles in figure 10-9.   The effect of the Pauli principle is particularly important for low-energy cascade nucleons and has already been invoked in chapter 9 as part of the justification for the success of the shell model.

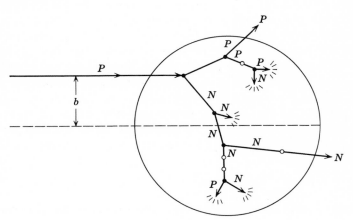

Fig. 10-9   Schematic diagram of intranuclear cascade generated by a proton with impact parameter b.   The solid circles indicate positions of collisions; the open circles indicate positions of collisions that were forbidden by the Pauli exclusion principle.   The short arrows ending within the nucleus indicate "captured" nucleons that contribute to the over-all excitation.

The cascade, then, is essentially a random-walk process in which each cascade particle has a mean free path for collision and each collision is characterized by a probability distribution for the various possible final states which is given by the characteristics of the collision in free space modified by the requirements of the Pauli exclusion principle.   Within this model the distribution of the properties of emitted nucleons and pi-mesons as well as those for the residual nuclei may be estimated by the so-called Monte Carlo method in which the selection of random numbers determines the locations and consequences of each nucleon-nucleon collision in the cascade.   A calculation of this type, carried out on a high-speed electronic computer, has considerable success when compared with experimental results (M2).

We have come to think, then, of high-energy reactions as occurring in two stages.

1. First there is the knock-on cascade in which several particles are directly ejected from the nucleus in a time of the order of 10^{-22} second, leaving behind the cascade product in an excited state.   The knock-on cascades are characterized by the probability that the cascade product have a particular mass, charge, and excitation energy.

2. The excited cascade products may subsequently de-excite in at least two ways:

a. The residual excited nucleus may evaporate nucleons or clusters of nucleons in a process similar to that of the evaporation of particles from compound nuclei, as mentioned in section B and elaborated in section G.   The products formed in this manner are called spallation products, from the verb "to spall"—to break up or reduce by chipping with a hammer.

b. The excited cascade product may divide into two roughly equal pieces in a manner that is analogous to the previously discussed fission process.   The products of such a reaction are considered to be high-energy fission products.

Spallation Products.   A large number of radiochemical studies have been made of spallation reactions in the bombarding-energy region up to about 30 GeV.   In summary of this rather complex field it is probably fair to say that essentially any spallation reaction which is energetically possible is believed to occur.   However, the cross sections (or at least the relative yields) of different products from a given bombardment are of interest.   In general the products in the immediate neighborhood of the target element, within perhaps 10 or 20 mass numbers on the low-mass side, are found in the highest yields.   The yields for lower mass numbers then drop off rather rapidly.   Thus the transfer of relatively small amounts of excitation, rather independent of incident energy, appears to be most probable, and this agrees with expectation from knock-on calculations (M2).

The spallation yields tend to cluster quite strongly in the region of stability in the case of medium-weight targets and somewhat more to the neutron-deficient side of stability for heavier elements.   This is just what one would expect from evaporation theory (see section G). The considerations here are the same as those encountered in lower-energy competition. It is interesting to note that the ratio of yields of two isobars in spallation usually remains almost the same regardless of target element and bombarding energy.

Excitation functions for spallation reactions divide into two general types.

1. If the threshold for the reactions is at an energy low enough for compound-nucleus formation still to be significant, the excitation function may go through the characteristic maximum but will then tend to level off rather than to continue to diminish with increasing energy. An example of this behavior is given by the excitation function for the $(p, p2n)$ reaction in figure 10-6.

2. If the threshold for the reaction is above the energy region in which compound nucleus formation is significant, the excitation function will rise to a value rarely exceeding 10 or 20 mb and then level off or diminish very slowly with increasing energy even up to energies in the GeV region. The energy at which the excitation function for a spallation reaction levels off increases with increasing mass difference between target and product nuclides.

High-Energy Fission. The characteristics of fission induced by high-energy particles differ markedly from those of thermal-neutron fission. The familiar double hump in the fission-product-yield curve (figure 10-7) is at this energy replaced by a single broad peak, centered around a value of A somewhat less than half the mass number of the target nuclide (cf. 400-MeV curve in figure 10-8). In contrast to low-energy fission, many neutron-deficient nuclides are found, especially among the heavy products. This observation has been interpreted as indicating that the excited nucleus formed after the knock-on phase of the reaction first evaporates a number of nucleons (mostly neutrons) and that after a sufficiently high value of Z^2/A is reached fission competes with further evaporation. In addition, there is evidence that many of the primary fission products, both heavy and light, from a given target and given bombarding energy have about the same neutron-proton ratio. This indicates that fission occurs too rapidly to permit redistribution of neutrons and protons and that it occurs from fissioning nuclei with about the same proton-neutron ratios that are found in the products.

Fission has been observed for many elements of Z greater than about 70. In these elements the fission product region is rather definitely separated from the spallation products, as shown in figure 10-8, by a region of A with very low cross section as long as the energy is not above several hundred MeV. This clean separation of the fission and spallation region is no longer evident when the energy is in the GeV region.

Fission has been reported for much lighter elements with cross sections of the order of microbarns, but here it is rather difficult to disentangle fission from spallation. The most persuasive evidence has come from the observation in photographic emulsions of tracks from two heavily ionizing particles going off in opposite directions.

Other Processes. It is doubtful whether all of the high-energy reactions can be explained within the framework of the two-step process: knock-on cascade followed by the de-excitation of the excited cascade products. The emission of high-energy (>50 MeV) particles such as He and Li ions in the forward direction and the formation, in high yield, of products with mass number between about 15 and 40 in the bombardment of heavy targets such as bismuth are examples of observations that do not appear to fit into this framework.

Serious consideration is being given to the idea that the He and Li nuclei may directly participate in the knock-on cascade, particularly in the diffuse edge of the nucleus in which there is a good possibility of the existence of clusters of nucleons. The high-energy He and Li products, then, would be considered as directly ejected after being struck by one of the cascade nucleons.

From the excitation functions for the formation of products such as F^{18}, Na^{24}, and P^{32} from heavy targets it is evident that they are largely produced in events in which a large amount of energy (several hundred MeV) has been transferred to the struck nucleus; occasionally the energy transfer approaches, or even exceeds, the total binding energy of the nucleus. Even if an excited intermediate nucleus is formed, its lifetime at these high excitation energies is no longer long compared to that of the cascade, a fact that immediately makes it difficult to invoke two separate steps. Efforts to understand these reactions, which are called fragmentation reactions, have been focused on what might occur as the damage done to the nucleus during the cascade is transformed into statistically distributed excitation energy; no significant progress on this question has yet been made.

Reactions with Pi-Mesons. The production of π-mesons (pions) in collisions between high-energy protons and complex nuclei (about 0.5 pion per collision with 1 GeV protons) has enabled studies to be made of nuclear reactions induced by pions. For high-energy pions, a knock-on cascade will be developed similar to that generated by protons. There is an additional important possibility in intranuclear cascades containing pions: the annihilation of the pion in a collision and the transfer of its total energy (kinetic plus rest mass) to cascade nucleons. Conservation of both momentum and energy demands that the annihilation process involve at least two nucleons which share the available energy between them. Thus it would be expected that the nuclear reactions induced by pions would be similar to those induced by high-energy protons, and this has indeed been found to be true.

8. Photonuclear Reactions

The excitation functions of γ, n and γ, p reactions look not unlike those of other low-energy reactions; with increasing energy, they rise rather steeply to a maximum and then drop again.   However, the drop in cross sections for these reactions is not accompanied by a corresponding rise in the cross sections for competing reactions.   In other words, the total cross section for photon absorption itself appears to show a resonance behavior.   This "giant-resonance" absorption is ascribed to the excitation of dipole vibrations of all the neutrons moving collectively against all the protons; on this basis the magnitude, energy, and A-dependence of the resonance absorption have been accounted for quite well.   Alternative mechanisms for the dipole absorption have also been proposed (W1).

The resonance peaks have widths of several MeV at half maximum, and the energy of the resonance peak decreases slowly with increasing A from about 24 MeV at O^{16} to about 14 MeV at Ta^{181}.   The integrated cross sections under the resonance peaks vary from about 0.05 MeV-barn for the lightest elements to 2 or 3 MeV-barn for heavy elements.   The energy of the dipole resonance is so low that mostly rather simple processes—such as γ,n, γ,p, and some $\gamma, 2n$ and photofission reactions—take place by this mechanism.   A few reactions have been observed with low cross sections at energies below the dipole absorption (presumably due to quadrupole effects).   The reaction cross section remains small ($\cong 10$ per cent of the peak cross section) above the energy of the resonance peak until meson production by high-energy ($\gtrsim 150$ MeV) γ rays enhances the reaction cross section and gives rise to processes similar to those discussed in section C7.

D. CENTER-OF-MASS SYSTEM

Before entering into a more detailed discussion of the optical, compound-nucleus, and direct-interaction models, we shall first enumerate some of the properties of the center-of-mass system (sometimes called the barycentric system), since this is the "natural" coordinate system in which to analyze the course of a reaction.

It has been pointed out in chapter 2, section F1, that because of momentum conservation all of the kinetic energy of a bombarding particle is not available for the production of a nuclear reaction in which energy is absorbed; part of the incident kinetic energy must go into kinetic energy of the products of the reaction.   Another way of looking at this problem is to consider the bombardment of a stationary target nucleus A of mass M_A, with projectile a of mass m_a which is moving with velocity v with respect to an origin O, fixed in the laboratory.   If we consider the problem nonrelativistically ($v/c \ll 1$), the kinetic energy of the incident particle is simply $\frac{1}{2}m_a v^2$.   Now, an apparent ambiguity arises when it is realized that all that the two particles a and A can possibly know is their relative velocity v.   Thus the bombardment described should be equivalent to one in which a stationary nucleus a is bombarded with nucleus A

moving with the same velocity v; but now the kinetic energy of the bombarding particle is $\frac{1}{2}M_A v^2$. The ambiguity appears when we ask which energy characterizes the bombardment, $\frac{1}{2}m_a v^2$ or $\frac{1}{2}M_A v^2$? The well-known answer is: neither one nor the other. The kinetic energy that characterizes the reaction is given by one half the product of the reduced mass of the system, $m_a M_A/(m_a + M_A)$, and the square of the relative velocity v of the two particles:

$$\epsilon = \frac{1}{2}\frac{m_a M_A}{m_a + M_A}v^2 \qquad (10\text{-}1)$$

This result is another way of expressing what was said in chapter 2: if a is the bombarding particle and A is stationary, then the fraction $M_A/(m_a + M_A)$ of the incident kinetic energy is available to make the reaction go; conversely, if A is the bombarding particle and a is stationary, the corresponding fraction is $m_a/(m_a + M_A)$.

The kinetic energy given by (10-1) is the kinetic energy in a coordinate system whose origin O' is at the center of mass of the two particles a and A and moves with a velocity v_{cm} with respect to the fixed laboratory origin O. The relationship between the velocities of a particle in the two coordinate systems is illustrated in figure 10-10 for the general situation in which both particles, a and A, are in motion in the laboratory. The velocities of the

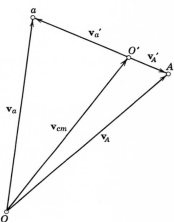

Fig. 10-10. Relationship between velocities of particles a and A in the laboratory system, v_a and v_A, and in center-of-mass system, v_a' and v_A'. The velocity of the center of mass with respect to the laboratory is v_{cm}.

two particles in the center-of-mass system must necessarily be in exactly opposite directions with a ratio of magnitudes given by the inverse of the ratio of their masses:

$$\frac{v_a'}{v_A'} = \frac{M_A}{m_a}.$$

The momenta of the two particles, then, are exactly equal in magnitude but opposite in direction in the new coordinate system and their vectorial sum vanishes. It is this property of the center-of-mass system that makes the energy given by (10-1) the energy "available" for the nuclear reaction since in this system the products of the reaction may have *zero* kinetic energy.

The over-all conservation of momentum is assured by the motion of the center of mass with respect to the fixed origin:

$$(m_a + M_A)\mathbf{v}_{cm} = m_a\mathbf{v}_a + M_A\mathbf{v}_A. \qquad (10\text{-}2)$$

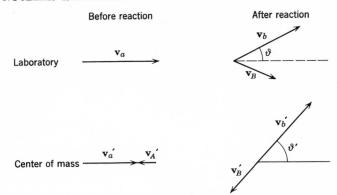

Fig. 10-11   Velocity diagram of nuclear reaction in laboratory system and in center-of-mass system.

All that this amounts to, then, is replacing the reacting system of two particles by a fictitious complex particle of mass $m_a + M_A$ moving with a velocity $\mathbf{v}_{cm}$ given by (10-2), and considering any interaction between a and A as occurring in the internal coordinates of the fictitious complex particle.   A velocity diagram for the nuclear reaction

$$a + A \rightarrow B + b,$$

as seen in the laboratory system and in the center-of-mass system is shown in figure 10-11.

Corresponding to the kinetic energy of relative motion given by (10-1) is the relative momentum of the two particles:

$$\mathbf{p} = \frac{m_a M_A}{m_a + M_A}\, \mathbf{v}, \tag{10-3a}$$

$$\mathbf{v} = \mathbf{v}_a - \mathbf{v}_A = \mathbf{v}_a{}' - \mathbf{v}_A{}'. \tag{10-3b}$$

The relative momentum of the two particles, $\mathbf{p}$, is the proper quantity to use in evaluating the density of translational quantum states for the reacting system discussed in footnote 13 of chapter 8.   The relation between the magnitude of the relative momentum and the momenta in the laboratory system may be obtained by a combination of (10-3a) and (10-3b).

$$p = \left| \frac{M_A}{m_a + M_A}\, \mathbf{p}_a - \frac{m_a}{m_a + M_A}\, \mathbf{p}_A \right|. \tag{10-4}$$

The magnitude of the relative momentum in the laboratory system is the same as the magnitude of the momentum of *either* particle in the center-of-mass system.

E. SOME GENERAL PROPERTIES OF NUCLEAR CROSS SECTIONS

There are a few important general predictions that can be made for the upper limit of the elastic and reaction[1] cross sections.   That this should be so is not surprising: it might be immediately expected that a nucleus that interacts with everything that hits it would have a reaction cross section of πR^2 where R is the sum of the radii of the two reacting particles.   As will shortly be seen, this is not quite correct; the wave nature of the incident particle causes the upper limit of the reaction cross section to be

$$\sigma_r = \pi(R + \lambda)^2, \tag{10-5}$$

where λ is the de Broglie wavelength of the incident particle in the center-of-mass system divided by 2π [$\lambda = \hbar/p$, where p is computed from (10-4)]. Equation 10-5 gives the expected answer of πR^2 at high energies, but even there the wave character of the incident particle makes the not-so-obvious requirement that there also be an elastic-scattering cross section of πR^2 (often called either diffraction scattering or shadow scattering) or a total cross section of $2\pi R^2$.   These limits on the cross section are properly derived by a quantum-mechanical analysis of the scattering event (cf. chapter 8 of ref. B1 or section 5 of ref. K1), which is sketched later, but the essence of the problem, at least for the reaction cross section, may be easily seen in a semiclassical treatment. Since, at this point, we are interested in the upper limit to the reaction cross section, we shall first consider incident neutrons, for which there are no Coulomb-repulsion problems.

Maximum Reaction Cross Sections for Neutrons.   A collision between a neutron and a target nucleus may be characterized classically by what would be the distance of closest approach of the two particles if there were no interaction between them.   This distance b, usually called the impact parameter, is shown in figure 10-12.   If we denote the magnitude of *relative* momentum of the two particles by p, the angular momentum of the system is normal to the relative momentum and of magnitude.

$$L = pb. \tag{10-6a}$$

From the de Broglie relation between the relative momentum and the wavelength in the center-of-mass system, $\lambda = h/p$, (10-6a) may be rewritten as

$$L = \frac{hb}{\lambda}. \tag{10-6b}$$

---

[1] From this point on we shall designate the cross section for all events other than elastic scattering as the reaction cross section.   Compound elastic scattering, which is negligible except for thermal neutrons, is included in the reaction cross section, although it cannot be distinguished experimentally from other elastic scattering.

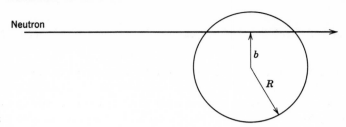

Fig. 10-12   Collision with impact parameter b between a neutron and target nucleus with interaction radius R.

As b may evidently assume any value between 0 and R, the relative angular momentum will vary continuously between 0 and hR/λ. We know, though, that this is not acceptable; quantum mechanics requires that the component of angular momentum in a particular direction be an integer when expressed in units of $\hbar$:

$$L = l\hbar, \quad \text{where} \quad l = 0, 1, 2, \dots . \tag{10-7}$$

Combination of (10-6b) and (10-7) gives

$$b = l\lambda \tag{10-8}$$

Equation 10-8 is not to be interpreted as meaning that only certain values of b are possible; such control over b would violate the uncertainty principle. Rather it means that a range of values of b corresponds to the same value of the angular momentum. In particular,

$$l\lambda < b < (l + 1)\lambda \tag{10-9}$$

corresponds to an angular momentum of $l\hbar$. This interpretation is illustrated in figure 10-13. From this figure it can be seen that the cross-sectional area which corresponds to a collision with angular momentum $l\hbar$ is

$$\sigma_l = \pi\lambda^2[(l + 1)^2 - l^2]$$
$$= \pi\lambda^2(2l + 1). \tag{10-10}$$

If it is assumed that each particle hitting the nucleus causes a reaction, then (10-10) gives the partial cross section for a nuclear reaction characterized by angular momentum $l\hbar$; and the reaction cross section may be obtained by summing (10-10) over all values of l from 0 to the maximum l_m.

$$\sigma_r = \pi\lambda^2 \sum_0^{l_m} (2l + 1). \tag{10-11a}$$

The summation in (10-11a) may be easily evaluated if it is recalled that the sum of the first N integers is equal to $[N(N + 1)]/2$. The expression for the reaction cross section becomes

$$\sigma_r = \pi \lambda^2 (l_m + 1)^2. \tag{10-11b}$$

The maximum value of l may be estimated from (10-8) by limiting the maximum impact parameter to the interaction radius R:

$$l_m = \frac{R}{\lambda}. \tag{10-12}$$

Substitution of (10-12) into (10-11b) yields the result already given in (10-5) for the maximum possible reaction cross section:

$$\sigma_r = \pi (R + \lambda)^2. \tag{10-5}$$

This result suggests the possibility of nuclear-reaction cross sections which are several orders of magnitude larger than the geometrical cross section of the nucleus, a possibility that is realized in slow-neutron reactions. (See section C2 and figure 10-5.)

It might appear that there is some sleight-of-hand involved in the derivation of (10-5); the area of a disk apparently turns out to be larger than the sum of the areas of the concentric rings contained in it (cf. figure 10-13). This result is a direct consequence of (10-12) for the maximum value of the angular momentum; a strictly classical approach to the neutron trajectory would have set $l_m = (R/\lambda) - 1$, in which case the upper limit to the cross section would be πR^2 and all quantal effects would vanish. This approach, though, would lead to an absurd result when $R/\lambda < 1$. Ambiguities of this sort are the usual unfortunate companions of classical calculations which are subject to quantum-

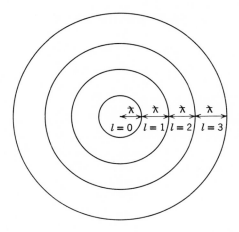

Fig. 10-13   The incident beam is perpendicular to the plane of the figure.   The particles with a particular l are considered to strike within the designated ring.

mechanical conditions.　The quantum-mechanical treatment of the problem (B1, R1) gives, instead of (10-11a)

$$\sigma_r = \pi\lambda^2 \sum_{l=0}^{\infty} (2l + 1)T_l,\qquad(10\text{-}13)$$

where T_l is defined as the transmission coefficient for the reaction of a neutron with angular momentum l and may have values between zero and one. Our semiclassical treatment assigns unity to T_l for all values of l up to and including l_m, as defined in (10-12); for all higher values of l, the transmission coefficients are zero.　The role of angular momentum here is analogous to the one that it plays in β and γ emission, discussed in chapter 8.　A further discussion of the evaluation of the transmission coefficients is postponed until later; but it should be mentioned here that the semiclassical result is quite right for $R/\lambda < 1$, where the only contribution comes from $l = 0$ and the reaction cross section has $\pi\lambda^2$ as its upper limit.

Reaction Cross Sections with Charged Particles.　The effect of the Coulomb repulsion on a reaction cross section may be easily estimated within the spirit of the semiclassical analysis.　The Coulomb repulsion will bring the relative kinetic energy of the system from ϵ when the particles are very far apart to $\epsilon - B$ when the two particles are just touching, where B is the Coulomb barrier:

$$B = \frac{z_a Z_A e^2}{R},\qquad(10\text{-}14)$$

and where z_a and Z_A are the atomic numbers of incident particle and target nucleus, respectively.　Further, the deflection of the particles causes the maximum impact parameter that leads to a reaction to be less than R, as illustrated in figure 10-14.　From this figure it is seen that the trajectory of the particle

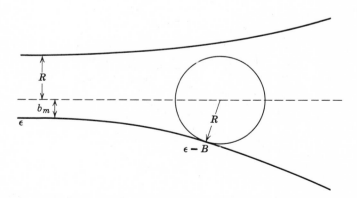

Fig. 10-14　Classical trajectories for charged particles with impact parameters R and b_m.

is tangential to the nuclear surface when it approaches with the maximum impact parameter b_m and that the relative momentum at the point of contact is

$$p = (2\mu)^{1/2}(\epsilon - B)^{1/2} = (2\mu\epsilon)^{1/2}\left(1 - \frac{B}{\epsilon}\right)^{1/2}, \qquad (10\text{-}15)$$

where μ is the reduced mass of the system. The magnitude of the maximum angular momentum is then obtained from the product of the interaction radius and the relative momentum:

$$L_m = R(2\mu\epsilon)^{1/2}\left(1 - \frac{B}{\epsilon}\right)^{1/2}. \qquad (10\text{-}16)$$

Recognizing that $(2\mu\epsilon)^{1/2}$ is the relative momentum of the two particles when they are far apart, we obtain from (10-16) in conjunction with (10-6a)

$$b_m = R\left(1 - \frac{B}{\epsilon}\right)^{1/2} \qquad (10\text{-}17)$$

for the maximum impact parameter. Equation 10-17 has meaning only for $\epsilon \geq B$; for lower ϵ the Coulomb potential, classically, prevents nuclear reactions. Thus the Coulomb barrier diminishes the l_m of (10-12) by a factor of $(1 - B/\epsilon)^{1/2}$. The upper limit for the capture of charged particles can be estimated as the area of the disk of radius b_m:

$$\sigma_r = \pi R^2\left(1 - \frac{B}{\epsilon}\right). \qquad (10\text{-}18)$$

It is to be noted that the method of estimating the upper limit to the reaction cross section for charged particles is different from that for neutrons, where the value of l_m was found and substituted directly into (10-11b). This procedure would not be appropriate for charged particles, as the Coulomb barrier has an important effect on the transmission coefficients of (10-13). In particular, it may be seen from (10-16) that $l_m \to 0$ as $\epsilon \to B$ for charged particles and from (10-12) that $l_m \to 0$ as $\epsilon \to 0$ for neutrons. However, the Coulomb barrier causes the transmission coefficient for charged particles to approach zero under these circumstances, whereas that for the neutron remains finite. The result is a vanishing cross section for charged particles of energies approaching that of the Coulomb barrier to be compared with an upper limit of $\pi\lambda^2$ for neutrons of very low energy. Further, since the Coulomb barrier is the most important factor in determining reaction cross sections with charged particles, (10-18) is an estimate of the reaction cross section rather than just its upper limit.

Equation 10-18, though approximate, has been useful for the estimation of reaction cross sections for charged particles, particularly when the Coulomb barrier is changed to an "effective" Coulomb barrier (D1) to allow for tunneling through the diffuse nuclear surface. Again, the complete analysis of the reaction cross section is properly carried out with (10-13), and the effect of the Coulomb interaction appears in the transmission coefficients (S1, H2).

Quantum-Mechanical Analysis of Cross Section. The general treatment of scattering and reactions between particles can be found in most standard texts on quantum mechanics; the particular analysis of *nuclear* scattering and reactions may be found in refs. B1 and R1, among many others. The following sketch of this treatment is meant to provide perspective both for the semi-classical analysis just described and the special models discussed in the next section.

The wave function that describes an undisturbed beam of particles moving in the z direction with momentum $p = k\hbar$ is just

$$\Psi_u = \exp{[ikz]}.$$

The problem in the quantum-mechanical analysis of scattering and absorption is to see what change occurs in the undisturbed wave function when an absorbing and scattering particle is put at the origin of coordinates and thus to find the disturbed wave function, Ψ_d. The method of analysis of this problem is suggested by the way that reactions and scattering are investigated experimentally. For scattering, a detector is placed at a distance r from the scattering center and at an angle θ with respect to the incident beam, and the flux of particles with and without the scatterer present is measured. This suggests that the wave function for the scattered beam is given by the difference of Ψ_u and Ψ_d:

$$\Psi_{sc} = \Psi_d - \Psi_u$$

and also that the wave functions should be expressed in terms of r and θ rather than z.

The relevant measurement for reactions is the determination of the rate of disappearance of particles with a magnitude of momentum $k\hbar$ (in the center-of-mass system). The straightforward way to carry out this measurement would be to surround the scattering and absorbing center with a spherical shell of radius r and to measure the net flux of particles of momentum $k\hbar$ through this shell. If there were only scattering and no absorption, the net flux would vanish, as there would be just as many particles coming in as going out; if there were reactions, there would be more coming in than going out. This suggests that the reaction cross section is determined by the net flux of particles corresponding to the disturbed wave function, Ψ_d. The program for the calculation, then, is first to express Ψ_u in terms of r and θ, and then to find the forms of Ψ_d appropriate to the scattering and absorbing center.

The mathematical problem of expressing the function $\exp{(ikz)}$ in terms of the product of a function of r times a function of θ (where $z = r\cos\theta$) is a very old one. The solution (see, for example, appendix A of ref. B1) involves spherical Bessel functions, $j_l(kr)$, and spherical harmonics $Y_{l,m}(\theta)$:

$$\exp (ikz) = \exp (ikr \cos \theta)$$

$$= \sum_{l=0}^{\infty} i^l \sqrt{4\pi(2l + 1)} \, j_l(kr) Y_{l,0}(\theta). \qquad (10\text{-}19)$$

The variable ϕ, which is normally found in the general solution to this problem, does not appear in (10-19) because the incident beam (and also the scattered beam) is taken to be cylindrically symmetric about the z direction, a condition which also causes the second subscript of the spherical harmonics to take on the value zero, corresponding to a zero component of angular momentum along the z axis for the incident particles.   Physically, each term of the summation in (10-19) corresponds to particles with angular momentum $l\hbar$ and so replaces a range of impact parameters given by (10-9) of our semi-classical treatment.

Now, measurements of either reaction or scattering cross sections are made at distances from the target which are enormous compared to the wavelength of the incident particles ($\cong 10^{-13}$ cm).   Since the momentum of the particle is $k\hbar$, and $p = h/\lambda$, we have that $kr = r/\lambda$, and so we are interested in the functions $j_l(kr)$ for enormous values of kr.   The form of the spherical Bessel function for $kr \gg l$, a condition that is certainly satisfied, is

$$j_l(kr) = \frac{\sin (kr - \frac{1}{2}l\pi)}{kr}, \qquad (10\text{-}20a)$$

$$= \frac{i\{\exp [-i(kr - \frac{1}{2}l\pi)] - \exp [+i(kr - \frac{1}{2}l\pi)]\}}{2kr}. \qquad (10\text{-}20b)$$

Substitution of (10-20b) into (10-19) gives the desired form for Ψ_u in terms of r and θ:

$$\Psi_u = \frac{\pi^{\frac{1}{2}}}{kr} \sum_{l=0}^{\infty} \sqrt{2l + 1} \, i^{l+1} \left\{ \exp \left[-i \left(kr - l\frac{\pi}{2} \right) \right] \right.$$

$$\left. - \exp \left[+i \left(kr - l\frac{\pi}{2} \right) \right] \right\} Y_{l,0} (\theta). \qquad (10\text{-}21)$$

The first exponential term represents particles coming toward the origin with momentum $k\hbar$ and angular momentum $l\hbar$; the second exponential term represents particles going away from the origin with just the same properties.   The fact that each term has the same absolute value for its coefficient (amplitude) and exponent (phase) means that nothing happened to the particle as it passed through the origin, and we indeed have an undisturbed wave.

The next problem is the form of the wave function when a scattering and absorbing center (a target nucleus) is placed at the origin of the coordinates.

We can answer this question in terms of (10-21) by asking the possible effect of scattering and absorption on the two exponentials in that equation.   First of all, there can be no effect on the first exponential, which represents particles coming toward the origin, since those particles do not yet even know that there is anything at the origin with which to interact; the whole effect must be on the second term, which represents those particles that are going away from the origin.   For that term there must be two effects: that which represents the elastic scattering of particles and that which represents the absorption of particles.   In elastic scattering, the relative momentum, $k\hbar$, must not change, and the particle does not disappear; so the coefficient (the amplitude) of the second exponential inside the bracket, which may be a complex number, must have an absolute value of unity.   Thus for scattering alone the coefficient of the second term is of the form $\exp(i\chi)$.   Absorption, on the other hand, means that there are fewer outgoing particles than incoming particles, and the coefficient of the outgoing wave inside the bracket must be diminished to a value less than one, which we shall call A_l.   The effect, then, of elastic scattering and absorption together is to multiply the second exponential by the term $A_l \exp(i\chi)$, which we call the amplitude η_l (a complex number).   This means that the wave function for the disturbed wave is

$$\Psi_d = \frac{\pi^{1/2}}{kr} \sum_{l=0}^{\infty} \sqrt{2l+1}\, i^{l+1} \left\{ \exp\left[-i\left(kr - l\frac{\pi}{2} \right) \right] \right.$$

$$\left. - \eta_l \exp\left[i\left(kr - l\frac{\pi}{2} \right) \right] \right\} Y_{l,0}. \quad (10\text{-}22)$$

The scattered wave, as mentioned before, is $\Psi_d - \Psi_u$.   The number of elastically scattered particles per unit incident particle, and thus the elastic-scattering cross section for a given l, can be computed from the flux of particles in the scattered wave.   The result has been shown to be

$$\sigma_{\mathrm{sc},l} = \pi\lambda^2(2l+1)\left|1 - \eta_l\right|^2. \quad (10\text{-}23)$$

The term $\left|1 - \eta_l\right|^2$ is the square of the modulus of the complex number $(1 - \eta_l)$ and is equal to $(1 - \eta_l)(1 - \eta_l)^*$, where $(1 - \eta_l)^*$ is the complex conjugate.

As mentioned earlier, the reaction cross section is obtained from the net flux of particles corresponding to the disturbed wave, Ψ_d:

$$\sigma_{r,l} = \pi\lambda^2(2l+1)(1 - \left|\eta_l\right|^2), \quad (10\text{-}24a)$$

$$= \pi\lambda^2(2l+1)(1 - A_l^2). \quad (10\text{-}24b)$$

Comparison between (10-13) and (10-24) shows the relationship between the transmission coefficient and the amplitude of the scattered wave.   The total cross section for a given l is the sum of (10-23) and (10-24)

$$\sigma_{t,l} = 2\pi\lambda^2(2l + 1)[1 - \mathrm{Re}(\eta_l)] \tag{10-25}$$

where $\mathrm{Re}(\eta_l)$ signifies the real part of the complex number η_l. Thus we see that complete information on the *total* reaction cross section and on the elastic scattering cross section is contained in the quantity η_l. The evaluation of η_l is usually accomplished by use of the optical model.

Two important aspects of these cross section expressions should be noted:

1. The maximum value of the reaction cross section for a given l is $\pi\lambda^2(2l + 1)$, in agreement with our semiclassical result (10-10), and occurs for $\eta_l = 0$, which means according to (10-23) that there must be a scattering cross section of equal magnitude. Indeed, nuclear reactions must always be accompanied by nuclear scattering; this is the source of the so-called shadow scattering.

2. The maximum scattering cross section is $4\pi\lambda^2(2l + 1)$. It occurs for $\eta_l = -1$, which means that the reaction cross section vanishes.

The expression (10-24) for the reaction cross section justifies the use of (10-10) in the semiclassical analysis of the reaction cross section; it remains to justify the cutoff at l_m, as given in (10-12). This justification may be presented in more than one way; but perhaps the most useful is the one that goes back to the Schrödinger equation for the two reacting particles. The particular functions chosen to represent the incident wave in (10-19) were not selected arbitrarily; they were chosen because they are the solutions to the Schrödinger equation that represents the relative motion of two noninteracting particles in spherical coordinates. In this equation, essentially because of the quantization of angular momentum, there appears a fictitious potential energy term which is

$$\frac{l(l + 1)\hbar^2}{2\mu r^2} \tag{10-26}$$

and is known as the "centrifugal potential." Its effect is to cause wave functions to be small at distances comparable to or less than l/k; classically, it implies that only particles with kinetic energy larger than the "centrifugal potential" at the point r can penetrate to that point:

$$\epsilon \geq \frac{l(l + 1)\hbar^2}{2\mu r^2}. \tag{10-27}$$

Since $\sqrt{2\mu\epsilon}$ is the relative momentum of the two particles, (10-27) becomes

$$\sqrt{l(l + 1)} \leq \frac{R}{\lambda} \tag{10-28}$$

as the condition on l for penetration to the interaction distance R.   Except for the usual confusion between $l(l + 1)$ and l^2 in hybrid classical-quantum analyses, (10-28) is the same as (10-12).   Equation 10-16 for the maximum angular momentum in the presence of a Coulomb barrier follows when (10-27) is changed to the condition that the particle must surmount both the Coulomb and the centrifugal barriers.

F. OPTICAL MODEL

An obvious nuclear model for the computation of the cross sections for nuclear processes is the one that was fundamental in chapter 9: the major part of the interactions among the nucleons in the system is replaced by an effective potential energy within which the undisturbed particles move; the residual interactions among the particles then appear as perturbations of this simple picture.   This model for nuclear reactions was, indeed, the first one considered.   It was dropped in the 1930's, when the earliest quantitative data on nuclear cross sections, information on slow-neutron resonances (section G), seemed to be directly opposed to it.   The model was revived in 1949 (F2) for the description of nuclear reactions at high energies (>100 MeV) and since that time has been used fruitfully in the interpretation of elastic-scattering and total-reaction cross sections at energies down to a few MeV.

The optical model, in its simplest form, represents the nucleus as a potential-energy well V_0 MeV deep and R fermis wide, which is illustrated in figure 10-4.   For $r > R$ the wave function of the particle is given by (10-22); for $r < R$ the wave function of the particle is given by the solution of the Schrödinger equation inside the potential well.   The quantities η_l are then evaluated through the condition that the two parts of the wave function join smoothly (equal value and slope) at the boundary.   This model, as discussed so far, must be incomplete; since the residual interactions play no role, there cannot be any nuclear reactions; there would be only elastic scattering.   This would appear formally as $|\eta_l| = 1$.   Further, the elastic-scattering cross section would have relatively sharp resonances at energies corresponding to single-particle states of the incident particle in the effective potential.   Both predictions are obviously at variance with fact, although it should immediately be stated that it *was* the observation of *broad* resonances in the scattering cross sections for neutrons of several MeV (B2) which led to the extension of the optical model to low energies (F3).   The residual interactions between the incident particle and the nucleons in the nucleus, which cause processes other than elastic scattering, are put into the optical model in the form of a complex potential instead of just the real one shown in figure 10-4:

$$V = -[V_o + iW_o] \qquad r < R$$
$$V = 0 \qquad\qquad\quad r > R \tag{10-29}$$

in analogy to the optical problem in which the absorption and refraction of a light beam is described by an index of refraction that has a real and an imaginary part. The introduction of the complex potential causes $|\eta_l| \leq 1$ which, as shown in section E, means that there will be a nonzero reaction cross section.[2] It has also been shown that the imaginary potential broadens the single-particle scattering resonances to a width of about W_o MeV. The optical model, then, gives information about the nucleus in terms of V_o and W_o from data on the elastic and reaction cross sections.

More detailed information about V_o and W_o may be obtained from the angular distribution in elastic scattering as exemplified in figure 10-3. Briefly, this information arises because the various peaks in the angular distribution depend on interferences between waves of various l values and are thus a more sensitive measure of η_l, and so of V_o and W_o, than are the reaction and elastic-scattering cross sections. The differential cross section for scattering into a unit solid angle at an angle θ in the center-of-mass system may be obtained from the flux at an angle θ corresponding to the difference in wave functions (10-21) and (10-22):

$$\frac{d\sigma_{sc}(\theta)}{d\Omega} = \pi\lambda^2 \left| \sum_{l=0}^{\infty} \sqrt{2l+1}\; Y_{l,0}(\theta)(1-\eta_l) \right|^2. \tag{10-30}$$

Analyses of angular distributions in elastic scattering in terms of (10-30) have shown that the square-well potential is much too simple: a potential well with rounded corners is required. Further, it is found that both V_o and W_o depend on the energy of the incident particle: V_o ranges from around 50 MeV at low energies (<10 MeV) to about 15 MeV at about 150 MeV; W_o varies from about 1 MeV at low energies to about 20 MeV in the 100-MeV range.

Since the interactions of the incident particle with the nucleons in the nucleus are the source of the optical potential, the latter should be calculable from the properties of the former. The relationship between the real part of the optical potential, V_o, and the interaction of the incident particle with the individual nucleons in the nucleus is obscured by the many-body aspect of the problem. Nevertheless, an approximate but still very complicated relationship between the two has been derived which is best at energies in excess of 100 MeV (G3).

The physical significance of the imaginary part of the potential, W_o, is more easily seen. The eigenfunctions of the Schrödinger equation containing a potential energy of the form (10-29) (note that V_o and W_o, as defined there, are usually positive quantities) are plane waves of the form (10-19), but the wave number, k, is complex. The imaginary part of k corresponds to absorption of

---

[2] The amplitude of the outgoing wave is often discussed in terms of the *phase shift* (δ_l): $\eta_l = e^{2i\delta_l}$. If there is only a real potential, $W_o = 0$, δ_l is real and, as can be seen from (10-22), the only effect is to change the phase of the outgoing wave but to leave $|\eta_l| = 1$. Introduction of the complex potential turns δ_l into a complex number and results in $|\eta_l| \leq 1$.

the particle and is equivalent to a mean free path Λ, within nuclear matter that turns out to be given by[3]:

$$\frac{1}{\Lambda} = \frac{2\mu^{1/2}}{\hbar} \{[(\epsilon + V_o)^2 + W_o^2]^{1/2} - (\epsilon + V_o)\}^{1/2}, \qquad (10\text{-}31a)$$

where ϵ is the relative kinetic energy of the system. For small values of $W_o/(\epsilon + V_o)$, (10-31a) reduces to

$$\Lambda \cong \frac{\hbar}{W_o} \sqrt{\frac{\epsilon + V_o}{2\mu}}. \qquad (10\text{-}31b)$$

The mean free path may also be immediately expressed in terms of the density of nucleons within the nucleus, ρ, and the effective average cross section, $\bar{\sigma}$, for the interaction of the incident particle with the nucleons within the nucleus:

$$\Lambda = \frac{1}{\rho\bar{\sigma}}, \qquad (10\text{-}32)$$

which establishes a relationship between optical-model parameters and the effective nucleon-nucleon interaction cross section within the nucleus.

The mean free path given in (10-31a) and (10-31b) is correct only for a uniform potential of the form (10-29); it may be used, nevertheless, at any particular point in a nonuniform potential as long as the fractional change of potential energy in a distance corresponding to the wavelength of the particle is negligible.

As mentioned at the outset of this section, the optical model is used in the calculation of η_l and therefore of the scattering cross section from (10-24) and of the angular distribution in elastic scattering from (10-30). It neither predicts the relative probabilities of the various possible reactions that may occur in the medium-energy region nor the resonances that are seen in slow-

---

[3] The complex wave number can be written $k = k' + ik''$; the wave function, then, is

$$\Psi = \exp(ikx) = \exp(ik'x) \exp(-k''x).$$

The probability of finding the particle between x and $x + dx$, which is $|\Psi|^2 \, dx$, becomes

$$|\Psi|^2 \, dx = \exp(-2k''x) \, dx.$$

Thus the probability of finding the particles diminishes exponentially with x and is characterized by a mean free path

$$\Lambda = \frac{1}{2k''}.$$

The quantity k'' can be obtained from the imaginary part of the complex quantity k, where

$$k = \frac{1}{\hbar} [2\mu(\epsilon + V_o + iW_o]^{1/2}$$

neutron reactions. These two topics are related through the compound-nucleus model which is discussed in the next section.

G. COMPOUND-NUCLEUS MODEL

The first model for nuclear reactions which enjoyed much success in the detailed interpretation of experimental data was the compound-nucleus model introduced by Bohr (B3) in 1936. As stated in section B, this model asserts that when an incident particle penetrates to the surface of a target nucleus, the incident particle is absorbed and a new nucleus is formed which is in an excited quasi-stationary state. This new nucleus is the compound nucleus, and it is said to be in a quasi-stationary state because the excitation energy makes it unstable with respect to the emission of particles, although its lifetime is long compared to the transit time of a nucleon across a nucleus.

Difficulties in the analysis of the compound-nucleus model arise primarily from a lack of detailed knowledge about this quasi-stationary state, which, because of its finite width, includes many excited states of the compound nucleus. This problem, however, is not serious for thermal neutrons because only a single excited state is involved.

(n, γ) Reaction with Thermal Neutrons. Three important characteristics of the excitation functions for (n, γ) reactions with low-energy neutrons (0.01 to several hundred eV) can be seen in figure 10-5:

1. The cross sections show enormous fluctuations over a very small energy range, that is, resonances are apparent.
2. The widths of the resonances are small ($\cong 0.1$ eV).
3. The spacing between the resonances is large compared to their widths; the spacings vary from the order of keV in the lightest elements to the order of eV for the heaviest).

The small widths of the resonances lead to the conclusion, by use of the Heisenberg uncertainty principle, that the compound nucleus has a lifetime of about 10^{-14}–10^{-15} second, which is long compared to the transit time of a thermal neutron across a medium weight nucleus, $\cong 10^{-18}$ second. This conclusion suggested the idea of the quasi-stationary state for the compound nucleus. Further, the observation that the average spacing between the resonances is 100 to 1000 times smaller than the average spacing between single-particle levels showed that the quasi-stationary excited state of the compound nucleus must involve the excitation of many particles. These conclusions were among those that eliminated serious consideration of the optical model for slow-neutron reactions. Classically, then, the compound-nucleus model envisages the incident particle as amalgamating with the target nucleus, with the excitation energy[4] randomly distributed among all the nucleons in the resultant compound nucleus. Some time later the compound nucleus is de-excited by the emission of one or more particles including photons.

Independence Hypothesis in the Resonance Region.   The formation of a compound nucleus by low-energy neutrons does not always lead to the emission of gamma rays; it is also possible for the neutron to be re-emitted or for an (n, p) or (n, α) reaction to occur.   For the heaviest elements fission is often the most probable process.   Since the compound-nucleus model divides the reaction into two parts—formation and decay of the compound nucleus—the relative probabilities of the various possible events should be completely determined by the quantum state of the compound nucleus.   In particular, if the resonances do not overlap, the behavior of the compound nucleus is essentially governed by the properties of a single quantum state (the resonant state) and should thus be independent of the manner in which the state was formed. This means, for example, that the relative amount of γ-ray emission and neutron emission will be the same when nucleus $_Z X^A$ is irradiated with neutrons and $_{Z-1} X^A$ is irradiated with protons as long as the energies of the particles are such that they form the same nonoverlapping resonant state.   This conclusion is known as the "independence hypothesis."   We return to it again in the more ambiguous situation of overlapping states.

Breit-Wigner Formula.   The rapidly varying cross section illustrated in figure 10-5 shows that the amplitudes of the outgoing waves in (10-22) are very sensitive functions of the energy in this low-energy region.   In the first solution of this problem by Breit and Wigner,[5] the quantities η_l were not directly calculated; rather, perturbation theory was used to solve the problem in the two steps suggested by Bohr involving the formation and decay of the compound nucleus.   It is useful to give the results of their important calculations for a general reaction

$$a + A \rightarrow C \rightarrow B + b, \tag{10-33}$$

going through a compound nucleus C which is in a *single* well-defined quasi-stationary quantum state.

The cross section for the particular reaction (10-33) would be written

$$\sigma_{A \rightarrow C \rightarrow B} = \sigma_{A \rightarrow C} W_B, \tag{10-34}$$

[4] From section D the excitation energy U is given by

$$U = \frac{M_A}{M_A + m_a} \epsilon_a + S_a,$$

where M_A and m_a are the atomic masses of the target and bombarding particles, respectively; ϵ_a is the laboratory kinetic energy of the bombarding particle; and S_a is the binding energy of particle a in the compound nucleus.
[5] See reference R1 for a complete discussion of theory and experiment with slow neutrons.

where $\sigma_{A \to C}$ is the cross section for forming the compound nucleus C and W_B is the probability that the compound nucleus decays in the particular manner prescribed by reaction (10-33) or goes into channel Bb.[6] Equation 10-34 explicitly presents the two-stage and the independence hypotheses. The Breit-Wigner treatment gives the important expression

$$\sigma_{A \to C} = \pi \lambda_{Aa}^{2} \frac{2I_C + 1}{(2I_A + 1)(2I_a + 1)} \frac{\Gamma_{Aa}\Gamma}{(\epsilon - \epsilon_0)^2 + (\Gamma/2)^2}, \qquad (10\text{-}35)$$

where λ_{Aa} is the relative wavelength in the entrance channel, ϵ_0 is the center-of-mass energy at which resonance occurs, Γ is the total width of the level, and Γ_{Aa} is the partial width of the level for decay into channel Aa. The meaning of "width-of-level" lies in the statement that $\Gamma_J/\hbar$ is the probability per unit time that the compound nucleus decays into channel J. This means that

$$\Gamma = \sum_J \Gamma_J \qquad \text{(summed over all channels)} \qquad (10\text{-}36)$$

and that

$$W_B = \frac{\Gamma_{Bb}}{\Gamma}. \qquad (10\text{-}37)$$

The substitution of (10-35) and (10-37) into (10-34) gives the famous Breit-Wigner one-level formula

$$\sigma_{A \to C \to B} = \pi \lambda_{Aa}^{2} \frac{2I_C + 1}{(2I_A + 1)(2I_a + 1)} \frac{\Gamma_{Aa}\Gamma_{Bb}}{(\epsilon - \epsilon_0)^2 + (\Gamma/2)^2}. \qquad (10\text{-}38)$$

For the (n, γ) reaction in particular,

$$\sigma_{(n,\gamma)} = \pi \lambda^{2} \frac{2I_C + 1}{2(2I_A + 1)} \frac{\Gamma_n \Gamma_\gamma}{(\epsilon - \epsilon_0)^2 + (\Gamma/2)^2}, \qquad (10\text{-}39)$$

where Γ_n and Γ_γ are the partial widths for neutron and gamma emission, respectively. Equation 10-39 is meant to describe the cross section at any particular resonance such as those shown in figure 10-5. For example, the first resonance seen in this figure is at $\epsilon_0 = 5.120$ eV and is characterized by $\Gamma_\gamma = 136 \times 10^{-3}$ eV and $\Gamma_n = 5.9 \times \epsilon^{\frac{1}{2}} \times 10^{-3}$ eV.

The resonant state need not correspond to a positive energy for the incident neutron; the state may be at an excitation energy that is below the binding energy of a neutron in the compound nucleus. Although under these circumstances the resonance will not be directly observable with neutrons, the resonance can cause large capture cross sections for neutrons of thermal energy

---

[6] The particular manner of the formation and the decay of the compound nucleus are often referred to as "channels"; reaction (10-33) would be said to go from channel Aa to channel Bb. The definition of a channel in general requires specification of the relative energy of the particles, the total angular momentum, and the internal quantum numbers, (excited states) of the particles.

(of the order of 0.025 eV) if the width of the resonance is not too small when compared with the energy difference between the position of its peak and the excitation energy of the compound nucleus produced in thermal-neutron capture.

It is clear, then, that the observation of neutron-capture resonances yields information about the energies of nuclear excited states and about their widths. Except for the lightest nuclei, this type of experiment is not possible with incident charged particles because the Coulomb barrier causes Γ_A to become vanishingly small at low energies. With slow neutrons, the centrifugal barrier causes Γ_n to be most important for $l = 0$; the spin of the compound-nucleus state, therefore, must be $I_A \pm \frac{1}{2}$.

It is of interest to examine the cross section of silver for neutrons (cf. figure 10-5) with energies below about 0.4 eV. In this region the dominant term in the denominator of (10-39) is clearly ϵ_0, and thus the denominator is essentially a constant.

The energy dependence of the cross section will depend on three factors:

1. $\lambda^2 \propto 1/v^2$, where v is the relative velocity of neutron and target nucleus.

2. $\Gamma_n \propto v$ because Γ_n is proportional to the density of final states for the system which (see footnote 13 in chapter 8) is proportional to the relative velocity of neutron and target nucleus.

3. Γ_γ is independent of changes in neutron energy of a few eV because the energy of the gamma ray is several MeV.

The result of these three factors is to make $\sigma_{n,\gamma} \propto 1/v$ in the region where $\epsilon \ll \epsilon_0$ ($\epsilon \ll |\epsilon_0|$ for negative-energy resonances). The $1/v$ dependence for the neutron-capture cross section of silver is shown as the dotted line in figure 10-5.

It is seen from the preceding discussion that the thermal-neutron capture cross section of any particular nuclide will depend critically on the energies and widths of its resonant states. In particular, if there is a resonant state at an energy within about 0.01 eV (either positive or negative) of the binding energy of the neutron, the capture cross section can be enormous. On the other hand, if there are no close resonances, the capture cross section may be quite small and follow the $1/v$ law. A compilation of thermal-neutron cross sections is given in appendix C. Because the values have been determined by a variety of experimental methods it is not easy to find a common basis for listing them. Many have been measured for the neutron spectrum present in a particular nuclear reactor. Some have been measured in a thermal-neutron flux characterized to good approximation by the velocity distribution at approximately 20°C. Others have been measured at particular neutron velocities by the use of neutron monochromators. Following the usual practice, we have tabulated all thermal-neutron reaction cross sections for the discrete neutron velocity of 2.20×10^5 cm sec^{-1} (which corresponds to the discrete energy 0.025 eV and is the most probable velocity in a Maxwellian distribution at 20°C).

The Statistical Assumption. As the energy of the bombarding particle is increased, two effects act in concert to make (10-38) increasingly difficult to use.

1. The width of each level, Γ, becomes larger and larger because more outgoing channels become available.
2. As is usually true in many-particle systems, the energy-spacing, D, between levels becomes smaller and smaller.

The net effect is that the resonances begin to overlap, and it is no longer possible, in general, even with ideal energy resolution of the incident beam, to excite but a single state of the compound nucleus. Under these circumstances the various states of the compound nucleus which enter into the reaction do not behave independently; interferences among them must be taken into consideration, and the cross section would *not* simply be given by a sum of terms, each of which has the form of (10-38). These interferences could have two important effects:

1. The angular distribution of the emitted particles would not be symmetric about a plane normal to the direction of the incident beam, as it must be if the compound nucleus is in a single nonoverlapping quantum state. The lack of symmetry can arise from interferences between particles emitted with, for example, $l = 0$ and $l = 1$ (s and p waves), for s waves are an even function of θ and p waves are an odd function of θ.
2. The relative values of the various interferences, which would affect the relative probabilities for the emission of various kinds of particles, would depend on how the compound nucleus was made, and the independence hypothesis would no longer be true.

Further, aside from the interferences, if each of the overlapping states had a different width for a particular mode of decay of the compound nucleus, then again the independence hypothesis would be invalid.

These problems, caused by the interferences and by the fluctuating partial widths, are removed if two assumptions are made which together are called the statistical assumption. It is first assumed that the interference terms, which may be either positive or negative, have random signs and thus cancel out; this reinstates a symmetrical angular distribution. It is further assumed that the overlapping states all have essentially the same relative partial widths for the various possible decay channels of the compound nucleus; this reinstates the independence hypothesis. The statistical assumption, then, allows the extension of the Bohr model to the region of overlapping energy levels and may be tested by the measurement of the angular distribution of evaporated particles and experimental tests of the independence hypothesis.

It has been observed, as exemplified in figure 10-2, that the angular distribution of most of the particles emitted by compound nuclei excited up to a few tens of MeV has the required symmetry, and thus the statistical assumption has some validity. However, some of the particles, usually of relatively high energy, tend to be preferentially emitted in the forward direction and thus represent a partial failure of the statistical assumption.

Unfortunately there have been very few tests of the independence hypothesis in the continuum region. The first, and most widely quoted, is that of Ghoshal (G1), and the test has come to be known as the "Ghoshal experiment." Ghoshal investigated the behavior of an excited Zn^{64} nucleus made in two different ways:

$$
\begin{array}{c}
_2He^4 + {}_{28}Ni^{60} \\
\\
_1H^1 + {}_{29}Cu^{63}
\end{array}
\searrow
Zn^{64}
\begin{array}{c}
\nearrow Zn^{63} + n \\
\longrightarrow Zn^{62} + 2n \\
\searrow Cu^{62} + H^1 + n
\end{array}
\tag{10-40}
$$

In light of (10-34), the independence hypothesis demands, for example, that

$$
\frac{\sigma(\alpha, pn)}{\sigma(\alpha, 2n)} = \frac{W(pn)}{W(2n)} = \frac{\sigma(p, pn)}{\sigma(p, 2n)},
\tag{10-41}
$$

where all of the cross sections are measured under the same condition for the compound nucleus. Results derived from Ghoshal's data are presented in figure 10-15, in which it is seen that the independence hypothesis seems to be confirmed. The situation can be less clear-cut for ratio curves which vary more rapidly with excitation energy than do those in figure 10-15; it is often found, for example, that a ratio curve for α-particle-induced reactions has a shape similar to that of the corresponding proton-induced reactions but is displaced on the energy scale, usually to higher energies. This shift in the energy is discussed in the paragraphs to follow on the subject of evaporation theory; it should be mentioned here that the comparison in figure 10-15 is made for

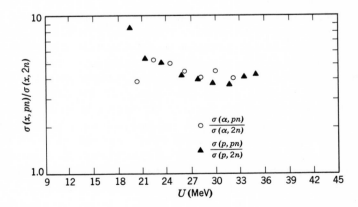

Fig. 10-15   Comparison of the behavior of an excited Zn^{64} compound nucleus made in two different ways: proton bombardment of Cu^{63} and alpha-particle bombardment of Ni^{60}. (Data are from reference G1.)

compound nuclei at the same excitation energy but not with the same distribution of angular momentum [cf. (10-16)], and so the compound nuclei are not, strictly speaking, identical. The distributions of angular momentum are different because the wavelengths of the protons and α particles are not the same when their kinetic energies are such that the excitation energies of the Zn^{64} are the same in the two bombardments.

The statistical assumption has proved to be successful for the description of most of the reactions induced by protons or neutrons with energies up to perhaps 40 MeV and complex particles with energies up to about 10 MeV per nucleon. This is not to say that it is not useful at even higher energies; unfortunately there are few data in the energy region between 40 and 100 MeV. At still higher energies the statistical assumption is known to fail.

The most powerful aspect of the statistical model lies in its ability to predict the energy spectrum of evaporated particles as well as excitation functions for various products, in terms of certain average nuclear properties, in a manner that is described in the following paragraphs.

Evaporation Theory. The excitation functions exhibited in figures 10-1 and 10-6 and the energy spectrum of emitted particles shown in figure 10-2 are typical of the data that can be interpreted within the compound-nucleus model extended up into the medium-energy region through the statistical assumption. Figures 10-1 and 10-6 show the competition among some of the various channels that are available for the decay of the compound nucleus. Explicit in the spectrum shown in figure 10-2 and implicit in the fairly sharp maxima shown by the excitation functions is the fact that most of the particles are emitted with considerably less than the maximum energy available. As discussed in section B, this would be qualitatively expected from the compound-nucleus model.

These qualitative remarks can be given quantitative expression because the statistical assumption implies that statistical equilibrium exists during a compound-nucleus reaction. The key to the solution is that the principle of detailed balance may be used in the following manner if, for simplicity, the effects of angular momentum and parity are ignored.

Consider a box of volume V containing compound nuclei C at an excitation energy U_C as well as all combinations of particles that correspond to the available decay channels of C. This means, for example, that there are complex nuclei B at excitation energy U_B and particles b with kinetic energy ϵ_{Bb} relative to B such that

$$U_C - S_b = U_B + \epsilon_{Bb}, \tag{10-42}$$

where S_b is the separation energy of particle b from the compound nucleus C. Equation 10-42 is evidently a statement of the conservation of energy for the decay:

$$C(U_C) \xrightarrow{\epsilon_{Bb}} B(U_B) + b, \tag{10-43}$$

in which it is assumed that particle b has no internal excitation energy. Statistical equilibrium in this instance means that the relative numbers of compound nuclei and of sets of particles that correspond to the various decay channels are determined by their relative statistical weights or state densities in the sense used in section C of chapter 8. The principle of detailed balance demands that the equilibrium be maintained by reactions such as (10-43) proceeding forward and backward at precisely the same rate. If we define $N_C(U_C)$ and $N_{Bb}(U_B, \epsilon_{Bb})\, d\epsilon_{Bb}$ as the number of compound nuclei and B-b pairs with appropriate energies, respectively, then statistical equilibrium gives

$$\frac{N_C(U_C)}{N_{Bb}(U_B, \epsilon_{Bb})\, d\epsilon_{Bb}} = \frac{\rho_C(U_C)}{\rho_B(U_B)(4\pi V p_{Bb}{}^2/h^3)(dp_{Bb}/d\epsilon_{Bb})\, d\epsilon_{Bb}}, \quad (10\text{-}44)$$

where $\rho_j(U_j)$ is the density of energy states at excitation energy U_j, that is, the reciprocal of the average spacing between states at U_j; p_{Bb} is the relative momentum of particles B and b, and the quantity in the denominator is the density of states when the relative kinetic energy of the pair B and b is between ϵ_{Bb} and $\epsilon_{Bb} + d\epsilon_{Bb}$ (cf. footnote 13 of chapter 8).

The law of detailed balance demands that

$$N_C(U_C)\, I(\epsilon_{Bb})\, d\epsilon_{Bb} = N_{Bb}(U_B, \epsilon_{Bb})\, W(U_B, \epsilon_{Bb})\, d\epsilon_{Bb}, \quad (10\text{-}45)$$

where $I(\epsilon_{Bb})\, d\epsilon_{Bb}$ is the probability per unit time that the compound nucleus emits particle b with kinetic energy (relative to the residual nucleus B) between ϵ_{Bb} and $\epsilon_{Bb} + d\epsilon_{Bb}$, and $W(U_B, \epsilon_{Bb})$ is the probability per unit time that particle b with relative kinetic energy in the stated interval reacts with the residual nucleus B at an excitation energy U_B to produce the compound nucleus C with excitation energy U_C. The quantity $W(U_B, \epsilon_{Bb})$ is given by

$$W(U_B, \epsilon_{Bb}) = \frac{\sigma_{Bb} v_{Bb}}{V}, \quad (10\text{-}46)$$

where σ_{Bb} is the cross section for the inverse of the process (10-43) and thus is the cross section for the particle b moving with relative kinetic energy ϵ_{Bb} to be captured by the nucleus B with excitation energy U_B to form the compound nucleus C with excitation U_C; v_{Bb} is the relative velocity of B and b. Equation 10-46 may be understood by realizing that it is the fraction of the volume of the box swept out in a unit time that corresponds to the desired reaction between B and b. Although σ_{Bb} is the cross section for the capture of particle b by the residual nucleus B in an *excited* state, lack of information about reaction cross sections of excited nuclei leads to the assumption that σ_{Bb} is the same as if B were in its ground state.

The substitution of (10-44) and (10-46) into (10-45) gives the desired expression:

$$I(\epsilon_{Bb})\, d\epsilon_{Bb} = \frac{4\pi p_{Bb}{}^2(dp_{Bb}/d\epsilon_{Bb})\sigma_{Bb} v_{Bb}\rho_B(U_B)}{h^3 \rho_C(U_C)}\, d\epsilon_{Bb}. \quad (10\text{-}47)$$

Recalling that $\epsilon_{Bb} = \frac{1}{2}\mu_{Bb}v_{Bb}^2$ and $p_{Bb} = \mu_{Bb}v_{Bb}$, where μ_{Bb} is the reduced mass, and using the conservation of energy as expressed in (10-42), we rewrite (10-47) as follows:

$$I(\epsilon_{Bb})\, d\epsilon_{Bb} = \frac{8\pi\mu_{Bb}\sigma_{Bb}\epsilon_{Bb}\rho_B(U_C - S_b - \epsilon_{Bb})}{h^3\rho_C(U_C)}\, d\epsilon_{Bb}, \qquad (10\text{-}48)$$

which is the expression that describes the spectrum shown, for example, in figure 10-2. The maximum at relatively low energies in the energy spectra of emitted particles occurs because, although ϵ_{Bb} obviously increases with kinetic energy, the state density $\rho_B(U_C - S_b - \epsilon_{Bb})$ decreases with increasing kinetic energy in what we shall see is approximately an exponential manner. The inverse cross section σ_{Bb}, as mentioned before, is usually approximated by an expression such as (10-18) and prevents the emission of charged particles with kinetic energies less than the height of the Coulomb barrier. The total probability per unit time for the emission of particle b is given by the integral over the whole spectrum:

$$\frac{\Gamma_b}{\hbar} = \int_0^{U_C - S_b} I(\epsilon_{Bb})\, d\epsilon_{Bb}. \qquad (10\text{-}49)$$

The cross section for the reaction (10-33) (induced by a charged particle a with Coulomb barrier B_{aA}) in the energy region in which the statistical assumption is valid may be written as suggested by (10-34):

$$\sigma(a,b) = \pi R_{aA}^2 \left(1 - \frac{B_{aA}}{\epsilon_{aA}}\right) \frac{\int_x^{U_C - S_b} I(\epsilon_{Bb})\, d\epsilon_{Bb}}{\sum_j \int_0^{U_C - S_j} I(\epsilon_j)\, d\epsilon_j}, \qquad (10\text{-}50)$$

in which the summation in the denominator is to be extended over all particles j which may be emitted by the compound nucleus C. The lower limit of the integral in the numerator of (10-50) is subject to the following conditions:

$$x = 0 \quad \text{if} \quad U_C - S_b - S' \leq 0, \qquad (10\text{-}51)$$

$$x = U_C - S_b - S' \quad \text{if} \quad U_C - S_b - S' > 0, \qquad (10\text{-}52)$$

where S' is the excitation energy above which the product nucleus B is unstable with respect to the emission of another particle. Under conditions (10-51, 52), one and only one particle is emitted from B. It is to be noted that the effects of angular momentum and parity are ignored in (10-50); we shall return to this point later.

The estimation of cross sections for reactions in which more than one particle is emitted becomes complicated but in essence involves the same formulation that leads to (10-50). It requires multiple integrals and is in general most expeditiously carried out on an electronic computer (D1).

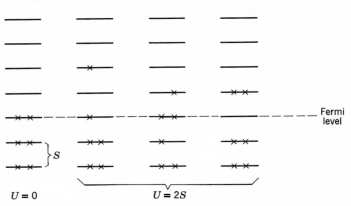

Fig. 10-16   Counting of states in simple independent-particle model with equal spacing for single particle states: $\rho(0) = 1$, $\rho(2S) = 3$, etc.

So far nothing explicit has been said concerning the state densities $\rho(U)$, which are evidently of great importance to calculations in evaporation theory. This quantity is discussed in the following paragraphs.

State Densities in Evaporation Theory.   The number of different states that are at the same excitation energy may be straightforwardly computed for the extreme single-particle model in which it is assumed that the excitation energy is the sum of the excitations of the individual particles.   The problem becomes one of counting the number of *different* ways in which the excitation energy may be distributed among the particles.   A simple illustration is given in figure 10-16 in which the single-particle states are taken to be equally spaced. A more realistic model, mentioned in section C of chapter 9, considers the nucleus as a mixture of a proton Fermi-gas and a neutron Fermi-gas.   For this model the dependence of the density of states on an excitation energy which is large compared to the spacing of single-particle levels in the region of the Fermi energy may be obtained by standard, although rather difficult, statistical mechanical methods.   The result of this analysis gives (E1):

$$\rho(U) = \frac{C \exp (2a^{\frac{1}{2}}U^{\frac{1}{2}})}{U^{\frac{5}{4}}}, \tag{10-53}$$

where a and C are constants that depend on the mass number of the nucleus; in particular, a is proportional to A.   Since this model views the nucleus as a mixture of proton and neutron Fermi-gases, it is possible to characterize the nucleus by the usual thermodynamic quantities such as the energy U and the temperature τ.   For this Fermi-gas the relationship between these two thermodynamic quantities is, to a good approximation, given by

$$U = a\tau^2. \tag{10-54}$$

It is to be noted that the expression for the density of states given in (10-53) includes all spins I and all orientations of the spins in space.[7]

The expression (10-53) is not meant to be correct at low excitation energies where the term in the denominator tends toward zero. Largely out of convenience rather than conviction, the relatively slowly varying term in the denominator of (10-53) is often ignored and the analysis of data has been carried out in terms of an approximate form of (10-53) (B1):

$$\rho(U) = C' \exp (2a^{\frac{1}{2}}U^{\frac{1}{2}}). \tag{10-55}$$

The use of expression (10-55) in (10-48) leads to an energy spectrum of the form

$$I(\epsilon) \propto \epsilon\sigma \exp [2a^{\frac{1}{2}}(\epsilon_m - \epsilon)^{\frac{1}{2}}], \tag{10-56}$$

where ϵ_m is the maximum kinetic energy with which the particle may be emitted. A plot, then, of $\ln [I(\epsilon)/\epsilon\sigma]$ versus $(\epsilon_m - \epsilon)^{\frac{1}{2}}$ should give a straight line and allow the evaluation of the level density parameter a, which is expected to be proportional to the mass number of the nucleus. Such plots are usually straight over only a limited portion of the spectrum, and the values of a that may be derived from them are thus ambiguous; further, they often do not exhibit the expected proportionality to mass number. These divergences are ascribed to the following:

1. The inadequacy of the approximate equation (10-55).
2. The ignoring of the effect of the angular momentum of the compound nucleus on the spectra of emitted particles.
3. The possibility that the statistical model is inadequate for the description of some of the reactions that occur, particularly those leading to the emission of high-energy particles.

Equation 10-56 often undergoes even further simplification in the treatment of spectral data for which ϵ/ϵ_m is small compared to 1:

$$I(\epsilon) \propto \epsilon\sigma \exp \left[2a^{\frac{1}{2}}\epsilon_m^{\frac{1}{2}}\left(1 - \frac{\epsilon}{\epsilon_m}\right)^{\frac{1}{2}}\right]$$

$$\left(1 - \frac{\epsilon}{\epsilon_m}\right)^{\frac{1}{2}} \cong \left(1 - \frac{\epsilon}{2\epsilon_m}\right) \quad \text{for} \quad \frac{\epsilon}{\epsilon_m} \ll 1$$

$$I(\epsilon) \propto \epsilon\sigma \exp (2a^{\frac{1}{2}}\epsilon_m^{\frac{1}{2}}) \exp \left(-a^{\frac{1}{2}}\frac{\epsilon}{\epsilon_m^{\frac{1}{2}}}\right),$$

[7] Since the angular momentum carried away by an evaporated particle is usually small, it is probably better to use the level density for a *given* spin in (10-48). This expression, as reviewed in refs. E1 and L1, is

$$\rho(U, I) = \frac{\text{constant } (2I + 1) \exp \left[-\dfrac{\hbar^2(I)(I + 1)}{2\mathscr{I}\tau}\right] \exp (2a^{\frac{1}{2}}U^{\frac{1}{2}})}{U^2}.$$

The quantity $\mathscr{I}$ is the effective moment of inertia of the nucleus and, unfortunately, may depend on the excitation energy.

recalling from (10-54) that $\epsilon_m^{1/2}/a^{1/2}$ is just the temperature τ that the residual nucleus would have at an *excitation energy* ϵ_m,

$$I(\epsilon) \propto \epsilon \sigma e^{-\epsilon/\tau}. \qquad (10\text{-}57)$$

Equation 10-57 is the source of the designation of the spectra of emitted particles as Maxwellian and their characterization by the temperature τ.

Angular-Momentum Effects. As mentioned in item (2) of the preceding paragraph and in connection with the Ghoshal experiment (p. 342), the angular momentum of the system should not be ignored. The proper treatment can become very complicated, as is nearly always true with angular-momentum considerations, although the essence of the problem is quite simple. The difficulty occurs because the compound nucleus can be formed in states of rather high angular momentum [cf. (10-16)], whereas the emitted particles, because they tend to be of low energy, do not carry away much of the angular momentum. Conservation of angular momentum demands, then, that the residual nucleus B contain the appropriate residual angular momentum and thus $\rho_B(U_B)$ in (10-47) should really be $\rho_B(U_B, I_B)$, where I_B is the required spin. The effect of this spin-dependent level density can be surmised without a detailed analysis: the important point is that for each spin I there is a corresponding excitation energy, $U(I)$, below which there are on the average no states of spin I or greater. To the approximation that the emitted particle is in an $l = 0$ state and spinless, the maximum kinetic energy of the particle emitted from a compound nucleus of spin I must change, then, from $U_C - S_b$ to $U_C - S_b - U(I)$, which means that there are proportionately more low-energy particles and that this effect, in turn, will be reflected as an increase in the apparent value of the parameter a, defined in (10-54). It is, of course, possible for $U_C - S_b - U(I)$ to be less than zero while $U_C - S_b$ is greater than zero, in which case the particle b will not be emitted even though the excitation energy of the compound nucleus is greater than the binding energy of the particle b. The compound nucleus will instead emit some other particle (including photons) or emit particle b into a state $l > 0$ with therefore a much reduced probability. The latter effect, contrary to that on the spectrum, is reflected in a smaller value for the effective a when it is determined from excitation functions (G2). The quantity $U(I)$ effectively raises the threshold for any given reaction and can thereby contribute to the energy shifts in the Ghoshal experiment.

Complete analysis of the problem requires the proper averaging over the spectra of angular momenta of the compound nuclei, consideration of the orbital and intrinsic-spin angular momentum carried away by the emitted particles, and an explicit expression (E1, L1) for the quantity $\rho(U, I)$.

Odd-Even Effects on State Densities. The dependence of the level density on the nuclear species appears in the parameter a, which was said to be proportional to the mass number of the nucleus. It is well known, however, that the pairing effect has an important influence on the binding energy of nuclei, as appears, for example, in (2-3), and on the spectrum of low-lying energy states, as discussed in section F of chapter 9; thus it might be expected that these effects would also affect level densities.   On the other hand, pairing effects are expected mainly for the lower energy states of nuclei because in the higher states, when many nucleons are excited, it is not likely that two particles will find themselves in paired states (in the sense of section F of chapter 9) because there are so many other states of essentially the same energy available to them.   To a first approximation, then, the pairing effect may be included in the level-density expression merely as an effect on the position of the ground state.   This is often accomplished by measuring the excitation energy, not above the ground state of the nucleus but above a fictitious ground state which it would have in the absence of enhanced stability from pairing.   This approach leads to an expression of the form

$$\rho(U) = c' \exp \left[2a^{\frac{1}{2}}(U - \delta_n - \delta_p)^{\frac{1}{2}}\right]. \tag{10-58}$$

in place of (10-55).   The quantities δ_n and δ_p are zero for odd neutron and odd proton number, respectively; they are positive for even neutron and proton number, respectively, with a numerical value that depends on that even number.   Thus the level density of an odd-odd nucleus at a given excitation energy is, in general, greater than that of an adjacent even-odd or odd-even nucleus, which, in turn, is greater than that of an adjacent even-even nucleus.   The quantities δ_n and δ_p are discussed in detail in ref. D1.

The effect of pairing on the level density of a nucleus can be quite important, as shown in figures 10-1 and 10-6.   The same surprising result is seen in both excitation functions: the probability of the evaporation of a proton and a neutron from an excited compound nucleus is considerably greater than that of the evaporation of two neutrons, despite the fact that the Coulomb barrier to proton emission, as reflected in the inverse-cross-section term in (10-48), serves to diminish proton emission.   This enhancement of proton emission occurs because the compound nucleus in both examples is an even-even nucleus which, on the evaporation of two neutrons, goes to an even-even product of low level density in relation to the odd-odd isobaric product formed by the emission of a neutron and a proton.   If the inverse cross section in (10-48) is taken to be of the form

$$\sigma = \pi R^2 \left(1 - \frac{B}{\epsilon}\right),$$

where B is the effective Coulomb barrier and the level density is given by (10-58), then (10-48) may be integrated in closed form to give a probability per unit time for the emission of particle j, which is an ever-increasing function of the quantity $(U_c - S_j - B_j - \delta_n - \delta_p)$ Thus, if the sum of the binding energy of a neutron and the δ_n and δ_p of the nucleus resulting from neutron emission is greater than the corresponding sum for proton emission plus the Coulomb barrier, then proton emission can be more probable than neutron emission. This is not an unusual situation for compound nuclei with atomic numbers up to about 30 or 40; above that the Coulomb barrier becomes so high that it is usually decisive and neutron emission predominates. Considerable success in the interpretation of excitation functions has been achieved with (10-58) (D1).

The physical interpretation of the quantities δ_n and δ_p is still somewhat obscure. This subject is discussed in refs. E1 and L1 and in the references to be found in them.

H. DIRECT INTERACTION

In our survey of nuclear reactions in section C we encountered two types of nuclear reactions which were attributed to direct interactions. The first type is best illustrated by high-energy reactions in which the incident particle is considered to interact with just a few of the nucleons in the nucleus; this we call a knock-on reaction. The second type is best illustrated by deuteron stripping in which only part of the incident particle strikes the target nucleus. More details concerning these two processes are given in the following paragraphs.

Stripping Reactions. The stripping reaction at medium energies (10–20 MeV) has been very useful for the determination of the energies as well as the spins and parities of excited states of nuclei (H3). To see how this is accomplished, consider, for example, the reaction $A(d, p)B^*$ which leaves the nucleus

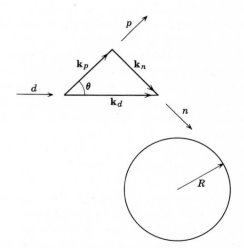

Fig. 10-17 Momentum diagram for (d,p) stripping reaction with proton emitted at an angle θ and neutron captured with impact parameter R.

B in a specific excited state whose energy, spin, and parity we wish to determine. A measurement of the energy spectrum of the emitted protons will, by conservation of energy, give the energies of the excited states of B, provided that the binding energy of the neutron in nucleus B is known. The spin and parity of a state may be estimated in the following manner from the angular distribution of the emitted protons corresponding to formation of that state. Consider a vector (momentum) diagram for the (d, p) reaction, as illustrated in figure 10-17: the deuteron approaches with momentum $\mathbf{k}_d\hbar$ and the proton goes off with momentum $\mathbf{k}_p\hbar$ at an angle θ with respect to the incident beam. The momentum of the captured neutron may be obtained from the conservation of momentum:

$$k_n{}^2 = k_d{}^2 + k_p{}^2 - 2k_dk_p \cos \theta. \qquad (10\text{-}59)$$

If the neutron is captured at an impact parameter R, *orbital* angular momentum carried in by the captured neutron is $l_n\hbar$, where

$$k_nR = l_n. \qquad (10\text{-}60)$$

Now, since k_d and k_p are fixed by energy conservation, (10-59) and (10-60) can be satisfied (if at all) only for some definite θ if l_n is to have a definite value. A quantum-mechanical analysis of this problem relaxes this condition somewhat, as, indeed, does the possibility of other impact parameters in this classical treatment. The result is that the position of a peak, or a pattern of peaks, in the angular distribution of the emitted protons yields the value of l_n. The spin of the excited state of B may then be bracketed by an inequality resulting directly from the conservation of angular momentum:

$$\text{minimum of } \left|\pm I_A \pm l_n \pm \tfrac{1}{2}\right| \leq I_B \leq I_A + l_n + \tfrac{1}{2}. \qquad (10\text{-}61)$$

Conservation of parity demands that A and B have the same parity if l_n is even, and opposite parities if l_n is odd.

Incident nuclei other than deuterons may participate in reactions in which nucleons (usually neutrons) are transferred between the target and bombarding nucleus during a grazing collision. The (α, d) reactions leading to low-lying states of the residual nucleus as well as $({}_ZX^A, {}_ZX^{A+1})$ reactions, where ${}_ZX^A$ may be an ion as heavy as Ne^{20}, are examples (B4). Again, it is occasionally possible to get information about excited states from a study of these reactions, although the analysis is considerably more complicated when more than one nucleon is transferred or when there is a large Coulomb repulsion between the nuclei, as is true in heavy-ion transfer reactions.

Knock-On Reactions. The extreme use of the impulse approximation for the analysis of high-energy reactions in terms of individual nucleon-nucleon collisions, discussed in section C, is questionable in the medium-energy region. Although the question is difficult to answer generally, it may be said that it is likely that this oversimplified model is useful even for medium-energy particles in the outer, very low-density regions of the nucleus (cf. p. 34), and it is there that most medium-energy direct interactions occur. Within this picture, then, compound-nucleus formation occurs with medium-energy particles when the incident particle penetrates to the denser parts of the nucleus and is captured; otherwise its collisions are made only in the diffuse edge of the nucleus, and there is a direct interaction.

Unfortunately, the quantitative details of the model are still to be worked out, and the knock-on type of direct interaction is usually invoked qualitatively to interpret any results that are not expected from either compound-nucleus or stripping reactions.

REFERENCES

*B1   J. Blatt and V. Weisskopf, *Theoretical Nuclear Physics*, Wiley, New York, 1952.

B2   H. H. Barschall, "Regularities in the Total Cross Sections for Fast Neutrons," *Phys. Rev.* **86**, 431 (1952).

B3   N. Bohr, "Neutron Capture and Nuclear Constitution," *Nature* **137**, 344 (1936).

*B4   G. Breit, "Theory of Resonance Reactions and Allied Topics," *Encyclopedia of Physics*, Vol. 41-1, p. 367 (S. Flügge, Editor) Springer-Verlag, Berlin, 1959.

B5   N. Bohr and J. A. Wheeler, "The Mechanism of Nuclear Fission," *Phys. Rev.* **56**, 426 (1939).

C1   B. L. Cohen and R.V. Neidigh, "Angular Distribution of 22-MeV Protons Elastically Scattered by Various Elements," *Phys. Rev.* **93**, 282 (1954).

D1   I. Dostrovsky, Z. Fraenkel, and G. Friedlander, "Monte Carlo Calculations of Nuclear Evaporation Processes. III. Applications to Low-Energy Reactions," *Phys. Rev.* **116**, 683 (1960).

*E1   T. Ericson, "The Statistical Model and Nuclear Level Densities," *Phil. Mag. Supp.* **9**, No. 36, 425 (1960).

*F1   F. L. Friedman and V. F. Weisskopf, "The Compound Nucleus," *Niels Bohr and the Development of Physics* (W. Pauli, Editor), McGraw-Hill, New York, 1955, p. 134.

F2   S. Fernbach, R. Serber, and T. B. Taylor, "The Scattering of High Energy Neutrons by Nuclei," *Phys. Rev.* **75**, 1352 (1949).

F3   H. Feshbach, C. E. Porter, and V. F. Weisskopf, "Model for Nuclear Reactions With Neutrons," *Phys. Rev.* **96**, 448 (1954).

G1   S. N. Ghoshal, "An Experimental Verification of the Theory of Compound Nucleus," *Phys. Rev.* **80**, 939 (1950).

G2   J. Robb Grover, "Effects of Angular Momentum and Gamma-Ray Emission on Excitation Functions," *Phys. Rev.* **127**, 2142 (1962).

G3   R. J. Glauber, "High-Energy Scattering," *Lectures in Theoretical Physics*, Vol. 1 (W. E. Britten and B. W. Downs, Editors), Interscience, New York, 1962.

*H1   B. G. Harvey, "Recoil Techniques in Nuclear Reaction and Fission Studies," *Ann. Rev. Nuclear Sci.* **10**, 235–258 (1960).

H2   J. R. Huizenga and G. Igo, "Theoretical Reaction Cross Sections for Alpha Particles with an Optical Model," *Nuclear Phys.* **29**, 462 (1962).

*H3   R. Huby, "Stripping Reactions," *Progress in Nuclear Physics*, Vol. 3 (O. Frisch. Editor) Pergamon, London, 1953, p. 177.

*K4   I. Halpern, "Nuclear Fission," *Ann. Rev. Nuclear Sci.* **9**, 245–342 (1959).

*K1   B. B. Kinsey, "Nuclear Reaction, Levels, and Spectra of Heavy Nuclei," *Encyclopedia of Physics*, Vol. 40 (S. Flügge, Editor) Springer-Verlag, Berlin, 1957.

K2   S. Katcoff, "Fission-Product Yields from Neutron Induced Fission," *Nucleonics* **18,** No. 11, 201 (March 1960).

L1   D. W. Lang, "Nuclear Correlations and Nuclear Level Densities," *Nuclear Phys.* **42,** 353 (1963).

M1   F. S. Houck and J. M. Miller, "Reactions of Alpha Particles with Iron-54 and Nickel-58," *Phys. Rev.* **123,** 231 (1961).

M2   N. Metropolis, R. Bivins, M. Storm, A. Turkevich, J. M. Miller, and G. Friedlander, "Monte Carlo Calculations of Intranuclear Cascades. I. Low-Energy Studies," *Phys. Rev.* **110,** 185 (1958).

*M3 J. M. Miller and J. Hudis, "High-Energy Nuclear Reactions," *Ann. Rev. Nuclear Sci.,* **9,** 159–202 (1959).

M4   J. W. Meadows, "Excitation Functions for Proton-Induced Reactions with Copper," *Phys. Rev.* **91,** 885 (1953).

O1   J. R. Oppenheimer and M. Phillips, "Note on the Transmutation Function for Deuterons," *Phys. Rev.* **48,** 500 (1935).

*R1 J. Rainwater, "Resonance Processes by Neutrons," *Encyclopedia of Physics,* Vol. 40 (S. Flügge, Editor) Springer-Verlag, Berlin, 1957.

S1   M. M. Shapiro, "Cross Sections for the Formation of the Compound Nucleus by Charged Particles," *Phys. Rev.* **90,** 171 (1953).

S2   R. Serber, "The Production of High Energy Neutrons by Stripping," *Phys. Rev.* **72,** 1008 (1947).

S3   R. Serber, "Nuclear Reactions at High Energies," *Phys. Rev.* **72,** 1114 (1947).

S4   R. Sherr and F. P. Brady, "Spectra of (p, α) and (p, p') Reactions," *Phys. Rev.* **124,** 1928 (1961).

*W1 D. H. Wilkinson, "Nuclear Photodisintegration," *Ann. Rev. Nuclear Sci.* **9,** 1–28 (1959).

W2   A. C. Wahl, R. L. Ferguson, D. R. Nethaway, D. E. Troutner, K. Wolfsberg, "Nuclear-Charge Distribution in Low-Energy Fission," *Phys. Rev.* **126,** 1112 (1962).

EXERCISES

1. Estimate the most probable charge for the fission product of mass number 140 formed in the slow-neutron fission of U^{235}.

2. Give the energy in the center-of-mass system in a collision between a 10-MeV proton and a 40-MeV O^{16} ion when (a) they are moving in the same direction, (b) they are moving in opposite directions.   *Answer:* (a) 2.4 MeV.

3. What are the de Broglie wavelengths in the center-of-mass system for the collisions described in exercise 2?

4. Prove (10-25) by adding (10-23) and (10-24a).

5. Prove that the magnitude of the relative momentum of two particles in the laboratory system is the same as the magnitude of the momentum of *either* particle in the center-of-mass system.

6. Prove (10-11b) starting from (10-11a).

7. (a) Derive (10-31a) by evaluating the complex part of the wave number k that is defined in footnote 3.   (b) Show that (10-31a) reduces to (10-31b) for small values of $W_o/(\epsilon + V_o)$.

8. Estimate the imaginary part of the potential energy felt by a 1-GeV proton in the center of a heavy nucleus.   Take the central density to be about 2×10^{38} nucleons cm^{-3} and the effective average cross section for nucleon-nucleon collisions to be 40 mb at this energy.   Assume that the real part of the potential energy is negligible compared to 1 GeV.   *Answer:* $\cong$ 115 MeV.

9. By the use of a cyclotron that accelerates He^{2+} ions up to 40 MeV, deuterons up to

20 MeV, and protons up to 10 MeV, suggest methods (target, type, and energy of incident particle) for the synthesis of (a) Bi^{210}, (b) Co^{58} (with a minimal contamination from Co^{56}), (c) Se^{75}, (d) Ag^{112}, (e) Ba^{140}. Consider targets of normal isotopic composition only. Radiochemical purity and good yield are goals to be kept in mind.

10. Show that the probability per unit time for the evaporation of particle j from a compound nucleus excited to an energy U_C is a function of the quantity $U_C - S_j - B_j - \delta_n - \delta_p$ if the inverse cross section for the evaporation is given by (10-18) and the state density by (10-58); note that (10-58) implies zero state density below an excitation energy $\delta_n + \delta_p$.

11. The neutron capture reaction of Au^{197} at neutron energies up to a few hundred eV is characterized by a number of resonances. The most prominent (and the one with the lowest energy) is at 4.906 eV and has $\Gamma_\gamma = 0.124$ eV and $\Gamma_n = 0.0071\epsilon^{1/2}$ eV. The compound nucleus formed in this resonance absorption has spin 2. From these resonance parameters estimate (a) the cross section of gold for 0.025-eV neutrons and compare your result with the experimental value given in appendix C; (b) the peak cross section of the 4.906-eV resonance.

Answer: (b) 3.3×10^4 barns (experimental value 3.0×10^4 barns).

12. Calculate and sketch the shape of the energy spectrum of protons evaporated from a Zn^{67} nucleus excited to an energy of 15 MeV. Take the level density parameter a as 6.6 MeV^{-1}. For the inverse cross section use the form $\sigma = \pi R^2 (1 - B/\epsilon)$, with the effective barrier B taken as 0.7 times the full electrostatic barrier.

11

Sources of Nuclear Bombarding Particles

A. CHARGED-PARTICLE ACCELERATORS

From the discovery of nuclear transmutations in 1919 until 1932 the only known sources of particles which would induce nuclear reactions were the natural α emitters. In fact, the only type of nuclear reaction known during that period of 13 years was the α, p reaction. The natural α-particle sources most frequently used in transmutation experiments were Po^{210} (5.30 MeV, $t_{1/2} = 138$ days) and RaC' (7.69 MeV, $t_{1/2} = 1.6 \times 10^{-4}$ sec) used in equilibrium with its β-emitting parent RaC. Today natural α-particle sources for nuclear reactions are chiefly of historical interest because of the much higher intensities and higher energies now available from man-made accelerators for charged heavy particles. The acceleration of ions to sufficiently high energies for the production of nuclear transmutations was first achieved with high potentials applied across an accelerating tube. Among the devices of this type are the voltage-multiplying rectifier of J. D. Cockcroft and E. T. S. Walton, still widely used for ion acceleration up to about 1 MeV, and the cascade transformer developed by C. C. Lauritsen and his co-workers at the California Institute of Technology.

Electrostatic (Van de Graaff) Generator. The adaptation of the electrostatic machine to the production of high potentials for the acceleration of positive ions was pioneered by R. J. Van de Graaff of the Massachusetts Institute of Technology, beginning in 1929. In the Van de Graaff machine a high potential is built up and maintained on a conducting sphere by the continuous transfer of static charges from a moving belt to the sphere. This is illustrated in figure 11-1. The belt, made of silk, rubber, paper, or some other suitable insulator, is driven by a motor and pulley system. It passes through the gap AB, which is connected to a high-voltage source (10,000 to 30,000 volts d-c) and adjusted so that a continuous discharge is maintained from the sharp point B. Thus positive (or negative) charges are sprayed from B onto the belt which carries them to the interior of the insulated metal sphere; there another sharp point or sharp-toothed comb C connected to the sphere takes off the

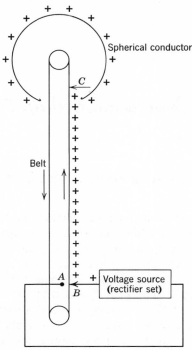

Fig. 11-1 Schematic representation of the charging mechanism of a Van de Graaff generator.

charges and distributes them to the surface of the sphere. The sphere will continue to charge up until the loss of charge from the surface by corona discharge and by leakage along its insulating support balances the rate of charge transfer from the belt. The continuous current that can be maintained with an electrostatic generator depends on the rate at which charge can be supplied to the sphere.

Van de Graaff's first installation consisted of two spheres, each 2 ft in diameter, one charged to a positive potential of about 750,000 volts above ground and the other to an equal negative potential. Most of the more recent electrostatic generators use a single electrode, with acceleration of the ions between that electrode and ground potential, because considerable practical advantages are gained if much of the auxiliary equipment can be operated at ground potential.

Since the voltage of an electrostatic generator is limited by the breakdown of the gas surrounding the charged electrode, it is desirable to use conditions under which the breakdown potential is as high as possible. The breakdown potential is a function of pressure and goes through a minimum at a rather low pressure (small fraction of an atmosphere).   It is therefore advantageous to operate an electrostatic generator either in a high vacuum, which presents formidable difficulties, or in a high-pressure atmosphere.   Most electrostatic generators are completely enclosed in steel tanks in which pressures of 10 or more atmospheres are maintained.   A further improvement is the use of gases which have higher breakdown potentials than air.   Nitrogen is most commonly used, and sulfur hexafluoride has been very successful.   These gases have the additional advantage over high-pressure air that they will not support combustion following a spark to some combustible material.   Pressure-type electrostatic generators capable of accelerating protons or other positive ions to energies of 2 to 6 MeV are in operation.

A variety of models for positive-ion and electron acceleration up to 5.5 MeV are commercially available.   Proton currents of up to 500 μa and even larger electron currents are common.   The chief application of electrostatic generators is in nuclear physics work requiring high precision because, unlike other machines such as cyclotrons, they supply ions of precisely controllable energies (constant to about 0.1 per cent and with an energy spread of about the same order of magnitude).

Accelerating Tubes.   Any machine for the acceleration of ions by the application of a high potential requires an accelerating tube across which the potential is applied.   A source of ions near the high-voltage end, a system of accelerating electrodes, and a target at the low-voltage end must be provided and enclosed in a vacuum tube connected to the necessary pumping system. The ion source is essentially an arrangement for ionizing the proper gas (hydrogen, deuterium, helium) in an arc or electron beam; the ions are drawn through an opening into the accelerating system.   In electron accelerators an electron gun is used as the source.

A typical accelerating tube (figure 11-2) is built of glass or porcelain sections S.   Inside this tube, sections of metal tube T define the path of the ion beam. Each metal section is supported on a disk which passes between two sections of insulator out into the gas-filled space to a corona ring R equipped with corona points P.   The purpose of the corona rings and points is to carry the corona discharge from the high- to the low-voltage end of the tube and to distribute the voltage drop uniformly along the tube.   Depending on the number of sections used, a potential difference somewhere between 10 and several hundred kilovolts exists between successive sections.   Each gap between successive sections has both a focusing and a defocusing action on the ions traveling down the tube.   The ions tend to travel along the electric lines of force (see figure 11-2 for the pattern of these lines between a pair of sections).   In entering the gap the ions are focused and in leaving it they are defocused; but because the ions move more slowly on entering the gap than on leaving it, the focusing effect is stronger than the subsequent defocusing.   Well-focused beams (cross-sectional area less than 0.1 cm^2) can be obtained.   It should be mentioned that

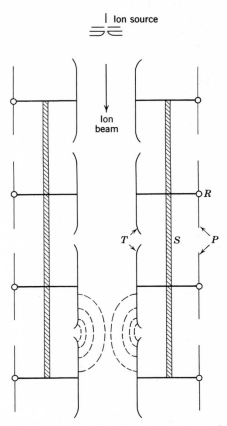

Fig. 11-2   Schematic diagram of a portion of an accelerating tube.

from hydrogen gas in an ion source not only protons but also hydrogen molecule ions (H_2^+) and H_3^+ ions are obtained; these are also accelerated in the tube but can be separated from the protons before striking the target by means of a magnetic analyzer.   One of the magnetically analyzed beams is usually used to obtain precise automatic energy control.   The position of this deflected beam along a slit system depends on its energy, and a signal from this slit system can be fed back to devices at the high-voltage terminal which will adjust the accelerating potential in the desired direction.

In a pressure electrostatic generator the charging system, high-voltage electrode, and accelerating tube are all enclosed in the steel tank containing the high-pressure gas.   The high-potential electrode is often more like a cylinder than a sphere.

Tandem Van de Graaff Machine. The energy attainable with electrostatic generators has been greatly increased by the application of the "tandem" principle, an ingenious idea first suggested in 1936 but not put into practice until more than 20 years later. In the two-stage tandem Van de Graaff negative ions (such as H^-) are produced by electron bombardment and are accelerated *toward* the positive high-voltage terminal which is located in the center of the pressure tank. Inside the terminal the negative ions, which now have an energy of 5 to 10 MeV, travel through a gas-filled channel and are thus stripped of electrons. The positive-ion beam so produced is further accelerated toward ground potential in the usual way. Two-stage tandems producing a few microamperes of 20-MeV protons are commercially available. The currents in tandem machines are generally lower than in ordinary Van de Graaff accelerators because of the difficulty of making efficient negative-ion sources. Acceleration of helium-ion beams in tandem machines has been reported, although the production of He^- or especially He^{2-} ions would certainly seem most surprising.

A further increase in energy can be achieved in the three-stage version of tandem Van-de-Graaff machine. This requires two separate tanks, one with a negative, the other with a positive high-voltage terminal. Figure 11-3 is a schematic drawing. Ions are produced in an ordinary positive-ion source, magnetically analyzed, and then neutralized by electron bombardment. The neutral beam so produced is allowed to drift to the negative high-voltage electrode in the first tank, where further electron addition produces negative ions which are then accelerated to ground potential. At this point the negative-ion beam passes from the first to the second tank and receives two additional stages of acceleration analogous to the two-stage tandem operation. A number of three-stage tandem Van de Graaff machines are under construction. Proton energies of up to 30 MeV are expected to be achieved.

Linear Accelerator. In all the devices for accelerating ions mentioned so far the full high potential corresponding to the final energy of the ions must be provided, and the limitations of this type of device are introduced by the insulation problems. These problems are very much reduced in machines that

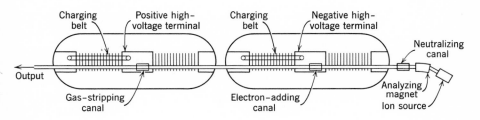

Fig. 11-3 Schematic sketch of a three-stage tandem Van de Graaff generator. (From reference B2.)

employ repeated acceleration of ions through relatively small potential differences. The linear accelerator was the first device developed in which advantage is taken of this possibility. In the early versions of this machine a beam of ions from an ion source was injected into an accelerating tube containing a number of coaxial cylindrical sections. (See figure 11-4 for a schematic diagram. Alternate sections were connected, and a high-frequency alternating voltage from an oscillator was applied between the two groups of electrodes. An ion traveling down the tube will be accelerated at a gap between electrodes if the voltage is in the proper phase. By choosing the frequency and the lengths of successive sections correctly one can arrange the system so that the ions arrive at each gap at the proper phase for acceleration across the gap. The successive electrode lengths have to be such that the ions spend just one half cycle in each electrode. Acceleration takes place at each gap, but the focusing action described for the accelerating tubes of direct high-voltage machines is for most types of linear accelerators replaced by a net defocusing effect because the rf field is rising while the particles cross the gap. Special focusing devices such as grids or magnetic lenses must then be provided.

The first linear accelerator on record was a two-stage device built in 1928 by R. Wideröe; it accelerated positive ions to about 50 keV. By 1931 E. O. Lawrence and D. H. Sloan in Berkeley had succeeded in accelerating mercury ions to 1.26 MeV in an accelerating system having 30 gaps. Intensive work on linear accelerators was carried out in many laboratories in the early 1930's. However, because the cyclotron was developed almost simultaneously and had obviously great advantages, the linear accelerator did not receive much further attention from about 1934 until after World War II, when the availability of high-power microwave oscillators made possible acceleration to high energies in relatively small linear accelerators. Various designs, some using traveling waves, some resonant cavities, have been explored and used in actual machines since 1946. Excellent discussions of "linacs," as linear accelerators are often called, may be found in refs. B3, L1, and M1. Of the several proton linacs in operation, that at the University of Minnesota has achieved the highest energy, 70 MeV. Linear accelerators are used to inject 50-MeV protons into the alternating-gradient synchrotrons at Geneva and Brookhaven (see later). A special design for the acceleration of heavy ions (He^4 to Ar^{40}) was developed

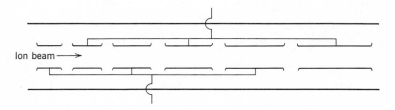

Fig. 11-4   Schematic diagram of the accelerating tube of a linear accelerator.

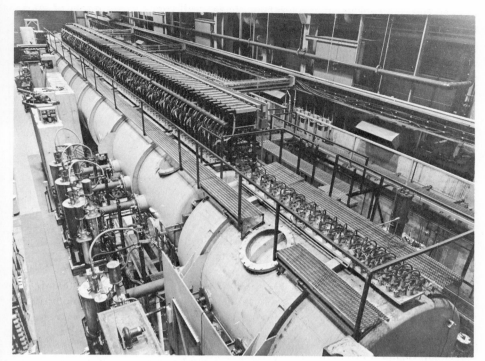

Fig. 11-5   Heavy-ion linear accelerator at Yale University.   (Courtesy Yale University.) In the short tank in the foreground, ions from the ion source (such as C^{+2}) are accelerated to about 1 MeV per nucleon.   In the region between the two tanks additional electrons are stripped off the ions by collisions with a gas jet; the resulting highly charged ions (for example, C^{+6}) enter the long main tank where they are accelerated to 10 MeV per nucleon.

jointly by groups at the University of California and Yale University, and "hilacs" (heavy-ion linear accelerators) producing heavy-ion beams with energies of 10 MeV per nucleon are in operation at both institutions.   The one at Yale is shown in figure 11-5.   A similar hilac is in operation at Kharkov, USSR.   Existing proton linacs and hilacs give average beam currents up to a few microamperes, but average proton currents in the milliampere range appear quite feasible.   The beams are always pulsed.

The operation of electron linacs is simpler than that of proton linacs because electrons at energies of even a few MeV move essentially with the velocity of light and therefore can travel down a waveguide with the accelerating wave without the need for drift tubes or other electrode structures (whose purpose is to shield particles during the "wrong" part of the rf phase).   On the other hand, the dimensional tolerances for the waveguide are extraordinarily exacting because the phase velocity of the traveling wave must be precisely maintained to keep the electrons at an accelerating phase at all times.   A number of linear accelerators for electrons with various maximum energies have been constructed both in the United States and abroad.   The largest electron linacs are the 1-GeV machines at Stanford University and at Orsay, France, and a 2-GeV linac nearing completion at Kharkov, USSR.   The Stanford accelerator is 220 ft long, and the radio-frequency power is fed into its waveguide by 22 klystrons, each delivering 17,000 kW at a frequency of about 3000 megacycles (Mc). A still larger machine, two miles long and expected to accelerate electrons initially to 20 GeV and eventually to 40 GeV, is under construction at Stanford.

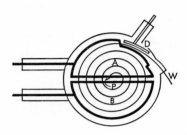

Fig. 11-6 Cyclotron vacuum chamber. The ions originate at the ion source P and follow a spiral path. The dees A and B, the deflector D, and the exit window W are shown. (Reproduced from E. Pollard and W. L. Davidson, *Applied Nuclear Physics*, 2nd ed., New York, John Wiley and Sons, 1951.)

For acceleration of electrons to the multi-GeV region the linear accelerator is undoubtedly the most suitable device because it avoids the huge energy losses by radiation common to all circular types of electron accelerators (see p. 370).   For proton acceleration to very high energies ($\gg$100 MeV), on the other hand, economic considerations favor the circular types of accelerators to be discussed.   In spite of this, the obvious advantages of well-focused straight-line beams have led several groups to propose construction of linacs for acceleration of protons to about 1 GeV.   Technically, such machines appear to be entirely feasible.

Cyclotron.   Perhaps the most successful device for accelerating positive ions to millions of electron volts is the cyclotron proposed by E. O. Lawrence in 1929.   A remarkable development has taken place (M1) from the first working model which produced 80-keV protons in 1930 to the giant synchrocyclotrons now in operation which accelerate protons to energies as high as 700 MeV.

In the cyclotron, as in the linear accelerator, multiple acceleration by a radio-frequency (rf) potential is used.   But the ions, instead of traveling along a straight tube, are constrained by a magnetic field to move in a spiral path consisting of a series of semicircles with increasing radii.   The principle of operation is illustrated in figure 11-6.   Ions are produced in an arc ion source P near the center of the gap between two hollow semicircular electrode boxes A and B called "dees."   The dees are enclosed in a vacuum tank, which is located between the circular pole faces of an electromagnet and is connected to the necessary vacuum pumping system.   A high-frequency potential supplied by an oscillator is applied between the dees.   A positive ion starting from the ion source is accelerated toward the dee which is at negative potential at the time.

As soon as it reaches the field-free interior of the dee, the ion is no longer acted on by electric forces, but the magnetic field perpendicular to the plane of the dees constrains the ion to a semicircular path.   If the frequency of the alternating potential is such that the field has reversed its direction just at the time the ion again reaches the gap between dees, the ion again is accelerated, this time toward the other dee.   Now its velocity is greater than before, and it therefore describes a semicircle of larger radius; however, as we shall see from the equations of motion, the time of transit for each semicircle is independent of radius.   Therefore, although the ion describes larger and larger semicircles, it continues to arrive at the gap when the oscillating voltage is at the right phase for acceleration.   At each crossing of the gap the ion acquires an amount of kinetic energy equal to the product of the ion charge and the voltage difference between the dees.   Finally, as the ion reaches the periphery of the dee system, it is removed from its circular path by a negatively charged deflector plate D and is allowed to emerge through a window W and to strike a target.

The equation of motion of an ion of mass M, charge e, and velocity v in a magnetic field H is given by the necessary equality of the centripetal magnetic force Hev and the centrifugal force Mv^2/r, where r is the radius of the ion's orbit:

$$Hev = \frac{Mv^2}{r} \quad \text{and} \quad r = \frac{Mv}{He}. \tag{11-1}$$

Remembering that the angular velocity $\omega = v/r$, we see that

$$\omega = \frac{He}{M}. \tag{11-2}$$

From this equation it is evident that the angular velocity is independent of radius and ion velocity and that the time required for half a revolution is constant for ions of the same e/M, provided that the magnetic-field strength is constant.   In practice, the magnetic field is kept constant, e/M is a characteristic of the type of ion used, and therefore ω is constant.   The radio frequency has to be chosen so that its period equals the time it takes for the ions to make one revolution.   For $H = 15{,}000$ gauss and e/M for a proton the revolution frequency $\omega/2\pi$, and therefore the necessary oscillator frequency, turns out (from 11-2) to be about 23×10^6 cycles per second (cps).   For deuterons or helium ions (He^{2+}) at the same H the frequency is half that value.   Most cyclotrons are operated as deuteron and helium-ion sources, and they often use rf oscillators tuned to about 11 or 12 Mc.

It is clear from (11-2) that in a given cyclotron both the magnetic field and the oscillator frequency can be left unchanged when different ions of the same e/M, such as deuterons and α particles, are accelerated.   Equation 11-1 shows that the velocity reached at a given radius is the same for ions of the same e/M; therefore α particles are accelerated to the same velocity, hence twice the energy, as deuterons. To accelerate protons in a cyclotron designed for deuterons either the frequency must be approximately doubled (which is usually impractical) or H must be about halved. Although the latter method makes inefficient use of the magnet, it is occasionally used, and the final velocity is again the same as for deuterons (11-1); therefore, protons are accelerated to half the energy available for deuterons.

By squaring (11-1) we get

$$r^2 = \frac{M^2 v^2}{H^2 e^2}$$

or

$$\frac{1}{2} M v^2 = \frac{H^2 e^2}{2M} r^2. \tag{11-3}$$

Thus the final energy attainable for a given ion varies with the square of the radius of the cyclotron. With $H = 15,000$ gauss, the deuteron energy $E = 0.035 r^2$ MeV if r is in inches. The size of a cyclotron is usually given in terms of its pole-face diameter.

From the equations of motion it is clear that an ion can reach the dee gap at any phase of the dee potential and still be in resonance with the radio frequency. As we have just derived, the final energy acquired by an ion is entirely independent of the energy increment the ion receives at each crossing of the dee gap. However, in practice only ions which enter the first gap in a favorable phase of the radio frequency (perhaps during about one third of the cycle) contribute to the beam current. To avoid difficulties due to excessive phase differences between beam and rf as well as to excessively long paths for the ions, rather high dee voltages (20,000 to 200,000 volts) are generally used.

A very important feature of the cyclotron is the focusing action it provides for the ion beam. The electrostatic focusing at the dee gap is entirely analogous to that in the high-voltage accelerating tubes. However, as the energy of the ions increases, this effect becomes almost negligible. Fortunately, a magnetic focusing effect becomes more and more pronounced as the ions travel toward the periphery. This can be seen from the shape of the

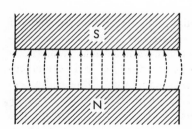

Fig. 11-7 Shape of magnetic field in the gap of a cyclotron magnet. The curvature of the lines of force gives rise to the focusing action.

magnetic field as shown in figure 11-7. Near the edge of the pole faces the magnetic lines of force are curved, and therefore the field has a horizontal component which provides a restoring force toward the median plane to an ion either below or above that plane. The focusing is so good that a cyclotron beam is generally less than 1 cm high at the target.

One difficulty we have so far neglected is presented by the relativistic mass increase of the ions as they reach high energies. This increase is about $\frac{1}{2}$ per cent for a 10-MeV deuteron and about 5 per cent for a 100-MeV deuteron. It is clear from (11-2) that if the revolution frequency is to be kept constant the increase in mass must be compensated by a proportional increase in field strength. When the relativity effects are small, this increase of the magnetic field toward the periphery can be readily achieved by slight radial shaping or shimming of the pole faces.[1] Notice, however, that this shaping of the field creates regions of magnetic defocusing. For moderate relativistic mass increases this difficulty has been overcome, mainly by the use of higher dee voltages and correspondingly shorter ion paths.

Almost thirty standard cyclotrons in the United States and an approximately equal number abroad are now operating; among them are several of about 60-in. pole-face diameter that accelerate deuterons to about 20 MeV and helium ions to about 40 MeV. The relativistic mass increase has limited the energies achieved with standard cyclotrons to about 25 MeV for deuterons and 50 MeV for helium ions. Present-day cyclotrons may have circulating beams of hundreds of microamperes; the deflected external beams are somewhat smaller. The large beam currents available have made target cooling a rather severe problem. The power dissipation in a target receiving 100 μA of 20-MeV particles is 2000 W, and even iron targets are melted unless water cooling is provided.

Thomas (Sector-Focused) Cyclotron. A method for overcoming the energy limitation on cyclotron acceleration was suggested as early as 1938 by L. H. Thomas (T1) but was not put to use until almost two decades later (see refs. B3 and L1). Thomas showed that azimuthal variations of the magnetic field can result in axial focusing (i.e., focusing in the direction perpendicular to the pole faces). It is thus possible to let the *average* field increase with radius (as required to compensate for the relativistic mass increase), yet to achieve focusing by means of azimuthal field variations. It has been shown that with this type of design, particles can be accelerated to a kinetic energy approximately equal to their rest energy. The periodic azimuthal field variations are obtained by the use of pole faces that have alternate "hill" and "valley" sectors (figure 11-8). In most, but not all, designs the sectors have spiral rather than radial contours. Therefore the name spiral-ridge cyclotron is frequently used, although the terms "sector-focused," "isochronous," or "azimuthally

[1] In a given cyclotron the field should be shaped slightly differently for protons and for deuterons because of the different relativity effects. For this reason deuteron cyclotrons do not give very good proton beams without major readjustments. Better proton beams can be obtained by acceleration of H_2^+ ions at full magnetic field; the required field shapes for H_2^+ and D^+ acceleration are almost identical. The final proton energy is the same whether H_2^+ ions are accelerated at full field or H^+ ions at half field.

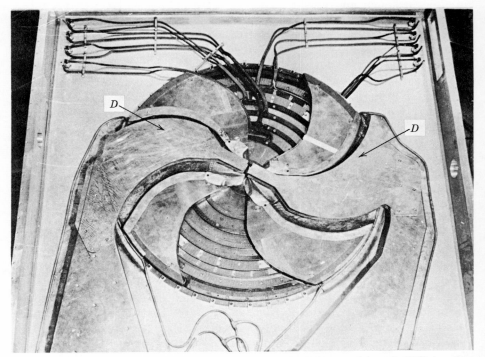

Fig. 11-8   View of the lower pole tip of the 50-MeV spiral-ridge cyclotron at UCLA.   The two "dees" (labeled D) are in place in two of the valley sectors.   (Courtesy J. R. Richardson.)

varying field" (AVF) cyclotron are more generally applicable.   The designation isochronous is meant to convey that in contrast to the synchrocyclotron (see later) the time per revolution (or the revolution frequency) stays constant just as in the ordinary cyclotron [cf. (11-2)], although both magnetic field H and ion mass M vary over the ion path.   Note that as in all cyclotrons the magnetic field stays constant with time.

Three or four pairs of hill and valley sectors are used in existing machines, although designs with six or eight pairs have been proposed.   The ratio of hill gap to valley gap varies widely in different designs but typically may be between 0.25 and 0.7.   The pole faces must be radially shaped also (usually in both hill and valley regions) to provide the radial rise of magnetic field required by the relativistic mass increase.   The resulting pole-face contours and field shapes can become quite intricate, as shown in figures 11-8 and 11-9.   A combination of model measurements and computer calculations is usually required to arrive at the final pole shapes.   The dee structures often take on odd shapes also, because they have to fit the pole designs (figure 11-8).

In 1964 some two dozen AVF cyclotrons were in operation or in advanced stages of construction in several countries. They range in size up to 90 in. in pole diameter and accelerate protons to various energies up to about 80 MeV. Circulating-beam currents of about 1 mA can be achieved. Great flexibility for the acceleration of various particles (H^2, He^3, He^4, and heavier ions) and for energy variation can be engineered into AVF cyclotrons if appropriate correcting coils are provided. The currents in these coils as well as in the main windings can then be programmed to give the field shape needed for acceleration of a particular ion to a particular energy. These advantages have prompted the conversion of a number of standard cyclotrons to sector-focused operation. Design studies are also underway in several laboratories for AVF machines for the acceleration of protons (and other ions) to much higher energies—up to about 800 MeV. Beam extraction from AVF cyclotrons is difficult, but it has been accomplished.

Synchrocyclotron. Another way of overcoming the relativity limitation in cyclotrons is by modulation of the oscillator frequency. Although this was, in a sense, an obvious solution that followed from the basic cyclotron equations, it was not seriously considered until about 1945 because the difficulty of main-

Fig. 11-9 Contour map of the magnetic field at the median plane of the 50-MeV sector-focused cyclotron at UCLA. The numbers on the contour lines are field strengths in kilogauss. [From D. J. Clark, J. R. Richardson, and B. T. Wright, *Nucl. Instr. and Meth.* **18-19,** 1 (1962).]

taining synchronism between oscillator frequency and revolution frequency seemed formidable. The discovery of the principle of phase stability in 1944–1945 (independently found by V. Veksler in the USSR and E. M. McMillan in the United States) led to the realization that, as McMillan put it (M1), "Nature had already provided the means" for overcoming this difficulty.

Without going into the quantitative theory, we can easily see how phase stability in any circular resonance accelerator comes about qualitatively.[2] Suppose a particle at a particular crossing of the accelerating gap is given more than its proper amount of energy; its next orbit will then have a larger radius than if the particle had just the "right" amount of energy, and the transit time for that revolution will be too long (longer than one rf wavelength). If the accelerator is designed so that the particles cross the gap when the sinusoidal rf voltage is in the 90°–180° phase—decreasing with time—the particle that has previously received too much energy and therefore arrives late at the gap now receives less energy. Conversely, a particle with too little energy follows an orbit of smaller than equilibrium radius, arrives at the gap early, and is given more energy than before. Thus the particles will perform phase oscillations (sometimes called synchrotron oscillations) around a stable phase. The frequency of these phase oscillations is usually much (perhaps several hundred times) lower than the revolution frequency.

As soon as phase stability was recognized, frequency modulation was applied to cyclotrons, first in Berkeley (see ref M1), then in many other laboratories. At least 18 frequency-modulated (FM) cyclotrons, or synchrocyclotrons as they are usually called, are in operation, eight of them in the United States. The largest of these machines are the ones at Berkeley, CERN (Geneva), and Dubna (near Moscow), all of which accelerate protons to 600 to 700 MeV. The 184-in. synchrocyclotron at Berkeley can also be used to accelerate deuterons (to 450 MeV) and helium ions (to 900 MeV); this involves interchangeable oscillators of different frequencies. For the acceleration of protons to 700 MeV the frequency has to be decreased during each acceleration cycle to about one half its initial value. In most synchrocyclotrons the frequency modulation is brought about by means of a rotating condenser in the oscillator circuit. Obviously, for successful acceleration, ions have to start their spiral path at or near the time of maximum frequency. Because ions are accepted into stable orbits only during about 1 per cent of the FM cycle the beam consists of successive pulses. Beam currents are therefore lower than in standard and sector-focused cyclotrons.

Average beam currents of 0.1 to 1 μA and pulse rates between 50 and 500 cps are rather common. Focusing presents less of a problem than in standard cyclotrons because the relativity effects need not be compensated for by the shape of the magnetic field. The magnetic field can actually be decreased near

---

[2] In modern linear accelerators for protons and other heavy particles phase stability is also important but the details are different; in linacs stability is obtained on the rising (0°–90°) rather than the falling (90°–180°) part of the rf sine wave.

the edges of the pole faces to increase the magnetic focusing. Most FM cyclotrons have but one working dee, and the dee voltage is relatively low, typically 10 to 30 kV. Acceleration to hundreds of MeV thus requires of the order of 10^4 revolutions, or a total path length of the order of 10^7 cm; an acceleration cycle then takes about 10^{-3} second.

Most experiments with synchrocyclotrons involve interception of the circulating beam with target probes; the bombarding energy is easily varied by variation of the radius of interception, as in standard cyclotrons. External beams with intensities up to about 10 per cent of the circulating beams have been extracted from a few synchrocyclotrons.

Betatron. The cyclotron which has been so successful for the acceleration of positive ions is not practical for the acceleration of electrons because of the high frequencies that would be required and because of the enormous relativistic mass increase of electrons even at moderate energies. (See appendix B; at 1 MeV the total mass is three times the rest mass.) The first device for producing electron energies above 2 or 3 MeV was the betatron suggested by a number of investigators and first developed by D. W. Kerst in 1940. The betatron may be thought of as a transformer in which the secondary winding is replaced by a stream of electrons in a vacuum "doughnut." The acceleration is supplied by the electromotive force induced at the position of the doughnut by a steadily increasing magnetic flux perpendicular to and inside the electron orbit. In order for the electrons to move in a fixed orbit, it is necessary that the field at the orbit change proportionally with the momentum of the electrons. This condition is fulfilled if the field at the orbit increases at just half the rate at which the average magnetic flux inside the orbit increases; this may be achieved by the proper tapering of the pole faces, as indicated schematically in figure 11-10. The radial variation of the field in the region of the electron orbit is important in the focusing problem. It turns out that if the field falls off as the inverse nth power of the radius in the region of the orbit and if $0 < n < 1$ the electrons will describe damped oscillations about the equilibrium orbit; if the revolution frequency is f, the frequency of the vertical "betatron" oscillations about the equilibrium orbit is $f_v = f \sqrt{n}$, that of the radial oscillations $f_r = f \sqrt{1 - n}$. Betatrons are generally designed

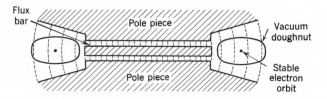

Fig. 11-10   Cross section through central region of a betatron (schematic), with the magnetic lines of force indicated.

with $n = 0.75$, and the excellent focusing permits the electrons to make hundreds of thousands of revolutions.

Electrons for acceleration are pulse-injected into the doughnut at 20 to 50 keV from an electron gun when the field at the orbit is very small. Alternating-current magnets constructed of laminated iron and operating between about 60 and 1800 cps are used in betatrons. The energy obtainable with a given betatron is limited by the saturation of the central flux bars. As they begin to saturate, the orbit begins to shrink. The electron beam can then be allowed to strike a target (often tungsten wire) mounted in the doughnut inside the equilibrium orbit. Or by the use of auxiliary coils the orbit can be expanded, raised, lowered, or contracted toward a suitably placed target at any electron energy.

Several 20- to 50-MeV betatrons and a few larger ones are in operation. The largest betatron built is a 300-MeV machine at the University of Illinois. The ultimate limitation on the energy attainable in a betatron or in any other circular electron accelerator is presumably set by the radiation of energy by electrons under centripetal acceleration; the radiative energy loss at a given radius increases with the fourth power of the electron energy and becomes quite prohibitive for betatrons above about 1 GeV.

It is generally conceded that synchrotrons and linear accelerators have many advantages over betatrons for electron acceleration.

Synchrotron. Both Veksler and McMillan, along with the discovery of the principle of phase stability (see the discussion of the synchrocyclotron above), proposed a new type of accelerator, called synchrotron by McMillan and synchrophasotron by Veksler. The two names have survived, one in Western countries, the other in the Soviet Union, and the device has been enormously successful for the acceleration of both electrons and protons.

In the synchrotron as in the betatron the radius of the orbit is kept approximately constant by a magnetic field which increases proportionally with the momentum of the particles. However, the acceleration (or rather the increase in energy, since the velocity must remain essentially constant at $v \cong c$) is provided not by a changing central flux but (more nearly as in the cyclotron) by a rf oscillator which supplies an energy increment every time the particles cross a gap in a resonator that forms part of the vacuum doughnut.

As in the synchrocyclotron, phase stability results when the acceleration takes place during the decreasing part of the rf cycle (phase between 90 and 180°). The phase stability considerations are qualitatively the same as for the synchrocyclotron, although the variation of the magnetic field with time makes the detailed phase relations somewhat more complex. As the magnetic field is steadily increased, a slight phase difference is maintained between the particle orbits and the resonator voltage, and on the average the particles gain some energy at each passage of the gap. As in the betatron, the particles perform radial and vertical oscillations around their equilibrium orbit. The radial field index

$$n = - \frac{dB/B}{dr/r} = - \frac{d \ln B}{d \ln r} = - \frac{dB}{dr} \frac{r}{B} \qquad (11\text{-}4)$$

has to be carefully chosen to avoid resonances between radial and vertical betatron oscillations or between either of these and the revolution frequency. At such resonances the beam would be lost.   From the relations between the oscillation and revolution frequencies (p. 369) it is clear, for example, that n values of exactly 0.2, 0.5, and 0.8 must be avoided because they would lead to ratios of vertical to radial oscillation frequencies of exactly 0.5, 1, and 2, respectively.   The actual choice of n is dictated by economic factors.   For a given angular deviation from the equilibrium orbit the amplitude of an oscillation is inversely proportional to its frequency, and therefore small n corresponds to large vertical, large n to large radial amplitudes.   A value of n between the 0.5 and 0.8 resonances results in the most economical shape of magnetic gap.

For electron acceleration the synchrotron is much more economical than the betatron because no large central magnetic flux is required and therefore magnet costs are much lower.   Most electron synchrotrons actually use betatron acceleration by means of small central flux bars until the electrons have almost reached the velocity of light ($v = 0.98c$ at 2 MeV).   Then the rf circuits are turned on for further acceleration by synchrotron action, and in this way no frequency modulation is required.   Injection, targeting, and beam ejection in electron synchrotrons present essentially the same problems as in betatrons.   At least 20 electron synchrotrons with maximum energies between 20 and 500 MeV are in operation.   The electron synchrotrons at Frascati, Italy, and at the California Institute of Technology produce electron beams of 1.1 and 1.5 GeV maximum energy, respectively.

For proton (or other positive-ion) acceleration to energies in the billion-electron-volt (GeV) range the synchrotron has proved to be the first practical device.   Because it requires a ring-shaped magnet only, a proton synchrotron is much cheaper to construct for these energies than a synchrocyclotron with its solid magnet structure.   The chief difference between proton and electron synchrotrons arises from the fact that protons do not approach the speed of light until they have energies of billions of electron volts ($v = 0.98c$ at 3.8 GeV). Therefore, for a constant orbit radius the revolution frequency of the protons changes by a large factor during the acceleration (a factor of 12 for acceleration from 4 MeV to the limiting velocity).   The frequency of the rf accelerating voltage must be modulated over this wide range, and in most of the present machines this is accomplished electronically rather than by rotating condensers, as in the FM cyclotrons.

Nine "ordinary" or constant-gradient proton synchrotrons were in operation in 1964: 1.0-GeV machines at Birmingham, England, and Delft, Holland,

3.0-GeV machines at Brookhaven National Laboratory (called the "Cosmotron"), at Princeton, New Jersey, and at Saclay, France (called "Saturne"), the 6.2-GeV "Bevatron" at Berkeley, the 7-GeV "Nimrod" at the Rutherford Laboratory in England, the 10-GeV synchrophasotron at Dubna, USSR, and the 12.5-GeV ZGS (zero-gradient synchrotron) at Argonne National Laboratory. These various proton synchrotrons have many characteristics in common, but some of them have interesting distinguishing features (B3). For example, although in most of the machines focusing is achieved by n values between 0.6 and 0.7, the Argonne and Delft synchrotrons have $n = 0$ (hence the name ZGS); in them focusing is accomplished by special field shapes at the ends of the magnet sectors. Protons are injected into the four highest-energy machines from linacs and into most of the others from Van de Graaff accelerators. All the proton synchrotrons except the one at Birmingham have at least four essentially field-free straight sections between magnet sectors; they greatly facilitate injection, rf acceleration, and targeting. The stored energy in the magnets of these accelerators is enormous—peak power inputs range from a few to more than 100 MW—and for this reason most of them use rather low pulse rates (5 to 30 pulses per minute) and have provisions for storage of most of the energy in flywheels between magnet pulses. The Princeton machine is unique in that it uses a choke-and-capacitor system for energy storage and is designed for about 20 pulses per second. Thus it is expected to achieve much higher time-average currents than the other proton synchrotrons.

Figure 11-11 shows a view of the Cosmotron, and we give here a few of its design features. The orbit radius is 30 ft, and the total magnet weight is about 2000 tons (to be compared with the 4300-ton magnet of the Berkeley 184-in. synchrocyclotron). The vacuum chamber is about 6 by 25 in. in cross section and has a total volume of about 300 ft³. Normal operation is at 12 pulses per minute, and the total acceleration time for each pulse is 1 second. Each proton travels about 150,000 miles to reach maximum energy. Beam intensities of a few times 10^{11} protons per pulse have been achieved. For radiochemical studies of nuclear reactions such intensities are quite ample, especially since the fractional energy loss of the protons in going through a thin target is so small that a given proton can make many target traversals in successive revolutions. The product of actual target thickness and number of traversals is approximately constant for all target thicknesses up to some maximum; in the circulating beam of the Cosmotron this limit, called the effective target thickness, is reached at 1 to 2 g cm^{-2}, the exact value depending on the atomic number of the target because protons are lost from the beam by multiple Coulomb scattering. Targeting in proton synchrotrons can be accomplished in a variety of ways: the target may be rammed into or flipped through the beam at the end of each acceleration cycle, or the beam orbit may be allowed to collapse into a target, which is achieved by switching off the rf accelerating field while the magnetic field is rising. The energy of the protons striking the target can be varied at will up to the maximum energy of the machine.

Fig. 11-11   General view of the Brookhaven Cosmotron.   The Van de Graaff injector is partially visible in the background.   Three of the vacuum pump stations and some concrete shielding can be seen in the foreground.   The two straight sections of the vacuum chamber are shown open to the atmosphere.   (Courtesy Brookhaven National Laboratory.)

For many types of experimentation it is advantageous to have external proton beams as well as the beams of secondary particles, such as neutrons, π mesons, K-mesons, and antiprotons, which originate in targets struck by the primary proton beam.   Several proton synchrotron installations have external-beam systems based on a scheme for beam extraction developed by O. Piccioni et al. (P1) which is capable of yielding external-beam intensities up to 50 per cent of circulating beams.   The beam is first made to go through an energy-loss target which causes the beam orbit to shrink so that on the next revolution the protons pass through a deflecting magnet placed at a smaller radius in a straight section; this magnet bends the beam into a trajectory which, at some distance downstream from the magnet, emerges from the periphery of the machine. The Cosmotron has three such external proton beams available.   It is characteristic for these large accelerators to have a multiplicity of complex configurations of experimental equipment (deflecting, analyzing, and focusing magnets, bubble chambers, spark chambers, counter telescopes) simultaneously set up in the experimental areas outside the main accelerator shielding.   The proton synchrotrons themselves and the external-beam facilities require very bulky shielding (L2) because the primary proton beams as well as some of the secondary particles are so penetrating.

Alternating-Gradient Synchrotron.   The weight and cost of the magnet for a synchrotron of a given orbit radius are largely determined by the size of the aperture in which the magnetic field has to be supplied.   This in turn depends on the radial and vertical oscillations of the particles around their equilibrium orbit.   These oscillations, caused by spread in angle and in energy at injection, by gas scattering at low energies, by inhomogeneities in the magnetic field, and by inadequate rf control, must be kept small enough in relation to the size of the vacuum chamber to avoid excessive loss of particles from the beam by collisions with the walls.   For economy it is clearly desirable to reduce the amplitudes of these oscillations and thereby reduce the required aperture. As we have mentioned before (p. 371), for a given angular deviation from the orbit the vertical amplitude is proportional to $n^{-1/2}$, the radial amplitude proportional to $(1 - n)^{-1/2}$.   Thus in a synchrotron the vertical dimension of the aperture can be reduced only at the expense of the radial dimension and vice versa.

From the foregoing discussion it may be seen that the amplitudes of vertical oscillations could be made very small in a magnet section of large positive $n(\gg 1)$, and the radial oscillations could be similarly reduced in a magnet section of large negative $n(\ll -1)$.   That a ring magnet design with alternate sections of large positive and large negative n values leads to strong focusing and therefore small aperture requirements, while still preserving phase stability for synchrotron acceleration, was first pointed out by N. Christophilos in 1950 and independently discovered by E. D. Courant, M. S. Livingston, and H. S. Snyder in 1952 (C1).   One undesirable consequence of the large n values is the existence of many resonances that can lead to beam loss.   The range of operating conditions under which stable operation is possible is therefore rather small, and very careful control of a number of parameters is necessary (B3, L1).

Alternating-gradient synchrotrons for electron acceleration are in operation at Cornell University, Ithaca, New York (1.3 GeV), at Cambridge, Massachusetts (6 GeV), at Lund, Sweden (1.2 GeV), at Bonn, Germany (0.5 GeV), and at Tokyo, Japan (1.3 GeV); additional 6-GeV accelerators are nearing completion at Hamburg, Germany and Yerevan, USSR.   The two large alternating-gradient proton synchrotrons at Geneva, Switzerland (CERN) and at Brookhaven have reached energies of 28 and 33 GeV, respectively. In spite of their almost incredible complexity, they have already proved to be extremely useful, reliable, and versatile research tools.   In the USSR a 7-GeV proton synchrotron is in operation, and a 70-GeV machine is under construction.

The Brookhaven AGS has an orbit radius of about 128 meters[3] and consists of 240 magnet sections weighing a total of about 4500 tons (about the same as the Berkeley 184-in. cyclotron!).   The magnets are arranged in pairs, two

---

[3] The radius of curvature in the magnet sections is 85 meters, but the many straight sections interposed between magnets increase the effective orbit radius to the value quoted.

with positive field index ($n = +357$), then two with negative field index ($n = -357$), etc. The magnet cross sections are shown in figure 11-12; a view of part of the magnet ring can be seen in figure 11-13. The aperture is 7 x 15 cm, and the vacuum chamber is made of Inconel. A pressure of about 10^{-6} mm Hg is maintained by 48 pumping stations. The protons gain about 100 keV per revolution, supplied at 12 accelerating stations. The total time for acceleration to 33 GeV is about 1 second, and the repetition rate is one pulse per 3 seconds at full energy, one pulse per 2.4 seconds at 30 GeV, and even higher at lower energies. The protons are injected from a 50-MeV linac which itself uses strong-focusing lenses. The beam intensity of the AGS is at present limited to about 5×10^{11} protons per pulse by the output of the linac. Targeting problems are similar to those in constant-gradient synchrotrons. But the strong focusing causes what is known as momentum compaction: a great reduction in the radius change caused by a given momentum change; this leads to large numbers of multiple traversals in targets. In fact, a light-element target tends to be traversed by the beam until a large fraction ($\geq \frac{1}{2}$) of the protons have made nuclear interactions.

At the present time there is no indication of technical barriers to the energy attainable with accelerators. The limitations may become largely economic. The alternating-gradient synchrotron principle can undoubtedly be used for acceleration of protons to 100–1000 GeV, and design studies for machines in that range have already been made. One of the problems encountered with these ever-increasing beam energies is that not nearly all of the energy of a

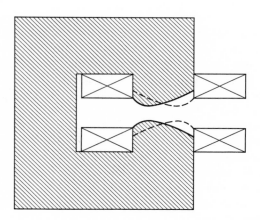

Fig. 11-12 Typical cross section for a magnet of an alternating-gradient synchrotron. Magnets with poles shown by the solid lines are alternated with magnets whose poles are shaped as shown by the dashed lines. The boxes with crosses indicate the positions of the magnet coils. In the Brookhaven AGS the external dimensions of the steel laminations are 33 in. wide by 39 in. high; the pole width is 12.5 in. and the gap height at the central orbit position is 3.5 in. (From reference B3.)

Fig. 11-13   View of a part of the Brookhaven Alternating-Gradient Synchrotron.   The vacuum chamber is visible in the pole gap.   (Courtesy Brookhaven National Laboratory.)

particle hitting a stationary target can be used for particle production or other nuclear processes; conservation of momentum requires that a certain fraction of the incident energy go to kinetic energy of the products, and this fraction increases rapidly with increasing incident energy, more so as the velocities approach the velocity of light.   For example, in the collision of a 33-GeV proton with a stationary proton, only 6.2 GeV are available in the center-of-mass system; if the incident energy in the laboratory system is raised to 500 GeV, the energy available in the center-of-mass system goes up only to 29 GeV![4]   These effects have led to serious consideration of various proposals for "colliding-beam" accelerators in which two proton beams would be made to collide head-on so that no energy would be lost to center-of-mass motion. The disadvantage of any such scheme is the relatively small number of interactions that can be achieved.

---

[4] The formula for computing E_c, the energy available in the center-of-mass system, from the *total* energy W of a particle of rest mass m_0 striking another particle of equal rest mass is

$$\frac{E_c}{m_0 c^2} = \left[2 \left(\frac{W}{m_0 c^2} + 1 \right) \right]^{\frac{1}{2}} - 2.$$

B. GAMMA-RAY AND X-RAY SOURCES

Radioactive Sources.   Since nucleons are bound in most nuclei with binding energies of 6 to 8 MeV, photons having energies less than 6 MeV cannot be expected to induce many nuclear reactions.   No γ rays emitted in radioactive processes (except from a few short-lived, low-Z nuclides such as N^{16}) have energies as high as that; nor do the X rays from conventional X-ray tubes. The only γ-ray- or X-ray-induced[5] nuclear reactions produced with such sources are therefore excitations of nuclei to isomeric levels and the "photo-disintegrations" of the deuteron (threshold 2.23 MeV) and of Be^9 (threshold 1.67 MeV).   Some of the radioactive γ-ray sources that have been used (W1) are listed in table 11-1.

Table 11-1   *Some typical radioactive gamma-ray sources*

Source	Half-life	Energy (MeV)	Number of Quanta per Disintegration
ThC″	3.1m	2.61	1
Na²⁴	15.0h	2.75	1
		1.37	1
Y⁸⁸	108d	0.90	1
		1.84	1

Nuclear Reactions as Sources.   In a number of nuclear reactions, especially with the lightest elements (which have large nuclear level spacings), very energetic γ rays are emitted.   These, in turn, have been used to induce other nuclear reactions.   Such sources are of particular value if the γ rays are monoenergetic.   The most important sources of this type are listed in table 11-2. With these reactions relatively high intensities of γ rays (of the order of 10^6 quanta per second) can be obtained from rather moderate high-voltage sets operating at 500 to 1000 keV.

The $H^3(p, \gamma)$ reaction can be used to obtain monoenergetic γ rays over a limited energy range by variation of the proton energy up to a few MeV.

Bremsstrahlung.   The continuous X rays produced when electrons are decelerated in the Coulomb fields of atomic nuclei are called bremsstrahlung (German for "slowing-down radiation").   This type of radiation is produced whenever fast electrons pass through matter, and the efficiency of the con-

---

[5] For convenience we shall speak of γ-ray-induced reactions even when the electromagnetic radiation used is not of nuclear origin but is produced by the deceleration of electrons in a target in the manner of continuous X rays.

Table 11-2   High-energy gamma-rays from nuclear reactions

Reaction	γ-Ray Energy (MeV)	Remarks
$Li^7(p, \gamma)Be^8$	17.6, 14.8	Resonance at 440 keV proton energy. Other components, at $\sim$15 MeV and at lower energies, also found.
$B^{11}(p, \gamma)C^{12}$	11.8, 16.6	11.8-MeV γ has 7 times intensity of 16.6-MeV γ; low-energy component ($\sim$4 MeV) also found.
$H^3(p, \gamma)He^4$	$19.8 + 0.75E_p$	Monoenergetic.

version of kinetic energy into bremsstrahlung goes up with increasing electron energy and with increasing atomic number of the material. In tungsten, for example, a 10-MeV electron loses about 50 per cent of its energy by radiation, whereas a 100-MeV electron loses more than 90 per cent of its energy by that mechanism (see chapter 4, section B).

The spectrum of bremsstrahlung from a monoenergetic electron source extends from the electron energy down to zero, with approximately equal amounts of energy in equal energy intervals; in other words, the number of quanta in a narrow energy interval is about inversely proportional to the mean energy of the interval.

The stopping of fast electrons in matter thus produces a continuous spectrum of X rays, and any electron accelerator also serves as an X-ray source. Van de Graaff machines, betatrons, and synchrotrons have all been used as sources of X rays for producing nuclear reactions. In fact, unless special devices are used to bring the electron beam out of betatron or synchrotron doughnuts, the X rays are the only radiation available outside the vacuum systems of these machines. The higher the energy of an electron producing bremsstrahlung, the more the X-ray emission is concentrated in the forward direction; with a 100-MeV betatron, for example, about half the intensity of the X-ray beam is contained in a 2° cone. The chief disadvantage of bremsstrahlung sources for nuclear work is their spectral distribution. However, they are capable of producing electromagnetic radiation in energy and intensity ranges not accessible by other means.

Two schemes have been used to obtain fairly monochromatic photon beams from bremsstrahlung sources. In one the electrons produce bremsstrahlung in a thin target, and the degraded electrons are then analyzed according to their momenta. The bremsstrahlung photons in coincidence with electrons in a particular energy range can thus be selected by a coincidence requirement for the pulses from the electron and photon detectors. This type of monochromator is useful only for experiments that can be gated by a coincidence pulse (but not, for example, for an activation experiment). In the other scheme (which does not suffer from this limitation) the primary electrons interact in a thick, high-Z target, producing there not only bremsstrahlung but also an appreciable intensity of electron-positron pairs. The positrons are momentum-analyzed in an electron spectrograph of high transmission, and those in a given energy band are allowed to strike a thin low-Z target. Again there will be bremsstrahlung (minimized by the low Z), but there will also be annihilation in flight, with one of the photons carrying most of the positron energy and going in the forward direction. With this method, intensities of the order of 10^6 photons per second of rather monoenergetic γ rays (half-width $\cong 250$ keV) of 10–25 MeV have been obtained.

C. NEUTRON SOURCES

Radioactive Sources. Our only sources of neutrons are nuclear reactions. There are several naturally occurring and several artificially produced α and γ emitters which can be combined with a suitable light element to make useful neutron sources (O1). Because of the short ranges of the α particles, α emitters must be intimately mixed with the light element (usually beryllium because it gives the highest yield). Such sources necessarily give neutrons with energies spread over a wide range. A γ emitter may be enclosed in a capsule surrounded by a beryllium or deuterium oxide target. (Only beryllium and deuterium have γ, n thresholds below 5 MeV.) Some of these sources can, in principle, give monoenergetic neutrons, but because of neutron and γ-ray scattering in targets of practical thickness the actual spectrum usually has an energy spread of about 30 per cent, with the average energy roughly 20 per cent below the expected maximum value. Some useful sources are listed in table 11-3.

Since spontaneous fission, like any fission process, is always accompanied by neutron emission, a sample of a nuclide which undergoes spontaneous fission can serve as a neutron source. At present by far the most practical nuclide for this purpose is Cf^{252}, which has a half-life of 2.55 years, decays about 97 per cent by α emission, 3.4 per cent by spontaneous fission, and emits 3.5 neutrons per fission. Thus Cf^{252} sources emit approximately 2.4×10^6 neutrons / μg sec, along with roughly 10 times that many α particles. In the future the use of other artificially produced spontaneous-fission neutron sources such as Cf^{254} may become practical.

Neutron-Producing Reactions with Accelerators. Much more copious sources of neutrons than can be obtained with radioactive α and γ emitters are available with ion accelerators. The reaction $H^2(d, n)He^3$ (often called a D, D reaction) is exoergic ($Q = +3.27$ MeV), and, because the potential barrier is low, good neutron yields can be obtained with deuteron energies as low as

Table 11-3   Alpha- and gamma-ray neutron sources

Source	Main Reaction	Q (MeV)	Neutron Energy (MeV)	Neutron Yield (per 10^6 disintegrations)
Ra + Be (mixed)	$\mathrm{Be}^9(\alpha, n)\mathrm{C}^{12}$	5.65	up to 13	460
Po + Be (mixed)	$\mathrm{Be}^9(\alpha, n)\mathrm{C}^{12}$	5.65	up to 11, av. 4	80
Ra + B (mixed)	$\mathrm{B}^{11}(\alpha, n)\mathrm{N}^{14}$	0.28	up to 6	180
RaBeF_4 (pure)	$\left\{\begin{matrix}\mathrm{Be}^9(\alpha, n)\mathrm{C}^{12}\\ \mathrm{F}^{19}(\alpha, n)\mathrm{Na}^{22}\end{matrix}\right\}$	5.65	up to 13	68.5
Pu^{239} + Be (mixed)	$\mathrm{Be}^9(\alpha, n)\mathrm{C}^{12}$	5.65	up to 11	60
Ra + Be (separated)	$\mathrm{Be}^9(\gamma, n)\mathrm{Be}^{8*}$	−1.67	<0.6	0.9†
Ra + $\mathrm{D}_2\mathrm{O}$ (separated)	$\mathrm{H}^2(\gamma, n)\mathrm{H}^1$	−2.23	0.1	0.03†
Na^{24} + Be	$\mathrm{Be}^9(\gamma, n)\mathrm{Be}^{8*}$	−1.67	0.8	3.8†
Na^{24} + $\mathrm{D}_2\mathrm{O}$	$\mathrm{H}^2(\gamma, n)\mathrm{H}^1$	−2.23	0.2	7.8†
Y^{88} + Be	$\mathrm{Be}^9(\gamma, n)\mathrm{Be}^{8*}$	−1.67	0.16	2.7†
Y^{88} + $\mathrm{D}_2\mathrm{O}$	$\mathrm{H}^2(\gamma, n)\mathrm{H}^1$	−2.23	0.3	0.08†
Sb^{124} + Be	$\mathrm{Be}^9(\gamma, n)\mathrm{Be}^{8*}$	−1.67	0.02	5.1†
La^{140} + Be	$\mathrm{Be}^9(\gamma, n)\mathrm{Be}^{8*}$	−1.67	0.6	0.06†
La^{140} + $\mathrm{D}_2\mathrm{O}$	$\mathrm{H}^2(\gamma, n)\mathrm{H}^1$	−2.23	0.15	0.2†
MsTh + Be	$\mathrm{Be}^9(\gamma, n)\mathrm{Be}^{8*}$	−1.67	0.8	0.9†
MsTh + $\mathrm{D}_2\mathrm{O}$	$\mathrm{H}^2(\gamma, n)\mathrm{H}^1$	−2.23	0.2	2.6†

* The product Be^8 is unstable and decomposes in less than 10^{-14} sec into two He^4 nuclei.
† The photoneutron yields are given for 1 g of target ($\mathrm{D}_2\mathrm{O}$ or Be) at 1 cm from the γ-ray source.

100 to 200 keV.   With thick targets of solid $\mathrm{D}_2\mathrm{O}$, the yields are about 0.7, 3, and 80 neutrons per 10^7 deuterons at 100 keV, 200 keV, and 1 MeV deuteron energy, respectively.   High-voltage sets and electrostatic generators are often used to produce the D, D reaction.   The neutrons are monoenergetic if mono-energetic deuterons of moderate energies (up to a few MeV) fall on a sufficiently thin target.

A very interesting reaction for neutron production is $\mathrm{H}^3(d, n)\mathrm{He}^4$.   The hydrogen isotope H^3 (tritium) is radioactive (half-life about 12 years) but has become available in useful quantities.   For use as a target it is usually adsorbed on zirconium or titanium.   Such targets have been reported to give about 150 neutrons per 10^7 deuterons at 200 keV.   The reaction has a strong resonance at 100 keV deuteron energy and can be a remarkable source of neutrons from very-low-energy deuterons.   The reaction is exoergic with $Q = 17.6$ MeV, and monoenergetic neutrons of about 14 MeV are produced from a thin target.

For a controlled source of monoenergetic neutrons of very low energy (down to about 30 keV) the $\mathrm{Li}^7(p, n)\mathrm{Be}^7$ reaction is suitable, especially when produced with the protons of well-defined energy available from electrostatic generators. The reaction is endoergic ($Q = -1.646$ MeV) and has a threshold of 1.88 MeV. Advantage may be taken of the differences in neutron energy in the forward and backward (and intermediate) directions.

With X rays from electrostatic generators, betatrons, and the like, neutrons can be produced by means of the $Be^9(\gamma, n)$ or $H^2(\gamma, n)$ reactions. The yields of these reactions go up quite sharply with energy. With an electrostatic generator operating at 2.5 MeV with 100 μA electron current the neutron yield per gram of beryllium is equivalent to that from about 4 g of radium mixed with beryllium; at 3.2-MeV energy the corresponding figure is 26 g of radium.

When high-energy deuterons are available, the most prolific neutron source is obtained by bombarding beryllium with deuterons. The reaction $Be^9(d, n)B^{10}$ has a positive Q value of 3.79 MeV, but the neutrons are far from monoenergetic. When a beryllium target is bombarded with deuterons of E MeV energy, neutrons with a distribution of energies up to about $(E + 3.5)$ MeV are emitted. The neutron yield goes up rapidly with deuteron energy; it is about 10^8 neutrons per second per microampere for 1-MeV deuterons, about 10^{10} neutrons per second per microampere for 8-MeV deuterons, and about 3×10^{10} neutrons per second per microampere for 14-MeV deuterons.

With a given deuteron source, a lithium target gives the highest neutron energies[6] because the $Li^7(d, n)$ reaction is exoergic by 15.0 MeV. The neutron yield is only about one third that of the $Be^9(d, n)B^{10}$ reaction. Neutrons are also obtained in the bombardment of almost any element with fast protons, deuterons, or α particles. The yields and energies vary from reaction to reaction, but if a neutron bombardment is needed for the activation of some substance it is often sufficient to place the sample near a cyclotron target that is being bombarded by deuterons, even if the target is not beryllium or lithium.

In the bombardment of targets with deuterons of much higher energy (for example the 460-MeV deuterons of the 184-in. synchrocyclotron in Berkeley) high-energy neutrons are emitted in a rather narrow cone in the forward direction as a result of the deuteron stripping mechanism discussed in chapter 10, section H. The energy distribution of these neutrons is approximately Gaussian, with the maximum at half the deuteron energy. The highest-energy machines in operation today accelerate protons to billions or tens of billions of electron volts. Such protons in striking target nuclei presumably will knock forward by essentially elastic collisons neutrons of about the same maximum energy. They can also be turned into neutrons by charge exchange in passing through a target. A 2.2-GeV proton beam of the Brookhaven Cosmotron was used to produce a neutron beam of average energy about 1.4 GeV.

Neutrons from all nuclear reactions initially are fast neutrons. Their slowing down and some properties of thermal neutrons have already been discussed in chapter 4, section D.

---

[6] An exception is the tritium target which gives slightly more energetic neutrons.

Nuclear Chain Reactors.   By far the most prolific souces of neutrons known are the nuclear chain reactors.   A nuclear reactor is an assembly of fissionable material (such as uranium, enriched U^{235}, Pu^{239}, or U^{233}) arranged in such a way that a self-sustaining chain reaction is maintained.   In each fission process some neutrons are emitted (for U^{235} the average number is 2.44, for Pu^{239} it is 2.89).   The requirement common to all reactors is that at least one of these neutrons must be available to produce another fission instead of escaping from the assembly or being used up in some other nuclear reaction.   Therefore, for a given type of reactor there is a minimum (or critical) size below which the chain reaction cannot be self-sustaining.   Actual reactors are always built with sufficient excess "reactivity" to make large amounts of neutrons available for other purposes.   We defer discussion of many aspects of nuclear reactors to chapter 14 and are here concerned only with their function as neutron sources (K1, C2).

Even in this more limited context, that is, excluding reactors intended primarily for production of power or fissionable materials, the variety of reactor designs is bewildering.   A 1962 catalogue of research and testing reactors (R1) listed well over 200 such devices in operation.   Almost all of them use uranium, either of natural composition or enriched in U^{235}, as fuel; but this may be in the form of plates, rods, bars, etc., of the metal or some alloy, in the form of uranium oxide pellets, or in the form of some salt in solution.   A few Pu^{239}-fueled reactors are in operation.   Since the great majority of research reactors use thermal neutrons for the propagation of the chain reaction, they contain moderators, that is, materials whose function it is to slow the fast neutrons emitted in fission (average energy $\cong 2.5$ MeV, most probable energy $\cong 0.6$ MeV) to thermal energies.   Low mass number and low neutron-absorption cross section are the desirable properties of moderators, and ordinary water, heavy water, graphite, and beryllium are the materials of choice.   The enormous heat produced by the fission chain reaction must be carried away, and all but the smallest reactors therefore must have efficient cooling systems.   Again, there is a wide choice of coolants—water, heavy water, air, carbon dioxide, and liquid metals.   Designs in which the same substance serves as coolant and moderator are often economical.

The maximum thermal-neutron fluxes available in research reactors range from about 10^{11} to about 10^{15} neutrons cm^{-2} sec^{-1} (with a few beyond either end of this range).   Some reactors such as the TRIGA reactors built by the General Atomic Division of General Dynamics are designed to be pulsed to give brief bursts with fluxes up to several times 10^{16} cm^{-2} sec^{-1}, although at steady operation they produce only about 10^{12} cm^{-2} sec^{-1}.   To put reactor fluxes somewhat in perspective, in table 11-4 we compare some typical examples with neutron fluxes from other sources.   It is also useful to recall the relation between neutron fluxes and energy production.   About 190 MeV per fission (all the energy released except for ~ 11 MeV carried off by neutrinos) is normally converted to heat in a reactor.   This corresponds to the produc-

Table 11-4   Thermal neutron fluxes

Source	Conditions	Thermal Flux $(\mathrm{cm}^{-2}\ \mathrm{sec}^{-1})$
1 g Ra mixed with Be	immersed in a large volume of water or paraffin; flux measured 4 cm from source	1×10^5
Po + Be, 3.7×10^{10} α particles sec^{-1}	immersed in large volume of water or paraffin; flux estimated 4 cm from source	2×10^4
Van de Graaff, 10 μA of 1-MeV deuterons on Be	target surrounded with paraffin; flux estimated in paraffin near target	1×10^7
Cyclotron, 100 μA of 8-MeV deuterons on Be	target backed up with large paraffin block; flux estimated in paraffin near target	1×10^9
Cyclotron, 100 μA of 14-MeV deuterons on Be	target backed up with large paraffin block; flux estimated in paraffin near target	3×10^9
Nuclear reactor, X-10 Oak Ridge, graphite-U	{in center of core {in hole in shield	2×10^{12} 1×10^7
Nuclear reactor, MTR, Arco, Idaho, $\mathrm{H_2O\text{-}U}^{235}$	in reflector (maximum)	7×10^{14}
Nuclear reactor, NRX, Chalk River, $\mathrm{D^2O\text{-}U}$	{in center of core {in hole in shield {in thermal column	6×10^{13} 4×10^7 1×10^9
Typical swimming pool reactor		2×10^{13}

tion of 1 MW of heat for 3×10^{16} fissions per second, which, for U^{235} fission, are associated with the emission of a total of about 7×10^{16} neutrons per second.   The neutron *flux* obtainable at a given power level depends, of course, on the size of the reactor core; in practice, the limiting problem in achieving the highest fluxes is the design of cooling systems adequate for the enormous specific powers involved.   Also the burn-up of fuel eventually becomes so high that replacement of fuel elements is required too frequently.   Even with the best reactor design the limit of tolerable fuel burn-up is about 30 to 40 per cent; beyond that the poisoning by neutron-absorbing fission products and the structural changes associated with replacement of uranium by fission-product elements become excessive.

Two general types of facilities for neutron use are desirable: (1) means (including pneumatic tubes) for the irradiation of samples in high-flux regions, (2) tubes and channels for bringing neutron beams outside the reactor for such purposes as structure studies by neutron diffraction, capture-γ-ray spectroscopy, or neutron cross section measurements.   Among the various reactor types, the large (and expensive) graphite-moderated, air-cooled uranium reactors (such as X-10 at Oak Ridge, BEPO at Harwell, BGRR at Brookhaven) have moderate fluxes (up to $\sim 10^{13}$ cm^{-2} sec^{-1}) but are particularly suitable for large numbers of simultaneous experiments because they can accommodate many beam pipes and irradiation facilities.   For smaller institutions the pool-type reactors using enriched U^{235} (20–93 per cent) as fuel and ordinary water as moderator, coolant, reflector, and shield are much more economical and suitable.   These popular research reactors give fluxes of about 10^{13} neutrons cm^{-2} sec^{-1} per megawatt and can be flexibly used for all types of irradiation studies but do not lend themselves well to the installation of beam tubes. Higher fluxes can be achieved if the reactor core, instead of being suspended in the bottom of an open pool, is enclosed in a sealed tank, with H_2O or D_2O under pressure serving as moderator and coolant.   Among the many reactors of this tank type are MTR (40 MW, H_2O, 5×10^{14} cm^{-2} sec^{-1}), ETR (175 MW, H_2O, 4×10^{14} cm^{-2} sec^{-1}), both at the National Reactor Test Station, Idaho; Pluto at Harwell, England (10 MW, D_2O, 2×10^{14} cm^{-2} sec^{-1}); BR2 at Mol, Belgium (50 MW, H_2O, 8×10^{14} cm^{-2}sec^{-1}).[7]

In most reactors the flux falls off from the center toward the outside, although some, such as MTR and the high-flux beam reactor at Brookhaven, are designed to have the maximum thermal fluxes in the reflector outside the core.   The neutron energy distribution can vary widely in different reactor types and also in different locations in a given reactor.   To provide pure thermal-neutron sources, so-called thermal columns are often attached to reactors.   A thermal column is a column of graphite (or some other moderator) of sufficient length to ensure a thermal-energy distribution for the neutrons which have passed through it.   The neutron flux at the end of a thermal column is, of course, several orders of magnitude smaller than that available inside the associated reactor.   Especially large ratios of fast-neutron to slow-neutron fluxes can be obtained inside uranium-walled containers placed in a reactor.

Neutron Monochromators.   A number of means have been devised for the conduct of experiments with neutrons selected to have a particular energy. One of these, the crystal spectrometer, is analogous to an optical-grating monochromator.   A thermal neutron with a velocity of 2.2×10^5 cm sec^{-1} (the most probable velocity at 20°C) has a wavelength $\lambda = h/mv = 1.8 \times 10^{-8}$ cm. This length is in the range of common X-ray wavelengths (1.54×10^{-8} cm for copper K_α radiation), and the spacing between crystal planes is about the proper "grating" spacing for slow-neutron diffraction as it is for X-ray diffraction.   Neutrons of considerably higher energy (i.e., shorter wavelength) may be diffracted successfully by the crystal at grazing incidence angles.   With an

---

[7] Reactors are usually designated by sets of initials.   The ones mentioned in this paragraph have the following meanings: BEPO is British Experimental Pile Zero, BGRR Brookhaven Graphite Research Reactor, MTR Materials Testing Reactor, ETR Engineering Test Reactor, BR2 Belgian Reactor 2.

intense source of slow neutrons available, such as a nuclear reactor, the crystal and slit system may be arranged to select neutrons from the spectrum with good resolution from about 0.02 to about 10 eV.

The other common means for selecting monoenergetic neutrons from a spectrum depend on control of the time of flight of the neutrons over a measured course.   A burst of neutrons, containing all energies in the spectrum of the source, may be selected mechanically by interposing in the beam a rotating "chopper."   Timing devices activate the detector circuits at a chosen time in each "chopping" cycle.   Thus neutrons which traverse the distance between chopper and detector in the chosen time lag and produce almost instantaneous response in the detector may be studied.   (Reactions involving only a delayed detector response, such as the production of radioactivity of conventional half-life periods, cannot be investigated by the time-of-flight technique.)   A neutron of energy $E = 0.025$ eV has a velocity $v = 1.38 \times 10^6 E^{1/2} = 2.2 \times 10^5$ cm sec^{-1} and so traverses a distance of 10 meters in 4.55×10^{-3} second. Monochromators with burst times and response times of a few microseconds give very good energy resolution for neutrons of this energy and are useful from about 0.001 to perhaps 10,000 eV when operated in conjunction with nuclear reactors.

If the source of neutrons is an accelerator, such as a cyclotron, the bursts of neutrons for time-of-flight studies can be generated directly by suitable modulation of the accelerator ion beam.   The modulated-cyclotron velocity selector at Columbia University is effective from 0.001 to about 1000 eV.   The pulsed electron beams from electron accelerators are well suited for this application; if the electron energy is about 15 MeV, the bremsstrahlung can give by γ, n reaction more than 100 neutrons per 10^6 electrons.   The Harwell electron-accelerator monochromator has a wide useful range, from 0.001 to several thousand eV.

An interesting possibility for future development is the use of precision neutron spectroscopy, in a way analogous to optical absorption spectroscopy, for qualitative and quantitative chemical analysis.   Elementary analysis for many elements may be possible without destruction of the sample.

REFERENCES

*B1   M. H. Blewett, "Low-Energy Sources" in *Methods of Experimental Physics*, Vol. 5B, *Nuclear Physics* (L. C. L. Yuan and C. S. Wu, Editors), Academic, New York, 1963, pp. 580–584.

*B2   M. H. Blewett, "The Electrostatic (Van-de-Graaff) Generator" in *Methods of Experimental Physics*, Vol. 5B, *Nuclear Physics* (L. C. L. Yuan and C. S. Wu, Editors), Academic, New York, 1963, pp. 584–590.

*B3   M. H. Blewett, "Medium- and High-Energy Sources" in *Methods of Experimental Physics*, Vol. 5B, *Nuclear Physics* (L. C. L. Yuan and C. S. Wu, Editors), Academic, New York, 1963, pp. 623–689.

C1   E. D. Courant, M. S. Livingston, and H. S. Snyder, "The Strong-Focusing Synchrotron—A New High Energy Accelerator," *Phys. Rev.* **88,** 1190 (1952).

C2   T. E. Cole and A. M. Weinberg, "Technology of Research Reactors," *Ann. Rev. Nuclear Sci.* **12,** 221–242 (1962).

*F1   B. T. Feld, "The Neutron" in *Experimental Nuclear Physics* (E. Segrè, Editor), Vol. II, Wiley, New York, 1953, pp. 208–586.

*K1   H. Kouts, "Nuclear Reactors" in *Methods of Experimental Physics*, Vol. 5B, *Nuclear Physics* (L. C. L. Yuan and C. S. Wu, Editors) Academic, New York, 1963, pp. 590–622.

*L1   M. S. Livingston and J. P. Blewett, *Particle Accelerators*, McGraw-Hill, New York, 1962.

L2   S. J. Lindenbaum, "Shielding of High-Energy Accelerators," *Ann. Rev. Nuclear Sci.* **11,** 213–258 (1961).

*M1   E. M. McMillan, "Particle Accelerators" in *Experimental Nuclear Physics* (E. Segrè, Editor), Vol. III, Wiley, New York, 1959, pp. 639–785.

O1   G. D. O'Kelley, "Radioactive Sources," in *Methods of Experimental Physics,* Vol. 5B, *Nuclear Physics* (L. C. L Yuan and C. S. Wu, Editors) Academic, New York, 1963, pp. 555–580.

P1   O. Piccioni, D. Clark, R. Cool, G. Friedlander, and D. Kassner, "External Proton Beam of the Cosmotron," *Rev. Sci. Instr.* **26,** 232 (1955).

R1   "Research and Testing Reactors the World Around," *Nucleonics* **20,** No. 8, 116 (August 1962).

T1   L. H. Thomas, "The Paths of Ions in the Cyclotron," *Phys. Rev.* **54,** 580, 588 (1938).

W1   R. West, "Low-Energy Gamma-Ray Sources," *Nucleonics* **11,** No. 2, 20 (February 1953).

EXERCISES

1. A standard cyclotron of 120-cm pole diameter is operated with a 10-Mc oscillator (a) What magnetic field is required for the acceleration of deuterons? (b) What will be the final deuteron energy? (c) With the same rf frequency, what is the maximum kinetic energy to which He^3 ions could be accelerated in this cyclotron? (d) Under the assumption that the magnetic field calculated in (a) is the maximum available, what oscillator frequency would be required to obtain the highest possible H^3 energy and what is that energy?
    Answers: (b) 14.8 MeV; (d) 6.67 Mc, 9.9 MeV.

2. A linac for the acceleration of protons to 45.3 MeV is designed so that, between any pair of accelerating gaps, the protons spend one complete rf cycle inside a drift tube (field-free region). The rf frequency used is 200 Mc. (a) What is the length of the final drift tube? (b) If the first drift tube is 5.35 cm long, at what kinetic energy are the protons injected into the linac? *Answer:* (b) 0.60 MeV.

3. Estimate (a) the percentage frequency modulation and (b) the pole diameter required for an FM cyclotron designed to accelerate protons to 350 MeV. Assume $H = 16,000$ gauss.

4. The H^+ and H_2^+ ions accelerated in a Van de Graaff generator to 5 MeV are to be magnetically separated from one another. Approximately over what distance must a 10,000-gauss field be applied if the two beams are to diverge by 20°?
    Answer: $\cong$ 18 in.

5. A synchrotron is to be designed to accelerate protons to 12 GeV kinetic energy. (a) Assuming a maximum field strength of 14,300 gauss, estimate the radius of curvature of the proton orbit. (b) If about 25 per cent of the protons' path is

spent in field-free straight sections, what is the final revolution frequency? (c) Assuming one revolution per rf cycle, at what kinetic energy must the protons be injected if the frequency over the entire acceleration cycle is to vary by a factor of 5? What type of device would you suggest for the injector?

Answers: (a) 30 meters; (b) 1.2 Mc.

6. In the synchrotron of the preceding problem the protons receive an energy increment of 7.5 keV per revolution. Approximately how long does it take to accelerate them from injection to full energy? What is the total path length?

Answer: $\cong$ 1.4 seconds.

7. A Cockcroft-Walton accelerator can be used to accelerate either deuterons or tritons to 500 keV. What is the maximum neutron energy attainable with (a) deuterons incident on a tritium target, (b) tritons incident on a deuterium target?

8. What would be the minimum n, γ cross section detectable by means of the product activity in a sample of 10 cm^2 area containing 1 mg-equivalent of target isotope, with a mixed Ra–Be source containing 1 g radium? Assume that the bombardment is continued to saturation and that 1 per cent of the neutrons emitted by the source strike each square centimeter of the target sample as slow neutrons. Consider 30 disintegrations per minute as the minimum detectable activity.

9. Suppose you want to prepare some 5.2-year Co^{60} with a cyclotron and have the choice of bombarding a cobalt sample directly with 14-MeV deuterons for 2 hours or of surrounding it with paraffin and placing it near a beryllium target bombarded with 14-MeV deuterons for a total of 20 hours. Which is more advantageous from the point of view of total activity obtained? Use data in appendix C and make reasonable assumptions about the solid angle subtended by the neutron-irradiated sample.

10. (a) What would be the approximate neutron energy from a beryllium target bombarded with Ni^{57} γ rays? (b) What would be the energy difference between neutrons in the forward and in the reverse directions?

Answer: (a) 0.21 MeV; (b) 9.1 keV.

11. If the fuel loading of a research reactor consists of 2 kg of uranium enriched to 90 per cent U^{235}, how long can the reactor be operated without reloading at a power level of 1 MW before burnup reaches 20 per cent?

12. In a neutron "chopper," a neutron-absorbing rotor with a narrow slit near its periphery rotates in front of a stationary collimator with a similar slit to produce short bursts of neutrons. The rotor is 20 cm in diameter and rotates at 15,000 rpm; the slits are 0.03 cm wide and 1 cm long. (a) What is the burst length? (b) With the neutron detector actuated for a time equal to the burst length, what flight path is required to achieve ± 10 per cent energy resolution for 5-keV neutrons?

Answers: (a) 4 μs; (b) $\cong$ 80 meters.

Techniques in Nuclear Chemistry

A. SOME GENERAL PRACTICES

The principal types of instruments used for detection and measurement of radiations from radioactive substances were discussed in chapter 5. Here we wish to review some of the techniques employed in the course of measurements of this kind. The choice of instruments and techniques will be determined in large part by the kinds of information desired. In a simple tracer application, employing one active isotope of favorable properties and available with ample activity and known purity, a single instrument (GM, proportional, or scintillation counter) is likely to be sufficient, and measuring techniques may offer no problems. The opposite extreme would be a large laboratory of nuclear chemistry devoted to the detailed study of the radiation characteristics of a variety of radionuclides, to the identification of new radioactive species, and to quantitative investigations of nuclear reactions produced in nuclear chain reactors and accelerators. Here a large number and wide variety of instruments, including highly specialized types, must be available, and the associated techniques and manipulations will be complex and ingenious. Most radiochemical laboratories represent intermediate situations. Even where tracer work only is carried out, a number of tracers are likely to be in use, calling for different choices of detectors and sample handling procedures. Frequently the desired radioactivity must be isolated and identified and checked for purity with regard to contamination by other radioactive substances. The typical laboratory may have thin-window proportional (or GM) counters for most β counting, sodium iodide scintillation counters for γ counting, semiconductor detectors, and perhaps a gas-flow proportional counter arranged for introduction of α or soft-β emitters into the counter. The associated equipment will certainly include amplifiers and scalers, very likely some forms of single-channel or multichannel pulse height analyzers, and possibly other types of spectrometers as well as coincidence circuits.

It is very convenient to have a standard arrangement for holding standard-size samples in various positions, the same arrangement being used for as many of the instruments as possible. A typical sample-holder arrangement, that used in the Chemistry Department of Brookhaven National Laboratory, is shown in figure 12-1.   This standard arrangement is fitted to all instruments (proportional and scintillation counters) which are used to measure externally placed solid samples.   The entire stand is made of lucite to minimize scattering of β and γ rays, and a lucite shelf (either solid or with a central hole) may be placed in any of the milled slots to hold samples at various distances from the detector.   The samples are mounted on cardboard or aluminum cards, $2.5 \times 3 \times \frac{1}{32}$ in., which slide onto the shelves.   The back wall of the stand serves as a stop and thus determines the position.   The machining of the stands is sufficiently precise and the clearance between cards and stand sufficiently small to allow placement of samples reproducible within a few thousandths of an inch.   For ease of handling each card has a tab along one of its 3-in. edges.

Absorbers may be placed directly on top of the sample or on a framelike

Fig. 12-1   Gas-flow proportional counter with lucite stand and lead shield.   The flow meter used to control and monitor the gas flow is mounted on the outside of the lead shield (to the left).   A sample is in place under the counter window on the top sample self.   An aluminum card for sample mounting (*right*) and a thin aluminum absorber in a frame (*left*) are shown in front of the counter assembly.   (Courtesy Brookhaven National Laboratory.)

shelf in a slot nearer the detector. Absorbers are cut to the same size as the sample cards. Sets of aluminum absorbers are the most useful. They should be available in an assortment of thicknesses so that absorption curves may include points from about 1 mg cm^{-2} to about 3 g cm^{-2}. The thinnest absorbers are compounded of thin aluminum foils, held in rectangular aluminum frames. A set of lead absorbers is convenient for the determination of γ-ray absorption curves, but it need not include very thin pieces. A few beryllium, paraffin, or polyethylene absorbers of the same shape are useful because of their very small absorption coefficients for all but the softest electromagnetic radiations.

All the measuring instruments should be checked routinely—preferably daily—with standard samples; ideally a standard should have radiations similar to those of the activity to be measured. With multipurpose counters this is not practical, and, in any case, other criteria for the choice of a standard source, such as long half-life and rugged physical form, may be of overriding importance. A very useful standard sample for routine checks of β counters can be prepared from Cl^{36} ($t_{\frac{1}{2}} = 3 \times 10^5$y, $E_{\max} = 0.7$ MeV), fused into a metal backing and covered with a thin evaporated layer of gold. Background rates should be measured at least daily. To reduce backgrounds due to cosmic rays and strong samples in the laboratory, most counters, including their stands, are enclosed in lead shields 1 to 2 in. thick. Voltage plateaus should be checked occasionally. Scintillation detectors for γ rays should have their energy resolution for a specific photopeak (e.g., that of the 0.66-MeV γ ray of Cs^{137}) determined from time to time. An intercalibration of various instruments for activities of interest can be useful but ordinarily should be depended on only for semiquantitative results. A knowledge of at least the relative geometry factors for samples placed on various shelves is often useful.

Each instrument must have its response to samples of different activity levels determined; outside the linear-response region it should be used cautiously, with calibrated corrections. This calibration can be made in several ways: (1) with samples of different activity carefully prepared from aliquot portions of an active solution; (2) by comparison of the decay curve of a very pure short-lived activity of known half-life with the exponential decay to be expected; and (3) by measurements of the separate and combined effects of samples located in reproducible assigned positions (see chapter 6, p. 187). With counters the failure of linearity at high counting rates is attributed to coincidence losses; the correction is known as a coincidence correction (see chapter 6, p. 186). Ordinarily the necessity for corrections amounting to more than a few per cent should be avoided.

B. TARGETS FOR NUCLEAR-REACTION STUDIES

Reactor Irradiations. The problems encountered in the preparation of samples that are to serve as targets in nuclear bombardments vary widely, depending on the purpose and degree of sophistication of the experiment. When one merely desires to prepare some radioactive nuclide for tracer applications or decay-scheme studies, target preparation is usually quite straightforward. Even then, special considerations sometimes enter in reactor irradiations. For example, containers for samples to be exposed in high-flux reactors have to be carefully chosen, with due regard to neutron flux, ambient temperature, and length of irradiation. Pyrex vessels should be avoided because of their high boron content (boron has a very high neutron-capture cross section). For irradiations of the order of minutes in the modest fluxes of many research reactors (10^{12}–10^{13} cm^{-2} sec^{-1}), plastic vials are often satisfactory, and they give rise to rather low activity levels. Aluminum-foil wrappers made of highest-purity aluminum are often convenient, if time for the decay of the 2.3-minute Al^{28} can be allowed. For longer irradiations samples are often sealed in evacuated quartz vials. However, these vials must generally be allowed to "cool" for some time after irradiation to let the intense Si^{31} activity ($t_{1/2}$ = 2.6 hours) decay; some thought must also be given to arrangements for breaking the seal without undue personnel exposure and contamination hazard. The thermal stability of the substance to be irradiated is, of course, a problem to be considered. The ambient temperatures in different types of reactors differ widely; water-cooled and water-moderated swimming-pool reactors are generally much more suitable for irradiation of organic materials than, for example, graphite reactors. Some reactors have special water-cooled or even liquid-nitrogen-cooled irradiation facilities. The irradiation of aqueous solutions creates special problems; for, even if cooling is adequate to keep them below the boiling point, the radiation decomposition of water can lead to the buildup of dangerous pressures unless provisions are made for venting or catalytically recombining the gases. Another problem encountered occasionally in reactor irradiations is the "self-shielding" of materials having high neutron cross sections. For example, a 0.1-mm layer of gold (whose absorption cross section for thermal neutrons is almost 100 barns) reduces a thermal-neutron flux by about 6 per cent, so that the interior of a cube of gold 1 cm on an edge would receive only a small fraction of the flux incident on its surface.

Thick-Target Accelerator Experiments. In accelerator bombardments the variety of possible targets and targeting problems is so large that only a few generalities can be mentioned. The simplest situation arises when production of a radionuclide is the goal, without the need for quantitative information about the reaction involved. Such problems are most frequently encountered in the energy range of fixed-frequency cyclotrons, and generally a thick target is adequate or even desirable, that is, a target in which the incident bombarding particles are appreciably degraded in energy. For example, if we wished to produce a radionuclide by α, n reaction and had 40-MeV He^4 ions available, we would probably use a target thick enough to degrade the He^4 ions to just a few MeV—approximately the α, n-threshold—to maximize the product yield.

On the other hand, even in a simple production problem there could easily be complicating circumstances that would dictate a different choice of bombarding conditions.   For example, if it were desirable to produce Rb^{84} with minimal contamination by Rb^{83}, we would not wish to use 40-MeV He^4 ions on Br but would first degrade them with absorber foils below the threshold of the reaction $Br^{81}(\alpha, 2n)Rb^{83}$, even though this would lower the yield of the desired reaction $Br^{81}(\alpha, n)Rb^{84}$ also.   Conversely, by use of the 40-MeV beam and proper choice of target thickness we could maximize the ratio of $(\alpha, 2n)/(\alpha, n)$ yield.

The principal problem in cyclotron irradiations for radionuclide production is one of cooling, since the energy dissipation in the target can become quite large—of the order of a kilowatt over an area of a square centimeter or two. Metal targets, bolted or soldered to water-cooled backing plates are most satisfactory.   However, it is, of course, frequently necessary to resort to the bombardment of nonmetallic elements or compounds; satisfactory targets can then often be made by pressing powders into grooves on a cooled target plate or wrapping them in metal-foil packages, which are clamped to a cooled plate. Fairly effective cooling can be achieved by flowing helium gas over the target surface.   Beam currents, of course, may have to be adjusted to the particular problem at hand; but it is usually possible to use many microamperes of particles with energies of tens of MeV.

Requirements for Thin Targets.   In a great variety of accelerator experiments thin targets are needed.   What constitutes a "thin" target depends very much on the particular information sought.   In any experiment designed for the measurement of a reaction cross section the target must be so thin that the energy degradation of the bombarding particle in its passage through the target will not cause a significant change in the cross section.   However, the implications of this general requirement may differ widely for different situations.   For example, a target that is "thin" for the study of p, xn reactions with 30–50 MeV protons may be thick indeed for the investigation of a narrow resonance in a p, γ reaction at 2 MeV.   If the spectra of particles produced in a reaction are to be measured, the criterion for maximum target thickness will likely be set by the interactions, not of the primaries, but of these secondaries in the target.   For example, if one wished to investigate the low-energy end of the α spectra produced in p, α reactions, the targets would have to be much thinner than those required for a study of total p, α cross sections measured via the activity of the reaction product (tens of micrograms versus milligrams per square centimeter).   Similarly, in any experiment designed for the determination of momentum and angular distributions of the recoil nuclei the targets need to be so thin that these recoiling reaction products will not undergo appreciable scattering or degradation on their way out of the target; this criterion may require targets of no more than a few micrograms per square centimeter.

Still another limitation on target thickness arises sometimes from the need to suppress secondary reactions caused by particles produced in the primary interactions, if the products of such secondary reactions interfere with the measurement at hand.   For example, the product of a p, $p\pi^+$ reaction is the same as that of an n, p reaction on the same target; thus in an attempt to measure the very low cross section ($\cong 10^{-4}$ barn) of a p, $p\pi^+$ reaction with high-energy protons the targets used must be thin enough so that n, p reactions caused by low-energy neutrons originating in the target will not swamp the sought-for effect.   The severity of this type of problem depends on the number of secondaries per primary interaction, on the ratio of primary to secondary cross section, and also on such factors as the angular distribution of the secondaries.   In practice, when the effect is likely to be of any significance, it is necessary to irradiate targets of several different thicknesses and to extrapolate the results to zero target thickness.

Techniques for Preparation of Thin Targets.   It is impossible to give, in a brief space, anything like a complete summary of the methods that have been used to prepare thin targets.   In a sense, each element presents a separate problem; different thickness ranges require different approaches; a method suitable when an abundant supply of target material is available may not be adaptable to a situation in which a few milligrams of a costly enriched isotope must be made into a target; whether or not a target backing can be tolerated and how big an area is required will strongly affect the method of preparation. Thus we can make only rather sketchy comments and refer the reader to the fairly comprehensive review of thin-target preparation methods by Yaffe (Y1).

Whenever suitable foils are commercially available, they, of course, offer the simplest solution to targeting problems.   However, not many metals can be purchased in thicknesses below a few milligrams per square centimeter; among those most readily available in the form of thin foils are aluminum, nickel, and gold.   Vacuum evaporation has been used to prepare targets of a large variety of metals, some nonmetallic elements, and some compounds, over a large range of thicknesses.   The method is generally wasteful of material, but has occasionally been used to make separated-isotope targets.   The evaporated films may be deposited on a variety of backing materials (metal foils, plastic films).   However, if an unsupported target is required, various techniques are available for stripping off or dissolving the backing; probably the most useful is evaporation onto a layer of water-soluble material (such as $BaCl_2$, $NaCl$, or glycerol) on glass, followed by gentle dissolution of this intermediate layer in a trough of water, whereupon the desired film is floated off. Self-supporting foils of various materials with thicknesses down to about 0.01 mg cm^{-2} have been prepared in this way.   For the deposition of small amounts of material with high efficiency cathodic sputtering is sometimes superior to vacuum evaporation.

Another important method for target preparation is electrodeposition.   This is not restricted to deposition of metals, but can be used, for example, for cathodic or anodic deposition of oxides and other compounds.   In most cases electrodeposition can be made nearly quantitative and is therefore suitable for use with enriched isotopes.   Removal of backing materials is generally more difficult than with vacuum-evaporated targets; but, if a very thin evaporated metal film on a plastic foil is used as the plating electrode, it may be possible to dissolve off the plastic.   Various forms of electrophoretic deposition of finely divided materials from suspensions have been successfully used for target preparation.   In a technique known as electrospraying the material to be deposited is dissolved in an organic solvent and the solution is placed in a capillary.   A high potential applied between the capillary and a metal foil placed opposite its tip causes the solution to leave the capillary in the form of a very fine spray, and a rather uniform deposit of the solute is obtained on the foil.   Electrospraying is useful for fairly quantitative deposition of very thin films ($\lesssim 0.1$ mg cm^{-2}) of compounds on backing foils.

Many other specialized techniques have been described.   Thermal decomposition of gases on hot surfaces is sometimes useful, for example, for the preparation of boron films from B_2H_6, nickel films from $Ni(CO)_4$, and carbon films from CH_3I.   The preparation of separated-isotope targets often presents problems; ideally the targets can be prepared directly in the isotope separator if the isotope of interest is collected on the target backing, but this technique can be used in a rather limited number of laboratories only.

If uniformity criteria are not too stringent, targets can often be successfully prepared by sedimentation from a slurry, perhaps with the use of some binder. A useful though tedious technique (D1), especially for use with enriched isotopes or other precious target materials, involves painting onto the target backing many successive portions of an alcohol solution of metal nitrate containing a small amount of Zapon lacquer.   After each application the deposit is ignited to remove most of the organic material and rubbed with tissue paper to improve uniformity and adhesion.   Very satisfactory targets of such materials as lanthanide and actinide oxides have been produced in this manner.

Measurement of Target Thickness.   Whenever a thin target is required, it is usually necessary to know what its thickness is.   Furthermore, there are generally some requirements for uniformity of thickness over some areas. Methods for determining target thickness and uniformity are varied.   Measurements with mechanical thickness gauges are rarely applicable.   Weighing an accurately measured area is often the method of choice for self-supporting targets; it can be applied to backed targets as well if the backing material is weighed separately before target deposition and if the ratio of target weight to backing weight is not too small.   For very thin targets a microbalance may be required.   Uniformity within the target area to be used cannot be established by this gravimetric method, but measurements on several neighboring areas can help establish the degree of uniformity on a slightly larger scale.   X-ray fluorescence spectrometry can be useful for thickness measurements on thin films.

Methods based on the absorption of α and β particles in matter have found widespread use in the determination of foil thicknesses and foil uniformity (Y1). Collimated beams of monoenergetic α particles or low-energy β particles are used, and the foil to be measured is interposed between source and detector. Alpha gauges are most sensitive if the α particles reaching the detector are near the end of their range; then slight changes in interposed thickness cause large changes in counting rate. Alpha and β gauges are particularly simple to use for relative measurements and uniformity checks; but, if calibrated carefully, they can also be used for absolute thickness measurements with accuracies of 1–2 μg cm^{-2}. In a particularly useful variant of the α gauge the well-collimated monoenergetic α beam is detected by a high-resolution spectrometer, such as a semiconductor detector with pulse height analyzer. The shift of the spectral line to lower energy when a foil is interposed is a measure of the average foil thickness over the area of the beam; the line broadening can give information on nonuniformities on a microscale. A monoenergetic accelerator beam can be substituted for the α source (R1).

Occasionally it may be most practical to determine target thickness after, rather than before, an irradiation. This may be done by dissolving an accurately measured area of the target and analyzing the solution or an aliquot of it for the target material.

C. MEASUREMENT OF BEAM ENERGIES AND INTENSITIES

In almost any investigation of a nuclear reaction it is necessary to know either the energy or the intensity of the bombarding particles or, more frequently, both. The accuracy with which these quantities are required can vary widely, depending on the problem at hand. The techniques available for energy and intensity determinations differ for different energy ranges. In the following paragraphs a brief account is given of the principal methods used by nuclear chemists for the measurement of these important parameters, and attention is called to some of the problems encountered.

Determinations of Beam Energy. In general, the most accurate methods for the determination of the energy of charged-particle beams use deflection in magnetic or electric fields. Beam-deflection equipment is commonly employed in the external beams of low-energy accelerators (such as Van de Graaff machines, cyclotrons, and linear accelerators) not only for energy determinations but also to achieve energy analysis of initially inhomogeneous beams. Such analyzed, highly monoenergetic particle beams (energy spread typically of the order of 0.1 per cent) are often essential for scattering experiments, nuclear spectroscopy, etc., but they are not practical for certain types of experiments of interest to nuclear chemists (such as excitation function determinations) because the magnetic or electrostatic analysis generally results in greatly lowered beam intensities.

In any circular accelerator the magnetic field of the accelerator itself accomplishes some energy analysis, and a knowledge of field strength and orbit radius in principle gives the beam energy. For many purposes this type of information on beam energy is sufficient. In both standard cyclotrons and synchrocyclotrons the maximum beam energies are usually known from the machine characteristics within perhaps 2 or 3 per cent. The actual inhomogeneity in beam energy is usually somewhat less (typically $\cong 1$ per cent) for the full-energy beam. But, since much of this inhomogeneity results from nonconcentric orbits (which, in turn, are brought about by ion source optics), the percentage energy spread increases with decreasing radius. Therefore excitation functions determined by beam interception at different radii in cyclotrons or synchrocyclotrons can be subject to rather serious distortion because of this energy spread. Yet at synchrocyclotron energies, radius variation is often the only practical method for selecting different energies.

In a synchrotron the particle energy at any particular time in the acceleration cycle is uniquely determined if the radio frequency at that time and the radius of the equilibrium orbit are known. Accurate frequency measurements can be made more easily than accurate magnetic-field measurements; therefore, particle energies in synchrotrons are readily known to about 1 per cent. Furthermore, energy variation is easily achieved by variation of the time within the acceleration cycle to which the rf field is turned off. A calibration of energy versus rf-turn-off time is usually available for a synchrotron.

At sufficiently low beam energies the use of solid-state detectors for accurate energy measurements is coming more and more to the fore. However, the method most frequently used by nuclear chemists at energies below 50 or 100 MeV is still the measurement of range, in conjunction with a range-energy relation. The range measurement can be done in a variety of ways but basically always involves the use of absorbers and of some detector. The detector may be a Faraday cup (particularly if a beam-intensity measurement is to be coupled with the energy determination) or almost any other radiation-sensitive device; the bleaching of blue cellophane gives a very convenient beam indication. It is extremely useful to be able to do the range measurements on a cyclotron beam remotely. A convenient device for this purpose is a wheel which, by means of a servomechanism, can be rotated to interpose, in the beam, absorbers of various thicknesses mounted on its periphery.

The accuracy of absolute determinations of beam energy by absorption methods depends on the range-energy relation used and at 40 MeV can probably not be expected to be much better than 0.5–1 per cent. Often, however, it is the relative accuracy of energy measurements at two or more energies which is important, and this can be much better than the accuracy of the absolute values. Absorption in foils is used not only to *measure* particle ranges, but also to *degrade* beam energies, as, for example, in the "stacked-foil" method for excitation function determinations. This important technique consists in the simultaneous bombardment of an entire stack of target foils of the same material, usually interspersed with appropriate degrading foils, for the purpose of obtaining, in a single experiment, cross sections at a number of bombarding energies. Even without intensity monitoring, this type of experiment gives the shape of an excitation function. However, to obtain the energy of the degraded beam at each target foil, one again has to depend on range-energy relations and usually in several materials. In addition, the secondary particles produced in the absorbers can cause trouble, and the beam energy becomes spread out by straggling (cf. chapter 4, p. 100). If reasonably monoenergetic degraded beams are wanted and a considerable loss in intensity (factor 10–100) can be tolerated, magnetic analysis after degradation is recommended. Energy spread and production of secondaries combine to make energy degradation by absorption less and less practical with increasing energy. Above about 100 MeV this method is rarely used.

Neutron Flux Measurement. The thermal-neutron flux, for example, in a reactor, is usually determined by the activation, under the exact conditions for which the flux is desired, of a substance of known activation cross section. The most frequently used flux monitor is gold (Au^{197}), from which, by the capture of thermal neutrons, 2.698-day Au^{198} is formed with a cross section $\sigma_a = 99$ barns. The number of Au^{198} atoms formed from W mg of Au^{197} at the end of an irradiation of t seconds in a flux nv of thermal neutrons is

$$N_{198} = nv \times \sigma_a \times \frac{W}{197} \times 6.02 \times 10^{20} \frac{1 - e^{-\lambda t}}{\lambda},$$

where λ is the decay constant of Au^{198} in sec^{-1}, nv is the neutron flux in $cm^{-2} sec^{-1}$ (n is the neutron density per cubic centimeter and $v = 2.2 \times 10^5$ cm sec^{-1}, is the most probable velocity of a Maxwellian distribution at 20°C). Since W and t are readily measured and σ_a is presumed to be known, the measurement of the thermal flux nv essentially reduces to the problem of determining N_{198}, the number of Au^{198} atoms, that is, to an absolute disintegration rate measurement. Fortunately, the decay scheme of Au^{198} is simple enough in its main features (cf. p. 435) to lend itself to the use of the coincidence method described in section F. Other capture reactions, for example, $Co^{59}(n, \gamma)Co^{60}$, can also be used for thermal-neutron flux measurements. A general requirement for a convenient thermal-neutron flux monitor is that the cross section follow a $1/v$ law, so that the Maxwellian velocity distribution can be replaced by the single velocity $v = 2200$ meters/sec (see introduction to appendix C).

Primary Monitoring of Charged-Particle Fluxes (C2). The most widely used instrument for absolute determination of charged-particle fluxes is the Faraday cup. This is essentially an insulated electrode designed to stop all the beam particles striking it as well as any charged secondaries produced in it by the beam. The total charge built up on the Faraday cup divided by the charge per particle (e for protons, $2e$ for α particles, etc.) thus gives the total number of particles which has fallen on the cup. The charge Q may be measured by the voltage drop ΔV developed across a condenser of known capacity C ($Q = C\,\Delta V$). More commonly a null method is used in which the voltage that must be applied across the condenser to bring the cup back to its original potential is accurately measured with a potentiometer. In the best devices electronic circuits are used to measure the current flowing to the cup and to keep the cup near ground potential at all times.

The design of Faraday cups presents a number of problems. Good insulation is essential, and the cup should be operated in a high vacuum, since gas ionization in the vicinity of the cup can lead to erroneous results. The principal concern in Faraday-cup design is usually the retention of charged secondaries, chiefly secondary electrons at modest beam energies. Proper design of the cup-shaped electrode, with an entrance aperture small compared to the depth of the cup, is used to minimize the solid angle for escape of secondary electrons. Magnetic fields of a few hundred gauss are also useful for preventing electron escape. On the other hand, secondary electrons coming from any other objects in the beam path (windows, collimators, etc.) must be prevented from reaching the cup. With increasing beam energy, the difficulty of retaining the charged secondaries increases, and Faraday cups for beams of several hundred MeV become quite unwieldy.

Another method which for practical reasons is limited to relatively low energies is calorimetry. If a beam and all its secondaries are completely stopped in a calorimeter, the product of beam energy and beam current is measured. One advantage of calorimetry over the Faraday-cup method is that it can be used for the circulating beams inside accelerators, where strong magnetic and electric fields would interfere with the operation of a Faraday cup. In external beams Faraday-cup measurements are to be preferred.

At higher energies ($\gg 100$ MeV) absolute measurements of beam intensities are usually based on counting individual beam particles by means of counter telescopes or nuclear emulsions (C3). Both methods are limited to use at fairly low beam intensities, and secondary devices suitable for higher intensities are usually calibrated in terms of these absolute monitors. Emulsion monitoring of near-minimum ionizing particles can be done with total intensities up to $\cong 5 \times 10^6$ particles cm^{-2} (independent of time distribution). Measurements in a counter telescope are limited by the requirement that the dead-time losses should be small; therefore the maximum average intensity that can be monitored with a given telescope and associated circuitry depends on the time distribution of the beam, which in synchrotrons tends to be strongly bunched.

Secondary Beam Monitors. Once a primary beam monitor is available, any other device can, in principle, be calibrated in terms of it. Caution is required in the use of secondary monitors in intensity ranges outside the range of primary calibration; linearity of response must as a rule be checked. This is particularly true for any instruments based on ionization measurements or on light collection (scintillation and Čerenkov counters). A large variety of secondary monitoring devices, each with its own virtues and shortcomings, has been described (C2).

Secondary beam monitors particularly useful for nuclear chemists and essentially free of nonlinearity problems are nuclear reactions of known cross section. The absolute cross section does not, in fact, have to be known, so long as the activity of the reaction product, preferably without chemical separation from the target foil, is measured in the same arrangement in which it was determined when its production was calibrated against an absolute beam monitor. Radioactivity induced by nuclear reactions can be useful as a beam monitor in almost any energy range but becomes of paramount importance at high energies (> 100 MeV) for several reasons.

1. Scattering and absorption in thin foils become relatively unimportant, and monitor foil and target foil can thus be made to intercept virtually the same number of beam particles of the same energy (which is not the case at lower energies).

2. In the circulating beams of synchrocyclotrons and proton synchrotrons the effective particle fluxes through targets may greatly exceed the circulating beam intensities (multiple traversals, cf. p. 372); since the number of traversals depends on particle energy, accelerator characteristics, and thickness and composition of target, activation of a monitor foil incorporated in the target stack is the only reliable method for the measurement of the effective particle flux through the target. In spite of this difficulty, circulating-beam irradiations are often attractive just because the multiple traversals raise the effective beam intensities by large factors over those that might be available in the more readily monitored external beams.

The nuclear reactions which have been found most useful for monitoring high-energy proton beams are shown in Table 12-1. The reaction $C^{12}(p, pn)C^{11}$ is the one that has been most thoroughly calibrated against absolute monitors over a wide energy range (50 MeV–30 GeV). However, because of the 20-minute half-life of C^{11}, this reaction is useful for relatively short irradiations only. For this reason, the production of Na^{24} from aluminum is the most widely used monitor reaction. The cross section of this reaction is almost independent of proton energy from 100 MeV to 30 GeV (cf. table 12-1), and the Na^{24} activity is readily measured in aluminum foils without chemical separation; about 24 hours after irradiation, when the shorter-lived activities have died out, Na^{24} can be measured without interference from other products. A disadvantage is the fact that Na^{24} can also be made by low-energy secondary neutrons produced in the target, according to the reaction $Al^{27}(n, \alpha)Na^{24}$. Production of F^{18} in aluminum is less sensitive to secondaries and may therefore be a preferable monitor under some circumstances. The last reaction listed in table 12-1, the production of the α emitter Tb^{149} in gold, has the

Table 12-1   Monitor reactions for high-energy proton beams

Reaction*	Product Half-Life	Principal Radiation Detected	Cross Section† in mb			Remarks
			300 MeV	3 GeV	30 GeV	
$C^{12}(p, pn)C^{11}$	20.4 min	β^+	35.8	27.1	26.8	Best monitor for short irradiations
$Al^{27}(p, 3pn)Na^{24}$	15.0 hr	β^-, γ	10.1	9.1	8.6	Sensitive to low-energy secondaries
$Al^{27}(p, spall)F^{18}$	110 min	β^+	6.6	6.8	6.2	
$C^{12}(p, spall)Be^7$	53.6 d	γ	10.0	10.3	9.2	Useful for long irradiations
$Au^{197}(p, spall)Tb^{149}$	4.1 hr	α	0.0	1.2‡	?	Threshold at $\cong$600 MeV

* The notation $(p, spall)$ merely indicates a spallation reaction.
† The cross section values are taken from ref. C3.
‡ The cross sections quoted for the production of Tb^{149} refer to the fraction of the nuclide which decays by α emission.

advantage of a very high threshold ($\cong$600 MeV); also, the α activity of Tb^{149} ($t_{1/2} = 4.1$ hours) is readily measured in irradiated gold foils after some short-lived α emitters have decayed out. All these monitor reactions are discussed in detail in ref. C3.

D. TARGET CHEMISTRY

We now turn to the chemical problems of purifying and isolating radioactive species following their production in nuclear reactions. In this connection the nuclear chemist or radiochemist may be confronted with one of two tasks. He may need to prepare a known reaction product free from other radioactive contaminants and sometimes free from certain inactive impurities and in a specified chemical form for use in subsequent experimentation or for determination of its yield in a nuclear reaction; or he may wish to identify a hitherto unknown or unidentified radioactive species by its atomic number, mass number, half-life, and radiation characteristics. In both cases chemical separations are usually required for two reasons.

1. In almost any nuclear bombardment more than one type of reaction occurs, and therefore the reaction products have to be separated.
2. Impurities present in the target material give rise to radioactive products.

Apart from the ordinary chemical impurities, further contamination is often introduced in the bombardment procedure; particularly in cyclotron bombardments, in which for cooling purposes the target material is soldered, pressed, or electroplated onto a metal backing or wrapped in a metal foil, transmutation products of the sample container, backing material, solder, and fluxes must be considered.

In a slow-neutron bombardment the only type of reaction produced in almost any target element is the n, γ reaction. Therefore, if an element of sufficient purity is bombarded with slow neutrons, chemical separations are often not required. However, in this case the radioactive product is isotopic with the target, and it is sometimes desirable to free the product isotope from the bulk of the target material in order to obtain high specific activities. Special techniques developed for this purpose are discussed in chapter 7, section E. Here we confine our attention to the separation of products not isotopic with the target material.

Comparison with Ordinary Analytical Practice. In many respects the chemical separations which the radiochemist carries out on irradiated targets are similar to ordinary analytical procedures. However, there are a number of important differences. One is the time factor which is often introduced by the short half-lives of the species involved. An otherwise simple procedure such as the separation of two common cations may become quite difficult when it is to be performed, and the final precipitates are to be dried and mounted, in a few minutes. When the usual procedures involve long digestions, slow filtrations, or other slow steps, completely different separation procedures must be worked out for use with short-lived activities. Ingenious chemical isolation procedures, taking as little as a few seconds, have been developed (K1, S1).

In radiochemical separations, at least those subsequent to bombardments with projectiles of moderate energies, we are usually concerned with several elements of neighboring atomic numbers. Thus the procedures given in complete schemes of qualitative analysis can often be modified and shortened. On the other hand, the separation of neighboring elements sometimes presents considerable difficulties, as can readily be seen by considering such groups as Ru, Rh, Pd, or Hf, Ta, or any sequence of neighboring rare earths. In very-high-energy reactions and in fission the products are spread over a wide range of atomic numbers. In these cases the separation procedures either become more akin to general schemes of analysis or, more frequently, are designed for the isolation of one or a few elements free from all the others; the latter type of procedure is required particularly when a short-lived substance is to be isolated, and for such cases many specialized techniques have been developed.

High yields in radiochemical separations are not always of great importance, provided that the yields can be evaluated. It may be more valuable to get 50 per cent (or perhaps even 10 per cent) yield of a radioactive element separated in 10 minutes than to get 99 per cent yield in 1 hour (this is certainly so if the activity has a half-life of 10 or 20 minutes).   High *chemical* purity may or may not be required for radioactive preparations, depending on their use. For identification and study of radioactive species and for many chemical tracer applications it is not important; for most biological work it is.   On the other hand *radioactive* purity is usually required and often has to be extremely good.

Hazards Encountered with Radioactive Materials.   Some specific effects of the radiations from radioactive substances on the separation procedures may be noted.   At very high activity levels (say 10^{12} β disintegrations per minute per milliliter of solution) the chemical effects of the radiations, such as decomposition of water and other solvents, and heat effects may affect the procedures. However, this is generally not so important as the fact that even at much lower activity levels, especially in the case of γ-ray emitters, the person carrying out the separation receives dangerous doses of radiation unless protected by shielding or distance.   At even lower activity levels, say in the handling of microcurie amounts, where the health hazards from radiation are minimal, special care is still required to prevent spread of radioactive contamination which could seriously raise counter backgrounds and interfere with low-activity experiments.   The degree of precaution needed, both to contain contamination and to prevent excessive radiation exposure, depends on many factors, including the amount of activity handled, the nature and energy of the radiation involved, the half-life of the active substance, and possibly its chemical properties.

Some general precautions should be observed in all work with radioactive sources.   Some survey instrument (see chapter 5, section D) should always be used to determine the actual radiation levels present, hence the type of protection required.   As many of the manipulations as possible should be carried out in hoods with adequate air flow or in dry boxes.   To contain possible spills, it is well to work in trays or on surfaces covered with absorbent paper.   Pipetting by mouth is to be avoided.   Even at the microcurie level, radioactive materials should never be handled with bare hands, but with gloves or tongs or in containers.   At somewhat higher levels of γ emitters (typically in the millicurie range) it becomes necessary to carry out separations behind lead shields, which are usually assembled from lead bricks to suit the particular purpose.   Operations are then performed with the use of tongs and other tools. For very high activity levels (say in excess of about 10^{12} γ quanta per minute) more elaborate remote-control methods are necessary.   It is obvious that separation procedures are more difficult under these conditions and in many cases have to be modified considerably to adapt them for remote-control operation.

More detailed discussions of the safe handling of radioactive materials and of appropriate health-protection measures may be found in refs. M1, B1, F1, G1, H1, U1.

Carriers. The mass of radioactive material produced in a nuclear reaction is generally very small. Notice, for example, that a sample of 37-minute Cl^{38} undergoing 10^8 disintegrations per second weighs about 2×10^{-11} g; a sample of 51-day Sr^{89} of the same disintegration rate weighs 1×10^{-7} g. Thus the substance to be isolated in a radiochemical separation may often be present in a completely impalpable quantity.[1] It is clear that ordinary analytical procedures involving precipitation and filtration or centrifugation may fail for such minute quantities. In fact, solutions containing the very minute concentrations of solutes which can be investigated with radioactive tracers behave in many ways quite differently from solutions in ordinarily accessible concentration ranges; adsorption on container surfaces, dust particles, and other suspended impurities may be important at these "tracer" concentrations.

Usually some inactive material isotopic with the radioactive transmutation product is deliberately added to act as a carrier for the active material in all subsequent chemical reactions. Most often it is not sufficient to add carrier only for the particular transmutation product to be isolated; frequently it is necessary to add carriers also for other activities which are known or assumed to be formed, including those which derive from target impurities.

It is in many cases not necessary to add carriers for all active species present, because several elements may behave sufficiently alike under given conditions so that traces of one will be carried by macroscopic quantities of another. For example, an acid-insoluble sulfide such as CuS can usually be counted on to carry traces of ions such as Hg^{2+}, Bi^{3+}, Pb^{2+}, which also form acid-insoluble sulfides. On the other hand, since many precipitates [such as $BaSO_4$ or $Fe(OH)_3$] tend to occlude or adsorb many foreign substances, it is usually necessary to add carriers not only for ions to be precipitated but also for ions to be held in solution when other ions are precipitated.[2] For example, if a zinc activity is to be separated from a ferric solution by ferric hydroxide precipitation with excess ammonia, all the zinc will not be left in solution unless zinc carrier is present. The carrier in such cases is sometimes referred to as holdback carrier. Later we shall discuss cases in which carriers are unnecessary.

We have mentioned before that extreme radioactive purity is often very important. Frequently the desired product has an activity that constitutes only a very small fraction of the total target activity; yet this product may be required completely free of the other activities. Such extreme purification is

---

[1] Actually the mass of an element formed in a nuclear reaction is often exceeded by that of the inactive isotopes of the same element present as an impurity in the target and in the reagents used in the separation procedure.

[2] In the early decades of radiochemistry there was a great deal of interest in the laws governing coprecipitation and adsorption and in the classification of various "carrying" phenomena (H2). These are no longer very active fields of research, and not much more than broad general guidelines are available for the prediction of coprecipitation behavior. A useful general rule formulated in 1913 by K. Fajans may be paraphrased as follows: Conditions which favor the precipitation of a substance in macroamounts also tend to favor the coprecipitation of the same material from tracer concentrations with a foreign substance.

usually quite readily attained by repeated removal of the impurities with successive fresh portions of carrier until the fractions removed are sufficiently inactive. This so-called "washing-out" principle may be illustrated by the separation of a weak cobalt activity from radioactive copper contamination. Cobalt and copper carriers are added to a $0.3M$ HCl solution of the activities, CuS is precipitated and filtered or centrifuged off, excess H_2S is removed by boiling, fresh copper carrier is added to the filtrate, and the procedure is repeated until a final CuS precipitate no longer shows an objectionable amount of activity. The same principle can be applied to other than precipitation reactions. Radioactive iron impurity might be removed by repeated extraction of ferric chloride from $9M$ HCl into isopropyl ether, with fresh portions of $FeCl_3$ carrier added after each extraction. In applying the washing-out method one must, of course, make sure that the desired product is not partially removed along with the impurity in each cycle. If the washing out works properly, the activities of successive impurity fractions should decrease by large and approximately constant factors, provided that the conditions in each step are about the same.

In order that an added inactive material serve as a carrier for an active substance, the two must generally be in the same chemical form. For example, inactive iodide can hardly be expected to be a carrier for active iodine in the form of iodate ion; sodium phosphate would not carry radioactive phosphorus in elementary form. The chemical form in which· a transmutation product emerges from a nuclear reaction is usually hard to predict and has been investigated in only a few cases. However, it is often possible to treat a target in such a way that the active material of interest is transformed to a certain chemical form. For example, if a zinc target is dissolved in a strongly oxidizing medium (say HNO_3, or $HCl + H_2O_2$), any copper present as a transmutation product is found afterward in the Cu^{2+} form. If there is any uncertainty about the chemical form of the transmutation product—its oxidation state or presence in some complex or undissociated compound, for example—the only method that can be relied on to avoid difficulties is the addition of carrier in the various possible forms and a subsequent procedure for the conversion of all of these into one form. To go through such a procedure prior to the addition of carrier may not be adequate. In fact, it appears that it may not always be sufficient to add the carrier element (say iodine) in its highest oxidation state (IO_4^-) and carry through a reduction to a low oxidation state (I_2). In the case of the iodine compounds this procedure does not seem to reduce all the active atoms originally present in intermediate oxidation states.

So far we have not spoken of the amounts of carriers used. For manipulative reasons it is often convenient to use between 2 and 20 mg of each carrier, and less than about 1 mg is used only rarely. In separations from large bulks of target material larger amounts of carrier (perhaps 100 to 500 mg) are sometimes useful.[3] The amount of carrier frequently has to be measured at various stages of a chemical procedure to determine the chemical yield in the different

[3] In a radiochemical laboratory it is convenient to have carrier solutions for a large number

steps or at least in the over-all process.   In such cases very small amounts of carrier are inconvenient.   On the other hand, it is necessary to keep the quantities of carrier small in the preparation of sources of high specific activity.   The specific activity of a sample of an element is sometimes expressed as the ratio of the number of radioactive atoms to the total number of atoms of the element in the sample; more conveniently, it is often expressed in terms of the disintegration rate per unit weight.   High specific activities are particularly essential in many biological and medical applications of radioactive isotopes and are often desirable in samples to be used in physical measurements or chemical tracer studies to ensure small absorption of the radiations in the sample itself or to permit high dilution factors.

It is often possible to prepare samples of very high specific activities by the use of a nonisotopic carrier in the first stages of the separation; this carrier may later be separated from the active material.   In the isolation of radioyttrium (108-day Y^{88}) from deuteron-bombarded strontium targets, ferric ion can be used as a carrier for the active Y^{3+}; ferric hydroxide is then precipitated, centrifuged, washed, redissolved, and, after the addition of more strontium as hold-back carrier, it is precipitated several more times to free it of strontium activity.   Finally the ferric hydroxide which carries the yttrium activity is dissolved in $9M$ HCl, and ferric chloride is extracted into isopropyl ether, leaving the active yttrium in the aqueous phase almost carrier-free.   The use of nonisotopic carriers became essential in early work with the artificially produced elements that do not occur in nature (chapter 7, section F).   For example, most of the chemical processes which were to be used during World War II for the large-scale isolation of plutonium from irradiated uranium were worked out on a tracer scale before any weighable amounts of plutonium were available.   A very rough rule governing coprecipitation of tracers with nonisotopic carriers is mentioned in footnote 2, p. 403.

Not all chemical procedures require the use of carriers.   In particular, procedures that do not involve solid phases may sometimes be carried out at tracer concentrations without the addition of carriers.   Because of the great importance of high specific activities, considerable work has been done on the preparation of carrier-free sources of many radioactive species (see for example, refs. G2, G3).   In the course of the following brief discussion of the various types of separation techniques we shall therefore point out those that lend themselves to the production of carrier-free preparations.   A rather complete collection of radiochemical procedures for the separation and isolation of every element (except H, He, Li, B) and including carrier-free procedures is available in a series of monographs (S1).

Precipitation.   In most radiochemical separations, as in conventional analytical schemes, precipitation reactions play a dominant role.   The chief difficulties with precipitations arise from the carrying down of other materials.

---

of elements on hand.   These may, for example, be made up to contain 1 or 10 mg of carrier element per milliliter.

Some precipitates such as manganese dioxide and ferric hydroxide are so effective as "scavengers" that they are sometimes used deliberately to carry down foreign substances in trace amounts. Other precipitates, such as rare-earth fluorides precipitated in acid solution, cupric sulfide precipitated in acid solution, or elementary tellurium brought down by reduction with sulfur dioxide, have little tendency to carry substances not actually insoluble under the same conditions and therefore can sometimes be brought down without the addition of hold-back carriers for activities that are to be left in solution. Most precipitates have an intermediate behavior in this regard. Precipitates previously formed in the absence of tracer may take up a tracer when added in suspension to the tracer solution. However, ordinarily such carrying on preformed precipitates is not so effective as coprecipitation.

A radionuclide capable of existence in two oxidation states can be effectively purified by precipitation in one oxidation state followed by scavenging precipitations for impurities while the element of interest is in another oxidation state. For example, a useful procedure for cerium decontamination from other activities uses repeated cycles of ceric iodate precipitation, reduction to Ce(III), zirconium iodate precipitation (with Ce(III) staying in solution), and reoxidation to Ce(IV).

The use of nonisotopic carrier has already been illustrated in the case of high-specific-activity yttrium carried on $Fe(OH)_3$. The same technique can, of course, be applied to many other di- and trivalent cations that coprecipitate with $Fe(OH)_3$, for example, Be^{2+}, Cr^{3+}, Bi^{3+}, rare earths, and even some anions such as phosphate [provided the activity of interest is the only one present in solution that will be carried on $Fe(OH)_3$]. Other coprecipitation reactions may often be found useful, especially if the subsequent separation of the radionuclide from the nonisotopic carrier can be accomplished by some technique other than precipitation.

Adsorption on the walls of glass vessels and on filter paper, which is sometimes bothersome, has been put to successful use in special cases. Carrier-free yttrium activity has been quantitatively adsorbed on filter paper from an alkaline strontium solution at yttrium concentrations at which the solubility product of yttrium hydroxide could not have been exceeded.

Ion Exchange. An exceedingly useful separation technique closely related to adsorption chromatography has been developed for use both with and without carriers. This technique involves the adsorption of a mixture of ions on an ion-exchange resin followed by selective elution from the resin. Both cation- and anion-exchange resins have been used very successfully (S2, S3, K2, K3, K4). Most of the cation resins (such as Amberlite IR-1 or Dowex-50) are synthetic polymers containing free sulfonic acid groups. The anion exchangers (such as Dowex-1) usually contain quaternary amine groups with replaceable hydroxyl ions. The distribution of any given element between a solution and the resin depends strongly on the particular ionic forms of the element present (either hydrated ion or various cation or anion complexes) and on their concentrations and therefore on the composition of the solution; for almost any pair of ions conditions can be found under which they will show some difference in distribution.

In practice a solution containing the ions to be separated is run through a column of the finely divided resin, and conditions (solution composition, column dimensions, and flow rate) are chosen so that the ions to be adsorbed will appear in a narrow band near the top of the column. In the simplest kind of separation some ionic species will run through the column while others are adsorbed. For example, Ni(II) and Co(II) may be separated very readily by passing a $12M$ HCl solution of the two elements through a Dowex-1 column; the Co(II) forms negatively charged chloride complexes and is held on the column, whereas Ni(II) apparently does not form such complexes and appears in the effluent.

More commonly, a number of ionic species may be adsorbed together on the column and separated subsequently by the use of eluting solutions differing in composition from the original input solution. Frequently complexing agents which form complexes of different stability with the various ions are used as eluants. There exists then a competition between the resin and the complexing agent for each ion, and if the column is run close to equilibrium conditions each ion will be exchanged between resin and complex form many times as it moves down the column.[4] The number of times an ion is adsorbed and desorbed on the resin in such a column is analogous to the number of theoretical plates in a distillation column. The rates with which different ionic species move down the column under identical conditions are different because the stabilities of both the resin compounds and the complexes vary from ion to ion. Separations are particularly efficient if both these factors work in the same direction; that is, if the complex stability increases as the metal-resin bond strength decreases. As the various adsorption bands move down the column, their spatial separations increase, until finally the ion from the lowest band appears in the effluent. The various ions can then be collected separately in successive fractions of the effluent.

The most striking application of cation-exchange columns is in the separation of rare earths from one another, both on a tracer scale and in gram or hundred-gram lots. The eluting solution in this case may be 5 per cent citric acid solution buffered with ammonia to a pH somewhere between 2.5 and 8, depending on the resin used and other conditions. More recently the use of α-hydroxy isobutyrate ion has been found to give faster and more efficient rare-earth separations (K5, p. 389; S1). The rare earths are eluted in reverse order of their atomic numbers, and yttrium falls between dysprosium and holmium. Very clean separations can be obtained, with impurities in some cases reduced

---

[4] Slow flow rate, high resin-to-ion ratio, and fine resin particle size favor close approach to equilibrium. In practice a compromise has to be made between high separation efficiency and speed.

to less than one part per million.   By continuously recording the specific activity of the effluent solution as a function of time one obtains separate sharp peaks for the activities of the various rare earths when a mixture of rare-earth radioactivities is run through the column.   This method led to the definite assignment of several decay periods to isotopes of element 61.   Cation exchange procedures (again with α-hydroxy isobutyric acid as one of the most effective complexing agents) have been equally successful for the separation of the actinide elements, as already mentioned in chapter 7, section F (see ref. K5, pp. 387–391).

For rather rapid target chemistry anion exchange is often more useful than cation exchange because larger flow rates can be used with anion columns.   A large number of elements form anionic complexes under some conditions, and available data make it appear likely that a general scheme of analysis based entirely on ion-exchange-column separations could be worked out for these elements.   If all the transition elements from manganese to zinc are present in a $12M$ HCl solution, all but Ni(II) are adsorbed on Dowex-1.   Then they may be successively eluted, Mn(II) with $6M$ HCl, Co(II) with $4M$ HCl, Cu(II) with $2.5M$ HCl, Fe(III) with $0.5M$ HCl, and Zn with $0.005M$ HCl. With milligram quantities of the elements and a column a few millimeters in diameter and about 10 cm long, this entire separation can be carried out in about half an hour.   Another interesting separation is that of palladium, rhodium, iridium, and platinum on Dowex-50 cation-exchange resin.   From a dilute $HClO_4$ solution free of halide ions, palladium, rhodium, and iridium are adsorbed while platinum runs through.   Subsequently palladium is eluted with $0.1M$ HCl, then rhodium with $2M$ HCl, and finally iridium with $5M$ HCl.   In this case cations are adsorbed, and the differences between the chloride complex constants are used to obtain the selective elution.

Most ion exchange separations work as well with carrier-free radioactivities as with carriers.   The remarkable performance of ion exchange resins in separating literally a few atoms of mendelevium has already been mentioned in chapter 7, pp. 216–17.

In addition to the organic ion exchange resins, some inorganic ion exchangers have come into use.   For example, excellent separations of alkali elements from one another have been obtained by elution with NH_4Cl solutions from columns of microcrystalline zirconium phosphate or zirconium molybdate.

Volatilization.   Other separation methods avoiding the difficulties inherent in precipitations have frequently been used in radiochemical work.   Among them are volatilization, solvent extraction, electrodeposition, and leaching. In special cases all of these techniques lend themselves to the preparation of carrier-free tracers.   Radioactive noble gases can be swept out of aqueous solutions or melts with some inert gas.[5]   The volatility of such compounds as

---

[5] This technique was, for example, successfully applied in the neutrino-capture experiment mentioned on p. 253 to sweep a few atoms of Ar^{37} out of several thousand gallons of carbon tetrachloride by a stream of helium gas.

GeCl$_4$, AsCl$_3$, and SeCl$_4$ can be used to effect separations from other chlorides by distillation from HCl solutions. Similarly, osmium, ruthenium, rhenium, and technetium can be separated from other elements and from one another by procedures involving distillations of their oxides OsO$_4$, RuO$_4$, Re$_2$O$_7$, and Tc$_2$O$_7$. Carrier-free palladium (Pd^{103}) has been prepared from a rhodium target by a method involving coprecipitation of palladium with selenium (by reduction of H$_2$SeO$_3$ with SO$_2$), followed by removal of selenium by a perchloric acid distillation.

Distillation and volatilization methods often give very clean separations, provided that proper precautions are taken to avoid contamination of the distillate by spray or mechanical entrapment. Most volatilization methods can be done without specific carriers, but some nonisotopic carrier gas may be required. Precautions are sometimes necessary to avoid loss of volatile radioactive substances during the dissolving of irradiated targets or during the irradiation itself.

Solvent Extraction (F2). Under certain conditions compounds of some elements can be quite selectively extracted from an aqueous solution into an organic solvent, and often the partition coefficients are approximately independent of concentration down to tracer concentrations (say 10^{-12} or $10^{-15} M$). In other cases, particularly if dimerization occurs in the organic phase (as in the ethyl ether extraction of ferric chloride), carrier-free substances are not extracted. Solvent extractions often lend themselves particularly well to rapid and specific separations. In most cases the extraction can be followed by a "back-extraction" into an aqueous phase of altered composition.

Extractions of the chlorides of Fe(III), Ga(III), and Tl(III) into various ethers are frequently used by radiochemists. The partition coefficients vary quite rapidly with HCl concentration. Extraction from $6M$ HCl into ethyl ether or from 8 to $9M$ HCl into isopropyl ether gives very good separations from nearly all other metal chlorides. The separation of gallium from iron and thallium can be achieved by ether extraction of GaCl$_3$ in the presence of reducing agents so that the reduced ions Fe(II) and Tl(I) are present.

Gold nitrate and mercuric nitrate can be extracted into ethyl acetate from nitric acid solutions. The extraction of uranyl nitrate by ethyl ether from a nitric acid solution of high nitrate concentration is sufficiently specific to serve as an excellent first step in the isolation of carrier-free fission products from the bulk of irradiated uranium. The extraction into ethyl ether of the blue peroxychromic acid formed when H$_2$O$_2$ is added to a dichromate solution is an excellent radiochemical decontamination step for chromium, although it tends to give low yields. Extraction of copper dithizonate into carbon tetrachloride, of cadmium thiocyanate into chloroform, of beryllium acetylacetonate into benzene, and many other examples could be cited. Judicious use of complexing agents such as ethylenediamine tetraacetate (EDTA) can often help to make extractions more specific for a particular element; the addition of EDTA is recommended, for example, in the extraction of beryllium acetylacetonate just mentioned because its complexing action prevents the extraction of some other ions that would otherwise accompany beryllium.

Several organic substances such as thenoyltrifluoroacetone (TTA) have been found to form chelate complexes with a large number of metal ions (P3). These complexes are preferentially soluble in nonpolar solvents such as benzene, and, since the dissociation constants of the different metal chelates show different pH dependence, specific separation procedures can sometimes be devised with several extraction steps at different pH values.

Occasionally it may be possible to leach an active product out of a solid target material. This has been done successfully in the case of neutron- and deuteron-bombarded magnesium oxide targets; radioactive sodium is separated rather efficiently from the bulk of such a target by leaching with hot water.

Electrochemical Methods. Electrolysis or electrochemical deposition may be used either to plate out the active material of interest or to plate out other substances, leaving the active material in solution. For example, it is possible to separate radioactive copper from a dissolved zinc target by an electroplating process. Carrier-free radioactive zinc may be obtained from a deuteron-bombarded copper target by solution of the target and electrolysis to remove all the copper.

In attempting to use electrode processes at tracer concentrations, one must keep in mind that the measured potential E for a reaction can deviate appreciably from the standard potential E^0, according to the Nernst equation:

$$ E = E^0 - \frac{RT}{nF} \ln Q, $$

where R is the gas constant, T is the absolute temperature, F is the Faraday, n is the number of electrons transferred in the reaction as written, and Q is the appropriate activity ratio for the reaction (product activities divided by reactant activities, each raised to proper power, as in an equilibrium constant). If the activity of tracer deposited on the electrode is taken as unity (which is by no means always a good assumption), Q can take on very large values. Measurements of the potentials needed to deposit a tracer (relative to some suitable reference electrode) have been used to estimate standard electrode potentials for some of the artificially produced elements before they were available in macroconcentrations.

Chemical displacement may sometimes be used for the separation of carrier-free substances from bulk impurities. The separation of polonium from lead by deposition on silver is a classical example. Similarly, bismuth activity obtained in lead bombardments may be separated almost quantitatively from the lead by plating on nickel powder from hot $0.5M$ HCl solution. This method for lead-bismuth separations is sufficiently rapid to permit isolation of the 0.8-second Pb^{207m} isomer from its bismuth parent.

E. PREPARATION OF SAMPLES FOR ACTIVITY MEASUREMENTS

Many points of experimental technique arise in the preparation of samples for activity measurements.   Most of them have to do with the attainment of a suitable and reproducible geometrical arrangement and with the scattering and absorption of radiations in the sample and in its support.   The difficulties encountered in sample preparation are greatest when absolute disintegration rates or energies are wanted, less when samples of different radiation characteristics are compared, and least when the relative strengths of several samples of the same kind, or of the same sample at several times, are to be determined. Fortunately the last-mentioned problem is probably the one most often met in radiochemical work.   However, even here adequate reproducibility may sometimes be troublesome to achieve.

Choice of Counting Arrangement.   Careful consideration must be given to the chemical and physical form in which samples are to be measured.   The radiations emitted by the substance and the available measuring equipment are among the determining factors.   Alpha emitters are usually counted in the form of thin deposits preferably prepared by electrodeposition or by distillation and placed inside a proportional counter or ionization chamber or near a solid-state detector.   Nuclides that emit primarily soft radiations (low-energy β rays, X rays, conversion electrons, or Auger electrons) may be very efficiently assayed for activity if they can be prepared in the form of a gas suitable as a component of a counter-filling mixture.   For example, C^{14}-labeled compounds may be burned to CO_2, which is then introduced into a proportional counter along with an appropriate amount of argon, methane, or argon-methane mixture; essentially 100 per cent counting efficiency and good counter behavior can be obtained over a fair range of CO_2 partial pressures (0.5 to 5 mm of mercury).   This technique requires the use of a good gas-handling and purification system.

Beta-active nuclides are most commonly prepared in the form of thin solid samples and measured with thin-window counters.   However, lack of reproducibility of absorption and self-absorption effects can be troublesome in this technique.   Therefore, if a sample emits both β and γ rays, the possible advantages of γ assay should be considered.   Absorption effects are generally much smaller for γ rays than for β rays, and variations in the thicknesses of samples and counter walls are usually negligible when γ rays are detected.   Particularly when scintillation counters with their high efficiency for γ-ray detection are available, γ counting will often be the method of choice.   Samples may be prepared as solids or in solutions which are placed under the counter in standard liquid cells.   The most convenient device for measuring γ activities of liquid samples is the well-type scintillation counter.   Even with GM or proportional counters γ assays may be preferable to β measurements, especially in experiments in which specific activity rather than total activity is a limiting factor; in γ counting a much larger sample can be used effectively.

In spite of the advantages of scintillation counters, much of the work of nuclear chemists almost inevitably involves measurement of β particles from solid samples. The following paragraphs therefore deal more specifically with problems encountered when solid samples are used for β counting.

Back-Scattering. The phenomenon of back-scattering of electrons has been described on p. 106. To achieve reproducibility in the measurement of β activities, it is clearly necessary to mount samples in such a way that the back-scattering from the mount is either negligible or constant from sample to sample. For determinations of relative counting rates it is probably best to mount all samples on thick supports of low Z (plastic or aluminum) and to count them in the same geometry. For accurate comparisons of different β emitters it will still be necessary to take into account possible differences between β^- and β^+ scattering and the dependence of back-scattering on energy for low-energy spectra. For work of the highest precision nearly weightless samples should be mounted on essentially weightless plastic films (<0.1 mg cm^{-2})[6] and assayed in a 4π counter (see section F).

Because the back-scattered electrons are degraded in energy an absorption curve for a given β emitter has a somewhat steeper initial slope when the sample is mounted on a thick backing than when it is deposited on a thin film. The end point is, of course, unaffected.

Self-Absorption and Self-Scattering. Whenever the β activities of samples of finite thickness are measured, consideration must be given to the effect of absorption and scattering of electrons in the samples themselves. To make corrections for these effects negligible, sample thicknesses have to be kept down to no more than a few tenths of a milligram per square centimeter, which is often impractical when chemical procedures requiring isotopic carriers are used.

Whenever it becomes necessary to do β measurements on thicker samples, it is advisable either to standardize the thickness at a fixed value—this is often adequate for relative measurements, for example in tracer applications—or to prepare an empirical calibration curve for different thicknesses. In either case careful attention must be given to a reproducible mechanical form for the sample, and reproducibility should be tested by experiment. The calibration curves obtained normally include the effects of back-scattering.

The need for empirical calibrations arises because the effects of self-scattering and self-absorption vary not only with the energy but also depend strongly on the chemical form of the sample (N1), on the backing used, and on the geometrical arrangement of sample and detector. In general, the counting rate

[6] Such films of VYNS, Formvar, Zapon lacquer, or nylon are readily prepared (Y1). The plastic is dissolved in a suitable solvent, a drop of the solution is allowed to spread on a clean water surface, the solvent evaporates, and the resulting film is lifted off the water by means of a metal frame or wire loop.

obtained with a given amount of activity does not by any means fall off uniformly or exponentially as the sample thickness is gradually increased. Under some circumstances, especially in materials of low Z and for moderate β energies ($E_{\max} \cong 0.2$–0.5 MeV), it may be almost independent of thickness over some range. Under other conditions, especially with high-Z sample materials and high-energy β emitters, an initial rise in counting rate is observed as the sample thickness is increased and is followed eventually by the drop to be expected from absorption effects. This increase, which is caused by the scattering of electrons by the sample material out of the sample plane into the counter, can raise the counting rate to as much as 1.3 or 1.4 times that observed with a weightless sample, and the maximum in the counting rate may typically occur at thicknesses of 1 to 10 mg cm^{-2}. Eventually, certainly by the time the thickness has reached about 10 per cent of the β-particle range, the self-absorption curve takes on the approximately exponential shape of an ordinary absorption curve (see p. 104).

When thicker and thicker samples are prepared from an active material, say, for example, $BaCO_3$ containing C^{14}, the measured counting rate at first increases because of the greater total activity in the sample and then approaches a constant value. This "saturation value" is clearly not a measure of the total activity of the sample but rather is related to the activity of the amount of sample material in an upper layer no thicker than the particle range R. It is thus a measure of the specific activity of the sample material. This fact is sometimes used to advantage in the measurement of the low-energy tracers; no correction for self-absorption is applied, and it is necessary only to measure the activities of thick samples of the same uniform area and the same chemical composition. Indeed in many tracer experiments the specific activity is more directly significant than the total activity. The minimum sample thickness required for this type of measurement is clearly not more than the range R, and for most practical purposes $0.75R$ is adequate because of the very small relative contribution of the lowest layers.

Useful Sample-Mounting Techniques. A large variety of methods is available for the preparation of solid samples for radioactivity measurements. The choice will depend on the type of measurement to be performed, the total as well as the specific activity available, the physical and chemical properties of the radioelement to be measured, the thickness and degree of uniformity desired, the need for quantitative or semiquantitative transfer, and so on.

One of the simplest techniques is the evaporation of a solution to dryness in a shallow cup or, in small portions, onto a flat disk. This procedure, best carried out under an ordinary infrared lamp, always leaves a very nonuniform deposit, with most of the residue in a ring around the edge. If the active substance is first precipitated and the slurry evaporated, preferably with stirring, a much improved deposit usually results. In another procedure the slurry is added and dried a little at a time. It is sometimes helpful to place a circular piece of cigarette paper, slightly smaller than the dish or disk, on the flat surface; the solution or slurry is allowed to spread over the paper and then to dry; when dry, this type of paper weighs about 1 mg cm^{-2}.   Other tricks are useful for obtaining fairly uniform deposits in well-defined areas.   A border (e.g., of Zapon lacquer) painted around the periphery of the sample disk prevents unwanted spreading, and the use of a wetting agent (such as tetraethylene glycol) can greatly improve the uniformity of deposits obtained by the evaporation of solutions (D1).

Other methods of preparing samples are usually preferable, especially if the bulk of the deposit or the volume of solution or suspension is large.   Filtration on a small Büchner or Gooch filter can give reasonably uniform and very nearly quantitative deposits of precipitates on the filter paper.   If the precipitate and filter are washed finally with alcohol or acetone,[7] the bits of precipitate creeping up the sides of the filter are likely to be washed down; also the paper is then more easily dried, either with or without an ether wash.   A glass chimney held tightly against the paper in a Büchner funnel with perforated surface ground flat will help to confine the precipitate to a definite area.   A particularly convenient arrangement for sample preparation by filtration is shown in figure 12-2.   The filter paper is supported on a sintered glass disk with a fire-polished rim, which is clamped between the thickened and ground ends of two glass tubes; the top tube serves as the area-defining chimney and the bottom tube is fitted into a rubber stopper on a filter flask.

Sedimentation of a precipitate followed by decantation or evaporation can be used to give uniform deposits.   Special cells with demountable bottoms are employed for this purpose so that the sample may be removed on its mounting for measurement.   Sedimentation cells of this type fitted into laboratory centrifuge cups may be used to give harder deposits in much less time.

Samples prepared in any of these ways should be thoroughly dry before measurement, otherwise the self-absorption and self-scattering will change with time as water evaporates.   Samples may be found subject to loss of precipitate through powdering when dry; this can be especially troublesome if the active dust should contaminate a measuring instrument.   To avoid this effect a few drops of a solution of zapon lacquer in alcohol or of collodion in acetone may be used to wet the sample when it is first dried; the concentration of the lacquer or collodion in the solution should be so small that only about 0.1 mg cm^{-2} of its dry residue will be left on the sample.

If the radiation to be measured is sufficiently penetrating, solutions of active

---

[7] The use of alcohol (presumably to lower surface tension) has other applications in the manipulation of samples.   For example, when a precipitate is centrifuged down in a semi-micro cone, some is almost always trapped in the liquid surface (meniscus); addition of a few drops of alcohol on top of the solution followed by another centrifugation will usually bring down most of this precipitate.   In the mounting of slurries, as preciously described, a few drops of alcohol as a wash will often clean the residual slurry from the transfer micro-pipet (a glass tube drawn down to have a long tip less than 1 mm in diameter and fitted with a rubber bulb, like an "eye dropper").

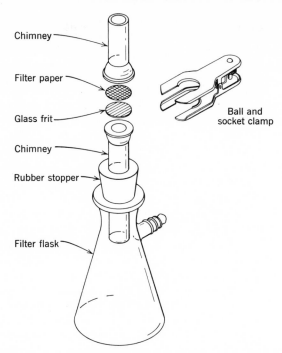

Fig. 12-2 Convenient filter apparatus for the preparation of radioactive samples for measurement. (Courtesy Brookhaven National Laboratory.)

materials may be placed in glass or plastic vials and assayed in well-type scintillation counters or they may be introduced into an outer jacket built as a part of a glass counter tube. A dipping counter is arranged to permit immersion of the whole counter (except an end for electric connections) in the solution to be measured. Radiations from relatively thick layers of solutions may be counted reproducibly only if the solution densities are comparable and if the relative composition of the solutions in terms of elements of various atomic numbers is kept approximately constant, especially for γ-ray measurements (see chapter 4, section C).

Many of the techniques mentioned for preparation of thin targets can also be used to prepare thin sources for measurement. For example, if the radioactive element is a metal like copper, excellent samples for measurement may be prepared by electrodeposition. Other types of ions may often be deposited by different electrode reactions under suitable conditions. For example, lead may be deposited on an anode as PbO_2 from alkali plumbite solutions. Insoluble hydroxides may be deposited from neutral solutions on cathodes because of the liberation of hydroxide ions there: $H_2O + e^- = \frac{1}{2}H_2 + OH^-$. Insoluble ferrocyanides may form on a cathode from the reduction of some metal ferricyanide solutions. A metal fluoride may be precipitated in adherent form if the metal can be oxidized or reduced at an electrode from a fluoride-soluble to a fluoride-insoluble state; for example, UF_4 may be deposited in this way and then ignited in air to U_3O_8.

"Weightless" Sources. The preparation of the extremely thin (sometimes loosely called weightless) sources required for α and β spectrometry and for 4π counting presents special problems. In order to prevent broadening of lines in α-particle or conversion-electron spectra, to minimize distortions of β spectra, and to ensure virtually 100 per cent efficiency in 4π measurements, such sources may have to be as thin as 1–10 μg cm^{-2}, and uniformity is important, insofar as the specification of maximum surface density, set by a given experimental situation, applies not only to the source as a whole but to any small portion of it. Samples for 4π counting and for investigations of β-spectral shapes must not only be thin themselves but they must be mounted on equally thin backings. The preparation of thin plastic films for this purpose has already been mentioned (footnote 6, p. 412) and is discussed in various review articles (Y1, S4). An insulating film with a radioactive source deposited on it can become highly charged as a result of the emission of charged particles from the source, and the source potential built up in this manner can seriously distort the spectrum of emitted particles. For this reason, films used for β-spectrometer or 4π sources should always be rendered conducting, usually by evaporation of a thin ($\cong 5$ μg cm^{-2}) metal coating, and grounded. A noble metal has obvious advantages since sources are often deposited from acid solutions. Gold coatings have been used most frequently but palladium is even more advantageous because its smaller infrared absorption (compared with gold) lowers the probability of film breakage when the source is evaporated under a heat lamp.

When quantitative deposition of a given amount of source material on a thin backing is required, as in absolute disintegration rate measurements by 4π counting, evaporation of a solution is the method of choice. Uniform spreading is usually ensured by use of a wetting agent such as insulin. An aqueous insulin solution (concentration $\cong 5$ per cent) is pipetted onto the spot to be covered by the source, then removed with the pipette; the residue may be dried, and the sample is then pipetted onto the spot and dried under a heat lamp. Successive portions of sample as well as washings may be added and evaporated.

When quantitative transfer is not essential, thin uniform samples may be prepared by one of the techniques discussed in connection with target preparation (pp. 393–394): volatilization, electrodeposition, electrophoresis, and electrospraying (Y1, D1, S4). Volatilization from a hot filament can be applied to most elements. Occasionally it can even be carried out in air, for example, for transferring such volatile elements as polonium and astatine from a metal holder to a counting disk placed above it. More often a simple vacuum system is used. By careful design of the filament and receiver assembly the evaporation can be made reasonably directional so that losses are not excessive. The catcher can even be a thin plastic film if heating by radiation from the filament can be kept from destroying the film. Whenever a source is prepared by volatilization, it is advisable to get rid of volatile impurities by heating the sample filament to a temperature just below that required for the evaporation of the desired material, and then bringing the source mount into position and raising the temperature to the required range.

A special technique is available for the preparation of thin samples of radionuclides which are themselves formed by radioactive decay, especially α decay. The recoil energy imparted by the α decay is used to carry the daughter atoms out of a deposit of the parent material and onto a nearby catcher plate. Similarly, the recoil energy imparted by a nuclear reaction can be used to transfer reaction products directly from a thin target deposit to a catcher foil placed downstream from the target in the ion beam. These techniques have been particularly useful in the investigation of short-lived transuranium nuclides produced in accelerator bombardments.

F. DETERMINATION OF ABSOLUTE DISINTEGRATION RATES

As mentioned before, the determination of absolute disintegration rates presents special problems in addition to those encountered in relative activity measurements. Yet a knowledge of absolute disintegration rates is sometimes required. Whenever a reaction cross section is to be determined, the number of product nuclei formed must be found, and for radioactive products this is best accomplished through knowledge of the decay constant and measurement of the disintegration rate. If the disintegration rates of a given nuclide are to be determined in many samples (e.g., in the study of an excitation function), relative measurements are quite adequate as long as the instrument used is calibrated with one sample of the nuclide whose absolute disintegration rate is known. Even in nuclear spectroscopy there is frequent need for what amounts to absolute measurements; in general, when the branching ratio of two decay modes is to be determined, measurements of two different types of radiation with different instruments are involved, and the absolute efficiency of each measurement must therefore be known. This applies to EC/β^+ ratios, to absolute conversion coefficients, and often even to measurements of the number of γ quanta per β decay.

Alpha Emitters. Given adequately thin, uniform samples, the determination of absolute α-disintegration rates is relatively simple. Either a proportional counter or an ionization chamber with linear amplifier can be readily arranged to count 100 per cent of the α particles entering the active volume, and 2π geometry is easily achieved if the sample is introduced into the counter or chamber volume. A correction must be applied for α particles backscattered from the sample mount; but in contrast to the β-particle case (p. 108) this correction is small (4 per cent for Pt, less for lower Z). With care, accuracies of ± 1 per cent can be achieved in measurements of α-disintegration rates by this method. The limitation is usually in sample preparation. The calorimetric method mentioned in chapter 3, section E, for the measurement of absolute α-disintegration rates is capable of at least comparable accuracy. It requires a larger sample activity but makes no demands on the sample's geometrical arrangement, thinness, and the like. It does require a knowledge of the α-particle energy; this is conveniently obtained from the range, or better from magnetic-spectrometer measurements.

4π Counting. For the determination of absolute disintegration rates there are obvious advantages to the use of 4π geometry, particularly if the counting efficiency is 100 per cent. Under these conditions every disintegration gives rise to one count, regardless of the decay scheme .(provided that there is included no state with lifetime comparable to or greater than the resolving time of the equipment). The observed counting rate then equals the disintegration rate.

A number of arrangements for 4π counting have been used. The introduction of a β-active sample in gaseous form inside a counter (usually a proportional counter) provides almost a 4π geometry; the end effects and wall effects can be made small and can be evaluated by experiments with different counters in which the ratio of sensitive to insensitive volume is deliberately varied. Gas counting is particularly useful for soft-β emitters (such as H^3, C^{14}, S^{35}, and Ni^{63}) and for low-Z electron-capture nuclides (such as Ar^{37}); in the latter case even the very soft Auger electrons may be counted quantitatively. In an analogous technique the sample may be dissolved in a liquid scintillator. For any but the lowest-energy β emitters and for emitters of electromagnetic radiation of less than about 200 keV every disintegration may again be counted. To minimize losses by edge effects, it is necessary to use scintillator dimensions that are large compared with the range of the radiations. One of the principal problems in 4π liquid scintillation counting is to prepare the active material in a form suitable for dissolution in the organic scintillator without adverse effect on the scintillation properties. Much work has been done along this line (B2).

With solid samples, 4π-β counting requires the use of extremely thin deposits on very thin supports. In the usual 4π-β counter such a sample is mounted in the central hole of a thin metal partition between two identical proportional counters connected in parallel. Each of the two counters may be in the shape of a hemisphere, a half cylinder, or a flat cylinder, with a straight or looped wire anode. One type of 4π-β counter is shown in figure 12-3. It has a simple slide between O-rings for introducing the sample. With careful source preparation on properly conducting films, accuracies of ± 1 per cent can be achieved in 4π

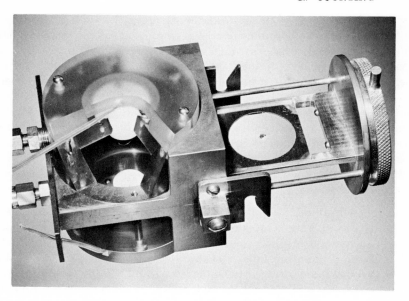

Fig. 12-3   A 4π counter with a section cut away.   The two wire loops of the upper and lower counter are visible.   The electrical leads and the tube connections for the counter gas are visible at the left.   The sample slide with a source mounted on a plastic film is in the "out" position.   The type of 4π counter shown is described by R. Withnell in *Nucl. Inst. and Methods* **14,** 279 (1961).   (Courtesy R. Withnell and Brookhaven National Laboratory.)

determinations of most β emitters.   It is advisable to take a voltage plateau with each sample, or at least for each nuclide to be determined.   The parameters affecting the performance of 4π counters have been carefully investigated (P1).

Complications arise in the use of a 4π-β counter when the source decays partially or entirely by electron capture.   Even the very thin sources and backings ordinarily used in 4π counters may not be thin enough to allow all the Auger electrons to escape from the source into the counting gas.   The K X rays of all but the light elements, on the other hand, will have an appreciable probability of crossing the active volume without making an ion pair.   The detection efficiency for electron capture decay will thus, in general, be neither zero nor 100 per cent and may be rather difficult to determine.   Other methods are described later on for the determination of electron-capture disintegration rates.

The usefulness of 4π geometry is not confined to the assay of β emitters. Various approximations to 4π counting of X rays and low-energy γ rays with scintillators have been described.   The growing of NaI crystals from melts containing the active material is very tedious.   More commonly, the sample is either sandwiched between two flat-faced NaI scintillators or placed in a well scintillator with a second scintillator covering the well.   In either case the outputs of the two photomultipliers are connected so that the pulse heights recorded in the two scintillators from a given event are added together.   Apart from losses in the crystal coverings, these arrangements can be made essentially 100 per cent efficient for the measurement of photons up to moderate energies, say $\cong 200$ keV.   The method is useful for the determination of the electron-capture disintegration rates of high-Z elements; here, if we denote by f_K the K-shell fluorescence yield, then in a fraction f_K of all the K-capture transitions K X rays are emitted, and these are detected whether or not γ rays are emitted also, so long as no X-ray-emitting delayed states are involved.   Since f_K is large and rather well known (see figure 2-8), the K-electron capture rate can be fairly accurately deduced.

Coincidence Method.   For nuclides with relatively simple decay schemes the absolute disintegration rates may be determined by coincidence measurements.   The method is easily understood for the simple case in which the emission of one γ quantum follows each β decay and the spectrum is simple. Consider two counters arranged to count β rays and γ rays, respectively, with measured counting rates $\mathbf{R}_\beta$ and $\mathbf{R}_\gamma$ and with β-γ coincidences also measured with rate $\mathbf{R}_{\beta\gamma}$; then $\mathbf{R}_\beta = \mathbf{R}_0 c_\beta$, $\mathbf{R}_\gamma = \mathbf{R}_0 c_\gamma$, where the coefficients c_β and c_γ may be thought of as defined by these equations and include all effects of solid angles, counting efficiencies and absorption corrections, and $\mathbf{R}_{\beta\gamma} = \mathbf{R}_0 c_\beta c_\gamma$. Now $\mathbf{R}_\beta \mathbf{R}_\gamma / \mathbf{R}_{\beta\gamma} = \mathbf{R}_0$, and the absolute disintegration rate is given very simply in terms of this ratio of three measured counting rates.   The contribution of γ rays to the counting rate in the β counter (and possibly to coincidence counts) must be measured in a separate experiment with an absorber that prevents the β rays from entering the β counter; this is essentially a background that must be subtracted from $\mathbf{R}_\beta$ (and from $\mathbf{R}_{\beta\gamma}$).   No complications result from a complex γ spectrum (two or more γ rays in cascade, possibly with cross-over transitions) as long as only one β transition is involved.   The coefficient c_γ then merely refers to the average over-all efficiency of the γ detector for the γ rays.

Many subtle effects arise in coincidence measurements (C1).   If an extended rather than a point source is used, the response of at least one of the detectors must be independent of the location in the source from which the detected radiation originates; otherwise the simple equations given do not hold.   This condition is usually more easily satisfied with the γ detector.   However, the use of a 4π counter as the β detector of virtually 100 per cent efficiency (for all parts of the sample and for different β branches if such are present) has also found extensive use in accurate standardization of samples (C1).   The validity of the equations given in the preceding paragraph also depends on the absence of angular correlations between the directions of emission of the coincident radiations.   If there is any suspicion that angular correlations do exist, measurements should be made at more than one detector-source-detector angle.

Even in simple $\beta\gamma$ measurements small errors can result from the detection in the γ counter of bremsstrahlung produced by β particles stopping in the absorber used to shield the γ counter from β rays.   This effect will increase $\mathbf{R}_\gamma$ without a corresponding increase in $\mathbf{R}_{\beta\gamma}$ because the particular β particles giving rise to this bremsstrahlung have little chance of being counted in the β counter.   This source of trouble is almost eliminated if the γ detector is a scintillation spectrometer operated with a narrow energy channel.   A more appreciable error of the same sort would be expected for any sample emitting positrons because of the annihilation radiation.   This is minimized if the absorber (just large enough to subtend the same angle at the sample as the counter) is placed near the sample rather than near the γ counter.   Again, energy discrimination can be helpful, unless the nuclear γ ray measured has an energy near 0.5 MeV.

Extreme caution is required when the coincidence method is to be applied to more complex decay schemes.   Scintillation spectrometry for γ rays can be used to great advantage, but does not, by any means, constitute a panacea. Some of the problems may be illustrated by the following example in which $\gamma\gamma$ rather than $\beta\gamma$ coincidence measurements are considered.   In principle, the two techniques are very similar, but some particular complications arising in $\gamma\gamma$ measurements are worth noting.   Consider a nuclide with the rather simple decay scheme shown in figure 12-4a and the problem of determining the disintegration rate of a sample of this nuclide by a $\gamma\gamma$ coincidence measurement. The two γ rays have energies E_1 and E_2, and the measurement is to be performed with two NaI(Tl) scintillation detectors, A and B, arranged as schematically shown in figure 12-4b; the pulses from the detectors are fed to pulse height analyzers, with a channel on detector A set to encompass the photopeak of γ_1 and a channel on detector B to encompass the photopeak of γ_2.   Figure 12-4c shows what the γ spectrum obtained with either detector might look like; we have taken $E_1 < E_2$ and have indicated on the graph where the energy channels on A and B might be set.

One of the problems encountered in $\gamma\gamma$ coincidence work (and also in many other attempts to use γ spectroscopy for quantitative intensity measurements —see later) is illustrated by the small peak at energy $E_1 + E_2$ shown in figure 12-4c.   This peak arises because there is a finite probability that a γ_1 quantum and a γ_2 quantum from the same disintegration undergo photoelectric absorption in the same detector.[8]

Similarly, there is a long "Compton" tail extending out to the energy $E_1 + E_2$ and including pulse addition events of three kinds: photoelectric absorption of γ_1 plus Compton effect of γ_2; Compton effect of γ_1 plus photoelectric absorption of γ_2; sum of two Compton events.   From the detection

---

[8] We neglect here the pulse addition that can result from "accidental" coincidences between pulses from two different decay events arriving within the resolving time of the coincidence circuit.   Such effects can be minimized by proper choice of sample strength, and their magnitude can usually be determined.

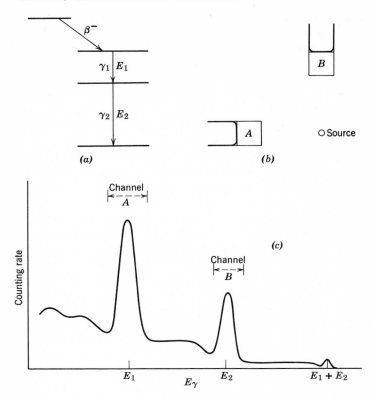

Fig. 12-4 Illustration of the use of $\gamma\gamma$-coincidence measurements for absolute disintegration-rate determination. (a) Decay scheme involving two γ rays in cascade, following β^- decay. (b) Schematic arrangement of scintillation detectors A and B for $\gamma\gamma$ measurement. (c) "Singles" spectrum observed with either detector.

efficiencies and spectral distributions for the two individual γ rays the intensity and spectral distribution of the sum spectrum can be computed in principle. However, since the intensity of the sum spectrum depends on the square of the solid angle subtended by the counter, whereas the main spectra have intensities proportional to the first power of the solid angle, it is in practice often advisable to work at solid angles small enough to make the summing effects negligible. Even when that is not possible, data taken in different geometries can be extrapolated to zero solid angle. The particular reason why pulse addition effects must be guarded against in coincidence measurements for absolute disintegration rate determinations is that they always alter the "singles" counting rates but leave the coincidence rate unaffected. In our example (although not necessarily in other situations) pulses are thrown out of the area under either of the two photopeaks in the "singles" spectrum by pulse addition with the pulse resulting from photo or Compton absorption of the other γ ray; this throw-out correction is only partly offset by the pulses thrown into the peaks by the addition of two Compton events (or, in the case of the peak at E_2, a photoelectric absorption of γ_1 coupled with an appropriate Compton scattering of γ_2). On the other hand, a coincidence count between counters A and B, with channels set on the two photopeaks, can result only when γ_1 undergoes photoelectric absorption in A, and γ_2 undergoes photoelectric absorption in B, since photoelectric absorption of a given γ ray in one detector makes it impossible for the same γ ray to deposit any energy in the other crystal.

From here on we shall assume that our illustrative measurement (figure 12-4) is made in geometry low enough to justify neglect of pulse addition effects. We now define the following quantities:

ϵ_{1A} is the efficiency for detection of γ_1 in detector A.

ϵ_{2A} is the efficiency for detection of γ_2 in detector A.

ϵ_{2B} is the efficiency for detection of γ_2 in detector B.

$\mathbf{R}_A$ is the measured counting rate (in channel) of detector A.

$\mathbf{R}_B$ is the measured counting rate (in channel) of detector B.

$\mathbf{R}_{AB}$ is the coincidence counting rate.

The efficiency ϵ_{1B} for detection of γ_1 in detector B is zero, since the channel of detector B is set at energy E_2 and we are neglecting pulse addition phenomena. If we denote the disintegration rate as $\mathbf{R}_0$, we can write

$$\mathbf{R}_A = (\epsilon_{1A} + \epsilon_{2A})\mathbf{R}_0,$$

$$\mathbf{R}_B = \epsilon_{2B}\mathbf{R}_0,$$

$$\mathbf{R}_{AB} = \epsilon_{1A}\epsilon_{2B}\mathbf{R}_0.$$

Therefore

$$\frac{\mathbf{R}_A\mathbf{R}_B}{\mathbf{R}_{AB}} = \left(1 + \frac{\epsilon_{2A}}{\epsilon_{1A}}\right)\mathbf{R}_0. \tag{12-1}$$

Equation 12-1 differs from the corresponding expression in simple $\beta\gamma$ coincidence measurements by the additive term $\epsilon_{2A}/\epsilon_{1A}$ and therefore does not immediately give the disintegration rate in terms of three measured counting rates. Although ϵ_{1A} is readily deduced from the measurements ($\epsilon_{1A} = \mathbf{R}_{AB}/\mathbf{R}_B$), the efficiency ϵ_{2A} for detection of the higher-energy γ ray (γ_2) in the detector set at the lower energy E_1 has to be obtained separately. If a nuclide emitting a single γ ray with energy near E_2 is available, the shape of its Compton spectrum in detector A in the region of energy E_1 can be experimentally determined. This measurement, together with the "singles" spectrum of the original source in detector A (figure 12-4c) gives essentially the ratio $\epsilon_{2A}/\epsilon_{1A}$ needed to evaluate (12-1). A cruder approximation can be obtained by assuming that the Compton distribution of γ_2 in the region of E_1 is flat.

The detailed discussion of the rather simple example given may serve to illustrate the care that must be taken in setting up the equations relating counting rates and efficiencies for any particular case under consideration. Additional problems arise when angular correlations exist. Without discussing them in detail, we merely call attention to the extreme case of angular correlation represented by the emission of the two 511-keV quanta emitted in opposite directions when a positron is annihilated. Measurement of coincidences between two annihilation quanta is not only a sensitive and selective method for the detection of positron emission (see p. 432) just because the extreme angular correlation is so characteristic, but it can also be used, in the form of a triple coincidence arrangement (with a third counter set to measure a nuclear γ ray that follows β^+ emission) to determine absolute disintegration rates of some β^+ emitters. In any measurement involving coincidences between two annihilation quanta it is well to consider intrinsic counter efficiencies and geometry factors separately. If each counter subtends a solid angle of $4\pi\omega$ steradians at the source and if each has an intrinsic efficiency ϵ for detection of 511-keV γ rays, the "singles" counting rate in each counter will be $2\omega\epsilon R_0$, where R_0 is the β^+-disintegration rate and where the factor 2 arises because there are two annihilation quanta per positron. The coincidence rate will be $2\omega\epsilon^2 R_0$, with only the first power of the solid angle appearing, since the emission of a 511-keV quantum into one counter ensures that its partner is emitted in the direction of the other counter.

Measurements at Known Solid Angle. When a detector of known (preferably 100 per cent) intrinsic efficiency is available, it is sometimes possible to obtain the absolute disintegration rate of a source by use of a defining aperture which makes it feasible to calculate the solid angle subtended by the detector at the source. Low-geometry arrangements of this type are often used in absolute α determinations. End-window proportional counters can be used in this way for determination of β-disintegration rates if the sample is placed at a distance of several centimeters from the counter window and a carefully machined defining aperture is put between sample and window. The solid angle is then readily calculated. A disadvantage of this arrangement is that absorption and scattering of the electrons in the air between sample and counter may require a sizable correction not readily evaluated. This difficulty is avoided if the space between sample and counter is evacuated or filled with helium. The self-absorption and back-scattering effects discussed in section E must be evaluated. To obtain results of higher accuracy, these corrections should be minimized by the use of very thin samples on very thin backings, much like the samples used in 4π counters. Once one has gone to such lengths in sample preparation, one might as well use a 4π counter which generally gives more accurate results.

The method of defined solid angles is well-suited to the determination of absolute X-ray emission rates; for example, from samples decaying by electron capture.   For all but the lowest X-ray energies absorption in air and in beryllium windows is quite small and the corrections for these effects can be accurately calculated.   For X-ray energies up to about 15 to 20 keV proportional counters filled to 1 to 3 atm with argon or krypton (with hydrocarbon admixtures) are quite suitable.   For X rays of higher energies thin wafers of NaI(Tl) crystals are more appropriate detectors.   Dimensions and, in the case of proportional counters, gas pressures can be chosen to assure almost 100 per cent absorption of the X rays.   The material surrounding the defining apertures must, of course, be of such a thickness that the X rays of interest are absorbed. With the use of pulse height analysis the desired X-ray counting rate can be determined even in the presence of other radiations.   To convert an absolute X-ray emission rate into an electron-capture rate, the fluorescence yield must be known (cf. figure 2-8).

Calibrated Detectors.   Once a source of a radionuclide of known disintegration rate is available, any detector may be calibrated in terms of this standard.   This calibration will be valid for other samples of the same nuclide, provided they are measured under precisely the same conditions.   The calibration may also be adequate for other radionuclides emitting radiations similar to those of the standard.

Standardized sources of a number of β and γ emitters are available in various forms (solutions and mounted solid samples) from the National Bureau of Standards (Washington, D. C.), from the International Atomic Energy Agency (Vienna, Austria), and from some commercial companies.

The efficiency of an end-window β counter for detection of a particular β emitter is usually best determined via a 4π-counter standardization.   The disintegration rate of a "weightless" sample of the nuclide in question prepared on a thin film is determined with a 4π counter.   Then aliquots of the same activity, mixed with appropriate amounts of carrier, are prepared and mounted in the desired manner for end-window counting.   The amounts of activity in the aliquots relative to the amount in the 4π sample may be determined by accurate pipetting and quantitative transfers or, more conveniently by comparison of the activities in a device whose response is not sensitive to sample thickness, backing, etc., such as a γ detector at a large distance.   This technique allows determination of the end-window counter efficiency directly for the sample thicknesses, sample backings, and geometrical arrangements of interest, without any need for separate corrections for self-absorption, self-scattering, back-scattering, air absorption, window absorption, etc.   Without too much difficulty, this method can usually be made to yield disintegration rates accurate to ± 5 per cent or less.   Attempts to determine absolute disintegration rates with end-window counters by evaluation of the separate correction factors mentioned (which are discussed in S5) were common before the use of 4π counters became widespread, but the errors involved can be sizable, and these methods are no longer considered satisfactory.

Absolute determination of γ-emission rates with NaI(Tl) scintillation counters is widely used. In its simplest form this involves merely the standardization of a scintillation counter with a source of known disintegration rate and use of the same detector for the assay of other samples of the same nuclide mounted in the same manner as the calibration standard. Pulse height analysis is not required in this application, and the method can be used with well-type scintillation crystals as well as with external source arrangements. The need for reproducible geometry cannot be overstressed; for example, the height to which the samples extend in a scintillator well must be carefully controlled. With proper precautions, the accuracy of the method is limited essentially by the accuracy with which the disintegration rate of the calibration standard is known.

Greater versatility in absolute γ-ray intensity measurements can be achieved with pulse height analysis. The emission rate of a particular γ ray is then usually inferred from the total counting rate in the photopeak. The photopeak efficiency of a given crystal for sources mounted in a particular geometry can be determined as a function of γ-ray energy by means of standard sources

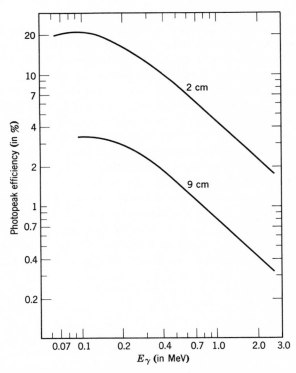

Fig. 12-5 Photopeak efficiency of a 3 × 3 in. NaI(Tl) scintillator as a function of γ-ray energy for two source-detector distances.

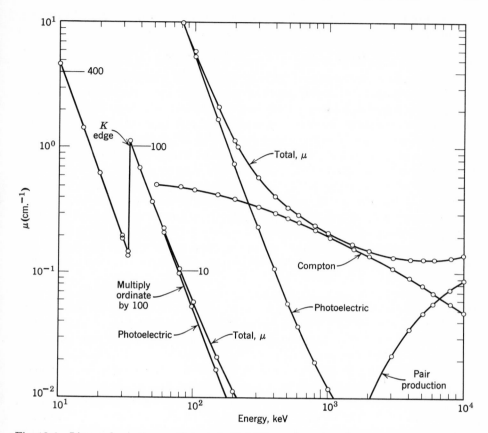

Fig. 12-6   Linear absorption coefficients in NaI (Tl) as a function of photon energy.   (Reproduced from reference J1.)

emitting γ rays of various energies.   Typical curves obtained by this procedure are shown in figure 12-5 for a 3 x 3 in. cylindrical crystal and for two source geometries.   As seen in the figure, a straight-line relationship is obtained over a wide energy range when the logarithm of the photopeak efficiency is plotted against log E_γ.

Most of the difficulties in absolute γ counting have to do with the problem of determining areas under photopeaks.   We have already discussed, in connection with coincidence counting (p. 422), the "throw-in" and "throw-out" corrections that can arise from pulse addition; as discussed there, these corrections can be evaluated by variation of source geometry.   Some of the other effects which can occur in scintillation detectors are most readily understood if the three basic processes for γ-ray interactions are briefly reviewed.   Figure 12-6 shows the linear absorption coefficient μ in NaI(Tl) as a function of γ-ray energy, along with its three component parts due to photoelectric absorption, Compton effect, and pair production (cf. chapter 4, section C for a discussion of the energy-dependence). Most of the interactions take place in iodine because of its high Z.

Gamma rays of low energy ($\lesssim 250$ keV) interact predominantly by photoelectric absorption. The photoelectrons produced have very short ranges in NaI and are essentially always absorbed. However, in a large fraction of the photoelectric interactions with γ rays above the K-absorption edge of iodine (33.2 keV), an iodine X ray (energy $\cong 28$ keV) is produced and, if the interaction took place near the surface of the crystal, this X ray has a finite chance of escaping without interaction. Therefore the photopeak of any low-energy γ ray is always accompanied by an "iodine escape peak" 28 keV below the photopeak. In general, the size of the escape peak in relation to the main peak increases with decreasing γ-ray energy because the lower the energy, the more interactions take place near the surface. Photopeak intensities must be determined consistently, either always including or always excluding the escape peak. An example of a pulse-height spectrum showing the iodine escape peak is given in figure 12-7.

Above about 250 keV the Compton process becomes dominant. The scattered photon may escape from the crystal or make another interaction (by photoelectric absorption or by a second Compton scattering). The relative probability of these two fates depends on the scintillator size; the larger the crystal, the greater the probability that even after an initial Compton scattering the entire energy of the γ ray will be deposited in the crystal, in which case the pulse height produced will be the same as that for an initial photoelectric absorption. This effect of crystal size is illustrated in figure 12-8, which shows the decrease in the height of the Compton distribution in relation to the photopeak when a Cs^{137} spectrum is measured with successively larger NaI crystals. The advantages of large crystal size for measurement of high-energy γ rays are evident. The figure also shows the prominent back-scattering peak which is usually observed (at ≤ 250 keV) in any high-energy γ spectrum because some of the γ rays from the source are Compton-scattered in the surrounding material (air, shield, sample support, etc.) and some of those that are scattered through about 180° reach the detector. Back-scatter peaks can sometimes be mistaken for or can mask photopeaks.

Pair production creates additional complications. The electron and positron formed have high probability of stopping in the crystal. However, the positron annihilates, and one or both of the annihilation quanta may leave the scintillator without interacting. Pair production thus results in a spectrum containing three peaks for a given incident γ-ray energy E: the full energy peak at E, the "single-escape" peak at $E - 0.511$ MeV, and the "double-escape" peak at $E - 1.02$ MeV. The relative magnitudes of these peaks are functions of crystal size and geometry in a manner analogous to the size effect discussed for the Compton continuum.

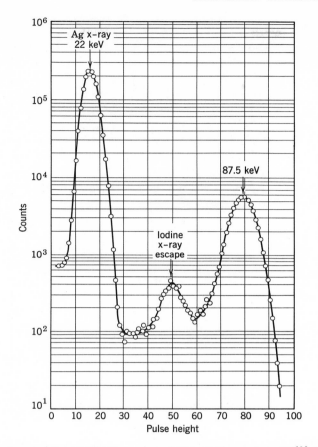

Fig. 12-7   Spectrum of the 87.5-keV γ rays and 22-keV X rays from Cd^{109} decay, observed with a 3 × 3 in. NaI(Tl) crystal.   In addition to the two photopeaks, the peak due to the escape of iodine K X rays is clearly seen.   (Reproduced from reference J1.)

From our brief discussion of the consequences of the various γ-ray interactions in NaI crystals it should be clear that the analysis of γ spectra when γ rays of more than one energy are incident on the detector can become a very complex problem.   Decomposition of a complex γ spectrum into its components can usually be accomplished fairly well if spectra of the individual components can be measured under the same experimental conditions.   The spectrum of the most energetic γ component is determined and normalized to the spectrum of the mixture at the full-energy peak, and this normalized component spectrum is subtracted from the mixed spectrum.   This procedure is repeated with the γ ray of next highest energy, etc., until all the components have been peeled off.   Some commercially available multichannel analyzers

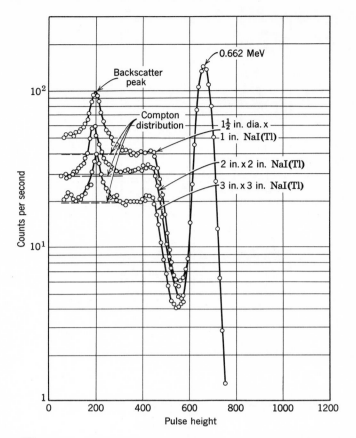

Fig. 12-8   Cs^{137} spectra obtained with NaI(Tl) crystals of three different sizes.   The spectra have been normalized to the same photopeak area.   (Reproduced from reference J1.)

have provisions for taking a standard spectrum while another spectrum is stored in another part of the analyzer memory and subtracting a certain fraction or multiple of the standard spectrum, channel by channel, from the stored spectrum.   In any attempt to decompose a γ spectrum by subtraction of standard spectra, care must be taken to avoid any shifts in pulse height scales between the spectra.   In most γ-ray spectrometers the gain varies somewhat with counting rate, especially at high rates.   Thus, not only the energies but also the intensities of the standards must be chosen to match those in the unknown to avoid gain shifts that require tedious corrections.

A standard spectrum which exactly matches a component of a complex mixture may, of course, not always be available.   Sometimes it is then possible to interpolate the desired spectrum from spectra of higher- and lower-energy γ rays, although this is by no means a simple matter because of the complex structure of Compton distributions and pair spectra. Some computer programs for γ-ray spectrum analysis have been developed (A1).

A simple and ingenious method has been suggested (B4) for the absolute determination of disintegration rates of nuclides emitting two (or three) γ rays in cascade. To illustrate this method, consider a sample of disintegration rate $\mathbf{R}$, emitting γ rays γ_1 and γ_2 in cascade. If ϵ_1 and ϵ_2 are the two photopeak efficiencies, ϵ_{1t} and ϵ_{2t} the total efficiencies (all including geometry factors), the areas A_1 and A_2 under the two photopeaks can be written as

$$A_1 = \mathbf{R}\epsilon_1(1 - \epsilon_{2t}) \tag{12-2}$$

$$A_2 = \mathbf{R}\epsilon_2(1 - \epsilon_{1t}). \tag{12-3}$$

The area under the sum peak is

$$A_{12} = \mathbf{R}\epsilon_1\epsilon_2. \tag{12-4}$$

Finally, the total counting rate under the entire spectrum can be expressed as

$$T = \mathbf{R}(\epsilon_{1t} + \epsilon_{2t} - \epsilon_{1t}\epsilon_{2t}). \tag{12-5}$$

Combining (12-2), (12-3), and (12-4), we obtain

$$\frac{A_1 A_2}{A_{12}} = \mathbf{R}(1 - \epsilon_{1t})(1 - \epsilon_{2t}) = \mathbf{R}(1 - \epsilon_{1t} - \epsilon_{2t} + \epsilon_{1t}\epsilon_{2t}),$$

and substituting (12-5) in this expression we have

$$\mathbf{R} = \frac{A_1 A_2}{A_{12}} + T. \tag{12-6}$$

To minimize the difficulties inherent in accurate determinations of photopeak areas and to avoid errors which arise if angular correlations exist, it is advisable to use large solid angles, thus making the first term in (12-6) relatively small compared to T. The method is, in fact, particularly useful in conjunction with well counters. Equation 12-6 is valid even if one of the γ rays is not 100 per cent abundant.

G. DECAY SCHEME STUDIES

A great deal of work by nuclear chemists and physicists has been directed toward the collection of data on decay schemes (disintegration schemes) of radioactive nuclides. A complete decay scheme includes all the modes of decay of the nuclide, their abundances, the energies of the radiations, the sequence in which the radiations are emitted, and the measurable half-lives of any intermediate states. When possible, spin and parity assignments of the various energy levels involved are included in the decay scheme. Elucidation of decay schemes usually requires careful and specialized measurements of the radiations, and, particularly if spins and parities are wanted, information is desirable on the shapes of β spectra and on angular correlations of various radiations. Some illustrative examples of decay schemes are shown in figures 12-9 to 12-12; these schemes are discussed individually in some detail in subsequent paragraphs.

The amount of detail known about any given decay scheme depends very strongly on the refinements in instrumentation and technique used in its investigation. Frequently, when a previously well-studied decay scheme is reinvestigated with instruments of improved resolution or sensitivity, new features such as branch decays of low abundance are discovered; it now appears that except among the lightest nuclei there are indeed few decay schemes that are truly simple (for example, consisting of a single β transition to a ground or first excited state).

The starting point of a decay scheme investigation depends, of course, on the information already available about the nuclide under study. If one deals with a newly discovered nuclide, the half-life is usually established first.[9] The nature of the radiations may be identified by measurements with different detectors—α counters, thin-window β counters, γ scintillation counters—and by some absorption measurements. Electron capture may be discovered by identification of the characteristic X rays emitted.

It is occasionally necessary to distinguish between β^- and β^+ particles. In the electron spectrograph this distinction is made easily according to the polarity of the magnetic field.[10] Without such an instrument a crude determination may be made to distinguish a predominantly positron from a predominantly negatron emitter. If the sample and detector are separated by a few inches and shielded from each other, a magnetic field from an electromagnet or sizable permanent magnet[11] may be arranged to bend particles of one sign around the shield toward the detector. A more sensitive method for the identification of positrons is based on the detection of the annihilation radiation (two γ rays of 0.51 MeV for each positron annihilated at the end of its range). The two γ quanta are emitted in opposite directions; therefore, if a positron emitter and sufficient material to absorb some or all of the positrons are placed between two γ counters, measurements of coincidences with the counters placed at 180° and at a different angle will give different coincidence rates, even if other coincidences due to nuclear γ rays are also recorded. If a calibra-

---

[9] Even when the decay scheme of a nuclide of well-known half-life is under investigation, the nuclide may not be available free from other radioactive isotopes. In that case measurements may have to be made as a function of time to sort out those radiations associated with the nuclide of interest.

[10] In a magnetic lens spectrometer electrons of either sign are focused unless special spiral baffles are incorporated which allow electrons of only one sign to reach the detector for a given polarity of the field.

[11] A field of 1000 to 2000 gauss over a volume of about 1 in.[3] is convenient and readily available with a permanent magnet such as those designed for radar magnetrons.

tion source of known β^+-emission rate is available, the method can be made into a convenient quantitative technique for β^+ measurements.

The determination of the energy spectra of the radiations emitted involves the use of the various types of spectrometers and spectrographs mentioned in chapter 5. The choice of instrument may depend on the information desired, on the resolution needed, on the source strength and specific activity available, and on the half-life. Source thickness and uniformity requirements also differ widely for different types of measurements, as discussed in section E. These requirements are least exacting for γ-ray measurements, and, partly for this reason, γ-ray spectrometry as outlined at the end of section F is probably the most widely used technique in decay scheme studies today.

Questions about the sequence in which various radiations are emitted and about the existence of alternative decay paths are usually answered by coincidence measurements. As already indicated in chapter 5, the more selectively each of the two detectors used in a coincidence study records one particular radiation, the more readily even very complex decay schemes may be disentangled. Since increased selectivity is almost always accompanied by decreased detection efficiency, some compromise usually needs to be made in practice. The development of the multiparameter, multichannel analyzers (see chapter 5) has enormously increased the scope of coincidence measurements that can be undertaken without excessive expenditures of time. Coincidence techniques are often very helpful in energy determinations also. For example, a low-intensity, low-energy β branch in the presence of an intense high-energy component may well escape detection in a β-spectrographic measurement. If, however, the high-energy β spectrum is not coincident with γ rays, but the low-energy branch is, then a β spectrum taken in coincidence with γ rays will show the low-energy component only, and precise measurement of its end point and spectrum shape thus becomes possible. With a β spectrometer and a scintillation spectrometer in coincidence, the β spectrum coincident with each of several γ radiations may be observed.

The following examples may serve to illustrate the techniques of decay scheme studies. Further details of these decay schemes and references to the experimental data may be found in refs. S6 and N2.

Gold-198. This nuclide of 2.698 days half-life was for a long time thought to have a simple disintegration scheme, decaying by the emission of a single β^- group of allowed spectrum shape and upper energy limit 0.96 MeV to the lowest excited state of Hg^{198} at 0.412 MeV above the ground state. This scheme was verified by numerous spectrometer and coincidence measurements.[12] The nuclide has, in fact, been frequently used as a standard source

[12] When reactor-produced Au^{198} sources became available, there was some confusion about the decay scheme because several workers found additional lower-energy γ rays in such sources. Subsequently, the relative intensity of these γ rays was found to depend on the neutron flux in which the gold had been irradiated, and they turned out to be associated with 3.15-day Au^{199} formed with very large cross section by neutron capture in Au^{198}.

for the calibration of spectrometers and coincidence circuits.   The energy of the γ ray has been determined with great precision in crystal spectrometers and by magnetic spectrometer measurements of the conversion electrons and is given as 0.411775 ± 0.000007 MeV.   The internal-conversion coefficients have been measured in magnetic spectrometers by comparison of the areas under the conversion electron peaks with the area under the entire β spectrum. The best values appear to be 0.028 for the K-conversion coefficient, 2.9 for the K/L conversion ratio, and $2.2/2.4/1.0$ for the $L_I/L_{II}/L_{III}$ ratio.   These data establish an $E2$ assignment for the 412-keV transition.   Since the ground state of even-even Hg^{198} is presumably $0+$, it thus appears that the 412-keV level is $2+$, in accord with the general rule for first excited states of even-even nuclei.   Since the shape of the 0.962-MeV β^- spectrum of Au^{198} corresponds to an allowed or (nonunique) first forbidden transition, the spin change for this transition must be 0 or 1.   The log ft value from (8-25) is 7.7, indicating most likely a first forbidden transition and therefore odd ($-$) parity for Au^{198}. A $1-$ assignment for Au^{198} is presumably excluded because it would make the β transition to the Hg^{198} ground state of the same order as that to the 412-keV state; this ground-state transition, however, is certainly not prominent and must therefore have a much higher log ft value than the observed 962-keV β transition.   The likely spin and parity designation for Au^{198} is therefore $2-$ or $3-$.

Since Pt^{198} is stable, Au^{198} might be expected to decay by β^+ emission or electron capture in addition to β^- emission.   However, searches for annihilation radiation by means of 180° coincidences, for platinum K X rays with a scintillation counter, and for platinum Auger electrons with a lens spectrometer have been negative; therefore upper limits have been set at 0.01 per cent of all disintegrations for K-electron capture and 0.003 per cent for positron emission.

With scintillation counters two additional γ rays of low abundance were found in Au^{198} decay in 1950.   Their energies have been determined as 0.675 MeV and 1.087 MeV; their intensities relative to the 0.412-MeV γ ray are 1.1×10^{-2} and 2.0×10^{-3}, respectively.   A $\gamma\gamma$ coincidence measurement with variable bias on one of the scintillation counters showed that the 0.675-MeV γ ray is in coincidence with the 0.412-MeV radiation and that the 1.087-MeV γ is not in coincidence with any other γ radiation.   Quite clearly the second excited state of Hg^{198} is at 1.087 MeV, and transitions from this state occur both to the 0.412-MeV level and to the ground state.   Measured internal conversion coefficients for the K shell and K/L conversion ratios indicate that the 0.675-MeV transition ($\alpha_K = 0.021 \pm 0.002, K/L = 5.7$) is $M1$ and the 1.087-MeV transition ($\alpha_K = 0.0045$ and $K/L = 6.3$) $E2$.   (More detailed information comes from a $\gamma\gamma$ angular-correlation experiment, which showed the 0.675-MeV transition to be a mixture of 60 per cent $E2$ and 40 per cent $M1$.)   The ground, first excited, and second excited states of Hg^{198} are therefore $0+$, $2+$, and $2+$.

Coincidence experiments with a lens spectrometer and a NaI scintillation

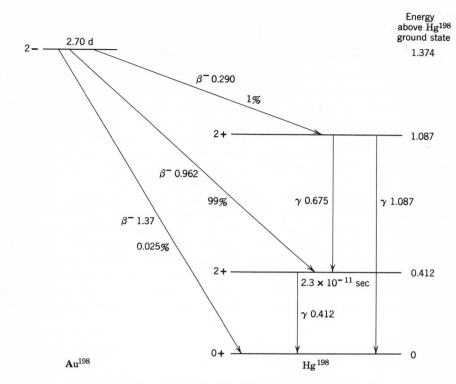

Fig. 12-9   Decay scheme of Au^{198}.   All energies are in MeV.

spectrometer as the two detectors showed the 0.675-MeV γ ray to be in coincidence not only with the conversion electrons of the 0.412-MeV γ ray but also with a β^- spectrum of upper limit 0.290 ± 0.015 MeV, which has the allowed shape and an intensity about 1 per cent that of the main β spectrum.   A third β^- transition, with an intensity 2.5×10^{-4} of the main spectrum and an upper limit of 1.37 MeV, was found in a magnetic-spectrometer study of a strong source and evidently represents the ground-state transition.   The spectrum shape identifies this transition as $\Delta I = 2$, yes; thus the spin and parity assignment of Au^{198} becomes $2-$.[13]   The log ft value of 11.8 for the ground-state transition [computed from (8-25)] is consistent with the assignment.

Finally, we mention that the half-life of the 0.412-MeV excited state has been measured by delayed coincidences.   It was found to be about 2.3×10^{-11} second, to be compared with the single-particle estimates (from table 8-5) of 5×10^{-10} sec and 3×10^{-13} second for $E2$ and $M1$ transitions, respectively.   The decay scheme deduced from all these data is shown in figure 12-9.

[13] The $I = 2$ assignment for Au^{198} has been independently established by an atomic-beam measurement.

Iodine-125.   An iodine isotope of about 60 days half-life produced by deuteron bombardment of tellurium has been assigned to mass number 125. Its decay proceeds entirely by electron capture to Te^{125}.   The low-lying levels of Te^{125} (including an isomeric level of 58 days half-life at 145 keV above the ground state) and the transitions between them were already well characterized when an investigation of the I^{125} decay scheme was undertaken.   This information is included in figure 12-10 without further discussion.

To check whether any I^{125} electron capture transitions go to the 145-keV level, tellurium and iodine carriers were added to an aged I^{125} sample and then separated.   The thoroughly purified tellurium fraction was found to be inactive, and from this experiment an upper limit of 5×10^{-4} can be set for the fraction of I^{125} decays that go to the 58-day Te^{125m} isomer.

When the radiations from I^{125} were allowed to enter a krypton-filled proportional counter through a beryllium window, the pulse height spectrum observed with a single-channel pulse height analyzer showed the presence of 35-keV γ rays as well as tellurium K X rays.   In another experiment a sodium iodide crystal containing I^{125} was grown from a melt and used as a scintillation detector; in this case every transition to the 35-keV level gives rise to a pulse height that corresponds to the sum of 35 keV and the (K or L) X-ray energy, whereas a transition to the ground state gives a pulse height corresponding to

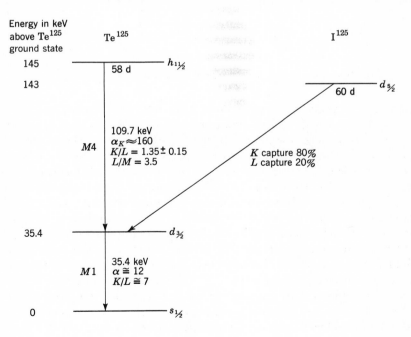

Fig. 12-10   Decay scheme of I^{125} and Te^{125m}.

the X-ray energy only. No ground state transitions were found, and it thus appears that all I^{125} decays go to the 35.4-keV level of Te^{125}. This result is consistent with a $d_{5/2}$ assignment for I^{125} and an allowed transition to the $d_{3/2}$ 35.4-keV level of Te^{125}; the unobserved transition to the $s_{1/2}$ ground state is then second forbidden.[14]

The ratio of K-electron to L-electron capture may be evaluated from the pulse height spectrum observed with the I^{125}-containing NaI scintillation crystal. The areas under the two peaks—one corresponding to the sum of 35 keV and tellurium K X rays, the other to the sum of 35 keV and tellurium L X rays—were compared, and an L-capture to K-capture ratio of 0.23 ± 0.03 was deduced. By (8-28) this ratio leads to a value of 103 keV for the transition energy of the electron capture decay to the 35.4-keV level. The corresponding log ft value [from (8-27)] is 4.5, corroborating the conclusion that the transition is allowed.[15] The I^{125} decay scheme is shown in figure 12-10.

Lead-204m$_2$. An interesting case of isomerism occurs in the even-even nuclide Pb^{204}. A 68-minute[16] isomer decaying with the emission of γ rays of about 1 MeV has been known for some time. It is formed in the electron-capture decay of Bi^{204} but not in the β^- decay of Tl^{204}. In 1950 an investigation of the electron spectrum of the lead isomer with a lens spectrometer showed K- and L-conversion lines of two γ rays, of energies 374 keV and 905 keV, with K/L conversion ratios of about 2.1 and 1.5, respectively. These K/L ratios suggest (cf. table 8-6) an $E2$ assignment for the 374-keV transition and an $E5$ assignment for the 905-keV transition. Approximate values of 0.05 and 0.1, respectively, for the total internal-conversion coefficients of the 374-keV and 905-keV transitions were indicated by absorption curves of the electrons and γ rays taken with a GM counter; these values, too, are compatible with the $E2$ and $E5$ assignments. According to table 8-5, the 68-minute half-life is compatible with a 905-keV $E5$ transition but not with an $E2$ transition of 374 keV. Thus it was concluded that the 905-keV step is the isomeric transition and is followed by the 374-keV transition.

Delayed coincidences between the two γ rays were found with scintillation counters as detectors. By variation of the delay time, the half-life of the 374-keV transition was determined to be 3×10^{-7} second. Later, a third γ ray of 899 keV was reported; it had been missed in the earlier measurements because its energy is so close to that of the 905-keV (now redetermined as 912-keV) γ ray and because careful relative intensity determinations were not made. This 899-keV γ ray is in prompt coincidence with the 374-keV

---

[14] According to shell structure, the ground state of I^{125} might be expected to be either $d_{5/2}$ or $g_{7/2}$. A $g_{7/2}$ assignment would make the transition to the 35.4-keV level of Te^{125} second forbidden and that to the 145-keV isomeric level first forbidden and thus appears to be ruled out by the experimental data. The $d_{5/2}$ assignment for I^{125} has more recently been corroborated by a direct microwave absorption measurement of $I = \frac{5}{2}$.

[15] More rigorous calculations lead to a decay energy of 108 keV and a log ft value of 4.8.

[16] The half-life has more recently been reported to be 66.9 ± 0.1 minute.

γ ray and delayed with a 3×10^{-7}-second half-life relative to the 912-keV transition. The order of emission of the 374- and 899-keV γ rays shown in figure 12-11 was originally inferred from indirect evidence, such as comparison of the 2.6×10^{-7}-second half-life with theoretical predictions (table 8-5) and absence of 374-keV γ rays in the β^- decay of Tl^{204} (a 2− state) and in the α decay of Po^{208} (0+). (A 2+ state in Pb^{204} at 374 keV would be expected to be appreciably populated in both decay modes.) More direct evidence for the γ-ray sequence 912-375-899 keV and unambiguous assignment of spins to the states involved has subsequently come through angular-correlation measurements for the pairs 375-899, 912-375, and 912-899. The results of these measurements proved that the spins 9, 4, and 2 previously assigned to the levels at 2.186, 1.274, and 0.899 MeV are correct (assuming that the ground state has 0 spin). The parity designations shown in figure 12-11 for these

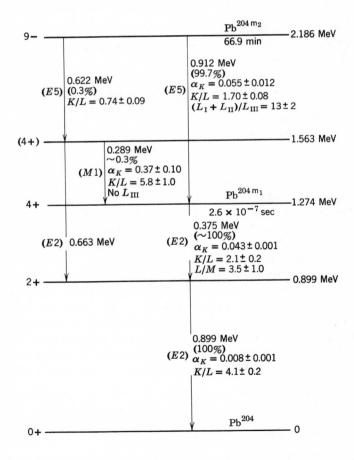

Fig. 12-11   Decay scheme of Pb^{204} isomers.

states then follow from the multipole character assigned to each of the three transitions, which, in turn, is solidly based on the rather detailed conversion-coefficient information shown in the figure. All of these deductions can be readily checked by reference to the information given in chapter 8.

Further investigations of the conversion electron spectrum of Pb^{204m_2} with magnetic spectrometers have revealed three additional weak transitions: two of them at 289 and 622 keV of about equal intensity (0.3 per cent of the intensity of each of the main transitions) and a still weaker transition at 663 keV. The most probable placement and multipole assignments for these transitions are shown in figure 12-11, along with some of the conversion coefficient evidence for these assignments. A second $4+$ level at 1.563 MeV thus appears in the level scheme. It should also be noted that the decay scheme of Pb^{204m_2} does not necessarily include all the Pb^{204} levels in the energy region covered. Three additional levels between the 1.563- and 2.186-MeV states appear to be populated in the electron capture decay of 12-hour Bi^{204}. Although they have not been completely characterized, they presumably have rather low spins and are therefore not measurably populated from the $9-$ isomer.

Pb^{204} is in a transition region between the spherically symmetric closed-shell configuration of Pb^{208} and a region of strongly deformed nuclei which begins around osmium. In this transition range spectra are difficult to interpret because states due to intrinsic particle excitations and vibrational states can occur in the same energy region (cf. chapter 9, p. 293). However, it appears from systematic trends among neighboring even-even nuclei (B3) that the first two excited states of Pb^{204} are most likely to be ascribed to the excitation of collective vibrations.

Iridium-186. As a final example, we consider briefly an exceedingly complex decay scheme which has not yet been completely unraveled but has been studied with highly refined techniques and may serve to illustrate how far it has been possible to push decay-scheme determinations. We can only sketch some of the highlights of this investigation of the levels of Os^{186} populated by the decay of Ir^{186}. For a detailed account the original paper (E1) on which the following paragraphs are based should be consulted. Prior to that study some information on the decay of 15-hour Ir^{186} was already available: Most of the decays were known to occur by electron capture, about 5 per cent proceeding by β^+ emission, with an upper energy limit of 1.92 MeV. Gamma transitions of 135, 297, 434, 625, 773, and 923 keV with intensities decreasing in the order given, had been reported, and the first three of these had been identified as $E2$ transitions by their K/L conversion ratios. From coincidence experiments the 135-, 297-, and 434-keV γ rays were known to be in cascade, and they were presumed to represent the transitions between the members of the ground state rotational band ($6+ \xrightarrow{434} 4+ \xrightarrow{297} 2+ \xrightarrow{135} 0+$). The location of the $2+$ first excited state of Os^{186} at $\cong 135$ keV was corroborated by the appearance of a 137-keV $E2$ transition in the β^- decay of Re^{186}. The only other γ rays reported in Re^{186} β^- decay were a 631-keV γ in coincidence with the 137-keV transition and a 768-keV crossover transition. Measurements of the 631-137-keV $\gamma\gamma$ angular correlation and a log ft value of 9 for the β^- branch from Re^{186} $(1-)$ populating the 768-keV level characterize that level as $2+$, and it was thought to be the lowest member of the γ-vibrational $(K = 2)$ band.

This was approximately the state of knowledge when the investigation described in ref. E1 was undertaken. The primary measurements in that study were made on the conversion electron spectra of Ir^{186} sources, and a double-focusing magnetic spectrometer with a resolution of about 0.3 per cent full width at half maximum for electron lines was used for this purpose. With an iridium source prepared by bombardment of enriched Re^{185} with 40-MeV He^4 ions, 235 electron lines with energies between 60 keV and 1800 keV were observed. Since the half-lives of Ir^{185}, Ir^{186}, and Ir^{187} are all very similar (in the study under discussion, they were determined to be 14.0 ± 0.9, 15.8 ± 0.3, and 10.5 ± 0.3 hours, respectively), the Ir^{186} lines could not be distinguished according to their rates of decay. Instead, a comparison of relative line intensities in sources produced from Re^{185} by 40- and 34-MeV He^4 ions allowed a clear-cut analysis because of the difference in energy dependence between the $(\alpha, 2n)$, $(\alpha, 3n)$, and $(\alpha, 4n)$ excitation functions in this energy range. Thus more than 150 electron lines could be ascribed to the decay of Ir^{186}, and from these lines possible transition energies corresponding to conversion in the various electron shells were calculated. Since most of the energies were measured to $\lesssim 0.1$ per cent, a transition energy corresponding to two or more electron lines could be assigned with considerable confidence; other transition energies could be associated with only one electron line (presumably the K conversion line) and were therefore less certain. A total of 101 transition energies were inferred. Whenever more than one electron line was observed for a given transition, the intensity ratio gave valuable and often decisive information on multipolarity of the transition.

The problem of fitting 101 transitions into a level scheme was a formidable one and was tackled with the aid of an electronic computer. All twofold energy sums of the 101 transition energies were formed and examined for relationships of the type $E_1 + E_2 = E_3$. In this manner, and starting with the known levels (now more accurately determined to be at 137.15, 433.91, 868.7, and 767.4 keV), it was possible to fit about two thirds of the 101 transitions into a scheme containing 23 levels. The usefulness of this procedure, which is analogous to the application of the Ritz combination principle in atomic spectroscopy, depends critically on the accuracy with which the transition energies are known; the poorer the accuracy, the greater the probability of spurious energy sum relationships.

Additional information regarding the level scheme was obtained from a variety of measurements. The positron spectrum was investigated with a lens spectrometer and was found to consist of components with end points at 1.94 and 1.37 MeV and a possible weak 1.0-MeV component.   The highest energy positron group was found to be in coincidence with 137-, 297-, and 434-keV γ rays (in equal intensity), which shows that that β^+ transition from Ir^{186} goes to the 6+ state at 868.7 keV.   Gamma-ray "singles" spectra taken with a NaI detector and multichannel analyzer yielded some useful intensity information for certain groups of γ rays, even though not many individual transitions were resolved.   Gamma-gamma coincidence measurements with two NaI scintillators and a two-dimensional 32- by 64-channel pulse height analyzer corroborated many features of the tentative level scheme.   Electron-electron coincidences measured with a double-lens magnetic spectrometer gave additional checks on level placements.   Detailed discussion of all these measurements and of the many considerations that led to spin and parity assignments of the 23 levels is beyond the scope of the present treatment and can be found in ref. E1.

A portion of the level scheme of Os^{186} is shown in figure 12-12.   This shows only those levels that are interpreted as members of the ground state rotational band (levels A–E) and of the $K = 2$ γ-vibrational band (levels F–K) and the transitions between them.   In figure 12-13 these level structures are compared with those of the other even-even osmium isotopes, and the systematic trends in both band structures are very evident: the energies of the ground state band increase and those of the $K = 2$ band decrease with increasing neutron number. This behavior represents a gradual transition from the pattern expected for rotational spectra of strongly deformed nuclei toward a near-harmonic vibrational pattern.

Detailed comparison of the energies and transition probabilities observed for the $K = 0$ and $K = 2$ bands with the predictions of the collective model reveals some discrepancies.   If the level energies in the two bands are calculated in terms of the energies of levels B and F by (9-5) (this is equivalent to using those two levels to determine the moments of inertia $\mathcal{I}_1$ and $\mathcal{I}_3$), energies higher than the experimental ones by up to 13 per cent are obtained.   Furthermore, various attempts to modify the simple strong-coupling model to include rotation-vibration interactions or to use more complex models such as asymmetric rotors have failed to improve the agreement between observed and calculated energy spacings significantly.   Transition probabilities can be used to obtain parameters in the nuclear models, such as the spheroidal-deformation parameter β and the asymmetric-rotor parameter γ.   Again, when all the transitions for which good data are available are considered, it has not been possible to arrive at a completely self-consistent set of parameters in terms of any of the models tried, although there is semiquantitative agreement.

Thus a complete theoretical interpretation of even the two well-characterized bands in Os^{186} is still lacking.   The nature of the other dozen states which have been located in the Os^{186} level scheme has not yet been characterized; and, finally, as indicated by the many observed transitions not fitted into the

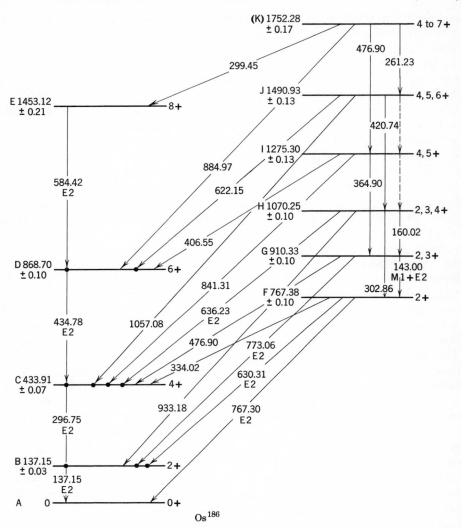

Fig. 12-12   A partial level scheme of Os^{186} showing those levels believed to be members of the ground-state rotational band and the gamma-vibrational ($K = 2$) band.   Transitions expected to occur which could not be observed because of interfering radiations are indicated by broken lines.   Energies (in keV) of transitions and their multipole orders (where established) are shown.   Observed coincidences of transitions are indicated by a solid dot at the terminus of the upper transition.   (Reproduced from reference E1.)

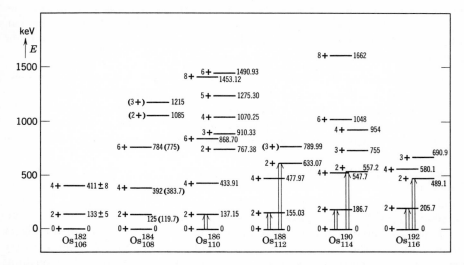

Fig. 12-13 Comparison of level structure of even-even osmium isotopes. Only the ground-state rotational and gamma-vibrational bands are shown. The double arrows indicate transitions observed in Coulomb excitation. (Reproduced from reference E1, in which the original sources of the data are given.)

level scheme, there must be additional levels which can be located only with the aid of additional experiments. The degree of complexity found in the decay of Ir^{186} is by no means unique; many decay schemes of comparable intricacy await elucidation.

H. MASS ASSIGNMENTS

The identification of new radioactive species in terms of their mass numbers and atomic numbers has historically been one of the chief problems of the nuclear chemist. By now almost all nuclides near the region of β stability have been identified. Apart from occasional isomeric states and a few other exceptions, only isotopes of the transcalifornium elements and nuclides far from the region of β stability (produced by bombardments with very-high-energy projectiles or heavy ions or by multiple neutron capture) remain to be discovered and characterized. Yet the nuclear chemist should be familiar with the techniques that have been used to assign the more than 1000 presently known radionuclides to the positions they now occupy in the nuclide chart. The principal problem has usually been the assignment of the mass number A, and in this section we briefly review the techniques developed for the solution of this problem. We assume that the atomic number of a nuclide can be determined by chemical analysis (see section D). This is, of course, not true for short-lived species (say $\lesssim 1$ second), but in these cases the assignment can often be made indirectly by identification of a longer-lived radioactive precursor or descendant.

Cross Bombardments. The only case in which the mass number of a reaction product is uniquely determined in a single bombardment is the slow-neutron activation of a single nuclear species. For example, the slow-neutron bombardment of arsenic (with the single stable species As^{75}) produces a 26.8-hour β-emitting arsenic isotope which is, therefore, readily assigned to As^{76}.

One can frequently make a mass assignment by determining whether the radioactive species being studied is formed in a number of different types of bombardments; this is called the method of cross bombardments. For this purpose target elements as well as projectiles may be varied. In each bombardment the possible products are limited by the stable isotopes of the target element and by the types of reactions possible with the projectile and energy used. As an illustration consider some radioactive isotopes of strontium; figure 12-14 displays the stable nuclides in the region of strontium. Slow-neutron activation of strontium produces strontium isotopes with 2.8-hour and 51-day half-lives; either or both of these activities might be assigned to any of the isotopes Sr^{85}, Sr^{87}, Sr^{88}, or Sr^{89}, since they are produced by neutron capture from the stable strontium isotopes. The fact that the same two activities are produced by fast neutrons (say 15 MeV) in zirconium, presumably by n, α reactions, eliminates Sr^{85}. The fast-neutron bombardment of yttrium (say with 10- or 15-MeV neutrons) produces only the 51-day strontium, presumably by n, p reaction, and this activity is therefore assigned to Sr^{89}. The 2.8-hour activity is then to be assigned to an isomeric state of Sr^{87} or Sr^{88}. The fact that 2.8-hour strontium is also produced by proton bombardment of rubidium by p, n reaction leads to its assignment to Sr^{87m}.

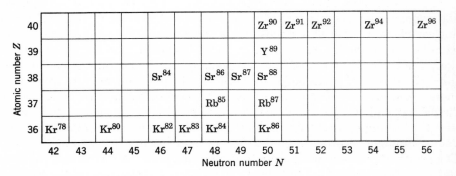

Fig. 12-14   Naturally occurring nuclides in the region of strontium.

In the interpretation of the results of cross bombardments the relative abundances of isotopes in the target elements often have to be considered. For example, failure to observe a certain reaction product may simply be due to low abundance of the isotope from which that product could have been made.

Use of Isotope Separations (H3).   In many cases the mass number of a radioactive species is uncertain because it is not known from which isotope of the target element the activity has been produced.   The bombardment of separated isotopes, or of isotopic mixtures sufficiently enriched in some isotope, is therefore of great advantage.   Enrichment or impoverishment of an isotope by a factor of 2 or even less may be sufficient for this purpose; by comparing the yield of the activity of interest from samples of normal and altered isotopic composition we can deduce the origin of the activity.   For a long time the assignment of the 37-minute chlorine activity produced by slow neutrons or by deuterons in chlorine (stable isotopes Cl^{35} and Cl^{37}) was uncertain and could not be readily determined by cross bombardments; but when it was shown that this activity was not produced by slow-neutron bombardment of a sample of almost pure Cl^{35} the 37-minute period could be assigned to Cl^{38} rather than Cl^{36}.   Electromagnetically enriched stable isotopes of most elements are available in milligram to gram quantities from the Isotopes Division of the United States Atomic Energy Commission at Oak Ridge, Tennessee, and from the Atomic Energy Research Establishment in Harwell, England.

Isotope separation can be used in an even more direct way for the identification of radioactive species: the reaction products themselves may be subjected to an isotope separation process.   The mass numbers of many radionuclides have been unambiguously established by mass-spectrographic identification; the technique has been successfully applied for half-lives as short as a few minutes and in special cases (rare-gas fission products) even down to 10 seconds. In some of these mass-assignment studies photographic detection in an ordinary mass spectrograph was used.   However, for identification of the mass-separated active species it is much more convenient to collect each isotope on a metal strip which can then be used directly as the source mount for activity measurements.   Isotope separators for this purpose are commercially available.   One of their important characteristics is high collection efficiency; over-all efficiencies (from source filament to collector) of the order of 10–20 per cent have been reported for some elements.   Mass-separated samples can often be used to great advantage in decay scheme studies.

Excitation Functions.   An excitation function for the production of the nuclide under study may help to determine the type of reaction involved by comparison with similar excitation functions for known reactions.   This technique becomes a particularly valuable extension of the cross-bombardment method at higher bombarding energies.   For example, in the unraveling of the exceedingly complex conversion electron spectra of the mercury isotopes formed in high-energy proton bombardments of gold the excitation functions were of great value.   The various electron lines, with characteristic energies and half-lives, were found to fall into three groups produced with maximum yields by protons of about 54, 63, and 72 MeV, respectively, and with corresponding thresholds at about 35, 45, and 55 MeV.   On this basis the reactions producing these activities could be identified as $p, 5n$; $p, 6n$; and $p, 7n$ reactions, respectively, and the activities themselves (including not only mercury isotopes but also their gold and platinum descendants) therefore are assigned to mass numbers 193, 192, and 191.

If a complete excitation function is not available, even a knowledge of the energy of the bombarding particle in a single experiment, together with an approximate yield of the product, can be of value for the mass assignment. For example, if a certain activity of element $Z + 1$ is produced in good yield in the bombardment of element Z with 5-MeV deuterons, the reaction can hardly be anything but a d, n reaction; if the deuteron energy had been 20 MeV, a $d, 2n$ reaction or perhaps a $d, 3n$ reaction might more likely have been responsible.   Referring again to figure 12-14, we see that the 2.8-hour strontium cannot be expected to result from the bombardment of rubidium with 5-MeV deuterons if it is Sr^{87}, but it could be formed in such a bombardment if it were Sr^{88}.

In the case of neutron-induced reactions the difference between the activities produced with and without cadmium shielding is often taken as an approximate measure of the thermal-neutron effect which may almost always be ascribed to n, γ reactions.   However, it should be noted that a cadmium shield does not completely eliminate n, γ reactions because they may occur with appreciable cross sections at energies above the cadmium resonance.

Beam Contamination and Secondary Reactions.   In connection with the subject of cross bombardments it is important to realize that extraneous bombarding particles or secondary particles produced by bombardment of the targets may sometimes give rise to nuclear reactions to an extent that can interfere with the recognition of the primary-reaction products.   For example, α-particle beams in cyclotrons are commonly contaminated with small amounts of deuterons (from residual deuterium in the ion source); this could be troublesome in the investigation of a relatively improbable α-induced reaction leading to a product that can be formed in high yield by deuterons (for example, $\alpha, p2n$ versus d, n).

The neutrons produced in the bombardment of targets with charged particles or γ rays usually give rise to nuclear reactions in the target material. The products of these reactions may be confused with those that are found or expected to be produced by the primary bombarding particle.   For example, in the deuteron bombardment of a thick sodium target to make Na^{24} by the d, p reaction the Na^{24} activity is found at depths beyond the range of the deuterons because it can be produced there by n, γ reaction.

In high-energy spallation reactions each primary interaction may give rise to a number of secondary particles—neutrons, protons, deuterons, helium ions, and even heavier fragments—which often are themselves energetic enough to produce further nuclear reactions. This is evidenced, for example, by the formation of products of atomic numbers $Z + 2$ to $Z + 4$ from a target element Z in very-high-energy proton bombardments. These secondary effects can usually be minimized by the use of targets that are thin compared to the ranges of the secondary particles.

Radiation Characteristics and Genetic Relationships. Half-life and atomic number are frequently not sufficient to characterize a radioactive species. It happens rather often that two isotopes of the same element have not very different half-lives. Then the isotopes can usually be distinguished by the types and energies of the radiations they emit. We may use again our example of the strontium isotopes (figure 12-14); an investigation of the radiations emitted by the slow-neutron-bombarded strontium sample after the 2.8-hour period has decayed reveals, in addition to the 1.5-MeV β^- particles of the 51-day Sr^{89}, some γ rays and characteristic rubidium X rays. On closer examination these last two radiations are found to follow a 65-day half-life. The 51- and 65-day periods could certainly not be resolved in a gross decay curve of the neutron-bombarded sample. The presence of rubidium X rays shows that the 65-day isotope decays to rubidium by β^+ emission or K capture; no positrons are observed, and the process must be an electron capture. From the evidence the assignment is probably to Sr^{85} or Sr^{87}; the latter is ruled out by the fact that the 65-day species is not formed by n, α reaction from Zr^{90}. Proton bombardment of rubidium produces the 65-day period (but not the 51-day Sr^{89}), as well as the 2.8-hour Sr^{87m} already discussed, and in addition a 70-minute strontium which also fails to appear in the fast-neutron bombardment of zirconium.

Using these four strontium activities as examples, we shall illustrate how a study of the mode of decay of an isotope helps in its assignment. The fact that the 51-day isotope emits β^- particles rules out its assignment to any mass number less than 89: a strontium nucleus of mass 88 or less would by β^- decay move away from rather than toward the stability region. A study of the radiations from the 70-minute and 2.8-hour activities reveals that both emit strontium X rays, indicating that these periods are associated with isomeric transitions. The electron capture decay of the 65-day isotope eliminates its assignment to any mass number greater than 87. From the facts listed in this and in the preceding paragraph the 65-day and 70-minute periods can both be assigned to Sr^{85}; the 70-minute period is associated with an isomeric transition to the lower state, which in turn decays by electron capture with a 65-day half-life to Rb^{85}.

On the basis of the systematics of α and β decay and of isomeric transitions (see chapter 8), it is often possible to correlate not only modes of decay but also half-lives and decay energies with the mass assignments of isotopes. Semiquantitative predictions of β-decay energies on the basis of binding-energy formulas like (2-3) or (2-8) are often very useful in this connection (although account may have to be taken of shell effects). For the identification of new nuclides in the transuranium region the predictions of α systematics have been particularly helpful.

If the product of a radioactive decay is itself radioactive, the genetic relationship may be investigated by studies of the growth and decay curves of fractions chemically separated at successive times. An understanding of genetic relationships helps in the assignment of the activities. This is particularly important in the fission product decay chains on the neutron-excess side of stability and in the decay chains of neutron-deficient nuclides produced in very-high-energy reactions. Nuclides far removed from stability are most commonly identified by determinations of genetic relationships with already characterized descendants near the stability valley. For example, the fission product chain of mass 89 was identified with that mass number because its last active member was shown to be the 51-day Sr^{89} already discussed. The decay chain containing 284-day cerium was assigned the mass number 144 by mass-spectrographic determination of the mass of that long-lived cerium.

Target Material and Effect of Impurities. When the product of a particular nuclear reaction is to be studied, it is important that the target not contain other elements from which the same product might result. As an obvious example, in a study of the production of 35-hour bromine (Br^{82}) by neutron bombardment of rubidium we would not wish to use rubidium bromide as a target. For the same reason impurities of neighboring elements, such as iridium impurity in an osmium target, may lead to misinterpretation of results. In general, it is advisable to use free elements as targets. However, sometimes the bombardment of a compound is indicated, for example, if the element is too reactive or in a physical state unsuitable for bombardment or if the dissolving of the element would make a slow step in the subsequent chemical separation procedure.

The presence of some impurities in a target is in most cases unavoidable. If the impurities are known, the chemical separation procedure can be designed to separate the products of interest from the products likely to result from the bombardment of the impurities. The purity requirements for the target material may become very exacting if a reaction of low cross section is studied, for then the expected product yield is small and care must be taken that comparable amounts of the same species are not produced from an impurity by a much more probable reaction. For example, the formation of Mg^{23} by the reaction $Al^{27}(\gamma, p3n)$ induced by bremsstrahlung of $E_{max} \cong 70$ MeV could be studied only with aluminum very free of magnesium because of the much larger yield of the $Mg^{24}(\gamma, n)Mg^{23}$ reaction.

REFERENCES

A1 *Applications of Computers to Nuclear and Radiochemistry*, Proceedings of a Symposium, Gatlinburg, Tennessee, October 1962 (G. D. O'Kelley, Editor), Report NAS-NS 3107. Available from Office of Technical Services, Department of Commerce, Washington 25, D. C., $2.50.

B1 C. B. Braestrup and H. O. Wyckoff, *Radiation Protection*, Charles C. Thomas, Springfield, Illinois, 1958.

B2 C. G. Bell and F. N. Hayes (Editors), *Liquid Scintillation Counting*, Pergamon, London, 1958.

B3 A. Bohr and B. R. Mottelson, "Collective Motion and Nuclear Spectra," in *Nuclear Spectroscopy* (F. Ajzenberg-Selove, Editor), Part B, Academic, New York, 1960, pp. 1009–1032.

B4 G. A. Brinkman, A. H. W. Aten, Jr., and J. Th. Veenboer, "Absolute Standardization with a NaI(Tl) Crystal—I.   Calibration by Means of a Single Nuclide," *Intern. J. Appl. Radiation Isotopes* **14**, 153 (1963).

C1 P. J. Campion, "The Standardization of Radioisotopes by the Beta-Gamma Coincidence Method Using High-Efficiency Detectors," *Intern. J. Appl. Radiation Isotopes* **4**, 232 (1959).

*C2 O. Chamberlain, "Determination of Flux of Charged Particles," in *Methods of Experimental Physics*, Vol. 5B, *Nuclear Physics* (L. C. L. Yuan and C. S. Wu, Editors), Academic, New York, 1963, pp. 485–507.

C3 J. B. Cumming, "Monitor Reactions for High-Energy Proton Beams," *Ann. Rev. Nuclear Sci.* **13**, 261–286 (1963).

D1 R. W. Dodson, A. C. Graves, L. Helmholz, D. F. Hufford, R. M. Potter, and J. G. Povelites, "Preparation of Foils," *Miscellaneous Physical and Chemical Techniques of the Los Alamos Project*, National Nuclear Energy Series Div. V, Vol. 3, McGraw-Hill, New York, 1952.

E1 G. T. Emery, W. R. Kane, M. McKeown, M. L. Perlman, and G. Scharff-Goldhaber, "Studies of Decay Schemes in the Osmium-Iridium Region.   III. Decay of 15.8-Hour Ir^{186}," *Phys. Rev.* **129**, 2597 (1963).

*F1 R. A. Faires and B. H. Parks, *Radioisotope Laboratory Techniques*, George Newnes, London, 1958.

F2 H. Freiser and G. H. Morrison, "Solvent Extraction in Radiochemical Separations," *Ann. Rev. Nuclear Sci.* **9**, 221–44 (1959).

G1 N. B. Garden and E. Nielsen, "Equipment for High Level Radiochemical Processes," *Ann. Rev. Nuclear Sci.* **7**, 47–62 (1957).

G2 W. M. Garrison and J. G. Hamilton, "Production and Isolation of Carrier-Free Radioisotopes," *Chem. Revs.* **49**, 237–72 (1951).

G3 I. J. Gruverman and P. Kruger, "Cyclotron-Produced Carrier-Free Radioisotopes. Thick-Target Yield Data and Carrier-Free Separation Procedures," *Intern. J. Appl. Radiation & Isotopes* **5**, 21–31 (1959).

H1 "Hot Labs—A Special Report," *Nucleonics* **12**, No. 11, 35–100 (November 1954).

H2 O. Hahn, *Applied Radiochemistry*, Cornell University Press, Ithaca, New York, 1936.

*H3 H. Hintenberger, "High-Sensitivity Mass Spectroscopy in Nuclear Studies," *Ann. Rev. Nuclear Sci.* **12**, 435–506, (1962).

*J1 N. R. Johnson, E. Eichler, and G. D. O'Kelley, *Nuclear Chemistry*, Wiley, New York, 1963.

K1 Y. Kusaka and W. W. Meinke, "Rapid Radiochemical Separations," *National Academy of Sciences—National Research Council, Nuclear Science Series* NAS-NS

3104 (1961). Available from Office of Technical Services, Department of Commerce, Washington 25, D. C. ($1.25).

K2   K. A. Kraus and F. Nelson, "Radiochemical Separations by Ion Exchange," *Ann. Rev. Nuclear Sci.* **7**, 31–46 (1957).

K3   R. Kunin, *Ion Exchange Resins*, 2nd ed., Wiley, New York, 1958.

*K4   R. Kunin, *Elements of Ion Exchange*, Reinhold, New York, 1960.

K5   J. J. Katz and G. T. Seaborg, *The Chemistry of the Actinide Elements*, Wiley, New York, 1957.

M1   K. Z. Morgan, "Techniques of Personnel Monitoring and Radiation Surveying," in *Nuclear Instruments and Their Uses* (A. H. Snell, Editor), Wiley, New York, 1962.

N1   W. E. Nervik and P. C. Stevenson, "Self-Scattering and Self-Absorption of Betas by Moderately Thick Samples," *Nucleonics* **10**, No. 3, 18 (March 1952).

*N2   Nuclear Data Group, National Academy of Sciences—National Research Council, "Nuclear Data Sheets," Washington, D. C., 1958–1963.

P1   B. D. Pate and L. Yaffe, "Disintegration-Rate Determination by 4π-Counting," *Can. J. Chem.* **33**, 610, 929, 1656 (1955); **34**, 265 (1956).

*P2   J. L. Putman, "Measurement of Disintegration Rate," in *Beta- and Gamma-Ray Spectroscopy* (K. Siegbahn, Editor), Interscience, New York, 1955, pp. 823–854.

P3   A. M. Poskanzer and B. M. Foreman, Jr., "A Summary of TTA Extraction Coefficients," *J. Inorg. Nuclear Chem.* **16**, 323 (1961).

R1   H. C. Richards, "Charged-Particle Reactions," in *Nuclear Spectroscopy* (F. Ajzenberg-Selove, Editor), Part A, Academic, New York, 1960, pp. 99–138.

*S1   Subcommittee on Radiochemistry, National Academy of Sciences—National Research Council, *Monographs on the Radiochemistry of the Elements*, NAS-NS 3001–3058. Available from the Office of Technical Services, Department of Commerce, Washington, D. C.

S2   O. Samuelson, *Ion Exchangers in Analytical Chemistry*, Wiley, New York, 1953.

S3   J. E. Salmon and D. K. Hale, *Ion Exchange: A Laboratory Manual*, Butterworth's, London, 1959.

S4   H. Slätis, "Source and Window Technique," in *Beta- and Gamma-Ray Spectroscopy* (K. Siegbahn, Editor), Interscience New York, 1955, pp. 259–272.

S5   E. P. Steinberg, "Counting Methods for the Assay of Radioactive Samples," in *Nuclear Instruments and Their Uses* (A. H. Snell, Editor), Wiley, New York, 1962, pp. 306–359.

*S6   D. Strominger, J. M. Hollander, and G. T. Seaborg, "Table of Isotopes," *Revs. Mod. Phys.*, **30**, 585 (1958).

*U1   U. S. Bureau of Standards, Handbook 42, "Safe Handling of Radioactive Isotopes." Available from Superintendent of Documents, U. S. Government Printing Office, Washington 25, D. C.

*Y1   L. Yaffe, "Preparation of Thin Films, Sources, and Targets," *Ann. Rev. Nuclear Sci.* **12**, 153–188 (1962).

EXERCISES

1. Some trace impurities in a silver metal sample are to be determined by neutron activation analysis in a reactor. What should be the maximum diameter of the sample (a sphere), if the variation of the thermal-neutron flux within the sample is to be held below 10 per cent?   *Answer:* $\cong$0.5 mm.

2. A 1-μA beam of 120-MeV C^{12} ions is incident on an aluminum foil 20 mg cm^{-2} thick. (a) Estimate the power dissipated in the foil. (b) If the foil is 20 cm^2 in area and mounted on an insulating frame in a high vacuum, how long an irradiation would raise its temperature to the melting point of aluminum (660°C)? Neglect heat losses by radiation and take the average specific heat of aluminum between room temperature and the melting point as 0.25 cal g^{-1} deg^{-1}. *Answer:* (b) $\cong 10$ seconds.

3. Cobalt foils are required for the following two types of experiments: (a) The excitation functions of α-induced reactions are to be measured by a stacked-foil experiment for He^4 energies down to the threshold of the reaction Co^{59} (α, n). The energy spread within a foil is not to exceed ± 5 per cent of the mean energy in the foil. (b) The proton spectra resulting from the (α, p) reaction are to be measured down to 2 MeV with an energy definition of ± 3 per cent and with protons being measured at angles as large as 30° with respect to the normal to the foil. What is the maximum thickness of cobalt foil you would use in each of the two experiments? (Assume that all the foils in a given experiment are to have the same thickness.) Suggest methods for preparing these foils and for measuring their over-all thicknesses as well as their uniformity.

4. A 0.70-mg sample of cobalt in the form of a fine wire is used to monitor the thermal-neutron flux in a reactor. It is exposed to the flux for exactly 5 minutes and a few hours later it is placed inside a well-type scintillation counter for a determination of the amount of Co^{60} formed. In virtually every Co^{60} disintegration two γ rays of 1.17 and 1.33 MeV are emitted in cascade. The γ spectrum of the cobalt sample, as measured with a multichannel analyzer, is found to contain 2175 counts per minute and 1700 counts per minute under the 1.17-MeV and 1.33-MeV photo-peaks, respectively, as well as 272 counts per minute under the "sum peak" at 2.50 MeV. The total γ-counting rate in the well counter is 180,800 counts per minute. What was the neutron flux to which the cobalt was exposed? Ignore any variation of flux within the small sample and consider the counting rates given as net counting rates after background subtraction and correction for coincidence losses. *Answer:* 1×10^{13} cm^{-2}sec^{-1}.

5. A 1-cm^2 area of 0.001-in. aluminum foil was exposed to a fast-neutron flux to produce Na^{24}. It was desired to determine the Na^{24} disintegration rate in this sample by a $\beta\gamma$ coincidence measurement. For the decay scheme of Na^{24}, refer to figure 8-8. The measurement was carried out with two scintillation counters, a plastic scintillator $\frac{1}{4}$-in. thick being used as the β detector and a 3 $\times$ 3-in. NaI(Tl) scintillator as the γ counter. The γ detector had a covering thick enough to prevent its response to the β particles; it was operated with a discriminator set to cut out all pulses corresponding to deposition of <1 MeV in the scintillator, thus making its response to bremsstrahlung negligible. Measurements were made (a) with sample in place, without additional absorbers, and without delay; (b) with sample in place, with a 0.7-g cm^{-2} aluminum absorber between sample and β detector, and without delay; (c) with sample in place, without absorber, and with the pulses from one of the detectors reaching the coincidence circuit after a 5 $\times$ 10^{-7} second delay; (d) with the sample removed and with no delay. The data taken over a period of several hours were as follows:

Experi-ment	Time at Mid-point of Counting Interval	Length of Count (min)	Total Counts Observed in		
			β Counter	γ Counter	Coincidence
(a)	11:00	10	1.830×10^6	3.63×10^5	9158
(b)	11:40	60	1.221×10^5	2.110×10^6	307
(c)	12:45	60	1.006×10^7	1.996×10^6	557
(d)	15:00	200	5800	6.00×10^4	0

What was the disintegration rate of the Na^{24} sample at 11:00? The response of the detectors may be considered to be the same over the entire sample area.

Answer: 7.20×10^6 min^{-1}.

6. The aluminum foil containing Na^{24} whose disintegration rate was determined in exercise 5 was used to calibrate an end-window proportional counter for Na^{24} radiations. A measurement taken with that counter at 14:00 on the same day as the measurements in exercise 5 gave 227,520 net counts per minute with the sample in a certain shelf position. Additional measurements in the same arrangement, but taken exactly 15.0 and 30.0 hours later, gave net counting rates of 114,880 and 57,720 per minute. Considering the statistical errors in these results as negligible and also neglecting the possibility of any activity other than Na^{24} in the sample, estimate (a) the dead time of the counting device, (b) the over-all efficiency of the counter for Na^{24} radiations in the particular geometrical arrangement used. *Answer:* (b) 0.037.

7. A 3-GeV proton bombardment is monitored by means of the Na^{24} activity induced in a 0.001-in. (6.85 mg cm^{-2}) aluminum foil (surrounded by two other aluminum foils in order to compensate any recoil losses of Na^{24} from the monitor by equal recoil gains). Exactly 20 hours after the end of the 15-minute irradiation the Na^{24} activity in the aluminum monitor is measured with the end-window counter of exercise 6 in the geometrical arrangement calibrated there. The net counting rate is 27,430 counts per minute. What was the average proton flux through the sample during the irradiation? *Answer:* 1.17×10^{14} min^{-1}.

8. Suggest methods for the chemical identification of (a) V^{52} produced in the fast-neutron bombardment of a chromate solution, (b) Mn^{52} produced in the deuteron bombardment of iron, (c) O^{14} produced in the proton bombardment of nitrogen gas.

9. A sample of sodium iodide is irradiated with fast neutrons to produce 105-day Te^{127m}. Suggest a chemical procedure for the isolation of the tellurium. How would you modify this procedure if you knew that the sodium iodide contained some sodium bromide impurity?

10. A 60-in. cyclotron capable of accelerating deuterons to 20 MeV and helium ions to 40 MeV is to be used for the preparation of the following nuclides: (a) 33-day Rb^{84} in high specific activity and as free as possible from other rubidium activities; (b) carrier-free Ce^{139}; (c) radiochemically pure As^{77}; (d) Ba^{140}. For each case outline a method of preparation, including a statement about target material, bombarding particle, and approximate bombarding energy to be used, and the chemical procedures following bombardment. Justify all your choices of conditions.

11. Calculate the electrode potentials corresponding to these half-reactions at the specified concentrations:
    (a) $Ag = Ag^+ + e^-$, with $(Ag^+) = 10^{-13}$ molar;
    (b) $Al = Al^{3+} + 3e^-$, with $(Al^{3+}) = 10^{-15}$ molar;
    (c) $2Hg = Hg_2^{2+} + 2e^-$, with $(Hg_2^{2+}) = 10^{-8}$ molar.
    (d) Would the shift in emf in (c) continue to be proportional to the logarithm of the concentration of mercury ions in solution as that concentration was indefinitely reduced? *Answer:* (a) -0.031 V.

12. (a) Verify the statement on p. 431 that (12-6) is valid even if one of the γ rays being measured is not emitted in 100 per cent of the disintegrations.
    (b) Derive a relation analogous to (12-6) for the case of three γ rays in cascade.

13. An end-window proportional counter is to be calibrated for the measurement of Sc^{46} in the form of $Sc_2(C_2O_4)_3 \cdot 5H_2O$ deposits of various thicknesses but all of 2-cm^2 area. The calibration samples are prepared by addition of various amounts of scandium carrier to aliquots of a Sc^{46} solution of high specific activity, followed by precipitation, filtration, drying, and mounting of the oxalate. A "weightless" source of Sc^{46} on a thin film is also prepared; it is found, by means of a 4π propor-

tional counter, to have a disintegration rate of 63,800 min^{-1}. After this disintegration rate determination the 4π source is mounted on an aluminum card in the same manner as the oxalate deposits. The relative Sc^{46} contents of all the samples are assayed with a NaI scintillation detector; the total scandium contents of the oxalate samples are determined analytically after completion of the activity measurements. From the following summary of data, construct a curve of counting efficiency versus sample thickness (in mg cm^{-2}) for the end-window counter measurements. All counting rates are net rates and have already been corrected for any decay during the course of the measurements.

Sample No.	Scandium Content (mg)	Net Counting Rate in Counts per Minute	
		on γ Counter	on End-Window Counter
"4π"	....	8140	....
1	0.42	6870	3257
2	0.91	7240	3415
3	1.38	7510	3530
4	1.90	7680	3558
5	2.95	7960	3560
6	3.84	7875	3295
7	4.77	7690	2982
8	5.86	7820	2645
9	7.72	7750	2110
10	9.81	7910	1705

Answer: For sample No. 10, efficiency is 0.0275.

14. A certain activity chemically proved to be associated with technetium (element 43) is produced in the bombardment of molybdenum with 12-MeV deuterons and in the bombardment of ruthenium with 12-MeV deuterons, but not in the bombardment of ruthenium with fast (up to 15 MeV) neutrons. To what isotope of technetium should the activity be assigned? What mode of decay would you expect?

15. Element Z has a single stable isotope of mass number A. In the bombardment of element Z with 28-MeV deuterons the following activities chemically identified with element $Z + 1$ were found:

A moderately strong 3-hour positron emitter.

A strong 2.6-day activity emitting mostly γ and X rays.

A weak 30-minute positron emitter.

The last activity was not produced when the deuteron energy was lowered to 20 MeV.

Critical-absorption measurements showed the X rays of the 2.6-day activity to be those of element Z. An 11-day isotope of element Z was shown to grow from the 2.6-day activity, whereas the 3-hour activity decayed to a 50-minute X-ray emitter chemically identified with element Z. The X rays from the latter activity were characteristic of element Z.

One-million-electron-volt neutrons produced in a target of element Z the 50-minute X-ray emitter, in addition to a 14-hour β^--emitting isotope of Z which had also been identified in the deuteron-bombarded Z samples. With 15-MeV neutrons the 11-day isotope of Z was also produced.

Make mass assignments for the various radioactive isotopes of elements Z and $Z + 1$, and indicate the most likely mode of decay for each. The stable nuclides of $Z + 1$ have mass numbers $A + 1$, $A + 2$, and $A + 3$; those of $Z - 1$ have mass numbers $A - 3$, $A - 2$, and $A - 1$.

16. The radiations emitted by a certain radioactive species were studied in a β-ray spectrometer. The β^- spectrum was resolved into two components of 0.61 ± 0.01 MeV and 1.438 ± 0.007 MeV maximum energies. The higher-energy component was about four times as abundant as the lower-energy one. When γ rays were allowed to strike a thin silver radiator placed in the source position of the spectrometer, the following five photoelectron energies were measured:

Energy in MeV	Intensity
0.216 ± 0.002	Strong
0.237 ± 0.002	Weak
0.801 ± 0.003	Weak
0.823 ± 0.003	Very weak
1.046 ± 0.005	Very weak

The K and L binding energies in silver are 25 and 4 keV, respectively. Draw a plausible decay scheme for the radioactive species under investigation.

17. The nuclide Z^A decays by β^- emission, largely to the first excited state of $(Z + 1)^A$; a small β^- branch (0.9 MeV maximum energy) goes directly to the ground state of $(Z + 1)^A$. The decay of $(Z + 2)^A$ proceeds entirely by K-electron capture to the first excited state. One sample of each of these two radionuclides is used in the following coincidence measurements with an anthracene (C_1) and a NaI (C_2) scintillation detector. The samples are placed in a standard position between the two counters, and a 0.5-g cm^{-2} copper absorber, sufficient to absorb the Z^A β particles and the K X rays of $(Z + 1)$, is placed between the sample and C_2 during all the measurements. The following data are obtained:

Sample	Delay between C_1 and C_2	0.5 g cm^{-2} Cu between C_1 and Sample	Counts per Minute in C_1	Counts per Minute in C_2	Coincidence Counts per Minute
Z^A	0	No	188,600	125,100	2928
Z^A	2 μs	No	188,300	125,600	237
Z^A	0	Yes	2,753	125,800	3.5
$(Z + 2)^A$	0	No	40,930	55,340	655
$(Z + 2)^A$	2 μs	No	41,070	55,090	23
$(Z + 2)^A$	0	Yes	1,216	55,510	0.7

(a) What is the disintegration rate of the $(Z + 2)^A$ sample? (b) What fraction of the Z^A decays go to the $(Z + 1)^A$ ground state directly? (c) What is the disintegration rate of the Z^A sample? (d) What is the coincidence resolving time of the circuit used? (e) If the K X rays of $(Z + 1)$ and the β particles emitted by Z^A are counted with the same efficiency in C_1, what is the K-fluorescence yield of $(Z + 1)$? Assume the two β^- groups of Z^A to be counted in C_1 with equal efficiency. Sample decay during the course of the measurements may be neglected.
Answers: (b) 0.09; (c) 8.66×10^6 min^{-1}; (d) 0.3 μs.

18. In an investigation of the decay scheme of 2.3-minute Ag^{108} the β^- spectrum was measured in an anthracene scintillation spectrometer and found to have an upper energy limit of 1.77 ± 0.06 MeV and a simple, allowed shape within the accuracy of the measurements.   Measurements with a proportional counter and pulse height analyzer showed that Ag^{108} emits Pd K X rays and that the ratio of the number of these X rays to the number of β particles emitted is 0.013 ± 0.001.   The gamma spectrum obtained with a NaI scintillation spectrometer showed weak gamma rays of 435, 510, and 616 keV with relative intensities 1.0, 0.27, and 0.27.   In a $\beta\gamma$ coincidence experiment, 616-keV γ rays were found to be in coincidence with β rays; however, these coincidences could be eliminated with an aluminum absorber of 480 mg cm^{-2} placed between sample and β counter.   Gamma-gamma coincidences were found between 435-keV and 602-keV γ rays, and between 510-keV and 510-keV γ rays, the latter, however, only when the two counters were 180° apart with respect to the sample.   The spectrum of γ rays in coincidence with X rays showed 435-keV and 602-keV γ rays in the intensity ratio 1.0:0.79.   The 602-keV peak in these coincidence spectra was definitely at a lower energy than the 616-keV peak found in the singles spectrum.   Additional experiments proved that 85 per cent of all the electron capture transitions lead to the Pd^{108} ground state.   Derive as much information as you can about the Ag^{108} decay scheme, including the intensities of the various β and γ transitions, and as many of the log ft values as possible. Discuss spin and parity assignments.

Most of the information in this exercise is based on a paper by M. L. Perlman, W. Bernstein, and R. B. Schwartz, *Phys. Rev.* **92,** 1236 (1953).

Nuclear Processes as Chemical Probes

Nuclear processes are largely unaffected by interactions with the chemical environment in which the nucleus exists. A new experimental probe for chemical structure is afforded by the few instances in which the nuclear processes *are* affected by their environment.

The most obvious candidates are electron capture and internal conversion because both processes directly involve orbital electrons. These two possibilities have been investigated, and it has been found that the half-life of Be^7 for electron capture is 0.08 per cent greater in BeF_2 than in Be metal and that the half-life for internal conversion in Tc^{99m} (largely in the M and N shells) is 0.27 per cent greater in Tc_2S_7 than in $KTcO_4$. The experimental techniques required for the investigation of chemical effects on even these two rather favorable decays were quite difficult, and further study of the environmental effects on these two processes has not been pursued.

Other more subtle, but more easily observed, effects of chemical environment on nuclear processes have been investigated. It is these processes that we discuss in this chapter, mainly from the point of view of the chemical information they can give. The processes include the Mössbauer effect, angular correlations of cascade radiations, annihilation of positrons, and the depolarization of μ mesons.

A. MÖSSBAUER EFFECT

The most thoroughly investigated nuclear process, which depends critically on the chemical environment, is recoilless nuclear resonance absorption or scattering (M3). To understand this phenomenon, consider the energy spectrum of γ rays emitted from a nucleus $_zX^A$ that goes from an excited state to its ground state with a transition energy E_r. The energy of the γ ray that is emitted, E_γ, is different from E_r for three reasons.

1. The emitting nucleus must recoil with a momentum that is equal and opposite to the momentum of the emitted γ ray; the energy associated with this nuclear recoil must come from E_r. The recoil energy, as discussed in section E of chapter 7, is given by

$$R \text{ (eV)} = \frac{537 E_\gamma{}^2}{M}, \tag{13-1}$$

where M is the atomic mass of the emitting nuclide and where E_γ is expressed in MeV. The recoil effect will generally lower the energy of low-energy photons by 10^{-2} to 10^2 eV in transitions that will be of interest to us.

2. The emitting nucleus is part of some chemical system and is in thermal equilibrium with it. The thermal motion causes the γ ray to be emitted from a moving source, and there is the consequent Doppler shift in the frequency of the emitted photon and the corresponding energy shift:

$$D = \frac{v}{c} E_\gamma \cos \vartheta, \tag{13-2}$$

where v and c are the magnitudes of the velocities of the nucleus and of light, respectively, and ϑ is the angle between the directions of motion of the emitting nucleus and the emitted γ-ray. Since $\cos \vartheta$ may vary between -1 and $+1$, the Doppler shift may either increase or decrease the energy of the emitted quantum and will cause the spectra of emitted quanta to show a distribution about the value $E_r - R$. The width of the distribution is about 0.1 eV at room temperature. It is important to realize that in Doppler broadening conservation of energy implies that either some of the energy of the chemical system goes into the γ ray or that some of the energy of the transition, E_r, in addition to the recoil energy mentioned in (1), goes into excitation of the chemical system.

3. Even if there were no Doppler broadening, the Heisenberg uncertainty principle implies that the finite half-life, $t_{1/2}$, of the excited state would cause a distribution in the energies of the emitted quanta. The width of that distribution would be

$$\Gamma \text{ (eV)} = \frac{4.55 \times 10^{-16}}{t_{1/2} \text{ (sec)}}. \tag{13-3}$$

[The numerical constant in (13-3) is the product of ln 2 and $\hbar$ in eV sec.] It is to be noted that this natural width, as given in (13-3), will exceed the room-temperature Doppler broadening only when the half-life of the excited state is less than about 10^{-15} second, which, for example, means a normal $E2$ transition greater than about 7 MeV (cf. chapter 8, table 8-5).

The foregoing three effects also apply to the inverse process, resonance

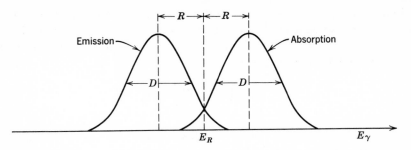

Fig. 13-1   The effect of recoil, R, and Doppler broadening, D, on spectrum of emission and resonant absorption of gamma rays.

absorption,[1] in which the nucleus $_zX^A$ in its nuclear ground state absorbs a photon and goes to the nuclear excited state E_r above the ground state. The recoil effect in this process requires that the energy of the incident photon be larger than E_r by an amount R; the Doppler broadening and the natural width will again cause a distribution about this value. An example of the two processes, emission and absorption, is shown in figure 13-1, in which the Doppler broadening is assumed to be large compared to the natural width. From figure 13-1 it is seen that the recoil energy prevents the γ ray emitted by nucleus $_zX^A$ in a direct transition from an excited state to ground state from being resonantly absorbed by the nucleus $_zX^A$ in its ground state. This remark is not completely true because of the small overlap brought about by the Doppler effect. While investigating the temperature dependence of this overlap, Mössbauer (M1) discovered that under particular circumstances a fraction of the γ rays emitted from a solid source show neither a measurable recoil energy loss nor any Doppler broadening; the energy of the γ ray was E_r and the line width approached the natural line width.

The explanation of this important observation lies in the tightness with which the emitting nucleus is bound in the chemical system. If the binding is tight enough, as it may be in crystalline systems, it is possible for the recoil momentum to be taken up immediately by the whole system, giving an effective value of M in (13-1) that approaches infinity and thereby reducing R to a vanishingly small value. To put it another way, the stiffness of the crystal makes it possible for the transition to occur without any excitation of crystalline vibrations. The Doppler broadening, which also involves the transfer of vibrational energy to or from the crystal, is quenched in a fraction of the decays for essentially the same reason.[2] This qualitative term "stiffness of

---

[1] The inverse process may be investigated experimentally either by observing the change in intensity of the transmitted beam or by observing the photons that are re-emitted at some angle with respect to the incident beam. In the latter experiment it is the resonant scattering that is being studied.

[2] Reviews of both theoretical and experimental methods are found in references F1 and M3.

the crystal" is roughly measured by the Debye temperature, θ, because θ is proportional to the highest fundamental vibrational frequency in the crystal, which in turn depends on the restoring forces for the atomic vibrations.   With this measure, the condition for the recoilless transition is

$$R < k\theta, \tag{13-4}$$

where k is Boltzmann's constant.   The fraction of the decays that occur without loss of recoil energy to the crystal increases with diminishing temperature and reaches a plateau value which depends on the relative magnitudes of the two quantities in (13-4).

When the nuclear recoil and the Doppler broadening are quenched, the emission and the absorption spectra should completely overlap, as both should peak at E_r and both should be characterized by the natural width Γ.   The condition given in (13-4) requires values of E_r less than about 100 keV (a condition that arises because θ generally lies between $\sim 100°$ and $\sim 1000°K$); this, in turn, implies half-lives greater than about 10^{-11} second or natural widths less than 10^{-5} eV.   For a decay energy of 100 keV a Doppler shift of 10^{-5} eV, that is, equal to one line width, is brought about by a velocity of only *3 centimeters per second* (13-2).   Thus a relative velocity of only a few centimeters per second between a source and an absorber for which the Mössbauer effect holds will cause the resonance absorption to vanish; energy shifts of the order of one part in 10^{10} may be relatively easily measured.   The sensitivity of the method is great enough so that the increase in the energy of a photon that had fallen less than 100 feet through the earth's gravitational field could be detected.

It would appear at this point that all effects of specific interactions between the nucleus and its environment have vanished because we now have an emission line with the energy and width determined by the characteristics of the nuclear states.   Actually, it is only at this point that the specific interactions may be detected.   They are of three kinds, and they all do the same thing; they cause small shifts in the energy E_r (of the order of 10^{-6} eV) but leave the width of the level unaffected and are thereby easily resolved.

1. *Isomer shift or chemical shift.*   The volume of a nucleus in an excited state is, in general, different from that in its ground state; as a result, the probability that the orbital electrons will be found inside the nucleus will be different for the two states.   This difference appears as a difference in the total binding energy of the electrons in the two states and contributes to the energy of transition:

$$E_r = \Delta E_{\text{nuc}} + \Delta E_{\text{elec}}, \tag{13-5}$$

where ΔE_{nuc} is the change in the nuclear binding energy and ΔE_{elec} is the change in the binding energy of the atomic electrons.   Now, if the emitting nucleus ($_ZX^A$ in an excited state) and the absorbing nucleus ($_ZX^A$ in the ground state) are in different chemical compounds, the distributions of the atomic electrons in space will be different, which will cause differences in ΔE_{elec} and therefore in E_r. This change in E_r is called the chemical shift. To a good approximation,

$$\Delta E_{\text{elec}} = \tfrac{2}{5}\pi Z e^2 (\overline{r_{\text{ex}}^2} - \overline{r_{\text{gr}}^2})[|\psi_e(0)|^2 - |\psi_a(0)|^2], \qquad (13\text{-}6)$$

where $\overline{r_{\text{ex}}^2}$ and $\overline{r_{\text{gr}}^2}$ are the mean square nuclear radii in the excited state and in the ground state, respectively; and $|\psi_e(0)|^2$ and $|\psi_a(0)|^2$ are the densities of electrons at the nucleus in the emitter and absorber, respectively.

Another contribution to the chemical shift, usually much smaller than the one just discussed, occurs because of the change in rest mass of the nucleus in the emission process which, in turn, causes a change in the zero-point vibrational energy.

2. *Magnetic hyperfine splitting.* If either the emitting or the absorbing nucleus has a spin $\geq \tfrac{1}{2}$, it will also have a magnetic moment; and, in the presence of a magnetic field, the energy of the nucleus will depend on its orientation with respect to that magnetic field. This means that, in general, another term must be added to (13-5):

$$E_r = \Delta E_{\text{nuc}} + \Delta E_{\text{elec}} + \Delta E_{\text{mag}}, \qquad (13\text{-}7)$$

where ΔE_{mag} is the change in the magnetic energy of the nucleus in the transition and is determined by the change in the magnetic moment and in the projection of the spin along the magnetic field and also by the strength of the magnetic field at the nucleus. Since the projection of the spin along the magnetic field may take on the usual $(2I + 1)$ values, the effect of ΔE_{mag} is not merely to shift E_r but to split it into several components. The splitting usually corresponds to Doppler shifts caused by relative velocities of the order of a centimeter per second.

3. *Quadrupolar hyperfine splitting.* If either the emitting or the absorbing nucleus has a spin $I \geq 1$ and is in an inhomogeneous electric field ($d^2V/dZ^2 \neq 0$, where V is the electrical potential at the nucleus and Z is the symmetry axis of the field), then, as in the magnetic interaction, E_r may be split into several lines because the interaction between the nuclear quadrupole moment and the inhomogeneous electric field causes the energy of the nucleus to depend on its orientation:

$$E_r = \Delta E_{\text{nuc}} + \Delta E_{\text{elec}} + \Delta E_{\text{mag}} + \Delta E_{\text{quad}}. \qquad (13\text{-}8)$$

Again, the splitting corresponds to Doppler velocities around a centimeter per second.

The experimental observation of the three interactions may be achieved with the simple technique shown schematically in figure 13-2. The emitter contains nuclei $_Z X^A$ in an excited state and the absorber contains $_Z X^A$ in the ground state. The intensity of the γ-ray beam in the detector is then determined as a function of the relative velocity of the emitter and absorber. The output from the detector is usually fed to a multichannel pulse height analyzer

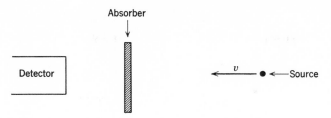

Fig. 13-2   Schematic representation of apparatus for observation of the Mössbauer effect. The source moves at velocity v with respect to the absorber.

to minimize effects from other γ-rays emitted from the sample and from background.   The experiment can become complicated when emitter and absorber must be cooled to very low temperatures or when minor fluctuations in their relative velocity are troublesome (very narrow lines).   The experimental results are more readily interpreted if the emitting nuclei are placed in a matrix in which there is no magnetic or quadrupolar splitting so that a single line is emitted; all of the splitting will then come from the absorption.

Applications.   A good example comes from the study of the Mössbauer effect with the 14.4-keV transition to the ground state of Fe^{57}.   The relevant nuclear information is shown in figure 13-3.   The emitter is prepared by diffusing Co^{57} into stainless steel in which there are evidently no electric or magnetic fields that will split the $\frac{3}{2}-$ or $\frac{1}{2}-$ states of Fe^{57} produced in the electron capture of Co^{57}.   In the absorber, on the other hand, Fe^{57} nuclei may be used to probe local electric and magnetic fields through the observed splitting patterns.   The effects of a magnetic field and of an inhomogeneous electric field on the ground state $(\frac{1}{2}-)$ and the first excited state $(\frac{3}{2}-)$ of Fe^{57}

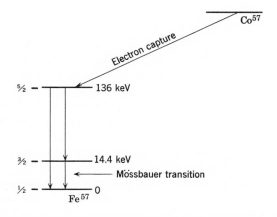

Fig. 13-3   Decay scheme of Co^{57} showing Mössbauer transition from 14.4-keV level in Fe^{57}.

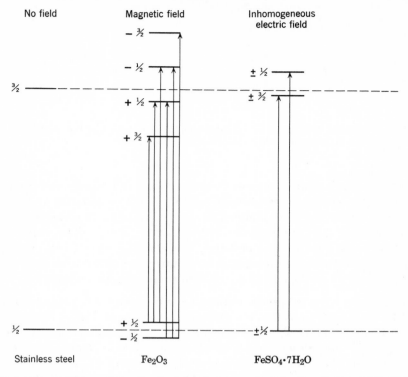

Fig. 13-4   The splitting of the $I = \frac{1}{2}$ ground state and of the $I = \frac{3}{2}$ first excited state of the Fe^{57} nucleus in a magnetic field (as in Fe_2O_3) and in an inhomogeneous electric field (as in $FeSO_4 \cdot 7H_2O$).   The lines between energy levels represent allowed transitions for γ-ray absorption; each level is characterized by its component of spin along the axis of symmetry of the field.   The spacings of the energy levels *are not* drawn to scale.

are shown schematically in figure 13-4.   It is to be noted that the center of gravity of the four levels into which the $\frac{3}{2}-$ state is split by the magnetic field and of the two levels in the inhomogeneous electric field does not coincide with the unsplit $\frac{3}{2}-$ state in stainless steel; this is an example of the chemical shift.

Experimental observations made with an Fe_2O_3 absorber are shown in figure 13-5 (K1).   The six lines expected from the magnetic splitting, as well as the lack of symmetry about zero velocity which is caused by the chemical shift, are seen.   From this spectrum it was deduced that the splitting is 2.9×10^{-7} eV for the $\frac{1}{2}-$ state and 1.6×10^{-7} eV for the $\frac{3}{2}-$ state.   From the known magnetic moments of the two states the splitting corresponds to a field of 5.2×10^5 oersteds at the Fe nucleus in Fe_2O_3 caused by the magnetic moments of the unpaired electrons in the Fe^{3+} ions.   The two lines that are expected from quadrupolar interactions are seen in figure 13-6, which shows the spectrum obtained with a $FeSO_4 \cdot 7H_2O$ absorber (D1).   The splitting is a

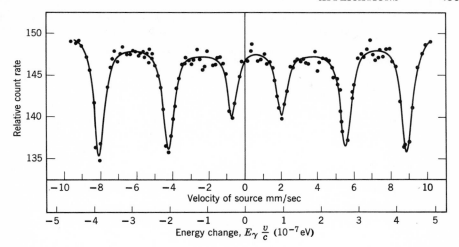

Fig. 13-5   The absorption in Fe^{57} (bound in Fe$_2$O$_3$) of the 14.4-keV γ ray emitted in the decay of Fe^{57m} (bound in stainless steel) as a function of the relative source-absorber velocity. Positive velocity indicates motion of source toward absorber.   (From reference K1.)

consequence of the inhomogeneous electric field produced by the sixth $3d$ electron in the Fe^{2+} ion; the other five give a spherically symmetrical electric field at the nucleus.   Again, the chemical shift causes a lack of symmetry about zero velocity.

The Mössbauer effect, then, can serve as a sensitive probe for atomic wave

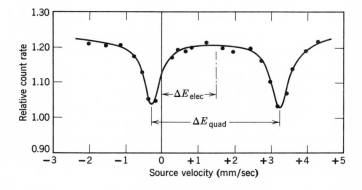

Fig. 13-6   Mössbauer spectrum of the Fe^{2+} ion in an absorber of FeSO$_4$·7H$_2$O at liquid-nitrogen temperature, taken with a room-temperature stainless-steel source.   The pattern exhibits the chemical shift ΔE_{elec} and the electric quadrupole splitting ΔE_{quad} of the excited state of the Fe^{57} nucleus.   The velocity is positive for the source approaching the absorber. (From reference D1.)

Table 13-1   Nuclides of interest as obsorbers in Mössbauer experiments

Fe^{57}	Cs^{133}	$Yb^{170,171,172,173,174}$
Ni^{61}	La^{139}	Lu^{175}
Zn^{67}	Nd^{145}	$Hf^{176,177,178,180}$
Kr^{83}	Sm^{152}	Ta^{181}
$Ru^{99,101}$	$Eu^{151,153}$	$W^{180,182,183,184,186}$
$Ag^{107,109}$	$Gd^{154,155,156,160}$	$Re^{185,187}$
$Sn^{117,119}$	Tb^{159}	$Os^{186,188,190,192}$
Sb^{123}	$Dy^{160,161,162,163,164}$	$Ir^{191,193}$
$Te^{123,125}$	Ho^{165}	Pt^{195}
I^{127}	$Er^{164,166,168}$	Au^{197}
$Xe^{129,131}$	Tm^{169}	$Hg^{199,201}$

functions, magnetic fields, and electric fields in the vicinity of the nuclei of atoms which are part of solid compounds.   It can also yield information about the chemical consequences of any nuclear processes which immediately precede the recoilless γ-ray.   For example, it has been found that electron capture of Co^{57} in Co(III) acetylacetonate leads to about twice as much Fe(III) as Fe(II) and that the higher oxidation states, which are expected from the Auger effect, must be reduced in a time less than 10^{-7} second (W1).

Table 13-1 lists nuclei for which the Mössbauer effect either has been observed or is expected to occur.   It is the nuclei with long-lived excited states $(t_{\frac{1}{2}} > 10^{-11}$ sec$)$ which emit low-energy γ rays $(<200$ keV$)$ that are the likely candidates.

B. POSITRON ANNIHILATION

It has been observed that the characteristics of positron annihilation (a process described in chapter 2, p. 51) depend on the chemical composition of the medium in which the event occurs (D2, D3).   In particular, the mean life of the positron as well as the fraction of the annihilations which emit three photons instead of two, quantities which we shall see are related to one another, can be grossly affected.   To understand the source of the effects, it is necessary to inquire further into the details of the annihilation process.

General Features.   The first important point is that nearly all annihilation events involve positrons which have been slowed down to thermal energies. Evidence for this statement comes, for example, from the experimental observation that positrons have a mean life of at least 1.5×10^{-10} second in condensed media as compared with the reliable calculation that the time required for slowing the positron down to thermal equilibrium is, at most, 5×10^{-12} second. These stopped positrons, then, will collide thermally with the electrons and will have a certain probability of being annihilated in each collision.

This annihilation probability depends on the relative orientations of the spins of the positron and of the electron: annihilation of the singlet state (opposed spins) is about 1100 times more probable per collision than annihilation of the triplet state (parallel spins). Further, because each photon must be emitted with at least $1\,\hbar$ of angular momentum and has only two states of polarization, annihilation in the singlet state gives two photons, whereas that in the triplet state gives three. It is this difference in multiplicity that causes the large difference between the probabilities of singlet and triplet annihilation. If the relative spins of the positron and electron were randomly oriented in each collision, the triplet state collisions would be three times as frequent as singlet state collisions and the ratio of two-photon to three-photon annihilation would be $1100/3 = 370$.

The mean life for annihilation of positrons depends on the density of electrons in the stopping material; for most metals the mean life is about 1.5×10^{-10} second.

Positronium. It is also possible for the collision between a slowed positron and an electron to result in the transient formation of a bound system, an atom of positronium $(e^{+}e^{-})$, before annihilation occurs. Positronium, which we denote by the symbol Ps, is a light isotope of atomic hydrogen with half the usual reduced mass and thus half the ionization potential (6.8 eV) and twice the Bohr radius. If the positronium were left undisturbed after being formed, there would be three times as many positronium atoms in the triplet state as in the singlet state, and thus there would be three times as many triple-photon as double-photon annihilations. Equivalently, since the mean life for annihilation of the singlet state is about 10^{-10} second, whereas that for the triplet state is about 10^{-7} second, one quarter of the annihilation would occur with a mean life of about 10^{-10} second and three quarters with 10^{-7} second; in the absence of positronium formation the mean life would be 10^{-10} second with no long-lived components. It was through the detection of the longer mean life for annihilation (with the delayed-coincidence technique) that Deutsch (D4) proved the existence of positronium.

Probability of Positronium Formation. It is not to be inferred that all positrons form positronium along their way to annihilation. Positronium is formed at most only about 20 per cent of the time; the rest of the time the positrons are annihilated in collisions as free positrons. This fact is qualitatively understood when it is realized that a positron must have an energy of at least $V - 6.8$ eV if it is to form positronium with an electron from a molecule of ionization potential V. If the energy is much in excess of the minimum value (say, about V), the collision is more likely to lead merely to ionization of the molecule without the formation of positronium. Since the previous collision is likely to leave the positron with any energy between zero and V, an upper limit to the probability of positronium formation may be estimated as $6.8/V$.

Reactions of Positronium. It is in the reactions of positronium that chemical information may be obtained about the medium in which the annihilation occurs. The relevant experimental observation is the rate of conversion of long-lived triplet positronium to either short-lived singlet positronium or free positrons.

The original work of Deutsch (D4) provides an excellent example of the conversion of triplet to singlet positronium. He found that the addition of a very small quantity of NO or NO_2, but not N_2O, to the stopping gas could cause the component of long half-life to vanish. Evidently an electron exchange between positronium and a molecule with an odd number of electrons can cause a spin flip, a process that is unfavorable in an electron exchange between positronium and a molecule with an even number of electrons because it would require the formation of the molecular triplet state.

Many interesting studies of the effects of the stopping medium on the annihilation process have been made. Among them is the study, by DeBenedetti and co-workers, of the rate of oxidation of triplet positronium in aqueous solution by several different oxidizing agents (M2). The resultant free positron is then annihilated with a mean life of about 5×10^{-10} second. It was found, for example, that the mean life of the triplet positronium decreased from about 1.8×10^{-9} second[3] in pure water to 0.7×10^{-9} second in $0.125M$ $HgCl_2$. The oxidation rate was observed to be proportional to the concentration of the oxidizing agent, and, for a given concentration, it seems roughly to increase with increasing standard potential of the oxidizing agent. This is also true for the more conventional electron transfer reactions mentioned in chapter 7, section B.

Studies of the chemical properties of the positronium atom are in an early state. The simplicity of the atom should encourage its future use as a chemical probe.

C. ANGULAR CORRELATIONS OF CASCADE RADIATIONS

It was pointed out in chapter 8 (p. 262) that when a nucleus emits two particles in sequence, such as a β particle followed by a γ ray, or two γ rays in succession, the angle between the two radiations is not, in general, expected to be randomly distributed. The deviation from a random distribution is measured

---

[3] The mean life is of the order of 10^{-9} second rather than the 10^{-7} second expected for isolated triplet positronium because of annihilation by the "pick-off" process in which a bound positron annihilates with an electron other than the one to which it is bound.

by the anisotropy defined as

$$A = \frac{W(180^0)}{W(90^0)} - 1,$$

where $W(\theta)$ is the number of events per unit solid angle which have an angle θ between the two radiations. Anisotropy can be observed only if the intermediate nucleus preserves its component of angular momentum along the emission direction of the first particle: the angular momentum vector must precess about that direction.

For any anisotropy to exist it is necessary for the intermediate nucleus to have a spin ≥ 1 (see p. 262). In general, then, the intermediate nucleus will have both a magnetic moment and an electric quadrupole moment. As discussed in section A, either or both of these nuclear moments will interact with the appropriate existing fields in the substance containing the intermediate nucleus and will cause the angular momentum vector to precess about the local field direction rather than the direction of emission of the first particle. This will diminish the anisotropy unless either the first particle is emitted along the local field direction or the mean life of the intermediate state is small enough so that the angular momentum vector will have moved only a small distance in its precession before the second particle is emitted. The latter possibility requires that

$$\tau \ll \frac{\hbar}{E},$$

where τ is the mean life of the intermediate state and E is the interaction energy of the nuclear moment with the local field (the splitting of the levels, discussed in section A). The critical lifetime for the intermediate state is of the order of 10^{-11} second. If the foregoing condition is not fulfilled and the precession diminishes the anisotropy, the quantity E may often be evaluated from the diminution of the anisotropy if τ is known; and, further, the magnitudes of the local fields may then be determined if the nuclear moments are known. The situation is more complicated in liquids because in them the local field varies randomly with time.

At this point it must be realized that the intermediate nucleus may find itself in very unusual chemical circumstances. This is best made clear by a particular example. Many studies have been made on the angular correlation of two cascade γ rays emitted from excited Cd^{111} produced in the electron capture by radioactive In^{111}. The Cd^{111}, then, finds itself in a chemical site previously occupied by an In^{111}, and, further, the electron capture process has just stirred up the electron shells. The latter effect will not be of much significance if, as is often true, the mean life for excited electronic states is considerably shorter than the mean life of the intermediate nucleus.

Details of the theory of angular correlations and their quenching by extranuclear fields are given in reviews by Steffen (S1) and Frauenfelder (F2, F3).

Some Experimental Results. Angular-correlation experiments are relatively easy to carry out. They only require measurement of coincidence rates as a function of the angle defined by the two detectors and the source. Our interest here lies in the effect of the chemical form of the source on this functional dependence (the anisotropy).

For very short mean life of the intermediate state ($<10^{-11}$ second), the local fields are expected to have but a negligible effect. For example, the angular correlation of two cascade γ rays in Ni^{60}, which result from the beta decay of Co^{60}, is essentially independent of the chemical nature of the source because the lifetime of the intermediate state in Ni^{60} is about 8×10^{-13} second.

The long-lived (1.25×10^{-7} second) intermediate state in the cascade of two γ rays that follow electron capture in In^{111} makes the angular correlation of the cascade γ rays very sensitive to chemical environment. Some typical results from the compilation of Steffen (S1) are given in Table 13-2. This kind of information has led to the following general conclusions concerning the chemical effects on angular correlations.

1. *Metallic sources.* The low anisotropies often seen in metallic sources usually arise from quadrupolar interaction with the strong inhomogeneous electric fields in those metallic structures that do not exhibit cubic symmetry. The low anisotropy in the thin silver film is expected because of the well-known structural distortions in thin films; the low anisotropy in gold, which has cubic symmetry, speaks for a serious structural distortion in the vicinity of the impurity atom. It is significant that although the anisotropy may become quite small it does not vanish; it approaches a minimum value expected for interaction with a static inhomogeneous electric field.

Table 13-2 Anisotropy values observed with In^{111} sources of different chemical composition and structure

Form of the Radioactive In^{111} Source	Anisotropy
In^{111} in Ag, thickness of Ag foil $> 10^4$ Å	-0.200 ± 0.006
In^{111} in Ag, thickness of Ag foil $< 10^2$ Å	-0.050 ± 0.021
In^{111} in Ag, formed in Ag by $(\alpha, 2n)$ reaction	-0.18 ± 0.01
In^{111} in Ag, same source as above but heated at 500°C for 6 hr	-0.05 ± 0.02
In^{111} in In metal powder	-0.095 ± 0.005
In^{111} in Au, electrodeposited	-0.072 ± 0.010
In^{111} in dry polycrystalline $InCl_3$ at room temperature	-0.012 ± 0.005
In^{111} in AgCl, double-stream evaporation	-0.004 ± 0.01
In^{111} in liquid indium metal	-0.21 ± 0.005
In^{111} in dilute aqueous solution of $InCl_3$	-0.221 ± 0.005
In^{111} in dilute aqueous solution of $In(NO_3)_3 \cdot 3H_2O$	-0.21 ± 0.02
In^{111} in xylene solution of InI_3	-0.22 ± 0.01
In^{111} in liquid InI_3 (220°C)	-0.19 ± 0.02

There is no evidence for interaction with an excited electron shell.   Any atomic excitation caused in the decay event must decay quite rapidly ($<10^{-12}$ second).

2. *Ionic Sources*.   The observation that the anisotropy can essentially vanish in ionic sources suggests an interaction with a time-dependent rather than a static field.   An appropriate field can be provided by electronic excitation of the ion in the decay process if the excited ion has a long mean life ($\cong 10^{-7}$ second).   A half-life of this order of magnitude is expected for excited impurities in an ionic lattice.

3. *Liquids*.   The anisotropy is found to remain large in nearly all liquids ranging from ionic solutions to molten metals.   Evidently this is a result of the rapid random fluctuations in any electric or magnetic fields that may be present in the liquid.   This explanation is supported by the observation that the anisotropy diminishes as the viscosity of the liquid increases, suggesting that the random fluctuation become much less rapid and the local structure approaches that of a solid.

The entry in Table 13-2 on the behavior of the In^{111} formed directly in a Ag lattice by the $(\alpha, 2n)$ reaction suggests that this technique might be particularly useful for learning about the state of recoil atoms in hot-atom chemistry.

D. MUON DEPOLARIZATION AND MUONIUM FORMATION

The fact that μ mesons interact with nuclei and electrons mainly through the electromagnetic field and the observation that their creation and decay occur through events in which parity is not conserved combine to make the μ meson potentially useful as a chemical probe.

Muon Polarization.   It was mentioned in chapter 9 that the muon (μ meson), a particle that does not interact strongly with nuclei, is formed in the decay of the pion (π meson), which does interact strongly with nuclei and is considered to be the quantum of the nuclear field.   It is the pion, not the muon, that is created in high-energy nuclear collisions; the muon appears only as a secondary particle resulting from the decay of a charged pion.

$$\pi^{\pm} \rightarrow \mu^{\pm} + \nu \qquad t_{1/2} = 2.6 \times 10^{-10} \text{ sec.}$$

The muon, in turn, is also unstable and decays into an electron, a neutrino, and an antineutrino:

$$\mu^{\pm} \rightarrow e^{\pm} + \nu + \bar{\nu}$$

The consequences of the nonconservation of parity in these two decay processes may be predicted from the discussion on p. 252 of chapter 8.

1. The muons produced in pion decay are polarized along their direction of motion: there are more muon spins (the spin of the muon is $\frac{1}{2}$) pointing in one direction than in the opposite direction.

2. In the β decay of the muon, as in the β decay of Co^{60}, the angular distribution of the emitted electrons is *not* symmetric about a plane perpendicular to the spin of the muon.

Because of these two consequences, parity nonconservation can be detected in the following manner. Consider a beam of muons resulting from pion decay; item 1 above means that the muon spins will be aligned along their direction of motion, which we shall take as the z direction. If the muon beam is then stopped in an absorber and measurements are made on the angular distribution of decay electrons, item 2 will require that the number of electrons observed at an angle ϑ with respect to the z axis be different from the number observed at the angle $\pi - \vartheta$. It was just this experiment, carried out by Garwin, Lederman, and Weinrich (G1), that demonstrated the violation of parity conservation in the two decay processes given above. Implicit in this experiment is the assumption that the muons are not depolarized while being stopped by the absorber and that they are not depolarized while they sit in the absorber waiting to decay. The magnetic moment of the muon will cause it to interact with any magnetic fields it may encounter in the stopping material, and the muon polarization can then be lost in the same manner as discussed for polarized nuclei in Section C. The occurrence of this depolarization was noticed in experiments in which the observed asymmetry in the β decay of the muon was found to decrease by about a factor of 2 when the stopping material was changed from graphite to a photographic emulsion (gelatin and silver bromide). It is the dependence of the depolarization on chemical environment that makes the muon potentially useful as a chemical probe.

Unfortunately, the depolarization of muons seems to involve several different processes in a rather complex manner so that the theory has not yet been developed to the point where quantitative information about the chemical environment can be extracted from the experimental results. The two different charge states of the muon present different problems which we shall sketch separately in the following paragraphs.

Depolarization of μ^+; Muonium. The magnetic interaction between a positive muon and the electrons in the absorber seems to be the predominant mechanism for depolarization. This magnetic interaction is assisted by the Coulomb attraction between the two particles which leads to the formation of muonium ($\mu^+ e^-$) in a manner analogous to the formation of positronium. The depolarization data, though, are not consistent with a one-time electron capture by the μ^+ to form muonium (F3, L1); evidently there must be repeated electron-capture-and-loss events which occur during the life of the μ^+.

A large, but not yet understood, chemical specificity is shown in the observations that there is essentially no depolarization in metals, semimetals and halocarbons, such as $CHBr_3$ and CCl_4, nearly complete depolarization in S, P, NaCl, and AgBr, and intermediate depolarization in hydrocarbons, H_2O, MgF_2, and teflon.   This curious grouping of substances does not lead to any immediate generalizations.

Depolarization of μ^-. The negative charge of the μ^- makes muonium formation impossible and, in general, diminishes depolarization of the μ^- by interactions with electrons in the stopping material.   It does, however, cause the formation of another interesting chemical substance in which the μ^- is captured into a stable atomic or molecular orbital: the μ-mesic atom or molecule.   Direct evidence for these new chemical species comes from the characteristic X rays emitted as the μ^- cascades down to the 1s state.[4]   The depolarization of the μ^-, then, can occur not only during this capture process but also through an interaction between the μ^- meson in the atomic 1s state and the nuclear magnetic moment if one exists (L2).   Unfortunately, little is yet known about the chemical specificity in μ^- depolarization, and its use as a chemical probe in this manner does not appear to be as promising as the use of the μ^+.

Mu-mesic Atoms and Molecules.   Aside from the depolarization process, the capture of the μ^- into molecular and, finally, atomic orbitals may also provide interesting chemical information.   For example, what determines the probabilities that a μ^- will be captured by the various kinds of atoms which may be present in the stopping material?   A theoretical treatment leads to the prediction that the relative probabilities are proportional to the product of the atom fraction and the atomic number.   This prediction has been verified for alloys but violated for other substances (S2).   Further experimental and theoretical study is required before the possible use of the process as a chemical probe can be evaluated.

REFERENCES

D1   S. DeBenedetti, G. Lang, and R. Ingalls, "Electric Quadrupole Splitting and the Nuclear Volume Effect in the Ions of Fe^{57}," *Phys. Rev. Letters* **6**, 60 (1961).

D2   S. DeBenedetti and H. C. Corben, "Positronium," *Ann. Rev. Nuclear Sci.* **4**, 191–218 (1954).

*D3   M. Deutsch, "Positronium," *Progress in Nuclear Physics*, Vol. 3, p. 131 (O. Frisch, Editor), Pergamon, New York, 1954.

D4   M. Deutsch, "Three Quantum Decay of Positronium," *Phys. Rev.* **83**, 866 (1951).

*F1   H. Frauenfelder, *The Mössbauer Effect*, W. A. Benjamin, New York, 1962.

F2   H. Frauenfelder, "Angular Correlation of Nuclear Radiations," *Ann. Rev. Nuclear Sci.*, **2**, 129–162 (1953).

*F3   H. Frauenfelder, "Angular Correlation," *Beta- and Gamma-Ray Spectroscopy* (K. Siegbahn, Editor), Interscience, New York, 1955, p. 531.

---

[4] The fact that the muon is 207 times as heavy as the electron causes the energy of a transition to increase by a factor of 207 and the radius of an orbit to decrease by a factor of 207.

G1   R. Garwin, L. Lederman, and M. Weinrich, "Observations of the Failure of Conservation of Parity and Charge Conjugation in Meson Decay: The Magnetic Moment of the Free Muon," *Phys. Rev.* **105,** 1415 (1957).

*G2   J. Green and J. Lee, *Positronium Chemistry,* Academic, New York, 1964.

K1   O. C. Kistner and A. W. Sunyar, "Evidence for Quadrupole Interaction of Fe^{57m}, and Influence of Chemical Binding on Nuclear Gamma-Ray Energy," *Phys. Rev. Letters* **4,** 412 (1960).

L1   G. R. Lynch, J. Orear, and S. Rosendorff, "Muon Decay in Nuclear Emulsion at 25,000 Gauss," *Phys. Rev.* **118,** 284 (1960).

L2   E. Lubkin, "Depolarization of a Muon by Hyperfine Interaction," *Phys. Rev.* **119,** 815 (1960).

M1   R. L. Mössbauer, "Kernresonanzfluoreszenz von Gamma-strahlung in Ir^{191}," *Z. Physik,* **151,** 124 (1958).

M2   J. D. McGervey, H. Horstman, and S. DeBenedetti, "Mean Lives of Positrons in Oxidizing Solutions," *Phys. Rev.* **124,** 1113 (1961).

*M3   R. L. Mössbauer, "Recoilless Nuclear Resonance Absorption," *Ann. Rev. Nuclear Sci.* **12,** 1–42 (1962).

*S1   R. M. Steffen, "Extranuclear Effects on Angular Correlations of Nuclear Radiations," *Advances in Physics,* **4,** 293 (1955).

S2   J. C. Sens, R. A. Swanson, V. L. Telegdi, and D. D. Yovanovitch, "An Experimental Test of the Fermi-Teller 'Z'-Law," *Nuovo Cimento* **7,** 536 (1958).

W1   G. K. Wertheim, W. R. Kingston, and R. H. Herber, "Mössbauer Effect in Iron (III) Acetylacetonate and Chemical Consequences of K Capture in Cobalt (III) Acetylacetonate," *J. Chem. Phys.* **37,** 687 (1962).

EXERCISES

1. What relative velocity is required to compensate a shift of 10^{-6} eV between the energy levels of Sn^{119} in an emitter and in an absorber (Sn^{119} emits a 24-keV γ ray)?
   Answer: 1.25 cm sec^{-1}.

2. The half-life of the 24-keV state of Sn^{119} is 1.9×10^{-8} second. What relative velocity will move the peak by an energy corresponding to the width of the level?

3. The Debye temperature of metallic tin is 195°K. Would a recoilless transition from the 24-keV state of Sn^{119} be expected in metallic tin?

4. Estimate the energies of the chemical shift and of the quadrupole splitting in $FeSO_4 \cdot 7H_2O$ from the information in figure 13-6.

5. Estimate the magnitude of the interaction energy between the excited Ni^{60} nucleus and its surroundings that would be required to destroy the angular correlation that exists between the two cascade gamma-rays that follow the beta decay of Co^{60}.
   Answer: $\cong 10^{-3}$ eV.

6. If the mean life of triplet positronium decreases from 1.8×10^{-9} second in pure water to 0.75×10^{-9} second in 0.10 M $HgCl_2$, what is the mean life for a reaction between the positronium and $HgCl_2$ in this solution? *Answer:* 1.3×10^{-9} second.

14

Nuclear Energy

A. BASIC PRINCIPLES OF CHAIN-REACTING SYSTEMS

Speculation about the possible use of nuclear processes for the large-scale production of power dates back to the early years of radioactivity. Only with the discovery of fission did such applications become a real possibility. The unique feature of the fission reaction which makes it suitable as a practical energy source is the emission of several neutrons in each neutron-induced fission; this makes a chain reaction possible.

Chain Reaction. The condition for the maintenance of a chain reaction is that on the average at least one neutron created in a fission process cause another fission. This condition is usually expressed in terms of a multiplication factor k, defined as the ratio of the number of fissions produced by a particular generation of neutrons to the number of fissions giving rise to that generation of neutrons. If $k < 1$, no self-sustaining chain reaction is possible; if $k = 1$, a chain reaction is maintained at a steady state; if $k > 1$, the number of neutrons and therefore the number of fissions increases with each generation, and a divergent chain reaction results. An assembly of fissionable material is said to be critical if $k = 1$ and supercritical if $k > 1$.

Since one neutron per fission is required to propagate the chain reaction, the number of neutrons increases by the fraction $k - 1$ in each generation. Thus the rate of change of the number of neutrons in a chain-reacting system is

$$\frac{dN}{dt} = \frac{N(k-1)}{\tau},$$

where τ is the average time between successive neutron generations. By integration we find that at time t the number of neutrons is

$$N = N_0 e^{(k-1)t/\tau}, \tag{14-1}$$

where N_0 is the number of neutrons at $t = 0$. If τ is very short (as it is when no moderator is used and fission takes place with fast neutrons) and if k is suddenly made to exceed unity by an appreciable amount, the chain reaction can proceed explosively as in a fission bomb.

In a nuclear reactor k is kept equal to unity for steady operation. However, a reactor must be designed in such a way that k can be made slightly larger than one (say 1.01 or 1.02) to allow the neutron flux and therefore the power to be brought up to a desired level. Control of the reactor, for example by the motion of neutron-absorbing control rods, is possible only if τ is not too short. Assume that $\tau = 10^{-3}$ sec (approximately the life expectancy of a thermal neutron in graphite or D_2O) and $k = 1.001$. Then, according to (14-1), $N = N_0 e^t$, where t is in seconds, and the neutron level will increase by a factor e every second or by a factor of about 20,000 in 10 seconds. This is too rapid an increase for safe and convenient control. Fortunately the fact that in thermal-neutron fission some neutrons are delayed with half-lives between 0.18 and 55 seconds increases the average time τ between neutron generations. As long as $k - 1$ is smaller than the fraction of delayed neutrons (0.0065 for U^{235}, 0.0021 for Pu^{239}, and 0.0026 for U^{233}), the effective time τ between generations is approximately

$$\tau = \tau_0 + \sum_i \left(\frac{f_i}{\lambda_i} \right),$$

where τ_0 is the generation time without delayed neutrons and the f_i's are the fractions of neutrons delayed with the decay constants λ_i. For the six known groups of delayed neutrons in U^{235} fission $\sum_{i=1}^{6} (f_i/\lambda_i) = 0.082$ second, which is large compared with τ_0. Hence τ is about 0.08 second. Then, with $k = 1.001$, the e-folding time or "period" of the system is about 80 seconds, and ample time is available for control.

Critical Size. The multiplication factor in a medium of infinite extent, denoted by k_∞, is given by the product of the number ν of neutrons emitted per fission and the fraction of the neutrons that produce another fission. This fraction is the ratio of the macroscopic fission cross section ($\sigma_f N_f$, where N_f is the number of fissionable nuclei per cubic centimeter) to the sum of this macroscopic fission cross section and all macroscopic capture cross sections:

$$k_\infty = \nu \frac{\sigma_f N_f}{\sigma_f N_f + \sum_i \sigma_{ci} N_i}, \tag{14-2}$$

where N_i is the number of nuclei of the ith substance per cubic centimeter and σ_{ci} is the (ordinary) capture cross section of that substance.[1] The nonfission capture of the fissile material used must be included in $\sum_i \sigma_{ci} N_i$.

---

[1] Since there is always a spectrum of neutron velocities present and cross sections generally vary with neutron velocity, the expression for k_∞ in (14-2) must in fact be appropriately averaged over the velocity spectrum.

Table 14-1   *Properties of moderators*

	L_S (cm)	L_0 (cm)	$\sigma_m N_m$ (cm^{-1})	Density (g cm^{-3})
H_2O	5.7	2.76	0.017	1.00
D_2O	11.0	100	0.000080	1.1
Be	9.9	23.6	0.0013	1.84
C	18.7	50.2	0.00036	1.62

In a reactor of finite extent the multiplication factor k is smaller than k_∞ because of the loss of neutrons by leakage through the surface; the smaller the reactor, the greater its ratio of surface to volume and therefore the greater the loss.   Quantitative estimates of neutron losses from a reactor surface are very complicated but are quite essential to any estimate of critical size.   As a rough approximation, the fractional loss of neutrons for thermal reactors is proportional to the sum $L_S{}^2 + L^2$, where L_S and L are the average (crow-flight) distances traveled in the moderating medium (of infinite extent) by a fission neutron before reaching thermal energy (L_S) and after reaching thermal energy (L).   For a spherical reactor of radius R the approximate relation is $k_\infty - k = \pi^2 R^{-2}(L_S{}^2 + L^2)$.   The critical radius R_c is that radius for which $k = 1$.   Thus

$$R_c = \pi(L_S{}^2 + L^2)^{1/2}(k_\infty - 1)^{-1/2}. \tag{14-3}$$

The slowing-down length L_S may be known from measurements on various moderators and generally is not appreciably altered by the addition of fuel to the moderator.   The diffusion length L will be smaller than that of the pure moderator (L_0) and is given by $L^2 = x L_0{}^2$, where x is the macroscopic absorption cross section of the moderator, $\sigma_m N_m$, divided by the macroscopic absorption cross section for the medium, $\sum_i \sigma_i N_i$.   Table 14-1 gives L_S, L_0, $\sigma_m N_m$ and the density for some popular moderators.

In most reactors x, which is the fraction of neutrons absorbed by the moderator, is kept rather small for reasons of neutron economy.   Therefore, a crude approximation to (14-3) which neglects L, namely, $R_c = \pi L_S (k_\infty - 1)^{-1/2}$, gives the right order of magnitude of the critical size for practical thermal reactors. For the same reason reactors often have k_∞ not much less than η (see p. 312). As an illustration we may estimate the critical radius for a solution of U^{235} in ordinary water:

$$R_c \cong \frac{\pi L_S}{(k_\infty - 1)^{1/2}} \cong \frac{\pi \times 5.7}{(2.1 - 1)^{1/2}} = 17 \text{ cm}.$$

Here we assumed that the concentration of the solution was great enough to ensure that neutron reaction with U^{235} was much more probable than capture by hydrogen.   The ratio of the two cross sections is $694/0.332 = 2100$, so that this condition is met if the concentration of U^{235} is a few tenths of a mole per liter.   The first homogeneous reactor put into operation, the Los Alamos Water Boiler (1944), consisted of a stainless-steel sphere, 30 cm in diameter, filled with a $\sim 1.2M$ solution of uranyl sulfate in H_2O, the uranium being enriched to 14.6 per cent in U^{235}.

In many reactors the fissionable material is not dissolved in the moderator but is separated from it in a heterogeneous arrangement.   All the reactors which have been constructed with ordinary uranium as the fuel have the uranium in lumps or rods arranged in a lattice embedded in graphite or heavy water.   The reason is that much of the loss of neutrons is due to several strong absorption resonances in U^{238} between 6 and 200 eV.   In a homogeneous mixture of uranium and moderator the probability that a neutron during the slowing-down process is absorbed by $U^{238}(n, \gamma)$ reaction in the resonance region is quite large.   If, however, the uranium is arranged in aggregates, a much greater fraction of the neutrons will be slowed down in the moderator to energies below the resonance region before encountering uranium nuclei. The optimum lattice spacing is approximately L_S for the moderator.   Without the advantage of this ingenious device, normal uranium-graphite assemblies would have k_∞ slightly less than unity; even with it, k_∞ cannot be greater than η, which is 1.3.   For normal uranium-graphite reactors k_∞ is commonly about 1.07, and, according to (14-3), the critical radius must be about $R_c = \pi \times 18.7 \times (0.07)^{-\frac{1}{2}} = 220$ cm.   For a cube-shaped assembly the length of an edge is approximately $\sqrt{3}\ R_c$, or about 13 feet.

In all practical reactors the core is surrounded by a neutron reflector which reduces neutron loss.   This makes the necessary size of the reactor core slightly smaller, but operating in the other direction are effects of impurities, provisions for cooling and for control, and overdesign.   Another important effect of the reflector in power reactors is the increase in neutron flux in the outer parts of the core; because the power level is likely to be limited by the temperature rise at the center this makes the outer parts contribute a better share to the overall power output.   In addition, the fuel lattice may be altered near the center to flatten the neutron- and power-flux distributions.

In practice, a reactor must always be designed with excess reactivity,[2] that is, in such a way that k can be made to exceed unity.   This is necessary not only to bring the reactor up to any desired power level (p. 474) but also to allow for some fuel burnup and for the buildup of neutron-absorbing fission-product "poisons" (cf. p. 483) and to make possible the introduction into the reactor of neutron-absorbing materials for irradiations.   Regardless of the reserve reactivity built into a reactor, for steady operation it must always be operated at zero reactivity.   This is usually achieved by control rods made of materials with large neutron-capture cross sections, such as boron, cadmium, or hafnium.   These control rods are moved in and out of the reactor to com-

---

[2] Reactivity is defined as $(k - 1)/k$; thus, when $k = 1$, reactivity is zero.

pensate for any changes in reactivity. Other methods of control use motion of fuel elements or of the reflector.

B. REACTORS AND THEIR USES

Reactor Types. Reactors may be classified in a variety of ways. We may consider the fuel used and distinguish among reactors using natural uranium, uranium enriched in U^{235}, Pu^{239}, or U^{233}. Reactors may also be grouped according to the neutron energy used to produce fission (fast, partly moderated or intermediate energy, or thermal), according to the moderator (water, heavy water, graphite, beryllium, or organic compounds), or the coolant (air, water, or liquid metal). Often reactors are classified according to whether fuel and moderator are homogeneously mixed or arranged in a heterogeneous manner. In terms of purpose, reactors may be designed primarily to produce fissionable material (Pu^{239} or U^{233}), to provide excess neutrons for the production of other nuclides (such as H^3), to produce useful power, to serve as research tools, or for a combination of these purposes. Some aspects of research reactors have already been discussed in chapter 11, section C. Table 14-2 lists some essential characteristics of a number of reactors now in operation. We devote a few paragraphs to several particular reactors, each representative of a different general type.

Oak Ridge Graphite Reactor. Built in 1943 and called the X-10 reactor, this was the world's first reactor operating at an appreciable power level. Because it is a natural uranium-graphite reactor it is a large one. Its core is a 24-ft cube of graphite blocks, provided with 1248 fuel channels each 1.75 in. square located in an 8-in. rectangular lattice. The fuel is 35 tons of uranium metal in the form of cylindrical slugs 1.1 in. in diameter, 4 in. long, and jacketed in aluminum for protection from oxidation. Because the fuel elements are round pegs in square holes, clear spaces are provided through which the coolant air is blown. At 3800 kW, with 100,000 cu ft of air per minute passing through the core channels, the metal slugs are kept below 245°C; the average graphite temperature is 130°C.

The fuel loading does not extend quite to the outer surfaces of the graphite, and the outer part of pure graphite serves as a neutron reflector. A concrete shield 7 ft thick surrounds the entire core, with a space left between for air passage. The fuel channels extend through the shield at one end, so that old slugs may be pushed out and new fuel inserted (figure 14-1). Of course, these openings in the shield are ordinarily kept plugged to reduce the leakage of dangerous radiations. Some of the channels serve for irradiations of other materials and for various experiments. There are also special channels for similar purposes and for the control rods. A reactor of this size is quite

Table 14-2   Properties of some representative reactors

Name and Location	Fuel	Moderator	Coolant	Thermal Power (MW)	Maximum Thermal Neutron Flux $(cm^{-2}sec^{-1})$	Purpose
BGRR, Brookhaven	93 % U^{235}	Graphite	Air	20	2×10^{13}]	Research
BR-5, Obninsk, USSR	PuO_2	None	Na	5		Experimental fast breeder
Calder Hall A, Great Britain	U	Graphite	CO_2	230		First British power station (1956): 45 MW electric power
Dresden Power Station, Dresden, Illinois	1.5 % U^{235} as UO_2	Boiling H_2O	Boiling H_2O	700	3×10^{13}	Electric power (200 MW)
EBR-II, National Reactor Testing Station, Idaho	48% U^{235}	None	Na	62.5	(3.7×10^{15}) fast	Experimental fast breeder
MTR, National Reactor Testing Station, Idaho	93 % U^{235} in U-Al plates	H_2O	H_2O	30	5×10^{14}	Materials testing; isotopes production
NRU, Chalk River, Canada	U	D_2O	D_2O	200	2.5×10^{14}	Research, materials testing; isotope production
OMR, Piqua, Ohio	1.9 % U^{235} in U-Mo alloy	Terphenyl	Terphenyl	45		Power (11.4 MW); demonstration of organic moderator design
PWR, Shippingport, Pennsylvania	Seeds: 93 % U^{235} in U-Zr. Blanket: normal UO_2	H_2O	Pressurized H_2O	231		Electric power (60 MW)
SGR, Hallam, Nebraska	3.6 % U^{235} in U-Mo alloy	Graphite	Na	240		Electric power (75 MW)
Swimming pool reactors	20 % U^{235} or 93 % U^{235} in U-Al plates	H_2O	H_2O	0.001–5.0	$\sim10^{13}$/MW	Relatively inexpensive research reactor
TRIGA	20 % U^{235} in U-Zr Hydride	U-Zr Hydride	H_2O	≤ 1 steady ≤ 1200	$\leq 5 \times 10^{12}$ steady $\leq 10^{16}$ pulsed	Research
VVPR1, Voronezh, USSR	1.5–2.5 % U^{235} in UO_2	H_2O	Pressurized H_2O	760		Electric power (210 MW)

expensive.   In round numbers, the value of the 620 tons of very pure graphite is about $2,000,000, and the value of the uranium is probably about $2,000,000. In addition, the costs of the building and of the air-handling system have been estimated at about $1,000,000 each.

Provided that all the aluminum jackets on the thousands of fuel slugs are intact, the exhausted cooling air should be free of fission product contamination.   Rupture of a slug jacket could cause contamination of the exhaust air with fission products and necessitate the removal of the slug.   At all times the cooling air will be made radioactive by passage through the reactor, chiefly by the reaction $Ar^{40}(n, \gamma)$ to produce β-emitting Ar^{41} of 109 minutes half-life. This thermal reaction cross section is approximately 0.5 barn, and from the

Fig. 14-1 Charging face of the Brookhaven National Laboratory's uranium-graphite reactor BGRR (similar to the X-10 reactor at Oak Ridge).   The uranium rods are inserted from this face.   (Courtesy Brookhaven National Laboratory.)

dimensions given we estimate that roughly 500 cu ft of air (containing $\sim$160 g argon) are exposed to the average thermal neutron flux, which is about 5×10^{11} cm^{-2} sec^{-1}. On this basis the steady-state activity of Ar^{41} present in the entire atmosphere as a result of the operation of this particular reactor is about $(160/40) \times 6 \times 10^{23} \times 0.5 \times 10^{-24} \times 5 \times 10^{11} = 6 \times 10^{11}$ disintegrations sec^{-1} or about 16 curies. Because of its radioactivity the air is discharged through a tall stack.

The reactor was originally built as a pilot plant for the plutonium-producing reactors at Hanford. Plutonium is produced by the neutron capture occurring in U^{238}:

$$_{92}\text{U}^{238} \xrightarrow{n,\,\gamma} {_{92}}\text{U}^{239} \xrightarrow{\beta^-} {_{93}}\text{Np}^{239} \xrightarrow{\beta^-} {_{94}}\text{Pu}^{239}.$$

As we have seen from the neutron material balance, very roughly one such capture occurs per fission of U^{235}. The total energy released in each fission (exclusive of neutrino energies) is 190 MeV so that the rate of fission at 3800 kW is about 10^{17} sec^{-1}. Thus Pu^{239} is produced at the rate of about 10^{22} atoms day^{-1} or 4 g per day. In recent years the reactor was operated primarily as a source of radioactive isotopes for distribution by the United States Atomic Energy Commission. Some of these isotopes are fission products, extracted from discharged fuel slugs. Others are produced by n, γ reaction on appropriate targets; for example, Na^{24} from Na$_2$CO$_3$, and I^{131} from tellurium metal by Te$^{130}(n, \gamma)$Te$^{131} \xrightarrow{\beta^-} $ I^{131}. A few are produced in special reactions, particularly C^{14} from N$^{14}(n, p)$, P^{32} from S$^{32}(n, p)$, and H^3 from Li$^6(n, \alpha)$.

After 20 years of almost continuous operation, the X-10 reactor, now made obsolete by more recent reactor designs, was shut down in 1963 and put on a standby basis.

Uranium-Heavy Water Reactors. The core of the experimental reactor (JEEP) at Kjeller, Norway, is a cylindrical tank 2 meters in diameter filled with 7 tons of D$_2$O. This volume of heavy water is expensive; the value is almost $2,000,000. But because of the very small thermal-neutron absorption cross section and the fairly small neutron-slowing-down length (L_S) for D$_2$O the reactor is much smaller than a uranium-graphite reactor and requires much less uranium. The fuel charge in this reactor is 2.5 tons of uranium metal (value about $200,000) in the form of 1-in. diameter slugs in 76 aluminum tubes arranged in a rectangular lattice. A graphite reflector about 3 ft thick surrounds the core. Control of the reactor is by insertion of cadmium plates into the space between core and reflector. (In an emergency the heavy-water moderator could be drained from the core.) A concrete radiation shield at least 2 meters thick gives biological protection.

Cooling is provided by forced circulation of the heavy water through an external heat exchanger. This originally permitted operation at 100 kW, giving a central neutron flux of more than 3×10^{11} cm^{-2} sec^{-1}. Later improvements in the heat exchangers have made it possible to operate the reactor at 450 kW and correspondingly higher flux. Several experimental holes into the core are used for experimental purposes and for isotope production.

The French uranium-heavy water reactor P-2 at Saclay (figure 14-2) is similar in size to the Kjeller reactor. A significant difference is that each fuel

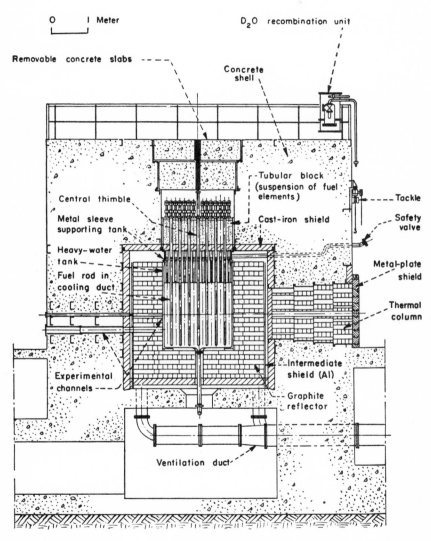

Fig. 14-2 Construction and experimental facilities of the heavy-water reactor P-2 at Saclay, France. (Reproduced from L. Kowarski, *Nucleonics* **12**, No. 8, 9 (1954), by permission of the McGraw-Hill Publishing Company.)

rod is cooled by a flow of nitrogen gas under 10 atm pressure through a surrounding annular space. The operational power level is about 2000 kW, and the maximum neutron flux is about 6×10^{12} cm^{-2} sec^{-1}. This reactor was designed primarily as an experiment and for radioisotope production.

Many other D_2O-moderated reactors fueled with natural uranium are in operation. The NRX research reactor at Chalk River, Canada, is unique in that it uses D_2O as a moderator but H_2O as a coolant. At 42-MW power it produces a maximum flux of 7.8×10^{13} cm^{-2} sec^{-1}. The larger NRU reactor built more recently at the same laboratory uses D_2O both as moderator and coolant, operates at 200 MW, and has a maximum thermal-neutron flux of about 2.5×10^{14} cm^{-2} sec^{-1}. Large uranium-D_2O reactors of still higher power ratings are used for plutonium production at Savannah River, Georgia.

Materials Testing Reactor. A number of reactors have been designed specifically for the purpose of providing facilities for testing the behavior of various materials at high radiation levels. The first such reactor built was the MTR (Materials Testing Reactor) at the National Reactor Testing Station near Arco, Idaho. Its maximum thermal-neutron flux is about 5×10^{14} cm^{-2} sec^{-1}. This reactor uses highly enriched (93 per cent) U^{235} as fuel and H_2O as moderator and coolant. The fuel elements are plates made of an aluminum-uranium alloy and are supported vertically in a lattice which is 40 by 70 cm and 60 cm high. A reflector of beryllium metal surrounds the fuel lattice and is contained in an aluminum and stainless steel tank 55 in. in diameter and about 30 ft deep. The tank is filled with water, and access to the reactor core for replacement of fuel elements, etc., is through the 20 ft of water, which is an adequate biological shield during shutdown. For cooling, water is forced downward through the tank between the fuel elements and through holes in the beryllium reflector at the rate of 20,000 gal per minute. The operating power level is 30,000 kW. Outside the tank is a secondary reflector of graphite about 4 ft thick, the inner part of which is made of loosely packed graphite spheres 1 in. in diameter. The construction is such that these spheres are free to expand upward without exerting undue stress on the tank wall or the solid graphite. Since the spheres are exposed to a higher neutron flux than the rest of the graphite, any radiation damage will affect them first. They can be replaced when necessary. The solid graphite is provided with holes for air cooling. Outside the reflector is a thermal shield consisting of two 4-in. thicknesses of steel, also air cooled. Outside this a 9-ft shield of special concrete reduces the radiation level during operation to less than 1 mr hr^{-1}. Many experimental holes penetrate the shields and reflectors. Some reach the maximum neutron flux of 5×10^{14} thermal and 1×10^{14} fast neutrons cm^{-2} sec^{-1}, which is in the beryllium reflector at the surface of the fuel lattice. Of course, these holes may not be opened during operation.

Reactor control is provided by two cadmium-loaded regulating rods (one is a spare) in the beryllium reflector. Automatic control circuits drive the rod in or out to maintain steady operation.   To ensure that no accidental rod with-drawal could make the reactor critical with regard to prompt neutrons only, each regulating rod controls no more than 0.5 per cent in k, which is less than the delayed-neutron contribution.   Further control of reactivity is provided by up to eight shim safety rods in the fuel lattice.   These rods contain cad-mium, and as the cadmium section is withdrawn a U^{235} fuel section is brought up into the lattice.   Safety circuits prevent too rapid removal of these shim rods, and if any minor malfunction or irregularity occurs rods are automatically reinserted until the power level is reduced.   One of the electronic circuits meas-ures the reactor period (or time rate of change of reactivity); if the period for any reason becomes shorter than one second, a complete "scram" results, with all shim rods dropped by their magnetic clutches.

One reason why a wide range of reactivity control is needed in a high-flux reactor is the effect known as xenon poisoning.   One of the fission products is Xe^{135}, a radioactive isotope of 9.2-hour half-life and with the astonishing thermal-neutron absorption cross section 2.7×10^6 barns.   In steady-state operation the presence of this poison reduces k by about 0.04.   On shutdown after steady operation the concentration of Xe^{135} increases because it is still being formed by decay of its parent, 6.7-hour I^{135}, and is no longer being con-sumed by n, γ reaction.   After a few minutes' shut-down the reactor cannot be started up again.   After about 10 hours the poisoning is a maximum, amounting, in the MTR, to a reduction of about 0.3 in k.   Thereafter the poison decays away, and the reactor can be made critical again after about 2 days.

Swimming-Pool Reactors.   In many respects similar to the MTR, but much less complex and less expensive, are the swimming-pool reactors which for some years have been the most popular research reactors (see chapter 11, section C, also refs. K2 and R2).   In a swimming pool, as the name implies, the reactor core is suspended in a pool of water at a depth of about 25 ft, and the water serves as moderator, coolant, reflector, and shield.   The fuel is enriched uranium, either 20 per cent or, preferably, 93 per cent U^{235}, alloyed with aluminum.   The U-Al alloy is in the form of plates typically 7.5 cm wide, 60 cm long, and 0.5 mm thick, clad in aluminum of the same thickness, and arranged in boxlike bundles with about 3-mm spacings between plates for water circulation.   The critical mass for pool reactors is approximately 2500 g of U^{235}, and the entire core occupies roughly 2 cu ft.   Control rods usually consist of aluminum boxes much like those containing the fuel element bundles but lined with cadmium and filled with boron carbide.   They are suspended from the same bridge that carries the reactor itself.

Several dozen swimming-pool reactors are in operation in many parts of the world.   Power levels and neutron fluxes vary over a considerable range.   With water passing through the fuel elements by convection only, swimming pools can be operated up to about 100 kW.   At higher power levels forced circulation is necessary and some provision has to be made to remove gaseous activities (mainly 7-second N^{16} formed by the O^{16} (n, p) reaction) produced in the water. Swimming-pool reactors have been operated at power levels up to 5 MW, with thermal-neutron fluxes up to $\cong 5 \times 10^{13}$ cm^{-2} sec^{-1}.

Breeding. Any reactor fueled with ordinary uranium makes new fissionable material as it burns up its U^{235}. When reactors (such as the X-10 reactor) make Pu^{239} at a rate that is not much less than the rate at which U^{235} is consumed, they are known as converters. This conversion can be very useful in extending the supply of fissionable fuel, especially if the yield of new fuel can actually exceed the consumption, in which case the process is called breeding. A reactor can operate as a breeder only if η, the number of neutrons produced per neutron absorbed in the fissionable material, exceeds two: one neutron to produce another fission so that the chain reaction can proceed and more than one neutron to produce more fissionable material than has been consumed. The values of η for the three important fissionable materials U^{235}, Pu^{239}, and U^{233} are shown in table 14-3 both for thermal neutrons ($v = 2.2 \times 10^5$ cm sec^{-1}) and for a fast reactor neutron spectrum; also given are the values of ν (average number of neutrons emitted per fission), σ_f (fission cross section), σ_c (n, γ capture cross section), and the ratio $\alpha = \sigma_c/\sigma_f$. As the data in the table show, η is always less than ν because all three of the fissionable nuclides have appreciable cross sections for n, γ reaction, particularly at thermal-neutron energies. The values of η for U^{235} and Pu^{239} fission by thermal neutrons exceed 2.0 by so little that no practical thermal breeder reactor appears likely with these fuels, as one cannot hope to reduce other neutron losses (absorption in other materials, resonance absorption, escape) to zero. The nuclide U^{233}, with $\eta = 2.28$, does appear to offer some possibility for thermal breeding. It can be produced from thorium, in a sequence of reactions com-

Table 14-3 *Some properties of fissionable materials at thermal-neutron and fast-neutron energies**

Properties	U^{235} Thermal	Fast	Pu^{239} Thermal	Fast	U^{233} Thermal	Fast
σ_f (in barns)	577 ± 5	1.44	742 ± 4	1.78	525 ± 4	2.20
σ_c (in barns)	101 ± 5	0.22	286 ± 4	0.15	53 ± 2	0.15
$\alpha = \sigma_c/\sigma_f$	0.18 ± 0.01	0.15	0.39 ± 0.03	0.086	0.101 ± 0.004	0.068
ν	2.44 ± 0.02	2.52	2.89 ± 0.03	2.98	2.51 ± 0.02	2.59
$\eta = \nu/(1 + \alpha)$	2.07 ± 0.01	2.18	2.08 ± 0.02	2.74	2.28 ± 0.02	2.42

* The thermal-neutron data are given for $v = 2200$ meters sec^{-1}; the fast-neutron data represent weighted averages over a typical reactor neutron spectrum.

pletely analogous to the production of Pu^{239} from U^{238}:

$$Th^{232}(n, \gamma)Th^{233} \xrightarrow[t_{1/2}=22 \text{ min}]{\beta^-} Pa^{233} \xrightarrow[t_{1/2}=27 \text{ d}]{\beta^-} U^{233} \xrightarrow[t_{1/2}=1.62 \times 10^5 \text{ y}]{\alpha}$$

Among the fuel-moderator systems considered for thermal U^{233} breeding are a D_2O solution of uranyl sulfate and a uranium solution in liquid bismuth with graphite moderator. Surrounding the reactor core would be a "blanket" containing thorium in an appropriate form. Some experimental reactors have been built to test the homogeneous aqueous solution system, but problems of container corrosion and solution instability are formidable and thermal breeders do not appear to be very promising (E1).

With the use of fast neutrons, much better neutron economy can be achieved (K1). Although fission cross sections decrease substantially with increasing neutron energy (aside from the region up to $\cong 100$eV, in which many resonances occur), capture cross sections decrease even more rapidly, as shown in table 14-3, and therefore α becomes small and η approaches ν (which is almost independent of energy in the region of interest). In a fast reactor, Pu^{239} breeding thus becomes a definite possibility—the higher value of η favors Pu^{239} over U^{233}—and several experimental fast breeder reactors have been constructed, two of which are listed in table 14-2. They are designed to take advantage of the U^{238}–Pu^{239} cycle (U^{238} may be present in the core as well as in the breeder blanket), although some of them initially used U^{235} rather than Pu^{239} fuel. The two experimental breeder reactors in Idaho, EBR-I and EBR-II, have achieved breeding ratios greater than unity. A full-scale power plant based on a fast breeder concept, the Enrico Fermi Power Plant, has been built at Monroe, Michigan. It is designed to deliver 90 MW electrical power when operating at a (thermal) power level of 300 MW. The core of its sodium-cooled reactor is about 30 in. in diameter and 31 in. high, and is initially fueled with a zirconium-clad uranium-molybdenum alloy containing uranium enriched to 25 per cent in U^{235}. The uranium in the blanket, on the other hand, is depleted to a U^{235} content of about 0.4 per cent.

Fast reactors have advantages as well as disadvantages relative to thermal-neutron reactors. The short lifetimes for prompt neutrons make control and safety more of a problem. The small fission cross sections for fast neutrons (table 14-3) dictate high fuel concentrations in the core, and the resultant high power densities lead to difficult heat-transfer problems. On the other hand, the choice of structural materials and fuel-alloy components is much wider in fast reactors because most elements have rather small neutron absorption cross sections in the energy range of interest. Similarly, the buildup of fission-product poisons is less serious.

With successful breeding, it is, in principle, possible to transform the world's supply of Th^{232} and U^{238} (so-called "fertile" materials) into the fissionable nuclides U^{233} and Pu^{239} and thus to increase the supply of nuclear fuel from only U^{235} to the total available uranium and thorium—a factor of perhaps 150.

However, the fuel materials will have to be chemically processed to separate fission products several times in each breeding cycle, and in the final analysis the efficiency of the chemical processing will determine the practicality of breeder reactors. The total potentially useful resources of uranium and thorium have been estimated as $\cong 3 \times 10^7$ tons. With breeding, this would yield almost 2×10^{21} Btu of energy, as compared with the world's estimated fossil fuel reserves of less than 10^{20} Btu.

Reactors for Power Production. The use of nuclear reactors as practical energy sources is no longer a dream but a reality. By 1963, less than 10 years after the first useful power was obtained from a reactor, some 4×10^3 MW of electrical energy were being produced from nuclear power plants the world over, about 20 per cent of this total in the United States (P1). These figures still represent only a very small fraction of the total installed electric generating capacity—about 0.4 per cent in the United States. But the share of nuclear plants in the total production of electricity is increasing rapidly, and all forecasts agree that it will be substantial by 1980, even in the United States, where supplies of fossil fuels are much less limited than in some other countries. At the present time power costs per kilowatt hour from nuclear reactors are beginning to be competitive with those from coal-fired plants in many areas, particularly where transportation costs for fossil fuels are an important factor. The relatively high capital cost of nuclear plants and the costs of chemical fuel processing and radioactive-waste disposal can be offset or outweighed by the lower fuel cost for nuclear fuels compared to coal (L1). With development of more efficient designs of nuclear plants, and with the progressive depletion of fossil fuels, the competitive position of nuclear power sources is bound to improve even further.

Quite a variety of reactor types are in use for power production, others are under investigation, and still different designs will undoubtedly be developed in the future (N1, P1). Representatives of some of the important types of "first-generation" power reactors are listed in table 14-2. In Great Britain natural-uranium-fueled, graphite-moderated, CO_2-cooled reactors have found greatest favor, whereas Canada has emphasized development of D_2O-moderated natural-uranium designs. In the United States and the Soviet Union a number of power plants are based on the use of enriched uranium, with either a pressurized-water cooling system (PWR, pressurized-water reactor) or with cooling effected by direct boiling of the water moderator, which thus produces steam for driving the turbines (BWR, boiling-water reactor). Other promising types, of which prototype plants have been built, are sodium-cooled, graphite-moderated reactors fueled with slightly enriched uranium (SGR), organic-moderated reactors, also with slightly enriched uranium fuel (OMR), and the fast reactors already discussed in connection with breeding.

Reactors for Propulsion. The use of nuclear power for propulsion purposes was suggested almost as soon as chain-reacting systems were developed. The small amount of fuel required per unit energy produced appears immediately as an attractive feature.   On the other hand, the necessity of a massive radiation shield around a nuclear reactor makes the weight-to-power ratio for nuclear engines so large that their use in small vehicles such as automobiles is ruled out.   Much preliminary work has been done on the development of nuclear-powered aircraft, particularly for military purposes, but no practical airplane driven by a nuclear engine has been built.

For ship propulsion, on the other hand, nuclear reactors have already proved advantageous.   The first nuclear-powered ships built were submarines, and, because of their ability to operate submerged for almost unlimited periods of time, their high speeds, and their extended ranges between refueling stops,[3] they are vastly superior to the conventional type.   The nuclear submarine fleet of the United States now numbers in the dozens, all driven by pressurized-water reactors and fueled with highly enriched U^{235} in the form of uranium-zirconium alloy plates.   Engines of the same general type are in use on an aircraft carrier, on frigates, and on cruisers which have joined the fleet. The United States Navy is, in fact, engaged in an extensive program of conversion to nuclear propulsion.   Other countries have also begun to use nuclear-powered ships.

The first nuclear-powered merchant vessel, the *N. S. Savannah*, is a 22,000-ton ship powered initially by a 75-MW (thermal) pressurized-water reactor with slightly enriched (4.2–4.6 per cent U^{235}) UO_2 fuel.   Capable of cruising at 24 knots, the *Savannah* is designed to carry 9250 tons of cargo, 60 passengers, and 110 crew members and will require refueling, that is, replacement of its reactor core, after three to four years.   The nuclear power plant of the 16,000-ton Russian ice-breaker *Lenin* also uses slightly enriched uranium as UO_2 with a pressurized-water coolant and moderator; it has three 90-MW (thermal) reactors, of which one is a spare.   This powerful ship, which went into service in 1959, is designed to cruise at 2 knots through ice that is 2 meters thick.

Aside from ship propulsion, the most promising area for the application of nuclear engines to propulsion is in the space program.   Intensive research and development work on nuclear rocket engines is underway, because, for a given weight, such engines can in principle produce much larger final velocities than chemical rocket engines.   This consideration will be particularly important in space missions undertaken from parking orbits.

C. ASSOCIATED PROBLEMS

Hazards and Safeguards (H2).   Even though reactors may contain sufficient amounts of fissionable material for an explosive chain reaction, there is

---

[3] The first United States nuclear submarine, the *Nautilus*, which went into service in 1955, operated for 60,000 miles and about two years before its first refueling.

very little possibility that one of them will explode like an atomic bomb.   The necessary conditions for such an explosion are very special indeed and are most unlikely to be achieved accidentally.   The possibility always exists that a series of malfunctions of controls and interlocks may permit a nuclear reactor to go out of control and exceed its design power level.   The probable consequences would be melting or vaporization of some core components and possibly initiation of chemical reactions such as burning.

The real hazard in a reactor is its content of fission products (M3).   The amount of these products stored in a reactor is determined by its power level and by the time its fuel has been in place without reprocessing.   For example, a reactor that has been operating steadily at 1000 kW for several months contains about 10^9 curies of fission products.   If any accident, nuclear or otherwise, should open the reactor and melt or vaporize the fuel elements, a fraction of these dangerously radioactive substances could be dispersed in the air, and a severe radiation hazard could be produced in the vicinity and possibly for a distance of several miles downwind.   For this reason, some reactors are enclosed in gastight and pressureproof buildings.   Also, the site, meteorological conditions, population distribution in the vicinity, and other factors which may affect the potential hazards are always carefully considered before a license for the construction of a reactor is issued.   In the United States the A.E.C. has the statutory responsibility for these safeguards.

Clearly, all reactors are not equally safe, although, of course, each contains a number of emergency controls, safety interlocks, and warning systems. Those reactor types with the most inherent safety features, chief among which is a sizable negative temperature coefficient of reactivity, as in the simple water boiler, are generally thought to be the safest.   Although the expansion caused by increase in temperature always tends to reduce the reactivity, effects due to changes in absorption and fission cross sections with neutron temperature and with Doppler broadening of resonance peaks need to be taken into account before a conclusion about the temperature coefficient is reached.   Another safety feature is present if any reasonable change in configuration can only reduce k.   Thus loss of D_2O from the Saclay reactor P-2, which very possibly could happen, would in itself stop the reaction.   On the other hand, losses of H_2O coolant from some reactor cores are expected to increase k.   Some reactors contain voids that might conceivably be closed by gross mechanical accident or even by earthquake, with corresponding increase in k just at a time when control rods might be jammed.   Large excess reactivity, especially in conjunction with sizable experimental holes in the core, would mark a reactor as slightly less safe than the average.   Apart from high power, very high specific power can introduce a special hazard in that the fuel elements may become so radioactive that they will melt from their own heat even after shutdown if the reactor should lose its coolant.   Dimensional changes in structural materials contribute to reactor hazards; for example, control rods might be inoperative.

The best reactor safeguards are careful design, inherent safety features, sound operating procedure, interlocks and controls subject to frequent operational tests, and some caution with regard to untried materials.   In the United States the A.E.C.'s Advisory Committee on Reactor Safeguards carefully scrutinizes any proposed reactor before its construction can be authorized.   A detailed hazards evaluation report is required, in which, among other things, all conceivable types of accidents, their possible consequences, and the precautions taken against them must be analyzed.

The actual safety experience with reactor operations is remarkably good in comparison with the record of conventional steam plants and other industrial enterprises.   A very few reactor incidents involving fuel melting and local radioactivity release have occurred; they have necessitated costly shutdowns and repairs but have not constituted major hazards.   In 1961 an unexplained, unscheduled sudden withdrawal of a control rod caused the explosion of the boiling-water reactor SL-1 in Idaho which killed three people.

Chemical Processes (P2).   The construction and operation of nuclear reactors have posed many new chemical problems.   Indeed, new reactor developments depend primarily on new techniques in chemistry and metallurgy.   As a random example, pure zirconium metal, until the early 1950's a very rare commodity, has become a widely used structural material for thermal reactors because of its low absorption cross section, and its exceptional corrosion resistance and strength at high temperatures.   The low cross section, 0.18 barn, was not recognized until special samples free of the usual 2 per cent hafnium impurity (with a cross section of 105 barns) were prepared.   Over a period of about 3 years a new technology of zirconium purification, reduction, and fabrication emerged.   Many other examples could be cited, but in this section we wish to consider only those chemical processes applied to fuel elements removed from reactors.

The nuclear fuel, whether ordinary uranium, U^{235}, U^{233}, or Pu^{239}, cannot simply be left in the reactor until all the fissionable material is consumed. Apart from the reduction in fuel mass, there are accumulated fission products which act as neutron-absorbing poisons; even more decisive in some types of reactors (especially those with metallic fuel elements) is the mechanical deterioration of the fuel elements with reduction in stability and in heat transfer.   Allowable burnup may be somewhere in the range of 1 to 40 per cent. After that, the fuel elements must be removed from the reactor.   If they are to be recovered, they may be stored for a few months while the radioactivity declines, then dissolved, purified, and decontaminated of radioactive products, reconverted to the desired chemical form (metal, alloy, oxide), and fabricated into useful shapes.   Any secondary fuel produced by nuclear reaction, such as Pu^{239} from uranium, is also separated in the chemical process.

The requirements put on the chemical and metallurgical processing are very severe.   Losses of fissionable nuclides must be kept exceedingly low.   For example, if the burnup per pass is 1 per cent and the recovery process after each pass is 99.9 per cent efficient, before all the valuable fuel is consumed 10 per cent of it will have been lost.   Moreover, the fission product radio-activity is so intense that the early process stages must be carried out entirely by remote control in heavily shielded, completely enclosed cells.   In these circumstances, for a complex process of many steps, an over-all yield of 99.9 per cent represents a minor miracle.   The advantages of increased burnup per pass in the reactor are obvious.

For the typical clad or canned solid fuel element the processing cycle always begins with the discharge of the "hot" element from the reactor, usually into a water channel sufficiently deep ($\gtrsim 20$ ft) to absorb all harmful radiations. After a "cooling" period the fuel elements are moved to the processing area, where the first step may be decladding, either by mechanical or by chemical means.   If this is not practical, cladding and fuel may be dissolved together; otherwise, the stripped fuel is dissolved separately.   Nitric acid is suitable for dissolving uranium, some of its alloys, and UO_2.   Other uranium alloys, for example, those containing molybdenum and zirconium, require other dissolving media.   In the dissolving process precautions must be taken to prevent escape of some volatile fission products such as iodine.

The subsequent separation processes may take on many forms (C2, P2). The uranium fuel from the wartime plutonium-producing reactors was proc-essed by cycles of precipitation reactions.   However, precipitations and filtra-tions must almost of necessity be carried out as batch processes and are there-fore not ideally suited for remote-control operations.   Solvent extraction separations in countercurrent columns are much more readily adapted to remote operation and have therefore largely replaced the precipitation pro-cedures.   A variety of processes is in use, and many more have been proposed and investigated.   The fuel solutions from different reactors may have widely differing compositions, and the objectives in the chemical processing may also differ.   In most of the processes the initial step involves extraction of uranium, plutonium, and, if present, thorium into an organic solvent, with most of the fission products and unwanted fuel alloy components left in the aqueous phase.   The organic phase is most often tributyl phosphate in some solvent such as kerosene, but various ethers, ketones, and alcohols have also been used as extractants.

The efficient extraction (> 99.9 per cent) of uranium and plutonium in the first step is based on the large partition coefficients for U(VI), Pu(IV), and Pu(VI) in the systems used.   At the same time, the accompanying total fission product activity (of the order of 10^6 curies per ton of fuel in the feed solution) is usually reduced by factors of several hundred, although some specific products, especially ruthenium, are partially extracted into the organic phase. For the subsequent separation of plutonium from uranium advantage is taken of the very low partition coefficient of Pu(III) for extraction into organic solvents.   Thus plutonium is reduced to Pu(III) under conditions that leave uranium in the $+6$ state—such reducing agents as Fe(II), SO_2, or hydrazine can be used—and another countercurrent solvent extraction then separates plutonium and uranium.   Further remote-control purification of the fractions by additional solvent extraction steps or by other techniques such as ion exchange is normally required before the material can finally be handled without γ-ray shielding.   Even then, some protective measures are always required when the highly α-active and toxic nuclide Pu^{239} is processed.

One of the problems in solvent extraction processes for irradiated fuels is the possibility of radiation decomposition of the organic reagents.   The initial aqueous feed solutions often contain hundreds of curies of β and γ activity per liter, and the effects of such radiation intensities on an organic material must be carefully studied before it can be selected as an extractant in a fuel processing plant.   The relatively high radiation stability of tributyl phospate has helped to make it a favorite reagent for extraction operations at high radiation levels.

Some significant advantages in efficiency and cost may accrue from the use of fuel-processing methods that do not involve aqueous solution chemistry. The initial dissolving step, for example, could then, be avoided and the final reconversion to metal or oxides could be greatly simplified.   Intensive development work has therefore been carried out on a number of nonaqueous processes. Among them are volatility methods which accomplish separation of uranium and plutonium from fission products by volatilization of UF_6 and PuF_6, and a variety of pyrometallurgical processes, one of which, a melt refining method, is actually being used to reprocess the fuel of the EBR-II reactor.   Here the uranium fuel elements are melted down in zirconium oxide crucibles at 1300°C in an inert atmosphere.   Many fission products, such as the rare gases, alkalis, alkaline earths, and cadmium, distil out; others form oxides (by reduction of the zirconium oxide to a suboxide) and separate in a layer of slag.   Still other fission products, such as the noble metals and molybdenum, remain alloyed with the uranium.[4]   This alloy, with the addition of some fresh fuel to make up for burnup, is recast into new fuel elements (by remote control) and returned to the reactor.   The relative simplicity of this type of process is clearly an asset.

An important concern of process designers and operators is that at no time must a critical assembly of fissionable material be allowed to accumulate. A plant might be designed to process daily thousands of gallons of solutions containing several times the minimum critical mass.   Unless this is done batchwise, it is not easy to guarantee that no divergent chain reaction can ever occur.   (In case of such an accident the attendant hazard would be serious but probably not widespread; certainly there would be no large atomic explosion.) The safeguards that have been considered for continuous operations include

---

[4] The equilibrium mixture of fission products which remains alloyed with uranium (or plutonium) has been termed "fissium."   Uranium-fissium alloys have been found to possess desirable mechanical properties for fuel-element use.

the use of neutron absorbers, unfavorable geometry such as tall narrow cylinders for all vessels, and, of course, careful material control in each stage.

Disposal of Radioactive Wastes (S2, W2).   The radioactive products made in nuclear reactors and discharged from their separations plants are essentially different from the wastes in ordinary industrial pollution problems.   The radio-activities are different substances not returned to natural ones by dilution, oxidation, precipitation, etc.   In a sense the earth's crust will not again in our time be quite the same as before.   If these products were mixed throughout all the volume of the oceans, the resulting concentration of radioactivity would be small, a fraction of that normally present in sea water.   Even so, this is not an altogether attractive prospect, with more and more reactors being built, and anyway the uniform mixing is easier said than done.

The major fission-product wastes have so far simply been stored in underground tanks, usually in aqueous solution evaporated to minimum liquid volume and adjusted to an appropriate pH.   Many millions of gallons of such waste solutions are now being stored.   This procedure cannot be permanently satisfactory; the possibility of eventual corrosion or rupture of the tanks is a hazard to the surrounding areas and to groundwaters.   Yet little progress has been made toward a safe, economical "ultimate-disposal" system, and tank storage appears to be destined to be used for some time for high-level wastes. It is hoped that it will eventually be possible to convert these liquid wastes into solid form, either by a calcining process or by incorporation of the fission products in clays, glasses, or synthetic minerals.   The relatively inert solid bodies formed could then be stored in mines or buried in other appropriately chosen locations.

The methods finally selected for waste disposal need to be inexpensive.   The cost simply of tank storage has been estimated as 44 cents per initial gallon and more if cooling coils are required to remove the radioactive heat.   These costs complicate the economics of nuclear-power production.   Clearly, chemical separations processes should give the radioactive wastes in the smallest volume of solution with the least bulk of added reagents.   It is conceivable that some of the fission products may find economically sound applications; this can hardly eliminate disposal but may help to pay for it.

D. THE FISSION BOMB

Nuclear technology has evolved because of its military applications.   The story of the so-called atomic bomb is well known (S3), but, in connection with our discussions of chain-reacting systems, it may be instructive to consider briefly the conditions under which a fission chain reaction can be made to proceed with explosive violence.

Explosive Chain Reaction.   If a fission chain reaction is to proceed to the intensity of a major explosion, several necessary conditions are immediately apparent.   The chain must operate on fast neutrons; otherwise the time τ between successive neutron generations (p. 473) will be determined by the time required to slow neutrons to thermal energies, which is of the order of 10^{-3} second, and the critical assembly will have moved apart in a small explosion before a really huge energy is generated.

A chain reaction based on fast neutrons requires no moderator, and the core is likely to be nearly pure fissionable material, U^{235} or Pu^{239}.   (Actually U^{238} is fissionable by reaction with neutrons of energy greater than 1.1 MeV, and although the raw fission-neutron spectrum has as its average energy about 1.5 MeV the losses of effective neutrons by inelastic scattering and by radiative capture prevent a fast-neutron chain reaction in this material.)   As in the case of the thermal-neutron chain reaction, there is a critical size or critical mass determined by the permissible surface losses of neutrons.   In order of magnitude the critical mass must correspond to that of a sphere of radius R_c given by the mean free path of the neutron before reaction.   The mean free path is $(n\sigma)^{-1}$, where n is the number of nuclei per cubic centimeter and σ is the neutron cross section in cm^2; then

$$R_c \cong (n\sigma)^{-1} = \left(\frac{6 \times 10^{23}}{235}\rho\sigma\right)^{-1},$$

where ρ is the density in grams per cubic centimeter.   The critical mass in grams is

$$M_c \cong \frac{4\pi}{3} R_c{}^3 \rho = \frac{4\pi}{3\rho^2}\left(\frac{235}{6 \times 10^{23}\sigma}\right)^3.$$

Taking $\rho = 19$ g cm^{-3} and guessing $\sigma \cong 2.5 \times 10^{-24}$ cm^2 (the geometrical cross section), we have $R_c \cong 8$ cm and $M_c \cong 40$ kg.   This is, of course, only a crude estimate.

If an object weighing 40 kg were the site of a fission energy release of 8×10^{20} ergs (equivalent to 20,000 tons of TNT or 1 kg of U^{235}) and all this energy were mechanical kinetic energy, the corresponding velocity would be about 2×10^8 cm sec^{-1}.   Then the time required for its parts to separate by a few centimeters (and so end the reaction) would be of the order of magnitude of 10^{-8} sec.   A fast-neutron generation can occur in this time.   We may estimate τ crudely as the time for a 1-MeV neutron, velocity 1.4×10^9 cm sec^{-1}, to travel a mean free path, 8 cm, which gives $\tau \cong 6 \times 10^{-9}$ second.   For the neutron intensity and the energy release to become very large the configuration of the fissionable material must be such that k is much greater than unity, say of the order of 2, and must remain so for a time t given by (14-1):

$$e^{(k-1)t/\tau} = \frac{N}{N_0}.$$

We estimate N/N_0 as the product of the number of atoms in the 1 kg U^{235} to be fissioned $[(1000/235) \times 6 \times 10^{23}]$ and the number of neutrons per fission ($\nu = 2.5$). Thus, with $k = 2$, we get $e^{t/\tau} \cong 6 \times 10^{24}$ and therefore $t \cong 60\tau \cong 4 \times 10^{-7}$ sec. Of course, as the nuclear explosion builds up, there is no possibility of providing a case strong enough to hold the critical assembly together. As assumed above, the only effective restraining force is that due to inertia. On this account the fissionable core may be enclosed in a layer of some dense material which adds to the inertia and which also may reflect neutrons into the core; this special reflector is called the tamper (S3).

Bomb Assembly Problem. The requirement for a reasonably efficient fission bomb that k be considerably greater than unity is met if the core contains, in good geometry, considerably more than one critical mass. The assembly of the bomb at the moment of detonation may be effected by bringing together two near-critical pieces of fissionable material. However, unless this is done in a time shorter than about 4×10^{-7} second the buildup of fission rate during assembly could lead to only a small premature explosion (a "fizzle"). If the parts are to move a distance of the order of 8 cm in 4×10^{-7} second, the required relative velocity is $\cong 2 \times 10^7$ cm sec^{-1}. This velocity is well above that even of artillery shells, which is $\cong 10^5$ cm sec^{-1}. Actually, in one type of bomb a self-contained cannon shoots one part of the fissionable material into a mating part, the latter being surrounded by tamper. This is successful only if the assembly phase, duration $\cong 10^{-4}$ second, is mostly completed before any stray neutron happens to start the chain reaction. Important sources of stray neutrons are spontaneous fission (0.7 sec^{-1} per kg of U^{235}) and α, n reactions on impurity elements of sufficiently low Z to permit penetration of the Coulomb barrier. If the total number does not exceed about 1000 sec^{-1}, the probability of a fizzle from this cause is small.

The bomb exploded at Hiroshima on August 6, 1945, is reported to have been of this gun-assembled type, using U^{235} as fissionable material. One feature of this type of bomb is that it can be made with a surprisingly small external diameter.

It has been known for many years that detonations of ordinary high explosive in particular shapes can by focusing effects concentrate energy in a portion of the mass so that very high velocities result. Shaped-charge weapons eject material with velocities in excess of 10^6 cm sec^{-1}. The implosion bomb uses this principle, in lenses of high explosive, to drive together very suddenly a mass of fissionable material. The Nagasaki bomb is reported to have been of this type.

A discussion of the effects of nuclear weapons is outside the scope of this book. Extensive information on this subject may be found in ref. U1.

REFERENCES

*C1   G. Cahen and P. Treille, *Nuclear Engineering*, Allyn and Bacon, Boston, 1961.
 C2   F. L. Culler, Jr., R. E. Blanco, L. M. Ferris, E. L. Nicholson, R. H. Rainey, and J.

Ullmann, "Proposed Aqueous Processes for Power Reactor Fuels," *Nucleonics* **20**, No. 8, 124 (August 1962).

E1   M. C. Edlund and P. F. Schutt, "The Future of Thermal Breeders," *Nucleonics* **21**, No. 6, 76 (June 1963).

*H1   J. F. Hill, *Textbook of Reactor Physics*, George Allen and Unwin, London, 1961.

H2   H. Hurwitz, Jr., "Safeguard Considerations for Nuclear Power Plants," *Nucleonics* **12**, No. 3, 57 (March 1954).

K1   L. J. Koch and H. C. Paxton, "Fast Reactors," *Ann. Rev. Nuclear Sci.* **9**, 457–472, (1959).

K2   H. Kouts, "Nuclear Reactors," in *Methods of Experimental Physics*, Vol. 5B, *Nuclear Physics* (L. C. L. Yuan and C. S. Wu, Editors) Academic, New York, 1963, pp. 590–622.

L1   J. A. Lane, "Economics of Nuclear Power," *Ann. Rev. Nuclear Sci.* **9**, 473–492, (1959).

*L2   S. E. Liverhant, *Elementary Introduction to Reactor Physics*, Wiley, New York, 1960.

*M1   G. Murphy, *Elements of Nuclear Engineering*, Wiley, New York, 1961.

*M2   F. S. Martin and G. L. Miles, *Chemical Processing for Nuclear Fuels*, Academic, New York, 1959.

M3   R. B. Mesler and L. C. Widdoes, "Evaluating Reactor Hazards for Airborne Fission Products," *Nucleonics* **12**, No. 9, 39 (September 1954).

N1   "Nuclear Power Plants—The Long View," *Nucleonics* **21**, No. 6, 59–82 (June 1963).

P1   "Power Reactors the World Around," *Nucleonics* **21**, No. 8, 112 (August, 1963).

*P2   "Processing Irradiated Fuels and Radioactive Materials," *Proceedings of the Second U. N. International Conference on the Peaceful Uses of Atomic Energy*, **17**, Geneva (1958).

*R1   "Research and Testing Reactors the World Around," *Nucleonics* **20**, No. 8, 116 (August 1962).

*R2   "Research Reactors," *Proceedings of the Second U. N. International Conference on the Peaceful Uses of Atomic Energy*, **10**, Geneva (1958).

*S1   R. Stephenson, *Introduction to Nuclear Engineering*, McGraw Hill, New York, 1958.

S2   K. Saddington and W. L. Templeton, *Disposal of Radioactive Waste*, George Newnes, London, 1958.

*S3   H. D. Smyth, *Atomic Energy for Military Purposes*, Princeton University Press, 1945.

U1   United States Defense Atomic Support Agency, *The Effects of Nuclear Weapons* (S. Glasstone, Editor), 1962 (available at $3.00 from Superintendent of Documents, U. S. Government Printing Office, Washington 25, D. C.).

*W1   A. M. Weinberg and E. P. Wigner, *The Physical Theory of Neutron Chain Reactors*, University of Chicago Press, 1958.

W2   "Waste Treatment and Environmental Aspects of Atomic Energy," *Proceedings of the Second U. N. International Conference on the Peaceful Uses of Atomic Energy*, **18**, Geneva (1958).

Numerous articles on many subjects mentioned in this chapter, including details of individual reactors, can be found in the journals *Nucleonics, Journal of Nuclear Energy,* and *Nuclear Engineering* and in the *Proceedings of the First* (1955) and *Second* (1958) *International Conferences on the Peaceful Uses of Atomic Energy* in Geneva.

EXERCISES

1. Assume that uranium metal is dispersed in heavy water at a concentration of 0.36 g per gram of D_2O.   (a) What is the value of k_∞ for this mixture?   (b) What is the radius of the sphere which is just critical?   *Answers:* (a) 1.3; (b) 85 cm.

2. The mixture in exercise 1 corresponds approximately to that in the Kjeller reactor. If that reactor is operated at 100 kW, what is the average thermal-neutron flux throughout the uranium?   *Answer:* 1.4×10^{11} cm^{-2} sec^{-1}.

3. Suppose a submarine reactor contains 40 kg U^{235} and operates at 50,000 kW. How long can the submarine run before 3 per cent of the fuel is burned up?

4. Estimate the neutrino (or, rather, antineutrino) flux 50 ft from the reactor of exercise 3. *Answer:* $\sim 3 \times 10^{11}$ cm^{-2} sec^{-1}.

5. A water-cooled uranium-graphite reactor operates at a power level of 200,000 kW. The reactor core is a cube 18 ft on the side. The cooling water enters at about 20°C, flows through the reactor at the rate of 20,000 gal min^{-1}, and resides in the reactor core for an average of 2 sec. Estimate (a) the exit temperature of the water, (b) its radioactivity (in curies per liter) as it leaves the reactor, assuming the water to be pure, (c) the radioactivity of the water 1 hour after leaving the reactor if it contains 1.2 ppm phosphorus, 1.8 ppm sodium, and 0.9 ppm chlorine impurities.

6. Verify the statement on p. 483 that Xe^{135} poisoning in a high-flux reactor such as the MTR reaches a maximum about 10 hours after shutdown.

7. After a reactor has operated for 1 year at 500,000 kW, its fuel is processed and the fission product wastes are stored. Estimate the number of curies of fission product activity remaining 10 years later.

8. Estimate (a) the equilibrium quantity of Ba^{140} present in a uranium-graphite reactor operating at 1000 kW, and (b) the total amount of Ce^{140} accumulated in the same reactor after 1 year's operation followed by 2 months' shutdown.
Answer: (a) 0.8 g.

9. From (14-2), estimate the minimum ratio of U^{235} atoms to moderator molecules that is required to make a thermal-neutron chain reaction possible in an infinitely large homogeneous mixture of (a) U^{235} and H_2O, (b) U^{235} and D_2O.
Answer: (a) 1:1100.

10. Why is Pu^{239} less suitable for use in the gun-assembled type of fission bom than U^{235}? Estimate the maximum concentration of Pu^{240} that could be tolerated in the plutonium used in such a bomb.

15

Nuclear Processes in Geology and Astrophysics

A. GEO- AND COSMOCHRONOLOGY

Radioactive Clocks. Not long after the discovery of radioactivity it was realized that radioactive decay constitutes a "clock" provided by nature. Until the beginning of this century geologists had no reliable absolute time scales for the geologic ages, although they had amassed a great deal of information on their relative sequence. The age of the earth was believed to be at most some tens of millions of years. The new knowledge about radioactive decay, the naturally occurring decay series, and the identity of α particles resulted in the first objective methods of geochronology, which, in turn, radically changed man's concepts of the earth's history. Rutherford was the first to suggest that α decay must lead to the buildup of helium in uranium minerals and that therefore the helium content of a uranium mineral could be used to determine the time elapsed since its solidification; he applied this method to the study of some mineral ages. Shortly afterward the realization that lead was the end product of uranium decay led to a study of the lead content of uranium minerals, and by 1907 it was correctly concluded from such investigations (B3) that geologic times had to be reckoned not in tens but in hundreds and thousands of millions of years.

Ages of Minerals. From these early beginnings radioactive dating techniques have been intensively developed and have become an increasingly valuable tool of the earth scientists (K1, A1). Some of the more important methods used to determine the time since the formation or solidification of particular minerals are discussed in the following paragraphs. As we shall see, none of these methods is free from difficulties, and the greatest reliance can therefore be placed on those ages that are determined by two or more independent techniques.

1. *Ratio of uranium to helium content.* Once an atom of U^{238} disintegrates, the chain of successive decays soon (say in less than about a million years) produces eight α particles. Because the ranges of these particles are very short in dense matter, most of the resulting helium atoms (the helium ions

497

at rest are easily capable of acquiring two electrons by oxidizing almost any substance) may be trapped in the interior of the rock. In favorable cases, with very impervious fine-grained rocks and a low helium concentration (pressure) from small uranium contents, this helium has been retained throughout the geologic ages and now serves as an indicator of the fraction of uranium transformed since the formation of the ore. The thorium in the rock also is a source of helium (six α particles per decay), and this must be taken into account. Very sensitive methods of assay for helium, uranium, and thorium are available and have permitted determinations on rocks with uranium and thorium contents below one part per million. Although the uranium-helium method was the first dating method based on radioactive decay, its applicability as a reliable technique is limited because of the possibility of helium leakage during geologic times. In general, uranium-helium ages can be considered as lower limits only. Application of the uranium-helium method to iron meteorites led to some much longer ages than seemed compatible with other data. These results were extremely puzzling until the method was shown to be invalid for these extraterrestrial bodies, since cosmic-ray-induced spallation reactions are an additional source of helium (see later); this interpretation was conclusively proved by mass spectrographic analyses of meteoritic helium which showed approximately 20 per cent of it to be He^3, whereas α decay can, of course, produce He^4 only.

2. *Uranium-lead and thorium-lead ratios.* The lead isotopes Pb^{206} and Pb^{208} are the stable end products of U^{238} and Th^{232} decay. Provided there is no other source of lead in a uranium or thorium mineral, the amounts of these lead isotopes present may be used as quantitative indicators of the amount of uranium and thorium decay that has taken place. This lead method might be expected to be more reliable than the helium method, since lead is not so likely to have been lost by slow diffusion; however, it is still quite possible that a lead-uranium or lead-thorium ratio has been changed by leaching or some other process. The distinction between these lead decay products (Pb^{206} from U^{238} and Pb^{208} from Th) and ordinary lead is made in a satisfactory way by mass-spectrographic analysis; it is usually presumed that absence of Pb^{204} establishes the absence of ordinary lead. If some Pb^{204} *is* found, a correction for the presence of nonradiogenic Pb^{206} and Pb^{208} must be made. Age determinations based on the uranium-to-Pb^{206} ratio are found (by comparison with other methods) to be more reliable than those made from the thorium-to-Pb^{208} ratio. Ages ranging up to roughly 3×10^9 years have been found by these methods.

3. *Ratio of uranium lead* (Pb^{206}) *to actinium lead* (Pb^{207}). Another method for very old rocks containing uranium involves determination of the ratio of Pb^{206} to Pb^{207}, the end products of the uranium (or U^{238}) and actinium (or U^{235}) series. This method should be free of many experimental errors and is less sensitive to chemical or mechanical loss of either uranium or lead than method 2. However, because of the very different half-lives of the radon

isotopes in the two decay chains (see figures 1-1 and 1-3) the results are affected by radon leakage from the minerals. Also, it is essential that the minerals be free of appreciable amounts of nonradiogenic lead; again absence of Pb^{204} is the criterion. The Pb^{206}/Pb^{207} ratio is an indicator of age because U^{238} and U^{235} decay at different rates (figure 15-1). From mass-spectrographic measurements of this ratio a convincing age scale of minerals more than 500 million years old can be established. Ages of about 2.6×10^9 years have been found by this method for samples of uraninites and monazites from several formations known from other evidence to be very old on a geologic time scale. Application of the Pb^{206}/Pb^{207} method to stone meteorites gives, for almost all specimens measured, ages since solidification of $(4.5 \pm 0.1) \times 10^9$ years.

4. *Ratio of K^{40} to Ar^{40} content.* In this method the argon produced in the rock by decay of the naturally radioactive K^{40}, half-life 1.27×10^9 years, is measured. One difficulty is that K^{40} decays both by electron capture to Ar^{40} and by β^- emission to Ca^{40}, and the fraction of decays that give Ar^{40} (about 0.110) has been hard to determine with great precision. Until fairly recently this uncertainty threw some doubts on the reliability of K-Ar dates. This method has been applied to a large number of potassium-bearing minerals. Comparison with the results of other dating methods (e.g., Pb^{206}/Pb^{207}) applied to the same minerals or to other minerals in the same rock formations indicates that the K-Ar method gives reliable ages for certain kinds of minerals, especially micas. In the feldspars, on the other hand, K-Ar ages are usually too low, and this is ascribed to argon leakage from these minerals during geologic time. Some K-Ar ages as large as 2.3×10^9 years have been found. In connection with the K-Ar dating method it is interesting to note that the large abundance of argon in the earth's atmosphere ($\cong 1$ per cent, almost all

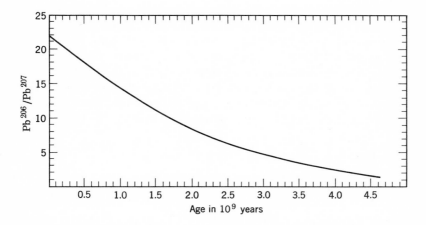

Fig. 15-1 Ratio of radiogenic Pb^{206} to radiogenic Pb^{207} in uranium-bearing minerals as a function of age.

Ar^{40}) presumably arises from the decay of K^{40} during the early history of the earth.

The K-Ar method has been applied to meteorites also (A2). For stone meteorites ages (i.e., values for the time elapsed since solidification) ranging from about 0.5×10^9 to 4.5×10^9 years are found. Some of the lower ages are undoubtedly to be ascribed to argon losses since the original solidification, probably during a high-temperature era in the life of the meteorite.

5. *Ratio of* Rb^{87} *to* Sr^{87}. These species are best determined mass-spectrometrically, although any good analysis for total rubidium should be satisfactory. Agreement of Rb-Sr and K-Ar ages for a given mineral is taken as strong evidence for the reliability of the determinations because the daughter products are chemically so different that any geochemical fractionation would have affected the Rb/Sr and K/Ar ratios quite differently. The Rb-Sr method is now regarded as one of the most reliable dating methods, although it reached this status only fairly recently, partly because the half-life of Rb^{87} was long uncertain. Like other techniques, the Rb-Sr method gives about (2.6 to 3.0) $\times 10^9$y for the ages of the oldest terrestrial rocks found and about 4.5×10^9y for the ages of a number of meteorites.

6. *Ratio of radiogenic lead to* Pb^{204}. This method differs from the other lead-uranium methods in that it is applicable only to lead minerals which contain no uranium or no thorium. Because of the production of Pb^{206}, Pb^{207}, and Pb^{208} by the decay of uranium and thorium the relative abundances of the lead isotopes throughout the world have changed continuously during geologic time. Lead ores deposited long ago contain a smaller proportion of radiogenic isotopes than does modern lead, and the Pb^{206}/Pb^{204} and Pb^{208}/Pb^{204} ratios particularly have been used to date galenas.

Age of the Earth. The terrestrial occurrence of the radioactive substances U^{238}, U^{235}, Th^{232}, and K^{40} gives some information about the time since the earth's genesis. Clearly, conditions as we know them today cannot have existed for a time very long compared to their half-lives, 4.5×10^9 years, 7.1×10^8 years, 1.4×10^{10} years, and 1.3×10^9 years. If the earth's crust is an isolated system, it should be possible to reach some conclusions about its age by extensions of the methods described for dating minerals. The following relations may be derived between the lead isotope abundance ratios, at times 0 and t, and the modern ratio of uranium to lead, $(U/Pb)_t$:

$$\left(\frac{Pb^{206}}{Pb^{204}}\right)_t = \left(\frac{Pb^{206}}{Pb^{204}}\right)_0 + \frac{(U^{238}/U)_t}{(Pb^{204}/Pb)_t}\left(\frac{U}{Pb}\right)_t (e^{\lambda_{238}t} - 1) \qquad (15\text{-}1)$$

$$\left(\frac{Pb^{207}}{Pb^{204}}\right)_t = \left(\frac{Pb^{207}}{Pb^{204}}\right)_0 + \frac{(U^{235}/U)_t}{(Pb^{204}/Pb)_t}\left(\frac{U}{Pb}\right)_t (e^{\lambda_{235}t} - 1) \qquad (15\text{-}2)$$

The isotopic abundances $(U^{235}/U)_t$, $(U^{238}/U)_t$, and $(Pb^{204}/Pb)_t$ are 0.00720, 0.9927, and 0.0148, respectively, and the modern ratios $(Pb^{206}/Pb^{204})_t$ and

$(Pb^{207}/Pb^{204})_t$ are measured on ordinary lead as 16.0 and 15.3. The primeval abundance ratios $(Pb^{206}/Pb^{204})_0$ and $(Pb^{207}/Pb^{204})_0$ are, of course, not known with certainty; but it may be assumed that they can be taken from the lead abundances found in certain iron meteorites (which have such a low uranium content that the radiogenic lead is negligible) as 9.4 and 10.3. Since the decay constants for the two uranium isotopes, λ_{238} and λ_{235}, are known, the two unknown quantities are t and $(U/Pb)_t$. Solution of (15-1) and (15-2) gives $t = 4.9 \times 10^9$ years. Had we assumed that primeval lead contained no Pb^{207} at all (it would still have contained Pb^{206}, since the buildup of Pb^{206} from U^{238} is much slower than that of Pb^{207} from U^{235}), the result would have been 5.5×10^9 years. A more detailed analysis involving Pb^{207}/Pb^{204} and Pb^{206}/Pb^{204} ratios in a whole series of meteorites [which allows (15-1) and (15-2) to be solved without *any* assumption about the primeval lead abundance ratios] led Patterson (P1) to a value of $(4.55 \pm 0.07) \times 10^9$ years for the time elapsed since the earth became an isolated system. The assumption is made that the meteorite parent bodies (see later) were formed at that same time. The meteorite solidification ages of $\cong 4.5 \times 10^9$ years determined by the various radioactive-decay methods support this assumption.

For the history of the earth from the time of its formation as a planet about 4.6×10^9 years ago to the date of formation of the oldest surface rocks, some 1.6×10^9 years later, we have to rely on indirect evidence. An earth originally at any reasonable temperature would have required a very short time to cool enough to form a solid crust, unless it were continuously supplied with heat, presumably from radioactive decay processes. Today the very crust of the earth, of average depth about 25 miles, contains enough uranium, thorium, and potassium to supply more than half of the average heat loss of the earth, about 10^{-6} calorie cm^{-2} sec^{-1}, and this heat source would have been several times greater $(3-4) \times 10^9$ years ago.

Very possibly the earth was formed at relatively low temperatures (as proposed by Kuiper, Urey, and others; see ref. U1 and chapter 10 of ref. F1) by gravitational accumulation of cosmic dust. At about the same time condensation of nearby dust eddies presumably produced the other planets, the moon, and the sun. (In this formative phase, the gaseous substances, such as He, Ne, Ar, Kr, Xe, Hg, H_2O, CH_4, NH_3, were mostly lost by the earth and the other inner planets.) The moon has such a large surface-to-volume ratio that it probably was at no time molten and is thought by some to have remained unchanged as a sample of the original cosmic accumulation; but the earth, because of internal radioactive heating, was melted and kept molten for almost 2000 million years. In this stage the metallic core was collected at the center and the metallic oxides and silicates formed the surrounding mantle. As the radioactivities decayed away, especially U^{235} and K^{40}, the planet cooled, and at some moment roughly 3000 million years ago, the first permanent solid crust appeared.

Evidence from Extinct Radioactivities. The dating methods discussed so far are all based on the decay of one of the long-lived radioactive nuclides now existing in nature. The suggestion has been made (see K1 and A2) that additional information on the early history of the earth and solar system could be obtained if evidence could be found for "extinct" radioactivities, that is, for nuclides with half-lives of the order of 10^7–10^8 years which might have existed at one time but would have decayed out. The nuclides I^{129} (1.6×10^7y), U^{236} (2.4×10^7y), and Pu^{244} (8×10^7y) are possibly of interest in this connection.

If a decay product of one of these extinct radioactivities were found in some system (meteorite, rock, etc.) and its origin from such decay could be established, this would show clearly that the time elapsed between element formation and the isolation of the system investigated could not have been very large compared to the half-life of the extinct radioactivity. Studies of the isotopic composition of xenon in meteorites, especially by J. H. Reynolds, have revealed abnormally high Xe^{129} abundances in a number of samples (A2), and this excess Xe^{129} has been attributed to decay of once present I^{129}. This hypothesis is greatly strengthened by the finding that the Xe^{129} occurs in the same phases of the meteorites as iodine (I^{127}). With the assumption that I^{127} and I^{129} were originally equally abundant (which is reasonable on the basis of presently accepted theories of nucleogenesis, see section B), the Xe^{129} data indicate a time interval of $(0.25 \pm 0.10) \times 10^9$ years between the end of the element-forming process and the time when the meteorite material could retain xenon. The uncertainty quoted arises partly from the spread in results for different meteorites, partly because the calculated time interval differs, depending on whether the relevant nucleosynthesis processes are assumed to have taken place suddenly or over an extended period of time. In any case the results seem to indicate that the time interval between nucleogenesis and formation of the planetary bodies in the solar system was rather short in relation to the time between then and the present.

Other Radioactive Nuclides in Nature. The so-called primary, long-lived radioactive nuclides which have survived since nucleogenesis, presumably without being replenished, are not the only radioactivities that occur in nature. In addition, there are first of all the secondary natural radioactivities, which are the short-lived descendants of the primary radionuclides U^{238}, U^{235}, and Th^{232}. Although some of these, such as ionium (Th^{230}, $t_{1/2} = 7.5 \times 10^4$y) and radium ($Ra^{226}$, $t_{1/2} = 1622$y) have been useful in geology, for example in dating ocean sediments (K1, S1), we shall not discuss them. Also, we merely mention that man-made radioactivities which have been introduced into the earth's atmosphere since 1945, largely as a result of nuclear bomb explosions, have, in addition to their well-known deleterious effects, made some interesting scientific studies possible; among the phenomena investigated with their aid are atmospheric mixing times between the Northern and Southern Hemispheres and residence times in the various vertical layers of the atmosphere (S1).

Of more widespread interest are the radionuclides (and stable products) formed by the interaction of cosmic rays with various objects.   The following paragraphs are therefore devoted to a brief discussion of these phenomena and their application.

Cosmic Rays.   Not long after the discovery of radioactivity it was known that detection instruments such as ionization chambers showed the presence of radiations even when not deliberately exposed to radioactive sources.   This background effect was attributed to traces of naturally occurring radioactive substances such as uranium and thorium and their decay products, and this assumption is, of course, partly correct.   Shielding the chambers with thick lead absorbers reduced but never eliminated the effect.   It was reasoned that, if the radiations resulted from radioactive contamination of the ground, elevation of the ionization chambers to 1000 meters or more should greatly reduce the effect because the ordinary β and γ rays would be strongly absorbed by the 100 g cm^{-2} or more of air.   In the period 1910 to 1913 several daring experimenters carried instruments consisting of ionization chambers and electroscopes aloft in balloons to altitudes as great as 9000 meters; surprisingly enough the background discharge rate at that height was about 12 times as great as on the ground.   The conclusion from this and other experiments was that a radiation of extraordinary penetrating power fell continuously upon the earth from somewhere beyond.   Since about 1925 this radiation has been known as cosmic radiation.[1]

Until the advent of high-altitude rockets and artificial satellites, cosmic-ray investigations were confined to the earth's surface or at least to relatively low altitudes, where the radiations observed are not the primary particles but rather are almost entirely secondary radiations produced by interactions of the primaries with the top of the atmosphere.   These interactions seem to be largely very-high-energy nuclear reactions resulting in the emission of many mesons (mostly π mesons) and nucleons, many of which undergo further nuclear reactions.   Many μ mesons, produced in flight by π-meson decay, are found lower in the atmosphere, and they constitute most of the "hard component" of the cosmic radiation.   With energies of many billions of electron

[1] The whole study of the cosmic rays was thus begun as a result of measurements on radiation detector backgrounds, and the knowledge gained has since been applied to the original problem.   For example, extraordinary reductions in counter background have been achieved by the use of electronic circuits which suppress all counts coincident with any count in nearby "shielding" counters.   The array of shielding counters surrounds the working counter and is almost certain to give a response to any cosmic-ray mesons, showers, etc. A counter about 2 in. in diameter and 8 in. long, filled with 10 cm of argon and 0.5 cm of ethylene, may have a natural background rate of about 500 counts per minute.   An 8-in. thick iron shield reduces this to about 100 min^{-1}.   (Iron is preferable to lead because it is less contaminated with natural radioactivities.)   Only extremely great thicknesses of shield could remove much of this residual background.   With a ring of about a dozen other Geiger or proportional counters placed around the first and operated as anticoincidence counters, the background can be reduced to 5 min^{-1}.

volts these mesons may penetrate a meter of lead or more and are observed to appreciable depths below water or ground. They have been best studied in cloud chambers triggered by counter telescopes.

The "soft component," readily absorbed in a few inches of lead, consists largely of photons, electrons, and positrons. It accounts for about 10 per cent of the cosmic-ray ionization at sea level but increases rapidly with altitude, constituting about 75 per cent of all the rays at an altitude of 10,000 ft. Most of the photons and electrons occur in showers of many particles of common origin. Initially, high-energy photons and electrons presumably result from meson decay, and subsequently positron-electron pair creation by photons, and ionization and bremsstrahlung emission by electrons tend to produce large cascades of these rays. These cascades are easily observed in cloud chambers and by arrays of coincidence counters. Some showers of particles resulting from a single primary have been found to trip simultaneously several coincidence counters located as much as 300 meters apart in a horizontal plane. Such an extensive shower may contain more than 10^5 individual particles, with a total energy of 10^{15} or 10^{16} eV.

The primary cosmic radiation arriving at the top of the earth's atmosphere consists predominantly, if not entirely, of positively charged particles, mostly protons. Their energy spectrum peaks in the vicinity of 1 or 2 GeV, but extends to extremely high energies, at least 10^{18} eV. Heavier nuclei are also present: per 1000 protons there are about 150 helium nuclei, about 8 nuclei in the carbon-nitrogen-oxygen range, and 3 or 4 heavier nuclei (N1). Their energy spectra, per nucleon, are about the same as the proton spectrum. The relative abundances of various nuclei in the primary radiation correspond roughly to the relative abundances of the elements in the universe.

A discussion of the origin of cosmic rays is beyond the scope of this book. Suffice it to say that most of the cosmic rays incident on the earth's atmosphere are believed to be of galactic origin and to have been accelerated to their present energies by interstellar magnetic fields. During solar flares the sun contributes significantly to the low-energy (mostly <1 GeV) cosmic-ray flux arriving at the earth. Another important effect of the sun is the decrease in galactic cosmic-ray flux reaching us during periods of intense sun-spot activity; this is most likely a magnetic phenomenon (N2).

Radionuclides from Cosmic Rays. The impact of the primary and very-high-energy secondary cosmic rays, mostly near the top of the atmosphere, produces violent nuclear reactions in which many neutrons, protons, α particles, and other fragments are emitted. Among the fragments in significant yield are H^3, Be^7, and Be^{10} nuclei, and because these nuclei are radioactive they have been found as natural products by radiochemical means.

The nuclide H^3 is formed not only directly by the spallation reactions but also by the action of resulting fast neutrons on nitrogen, $N^{14}(n, H^3)C^{12}$. The total yield is estimated as about 0.4 cm^{-2} sec^{-1}, but because of dilution by ordinary hydrogen its detection is difficult without deliberate isotopic enrichment. Careful investigations of world-wide tritium distribution in surface waters and rain (prior to 1954, when serious perturbations by bomb-produced tritium set in) appear to indicate a steady-state concentration several times that calculated from production by cosmic-ray neutrons and by spallation. The discrepancy is probably explained by the inflow of tritium from the sun and possibly from other extraterrestrial sources. Direct evidence for tritium of solar origin was obtained by the discovery of substantial tritium concentrations in the shell of a recovered satellite which had been in orbit during an intense solar flare.

Most of the neutrons produced by the cosmic rays are slowed to thermal energies and, by n, p reaction with N^{14}, produce C^{14}, the 5720-year β^- emitter. From cosmic-ray data the production rate averaged over the whole atmosphere is calculated to be approximately 2.4 per second per cm^2 of the earth's surface. The lifetime of C^{14} is long enough for the radioisotope to become thoroughly mixed with all the carbon in the so-called exchangeable reservoirs (S1): atmospheric CO_2 (1.6 per cent of the exchangeable carbon), dissolved bicarbonate in the oceans (88.1 per cent), living organisms (0.8 per cent), dissolved organic matter (6.8 per cent), and humus (2.7 per cent). The total carbon content of these reservoirs is estimated as 7.88 grams per cm^2 of the earth's surface. Therefore the specific activity of C^{14} in all of this carbon is expected to be $2.4/7.88 = 0.30$ disintegrations sec^{-1} g^{-1} or 18.3 dis min^{-1} g^{-1}. Actual measurements give values of about 16 dis min^{-1} g^{-1}, in satisfactory agreement with the predicted result. This radioactivity is readily detected in counters, provided that the ordinary high background is greatly reduced by shielding and by anticoincidence counters as mentioned in footnote 1, p. 503.

Radiocarbon Dating. The discovery that all carbon in the world's living cycle is kept uniformly radioactive through the production of C^{14} by cosmic rays led Libby to propose and pioneer the C^{14} dating method that has become such a powerful and widely used technique for determining ages of carbon-containing specimens (L1). The underlying assumption is that cosmic-ray intensity has been constant (apart from short-term fluctuations such as those associated with solar activity) over many thousands of years. Then the specific activity of C^{14} in the exchangeable reservoir has also been constant, and the time elapsed since a specimen was removed from the exchange reservoir can be determined from its C^{14}/C ratio. The constancy of the average cosmic-ray flux has been checked back to about 5000 years ago by means of specimens whose ages are known from historical records. The C^{14} method has been used fairly routinely in many laboratories to date such diverse samples as wood, charcoal, peat, grain, shells, bone, cloth, beeswax, and corncobs, gathered from all over the world. Radiocarbon dating has certainly become a most important tool for archaeologists and geologists. The range of the method is, of course, limited by the lower limit of detectable carbon activity. An age of

50,000 years corresponds to a specific activity of 0.03 dis min^{-1} g^{-1} of carbon, which is probably about as low a level as one can hope to measure.

In addition to its applications for dating archaeological objects and recent geological events, the C^{14} method has yielded other interesting results. Some years ago a puzzling problem was the apparent absence of isotope effects evidenced by the virtually identical C^{14}/C^{12} ratios found in sea shell carbonates and in wood. The C^{13}/C^{12} ratio in sea shells is about 1.025 times that in wood, presumably as a result of the isotope effect in the exchange equilibrium between CO_2 and HCO_3^-. The C^{14}/C^{12} ratios are therefore expected (chapter 7, section C) to differ by a factor of about 1.05. The experimentally found equality of the C^{14}/C^{12} ratios in shells and wood is interpreted as an apparent age of 400 to 500 years (C^{14} decay by about 1.05) for sea shells. This, in turn, means that the average residence time of dissolved carbon in ocean surface water is about 400 to 500 years. The almost exact cancellation between isotope effect and ocean residence time is fortuitous.

The residence time of carbon (as CO_2) in the atmosphere can be estimated from other data. The CO_2 from combustion of fossil fuels (which are old and therefore contain no C^{14}) has been "diluting" the C^{14} concentration in the atmosphere. By 1950 the total amount of "dead" CO_2 added to the atmosphere from this source (mostly since about 1900) amounted to about 12 per cent of the total atmospheric CO_2. Yet plants grown in 1950 show a specific C^{14} activity not 12 per cent, but only 1.75 per cent lower than wood from the nineteenth century (after correction for decay). Thus it is clear that any given carbon atom remains in the atmosphere for a time short compared with 50 years, and from the data the average residence time of carbon in the atmosphere has been estimated as 5 to 10 years. Exchange with the oceans is presumably the principal mechanism for the removal of CO_2 from the atmosphere. Since about 1950 the dilution of atmospheric C^{14} by "dead" CO_2 has been overshadowed by an effect in the opposite direction: the increasing concentration of C^{14} brought about by the neutrons released in nuclear bomb tests. By the early 1960's the specific activity of atmospheric carbon had more than doubled as a result of bomb-produced C^{14}. These man-made effects will surely cause trouble in future C^{14} dating.

By determination of C^{14} in the wood from individual tree rings of old trees it has even been possible to obtain information on short-term (year-to-year) fluctuations in cosmic-ray intensity during specific periods in past centuries, presumably associated with solar activity.

Cosmic-Ray Effects in Meteorites (A2, A3, S4). The important cosmic-ray interactions on our planet are largely confined to the atmosphere because (1) the surface of the earth is effectively shielded from the primary cosmic rays and (2) any reaction products that *are* formed on the surface are likely to be removed by weathering in times that are short on a geological scale. Some attempts have been made to use Cl^{36} ($t_{1/2} = 3 \times 10^5$y), produced by the reaction $Cl^{35}(n, \gamma)$, to determine the length of time during which chlorine-containing formations have been exposed on the surface (e.g., since being covered by glaciation), but such techniques have not been widely applied.

On the other hand, much interesting information has been obtained in recent years from the study of cosmic-ray-induced nuclear reactions in meteorites. Since a meteorite, prior to its arrival on the earth, has presumably spent some time in an interplanetary orbit, its surface has been bombarded by cosmic rays during that time, and the products of this bombardment should give some clues to its history. The interpretation of the data depends strongly on laboratory measurements of cross sections for nuclear reactions by multi-GeV protons (chapter 10, section C7) and is greatly aided by the fortunate circumstance that most such cross sections and certainly the ratios of cross sections for similar reactions are hardly energy-dependent above about 1 GeV, so that details of the cosmic-ray spectrum are usually not needed for the interpretation of results.

Some two dozen radioactive products, ranging in half-life from a few days to millions of years, and a number of stable reaction products have been identified in various meteorites. In space, any radioactive product in a meteorite should be at saturation (as many atoms decaying per unit time as are being formed), provided it has been in a constant cosmic-ray flux for a period long compared to its half-life. Typical activity levels of radionuclides in meteorites are in the range 10–100 dis $min^{-1} kg^{-1}$. Comparison of the ratios of activities of different half-lives in recently fallen iron meteorites with the ratios of saturation activities of the same products made in accelerator bombardments of iron will thus give information on the constancy of cosmic rays in time. Such comparisons of relative production rates for pairs like $Ar^{37}(35d)$-$Ar^{39}(270y)$, Ar^{37}-$Cl^{36}(3 \times 10^5 y)$, $Na^{22}(2.6y)$-$Al^{26}(7.4 \times 10^5 y)$, $Mn^{54}(280d)$-$Mn^{53}(\cong 2 \times 10^6 y)$ have all indicated little or no variation in cosmic-ray intensity (A3, S4). The method is, of course, not sensitive to variations over times short compared to the shorter of the two half-lives of the species studied.

With the assumption of constant intensity, the length of time during which a meteorite has been exposed to cosmic rays can be determined if the cosmic-ray intensity is measured by way of the saturation activity of a radioactive spallation product and the integrated flux is obtained from a measurement of a stable product which has accumulated throughout the exposure. Isobaric pairs such as Cl^{36}-Ar^{36} are probably most reliable, but other pairs, such as Ar^{39}-Ar^{38} and even Cl^{36}-Ne^{21}, have also been used, although the assumption that a ratio of laboratory cross sections determined at a particular bombarding energy is applicable for the cosmic-ray spectrum becomes less secure when the mass numbers of the products are far apart. Greatest reliance can, of course, be placed on exposure ages determined by more than one method. Reliable results range from about 2×10^6 to about 1.5×10^9 years, with appreciable clustering of values in two or three regions. The ages of many stone meteorites are in the vicinity of 30×10^6 years, and in general the iron meteorites are much older than the stones.

We recall that the solidification ages of meteorites are in the neighborhood of 4.5×10^9 y (pp. 499–501).   The interpretation of the much shorter and scattered exposure ages then must almost certainly involve the idea that the meteorites were, for a long part of their history, shielded from cosmic rays. Breakup of one or more larger parent bodies, probably as a result of collisions, is indicated; the similarity in composition between iron and stone meteorites on the one hand and the earth's core and crust on the other suggests that these parent bodies were of planetary (rather than lunar) size and had, at one time, undergone melting and differentiation into different phases.   However, the solidification age of 4.5×10^9 years found for stone meteorites (p. 499) shows that the parent bodies remained molten for a much shorter time than the earth did, and this conclusion sets an upper limit to the sizes of these bodies considerably smaller than the earth.   The best evidence seems to point to more than one parent body and to a series of collisions.   What little information is available on meteorite orbits indicates that they are very eccentric, with aphelia between the orbits of Mars and Jupiter, in the region of the asteroids; asteroids are probably fragments of the same former planets from which the meteorites originated.

Still other types of information can be obtained from data on cosmic-ray-induced reactions on meteorites, for example, regarding the amount of material ablated during the meteorite's trip through the atmosphere and the length of time a meteorite has been on the earth's surface.   For a discussion of these and other questions, the reader is referred to A2 and S4.

B. NUCLEAR REACTIONS IN STARS

The emission of radiation from stars at the observed rates, which has presumably continued over the last several billion years, requires enormous sources of energy.   Chemical reactions cannot possibly be of significance as stellar energy sources.[2]   Although the gravitational energy of contraction may in certain stars play an important role, it is now generally recognized that most stars derive the energy they radiate from exothermic nuclear reactions occurring in their interiors.   These reactions not only account for the vast amounts of energy radiated by stars but also constantly change the elementary and isotopic composition of matter in the universe, and detailed knowledge of these processes is therefore important to an understanding of the abundance distribution of the elements as we know it in our solar system (B1).   Before discussing the particular reactions believed to be important in some types of stars, we state briefly what distinguishes the different types from each other.

---

[2] The energy release in chemical reactions is only of the order of 10^{13} ergs g^{-1} (compared to $\cong 10^{19}$ ergs g^{-1} for nuclear reactions; see p. 59).   Thus the sun, which radiates 4×10^{33} ergs sec^{-1} would use up about 10^{20} g sec^{-1} if chemical reactions were the source of its energy. The total mass of the sun is 2×10^{33} g.

Characteristics of Stars. The observations made by astronomers show that stars differ widely in properties.   Measured masses range over a factor of about a hundred.   Surface temperatures, deduced from analyses of emission spectra, vary by a factor of about 25.   The absolute luminosities, which depend only on surface temperature and area, range from about 10^{-4} to about 10^4 times that of the sun.   Even so, certain distinct classifications have been made. In the *main sequence stars*, and these include the sun, the luminosity is a definite function of the surface temperature; they are probably alike in structure but of various sizes.   A class of stars too cool for their luminosity to fit into the main sequence are the *red giants*.   The *white dwarfs* have small luminosities compared to main-sequence stars of the same surface temperature and therefore must be smaller in size.   In addition there are *variable stars, super giants,* the rare *supernovae,* etc.

We may take the sun as representative of the main sequence stars.   Its mass is 2.0×10^{33} g; its mean density 1.4 g cm^{-3}.   The measured surface temperature is 6000°K, the estimated central temperature, 15,000,000°K, and the rate of energy loss, 4.0×10^{33} erg sec^{-1}.   The sun is believed to contain approximately 80 per cent hydrogen, 20 per cent helium, about 1 per cent carbon, nitrogen, and oxygen, and smaller amounts of all other elements.   The extent to which its composition is kept uniform by mixing between the interior and the surface is an important uncertainty.

Carbon-Nitrogen Cycle. Without doubt nuclear reactions are the source of the sun's energy.   At 15,000,000°K the average kinetic energy of thermal motion is of the order of $kT = 1.3$ keV.   This is much smaller than the height of Coulomb barriers between any nuclei; but rare collisions of light nuclei with energies far above the average (in the "tail" of the Maxwellian distribution) may give very small reaction rates.   H. Bethe in 1938 first proposed a sequence of reactions which is able to supply the sun's energy (B4):

$$C^{12} + H^1 \rightarrow N^{13} + \gamma,$$
$$N^{13} \rightarrow C^{13} + e^+ + \nu \ (\beta \text{ decay}),$$
$$C^{13} + H^1 \rightarrow N^{14} + \gamma,$$
$$N^{14} + H^1 \rightarrow O^{15} + \gamma,$$
$$O^{15} \rightarrow N^{15} + e^+ + \nu \ (\beta \text{ decay}),$$
$$N^{15} + H^1 \rightarrow C^{12} + He^4.$$

The net result of the cycle is the conversion of four protons into one helium nucleus, aside from the positrons (which are annihilated with two electrons) and the neutrinos.   The total energy release per He^4 formed is $\cong$26 MeV, of which only 2 per cent goes into neutrino energy.   The carbon serves only as a catalyst, entering into the rate-determining step and being later regenerated.   A calculation of the rate of the over-all reaction, based on extrapolations of the reaction cross sections into the "thermal" energy range, gives as the best approximation to the energy production (S2):

$$700\rho[\mathrm{H}][\mathrm{C}]\left(\frac{T}{15\times10^6}\right)^{20} \mathrm{erg\ g^{-1}\ sec^{-1}},$$

where ρ is the density of the reacting matter in grams per cubic centimeter, [H] is the weight concentration of hydrogen, and [C] is the weight concentration of carbon and nitrogen. Because the rate of this reaction is extremely sensitive to temperature, the proposed cycle can be important only in the very hottest region of the sun's interior.

Proton-Proton Chain. In 1952 E. Salpeter directed attention to an alternate sequence of nuclear reactions which appears more probable as a result of new knowledge of β-decay selection rules (S2). Among the available nuclei by far the lowest Coulomb barrier is for the reaction of two protons. But because He^2 cannot exist the only possible reaction requires that a relatively slow β decay occur during the collision:

$$\mathrm{H}^1 + \mathrm{H}^1 \rightarrow \mathrm{H}^2 + e^+ + \nu.$$

After this step relatively fast reactions complete a cycle which in effect transforms four protons into one helium nucleus:

$$\mathrm{H}^2 + \mathrm{H}^1 \rightarrow \mathrm{He}^3 + \gamma,$$
$$\mathrm{He}^3 + \mathrm{He}^3 \rightarrow \mathrm{He}^4 + 2\mathrm{H}^1.$$

The energy production rate has been calculated to be

$$0.50\rho[\mathrm{H}]^2\left(\frac{T}{15\times10^6}\right)^4 \mathrm{erg\ g^{-1}\ sec^{-1}}.$$

The best evidence available at present indicates that in the very center of the sun the carbon-nitrogen cycle is the faster reaction and that in a larger interior region at slightly lower temperatures the proton-proton chain is more important because of its smaller temperature dependence. Those main sequence stars that are smaller and cooler than the sun are probably heated by the direct proton-proton reaction; in those much larger than the sun the carbon-catalyzed reaction almost surely contributes most of the energy.

Laboratory experiments on reactions of light nuclei have been extremely important in establishing details of the reaction sequences in stellar interiors; however, to get measurable rates in the laboratory, the experiments always have to be done at much higher kinetic energies than are available in stars, and the cross section data must then be extrapolated. From such work it has become clear that some side reactions may be important in the proton-proton chain (B2). Among these is the following sequence, thought to contribute significantly at $T > 1.3 \times 10^7$ °K:

$$\text{He}^3 + \text{He}^4 \rightarrow \text{Be}^7 + \gamma,$$

$$\text{Be}^7 \xrightarrow{\text{EC}} \text{Li}^7 + \nu$$

$$\text{Li}^7 + \text{H}^1 \rightarrow 2\,\text{He}^4.$$

Stellar Evolution (B1, C1, B2).   It is evident both from observational data and from theoretical considerations that the rate of hydrogen burning in the main sequence stars depends strongly on their sizes.   The sun's supply of hydrogen is sufficient for it to radiate at about the present rate for several billion years (and, as we have seen, other evidence shows that the sun is already several billion years old).   But the very large stars at the upper end of the main sequence are burning hydrogen at a rate that will exhaust their supply in about 10^6 years.   Thus it is clear (1) that the stellar population of our galaxy (and of other galaxies) must consist of stars of various ages and (2) that stars cannot continue indefinitely to produce energy by the hydrogen-burning reactions.

A star that has exhausted an appreciable fraction of its hydrogen supply will eventually have a central region that consists largely of helium.   Model calculations show that the helium core of such a star will contract, whereas the hydrogen-containing shell may expand considerably.   The surface will thus increase substantially and become much cooler, so that the star leaves the main sequence and becomes a red giant.   In the interior, in the meantime, the gravitational contraction leads to higher densities and higher temperatures which eventually make reactions between helium nuclei possible.   When densities of $\sim 10^5$ g cm^{-3} and temperatures in the neighborhood of 10^8 °K are reached, the reaction

$$\text{He}^4 + \text{He}^4 = \text{Be}^8$$

produces a steady-state concentration of about 10^{-9} Be^8 atoms per He^4, in spite of the short half-life of Be^8 ($\cong 10^{-16}$ second), and this concentration is sufficient to support the reaction

$$\text{Be}^8 + \text{He}^4 \rightarrow \text{C}^{12} + \gamma$$

at an appreciable rate.   In other words, the helium-burning reaction

$$3\,\text{He}^4 \rightarrow \text{C}^{12} + \gamma$$

becomes significant and may be an important energy source in red giants. After the C^{12} concentration has been built up sufficiently, additional helium-burning may proceed by

$$\text{C}^{12} + \text{He}^4 \rightarrow \text{O}^{16} + \gamma.$$

Eventually (in a time of the order of 10^7–10^8 years), most of the helium in the stellar core will be used up, and further contraction and increase in temperature is expected, the details of the evolutionary process again depending on the size of the star.   Carbon-carbon reactions such as

$$C^{12} + C^{12} \rightarrow Mg^{24} + \gamma,$$

$$C^{12} + C^{12} \rightarrow Na^{23} + p,$$

$$C^{12} + C^{12} \rightarrow Mg^{23} + n,$$

$$C^{12} + C^{12} \rightarrow Ne^{20} + \alpha,$$

as well as $C^{12} + O^{16}$ reactions, now become possible ($T \gtrsim 6 \times 10^8$ °K), and at still higher temperatures further buildup to silicon, phosphorus, and sulfur may occur.   Above $\cong 10^9$ °K, (γ, α) reactions are expected with nuclei such as Ne^{20} and S^{32}, in which the α particles are relatively loosely bound.   The importance of these reactions lies in the production of α particles with kinetic energies considerably in excess of those of "thermal" He^4, even at 10^9 °K; these α particles will, on occasion, collide with other Ne^{20} nuclei to produce the reaction $Ne^{20}(\alpha, \gamma)Mg^{24}$.   This process, on a time scale of perhaps thousands of years, is believed to lead to the buildup of Mg^{24}.   In analogous fashion further synthesis by α, γ reactions is thought to account for the formation of Si^{28}, S^{32}, Ar^{36}, Ca^{40}, Ca^{44}, and Ti^{48}, in decreasing abundance.   Whatever the details, it is clear that nucleosynthesis by exothermic fusion reactions can certainly not proceed beyond the maximum in the binding-energy curve in the region of iron.

Element and Isotope Abundances.   Before proceeding further to trace possible transformations of matter in stellar interiors, we should briefly consider the available evidence concerning the distribution of elements and their isotopes in the solar system and in other parts of the universe.   The important observations may be summarized as follows:

1. The isotopic composition of all elements (except those affected by formation in radioactive decay processes, as discussed in section A) in terrestrial and meteoritic material has been found to be independent of source.[3]

2. From direct chemical analyses of terrestrial and meteoritic samples, from geologic data, from spectral analyses of the sun, and from information on the densities of the planets, the relative abundances of the elements in the solar system are rather well determined (see, e.g., S3).   Figure 15-2 is a plot of typical results.   The nearly exponential decrease up to $Z \cong 35$, the near-constancy for $Z > 35$, the relatively high abundances of iron and nickel, the very low abundances of Li, Be, and B, the general tendency toward even-odd alternation, and many finer details may be noted as characteristic.

3. Additional systematic trends are found when the relative abundances of

---

[3] Slight fluctuations in isotopic composition of some light elements such as hydrogen, carbon, and oxygen are explainable in terms of isotope fractionation in chemical reactions (see pp. 24–25 and chapter 7, section C).

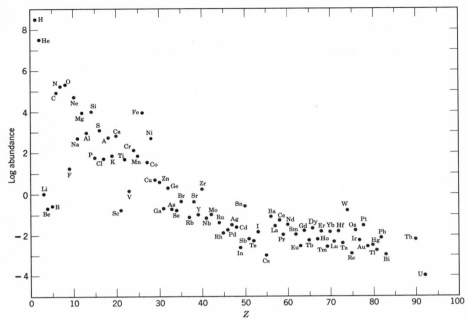

Fig. 15-2 Cosmic abundances of the elements. The ordinate is for each element the logarithm of the number of atoms per 10,000 atoms of silicon. (Data are from V. M. Goldschmidt, with added values from H. C. Urey and from H. Brown; see *Phys. Rev.* **88**, 248 (1952).)

individual isotopes are considered. (Direct observations on isotope abundances are, of course, confined largely to terrestrial and meteoritic material). The most striking feature is that the distribution of relative isotopic abundances for individual elements of even Z is generally different for the lighter and heavier groups. Among the lighter elements the tendency is for the lightest isotope to be an abundant one. The opposite is true for the heavier group; for each of the elements of even Z above Z = 33 the lightest isotope is rare and the heaviest isotope is always abundant.

4. The solar-system abundances do not by any means apply to the universe as a whole or even to our galaxy, as was once assumed. It is by now well established by spectroscopic observation that wide variations in composition occur among the stars of our galaxy (B2). Abundances of heavy elements (Z ≥ 6) relative to hydrogen are in some stars factors of hundreds or thousands lower than in the sun. From other evidence these stars appear to be very old, formed $(10-20) \times 10^9$ years ago. Other stars with the opposite anomaly— very large heavy-element abundances—have been found also and some of them are known from their rate of hydrogen burning to be very young. Stars are known with He/H ratios as large as several hundred and C/H ratios of about

10. Even abnormal isotopic abundance ratios such as $He^3/He^4 \cong 5$ and $C^{12}/C^{13} \cong 4$ have been established (from intensities in band spectra and isotope shifts of spectral lines). Finally we mention that spectral lines of the unstable element technetium (half-life of longest lived isotope, Tc^{97}, is 2.6×10^6y) have been discovered in the spectra of certain stars.

Nucleosynthesis in Prestellar Universe. The observations on element and isotope abundances, sketched in the preceding paragraphs, together with our earlier discussion of the chronology of the solar system, indicate that the present abundance distribution in the solar system must have been established before the differentiation into sun and planets about 5×10^9 years ago and has remained unchanged since then except for the alterations resulting from conversion of hydrogen to helium in the sun and from decay of the radioactive nuclides. On the other hand, it is clear from the widely varying compositions of different types of stars that a common origin for the present matter composition of our entire galaxy cannot be assumed. In fact, the observation of technetium in stars proves rather conclusively that, in addition to the hydrogen- and helium-burning and other light-element reactions already discussed, other nuclear processes which can lead to heavy-element synthesis must be taking place in stellar interiors essentially right up to the present time. These conclusions from fairly recent astronomical evidence dealt a blow to the previously rather widely accepted and in many respects very attractive theories of element synthesis by successive neutron captures in the earliest stages (first hour or so) of the expanding universe.

These so-called "big-bang" theories (A4) pioneered by G. Gamow received much impetus from the near-agreement between the age of the solar-system elements and what was then believed to be the time since all the galaxies, according to their measured recession velocities (spectral red-shifts), started from a common origin ($\cong 5 \times 10^9y$). The other significant observation was the behavior of fast-neutron capture cross sections as a function of atomic number:[4] they show a nearly exponential increase up to $Z \cong 35$ and then stay roughly constant at higher Z, mirroring the shape of the element abundance curve (figure 15-2). It was thus attractive to postulate that the entire universe began some 5×10^9 years ago as a primordial substance (called the ylem) consisting of neutrons at an extraordinarily high temperature and density, dissolved in electromagnetic thermal radiation of even higher density. Decay of neutrons to protons and electrons and expansion and cooling of the ylem were thought to bring about buildup of the entire periodic table of the elements, largely by successive neutron capture reactions, within about an hour, that is, before too many neutrons had disappeared by decay and before temperature and density had decreased too far. Subsequent adjustment of

---

[4] By "fast" neutrons we here mean neutrons with kinetic energies of tens of keV. Note that a temperature of 10^8 °K corresponds to a kinetic energy of about 10 keV for "thermal" neutrons.

element abundances by β^- decay was, of course, expected, since the nuclei initially formed would have been far on the neutron excess side of stability. This idea accounted well for the prevalence of the heavy isotopes among heavy elements.

In addition to the evidence cited earlier, other serious difficulties developed for this theory.   (1) It was hard to invent a mechanism for carrying the element synthesis past mass numbers 5 and 8.   (2) A redetermination of the astronomical distance scale showed that the galaxies have been receding from one another not for $(5–6) \times 10^9$ years, but for twice that long—much longer than is compatible with the age of the elements in the solar system.[5]

Further Nucleosynthesis in Stars.   Some significant contribution to element building in a rapidly evolving prestellar stage of the universe cannot be excluded.   But it is now generally believed that most of the important nucleosynthesis processes go on in stellar interiors, that these processes can account for most of the features of observed abundance distributions, and that they are compatible with what is known about stellar evolution.   Whether the universe is finite or infinite in extent and whether matter is being created continuously as postulated by some astrophysicists are questions that are not directly relevant in this context, although, of course, extremely interesting in themselves.

A detailed discussion of the connection between various observable stellar types, their presumed evolutionary stages, and the reactions believed to proceed in them is entirely beyond the scope of the present treatment.   Nor is it possible here even to indicate the remarkable quantitative work that has been done on stellar models and their hydrodynamic behavior, on calculations of nuclear reaction rates and their detailed effects on nuclide abundances, and many other fascinating aspects of recent research in this field of nuclear astrophysics.   However, it may be of interest to trace stellar evolution in very general outline somewhat beyond the point at which we left off on p. 512 and to see how the solar-system abundance distribution is believed to have come about.

We sketched how hydrogen burning, helium burning, carbon burning, and α, γ reactions are believed to follow each other in orderly progression, with the temperature rise required for the onset of each type of reaction provided by conversion of gravitational to thermal energy when the fuel of the preceding stage has been exhausted.   Other types of reactions which we have hitherto neglected can take place when products formed in the hotter interior, such as C^{12}, O^{16}, Ne^{20}, become mixed with hydrogen still present in the outer regions. Such reactions as $C^{12}(p, \gamma)N^{13} \rightarrow C^{13}$, $O^{16}(p, \gamma)F^{17} \rightarrow O^{17}$, $O^{17}(p, \gamma)F^{18} \rightarrow$

[5] In addition to the evidence from extinct I^{129} (p. 502), there are other strong arguments for an age of the solar-system elements not in excess of about 6×10^9.   At that time U^{235} and U^{238} would have had about equal abundances.   The systematics of other heavy-element isotope abundances make it highly unlikely that the odd-A U^{235} should have been more abundant than the even-A U^{238}.

O^{18}, and $O^{18}(p, \alpha)N^{15}$, when added to the reactions already discussed, account well for the relative abundances of all the nuclides up to the region of magnesium and in particular can produce all the species involved in the carbon-nitrogen cycle.

As mentioned on p. 512, a star that condensed out of a mass of pure hydrogen presumably cannot synthesize elements beyond the iron-nickel region. The observed absence or at least abnormally low abundance of heavy elements in some old stars is in agreement with the idea that they are first-generation stars. The question then arises, how do stars, such as our sun, which are still in the hydrogen-burning stage but contain appreciable abundances of heavy elements, come about? The suggested explanation involves their condensation out of interstellar gas clouds which themselves are the "debris" ejected from older stars. Various types of stars are known to emit material into interstellar space. However, the most important mechanism for large-scale ejection of matter appears to be in supernova explosions. Of the two types of supernovae known, the so-called Type II, which appear to occur among stars with low heavy-element content and have rather quickly decreasing light output after their eruption, are thought to represent the ultimate fate of the first-generation stars we have been discussing.

After a substantial fraction of the star's core has been converted to iron-group elements, further gravitational contraction is no longer held in check by energy-producing reactions. Densities of about 10^8 g cm^{-3} and temperatures close to 8×10^9 °K will be reached, and under these circumstances an equilibrium between iron (mostly Fe^{56}) and He^4 is set up. A relatively small temperature rise ($\cong 0.5 \times 10^9$ °K) can bring about almost complete dissociation of iron into helium and neutrons, but the energy for this process can be supplied only by gravitation. The result is a gigantic implosion which takes only a fraction of a second. As a consequence, the outer regions of the star, still consisting largely of light elements (H, He, C, O, etc.), experience a sudden heating, which, under appropriate conditions, can set off a huge thermonuclear explosion accompanied by the ejection of a large amount of the star's mass into interstellar space: a supernova. Such explosions of Type II supernovae are believed to occur roughly at the rate of one every 50 years per galaxy; each is expected to eject in the order of a solar mass into space. Thus in the course of a few billion years the galaxy has presumably accumulated a substantial mass of interstellar gas composed of material previously processed in stellar interiors. Out of this matter new second-generation stars will have formed all along.

In a second-generation star the basic sequence of energy-producing reactions will at first be the same as discussed before: hydrogen burning, followed by helium burning, etc. However, the presence of all the stable nuclides up to iron in appreciable concentrations makes additional reactions possible in the helium-burning stage, among them a group of α, n reactions such as $C^{13}(\alpha, n)O^{16}$, $O^{17}(\alpha, n)Ne^{20}$, and $Ne^{21}(\alpha, n)Mg^{24}$, which can set in at about

10^8 °K.   These reactions furnish a source of neutrons which can continue the element-building process beyond Fe by successive n, γ reactions, interspersed with β^- decays.   The abundance distribution of the majority[6] of nuclides up to bismuth is accounted for in considerable quantitative detail by such a process of successive neutron captures on a slow time scale, which has been named the s process.   This is also the process that accounts for the presence of technetium in stars.   However, since it does take place on a relatively slow time scale ($\cong 10^2$–10^5 years per neutron capture step), the s process cannot possibly carry the synthesis beyond bismuth to thorium and uranium because of the intervening short-lived species.

Uranium and thorium are believed to owe their existence to a neutron-capture chain that takes place on a much shorter time scale in an enormous neutron flux, so that many neutrons can be captured successively without intermediate α or β decays.   Again, supernova explosions are invoked.   The exact mechanism operative in these so-called Type I supernova outbursts is still under discussion; in any case, it is believed to involve gravitational collapse leading to a thermonuclear explosion which in turn produces the neutron flux that, in these second- (or later-) generation stars, can build nuclides up to $A \cong 270$ in a matter of seconds.   This rapid neutron capture sequence is known as the r process.

An extremely interesting result in this connection was the discovery, in the debris from man-made thermonuclear explosions, of Cf^{254}, produced from U^{238} by multiple-neutron capture.   This nuclide decays by spontaneous fission with a half-life of 56 days[7] and, after some shorter-lived species have decayed, the energy given off in the decay of this one nuclide greatly outweighs that emitted by all the other heavy-element isotopes formed in the multiple-capture processes because the energy release in spontaneous fission is so much greater than that in α or β decay.   These observations have tended to support the suggested origin of the heaviest elements in supernova explosions.   For the light emission of supernovae of Type I, after initial rapid decrease for about a month or two, decays with a half-life of about 55 days; this not only has been observed by modern astronomers for supernovae in other galaxies but also follows from the records kept on the three supernovae seen in our own galaxy in historic times: the one observed in China in 1054 and the ones recorded by Tycho Brahe (1572) and Johannes Kepler (1604).   The light curves are shown in figure 15-3.

The matter ejected from supernovae of Type I, possibly mixed with other interstellar gas, can presumably again serve as the raw material for the forma-

---

[6] Some nuclides, especially the relatively neutron-deficient and shielded ones, require still another mechanism which presumably involves p, γ reactions.   The high temperatures ($\cong 2 \times 10^9$ °K) and large hydrogen concentrations needed for such reactions suggest the envelopes of supernovae as a likely site.

[7] A recent redetermination of the Cf^{254} half-life as 60 days has cast some doubt on the specific interpretation of supernova light curves in terms of Cf^{254} decay; however, the general explanation of heavy-element buildup in supernovae explosions is believed to be valid.

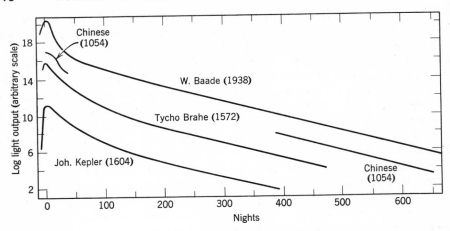

Fig. 15-3   Light curves of supernovae observed by Baade (1938), Johannes Kepler (1604), Tycho Brahe (1572), and by Chinese astronomers (1054). The abscissa gives the number of nights after maximum; the ordinate is the logarithm of the light output in arbitary units; the vertical positions of the different curves are arbitrary. (Figure adapted from reference B1.)

tion of stars.   Our sun with its planets is, according to this picture, a third- (or later-) generation star formed from the debris of one or more supernova explosions, the last of which apparently occurred only about 5×10^9 years ago. Within a few hundred million years after that cataclysmic event, to which we owe our uranium and thorium supplies and some other details of our element abundance distribution, the sun and its planetary system began to form.

REFERENCES

*A1   L. T. Aldrich and G. W. Wetherill, "Geochronology by Radioactive Decay," *Ann. Rev. Nuclear Sci.* **8**, 257–298 (1958).
*A2   E. Anders, "Meteorite Ages," *Rev. Mod. Phys.* **34**, 287–325 (1962).
A3   J. R. Arnold, "Nuclear Effects of Cosmic Rays in Meteorites," *Ann. Rev. Nuclear Sci.* **11**, 349–370 (1961).
A4   R. A. Alpher and R. C. Herman, "The Origin and Abundance Distribution of the Elements," *Ann. Rev. Nuclear Sci.* **2**, 1–40 (1953).
*B1   E. M. Burbidge, G. R. Burbidge, W. A. Fowler, and F. Hoyle, "Synthesis of the Elements in Stars," *Rev. Mod. Phys.* **29**, 547–650 (1957).
*B2   G. Burbidge, "Nuclear Astrophysics," *Ann. Rev. Nuclear Sci.* **12**, 507–576 (1962).
B3   B. B. Boltwood, "On the Ultimate Disintegration Products of the Radioactive Elements," *Am. J. Sci.* **23**, 77 (1907).
B4   H. A. Bethe, "Energy Production in Stars," *Phys. Rev.* **55**, 534 (1939).
*C1   A. G. W. Cameron, "Nuclear Astrophysics," *Ann. Rev. Nuclear Sci.* **8**, 299–326 (1958).
*F1   H. Faul (Editor), *Nuclear Geology*, Wiley, New York, 1954.
K1   T. P. Kohman and N. Saito, "Radioactivity in Geology and Cosmology," *Ann. Rev. Nuclear Sci.* **4**, 401–462 (1954).

*L1   W. F. Libby, *Radiocarbon Dating*, University of Chicago Press, 1955.

*N1   H. V. Neher, "The Primary Cosmic Radiation," *Ann. Rev. Nuclear Sci.* **8**, 217–242 (1958).

N2   E. P. Ney, "Experiments on Cosmic Rays and Related Subjects During the International Geophysical Year," *Ann. Rev. Nuclear Sci.* **10**, 461–488 (1960).

P1   C. Patterson, "Age of Meteorites and the Earth," *Geochim. Cosmochim. Acta.* **10**, 230 (1956).

S1   H. E. Suess, "The Radioactivity of the Atmosphere and Hydrosphere," *Ann. Rev. Nuclear Sci.* **8**, 243–256 (1958).

S2   E. E. Salpeter, "Energy Production in Stars," *Ann. Rev. Nuclear Sci.* **2**, 41–62 (1953).

*S3   H. E. Suess and H. C. Urey, "Abundances of the Elements," *Rev. Mod. Phys.* **28**, 53–74 (1956).

S4   O. A. Schaeffer, "Radiochemistry of Meteorites," *Ann. Rev. Phys. Chem.* **13**, 151–170 (1962).

U1   H. C. Urey, "The Origin and Development of the Earth and Other Terrestrial Planets," *Geochim. Cosmochim. Acta* **1**, 209 (1951).

EXERCISES

1. A 1-kg sample of an ocean sediment is found to contain 1.50 mg of uranium, 4.20 mg of thorium and 6.0×10^{-3} cm^3 of helium (at standard temperature and pressure). Estimate the age of the deposit, assuming complete helium retention.
Answer: 2.0×10^7 years.

2. The lead isolated from a sample of uraninite is mass-spectrometrically analyzed and found to contain Pb^{204}, Pb^{206}, and Pb^{207} in the ratio 1.00:914.7:92.6. Estimate the time since the formation of the uraninite ore. *Answer:* 1.50×10^9 years.

3. From (15-1) and (15-2) derive an expression relating the common age of two meteorites to the experimentally determined Pb^{207}/Pb^{204} and Pb^{206}/Pb^{204} ratios in them and to the present-day U^{238}/U^{235} abundance ratio. Assume that the original isotopic composition of lead in the two meteorites was the same, that they have existed since their formation as isolated systems, and that the present U^{238}/U^{235} ratio in the meteorites is the same as in terrestrial uranium. Verify that your expression gives the meteorite age quoted on p. 501 when you use the following experimental data from reference P1 for the lead composition in three meteorites:

Meteorite	206/204	207/204
Nuevo Laredo	50.28	34.86
Modoc	19.48	15.76
Canyon Diablo	9.46	10.34

4. From data in chapter 1 and appendix E calculate the heat generated per gram of normal uranium per year (a) by U^{238}, (b) by U^{235}, each in secular equilibrium with all its decay products. Assume the uranium is embedded in a massive body so that all α, β, and γ rays are absorbed. *Answers:* (a) 0.7 cal; (b) 0.03 cal.

5. Assuming the earth's crust to contain uranium, thorium, and potassium in the weight ratio 1:4:10^4, estimate the relative contributions of U^{238}, U^{235}, Th^{232}, and K^{40} to the heat production in the earth's crust (a) now, (b) 2×10^9 years ago, (c) 4.5×10^9 years ago. Use the results of the preceding exercise and the additional information that the corresponding heat production rates (in calories per

year per gram of element) are 0.20 for thorium in equilibrium with its decay series and 2.6×10^{-5} for potassium.

6. A proportional counter of about 6-liter volume, filled to 3 atm with pure CO_2 and surrounded with heavy shielding and anticoincidence counters, is used for C^{14}-dating measurements. Three samples of CO_2 are measured: sample A is "dead" CO_2 (age greater than 75,000 years) from coal, sample B is from contemporary wood (grown 50 years ago according to a tree-ring count), and sample C is of unknown age. The following counts are accumulated: 11,808 counts in 960 minutes for sample A, 21,749 counts in 180 minutes for sample B, and 20,583 counts in 480 minutes for sample C. Great care is taken to introduce the same weight of CO_2 into the counter at each filling. What is the age (and its standard deviation) of sample C? Take the half-life of C^{14} as 5720 ± 15 years.

7. At sea level the cosmic radiation produces about 2 ion pairs per second per cm^3 of air. At higher altitudes the intensity depends on the latitude, but for much of the United States it is about 10 ion pairs sec^{-1} cm^{-3} (of sea level air) at 10,000 ft and about 200 ion pairs sec^{-1} cm^{-3} (of sea level air) at 40,000 ft above sea level. Estimate the radiation dosage received per 24 hours in r units (probably we should say in rep units) from this source at (a) sea level, (b) 10,000 ft, (c) 40,000 ft.

8. A 1-kg sample of a recently fallen iron meteorite is found to contain 16 disintegrations per minute (dpm) of Cl^{36}, 14 dpm of Ar^{39}, and 1.88×10^{-4} cm^3 of Ar^{36} at STP. From bombardments of iron targets with high-energy protons it is known that the cross sections for formation of Cl^{36}, Ar^{36}, and Ar^{39} in such bombardments are in the ratio 1:0.2:0.9, rather independently of proton energy above $\sim$400 MeV. (a) What is the cosmic-ray exposure age of the meteorite? (b) What can you conclude about cosmic-ray intensity as a function of time? (c) How would you interpret the finding that another iron meteorite has nearly the same Cl^{36} and Ar^{36} contents as the first but contains only 3.6 dpm of Ar^{39} per kg? *Answer:* (a) 5×10^8 years.

9. At approximately what stellar temperature would you expect the carbon cycle and the proton-proton cycle to proceed at equal rates, assuming the weight ratio of hydrogen to carbon to be 1000? *Answer:* 1.5×10^7 °K.

10. The radiation from the sun at normal incidence at the earth (93 million miles away) amounts to 0.135 joule cm^{-2} sec^{-1}. To produce this energy, at what rate is hydrogen being consumed in the sun, in grams per second? *Answer:* 5.9×10^{14}.

11. If the sun's energy comes predominantly from the proton-proton chain, and if the neutrinos produced are abosrbed only negligibly in the sun, what is the flux at the earth of neutrinos from this source? *Answer:* 6×10^{10} cm^{-2} sec^{-1}.

12. (a) If the s process (neutron capture on a slow time scale) in a star starts with a mixture of Fe^{56} and Ni^{58}, which of the stable nuclides with mass numbers between 56 and 85 do you expect to be formed by this process? (b) Which stable nuclides in this mass range were presumably formed predominantly by the r process (p. 517)? (c) Which nuclides in the same mass region cannot be accounted for by either of these neutron capture processes, and what type of reaction might be invoked for their synthesis? *Answer:* (b) Zn^{70}, Ge^{76}, Se^{82}.

Appendix A. Physical Constants and Conversion Factors

The values given are taken from the consistent set of physical constants recommended in 1963 by a committee of the National Academy of Sciences-National Research Council and adopted by the National Bureau of Standards [*NBS Tech. News* **47,** 175 (1963)]. Where pertinent, the C^{12} atomic-mass scale is used (C^{12} mass $\equiv$ 12 amu).

Quantity	Symbol	Value	Error Limit*
Velocity of light	c	2.997925×10^{10} cm sec^{-1}	3
Planck's constant	h	6.6256×10^{-27} erg sec	5
	$\hbar = h/2\pi$	1.05450×10^{-27} erg sec	7
Boltzmann constant	k	1.38054×10^{-16} erg deg^{-1}	18
Electronic charge	e	4.80298×10^{-10} esu	20
		1.60210×10^{-20} emu	7
Avogadro's number	N	6.02252×10^{23} mol^{-1}	28
Faraday constant	$F = Ne$	9.64870×10^{4} absolute Coulombs mol^{-1}	16
		2.89261×10^{14} esu mol^{-1}	5
Electron mass	m_e	9.1091×10^{-28} g	4
		5.48597×10^{-4} amu	9
Neutron mass	M_N	1.67482×10^{-24} g	8
		1.0086654 amu	13
Hydrogen atom mass	M_H	1.67343×10^{-24} g	8
		1.00782522 amu	24
Energy equivalent of one atomic mass unit (1 amu)		931.478 MeV	15
electron mass		0.511006 MeV	5
neutron mass		939.550 MeV	15
proton mass		938.256 MeV	15
hydrogen atom mass		938.767 MeV	15
Energy conversion factors: 1 eV =		1.60210×10^{-12} erg	7
		3.8291×10^{-20} cal$_{th}$	2
Temperature corresponding to 1 eV		1.16049×10^{4} deg	16
Photon wavelength associated with 1 eV		1.23981×10^{-4} cm	4
Fine structure constant	$e^2/\hbar c$	1/137.0388	19
Bohr magneton	$\hbar e/2m_e$	9.2732×10^{-21} erg gauss^{-1}	6
Nuclear magneton		5.0505×10^{-24} erg gauss^{-1}	4
Number of seconds per day		86,400	
Number of seconds per year		3.1558×10^{7}	

* The error limits shown are based on three standard deviations of the least-squares-adjusted data and apply to the last digit of the values given.

Appendix B. Relativistic Relations

Consider a particle of rest mass m_0 moving at a velocity v.   If $\beta = v/c$, where c is the velocity of light, then the particle has

$$\text{mass} \equiv m = \frac{m_0}{\sqrt{1 - \beta^2}}, \tag{B-1}$$

$$\text{momentum} \equiv p = mv = \frac{m_0 v}{\sqrt{1 - \beta^2}} = \frac{m_0 \beta c}{\sqrt{1 - \beta^2}}, \tag{B-2}$$

$$\text{kinetic energy} \equiv T = m_0 c^2 \left(\frac{1}{\sqrt{1 - \beta^2}} - 1 \right) = mc^2 - m_0 c^2, \tag{B-3}$$

$$\text{total energy} \equiv E = mc^2 = \frac{m_0 c^2}{\sqrt{1 - \beta^2}}. \tag{B-4}$$

A useful relation between momentum and total energy may be obtained by squaring and rearranging (B-1):

$$m^2 c^2 - m_0{}^2 c^2 = m^2 v^2 = p^2.$$

Dividing by $m_0{}^2 c^2$, we have

$$\left(\frac{p}{m_0 c} \right)^2 = \left(\frac{mc}{m_0 c} \right)^2 - 1 = \left(\frac{mc^2}{m_0 c^2} \right)^2 - 1 = \left(\frac{E}{m_0 c^2} \right)^2 - 1. \tag{B-5}$$

Thus, if momentum is expressed in units of $m_0 c$ $(\eta = p/m_0 c)$ and total energy in units of $m_0 c^2$ $(W = E/m_0 c^2)$, the following relation holds:

$$\eta^2 = W^2 - 1. \tag{B-6}$$

Equation B-5 may also be rearranged to yield the relation

$$E^2 = E_0{}^2 + p^2 c^2, \tag{B-7}$$

where $E_0 = m_0 c^2$.

If a system A moves with a velocity v $(= \beta c)$ relative to another system B, a time interval Δt_A measured in system A will, in system B, appear as

$$\Delta t_B = \frac{\Delta t_A}{\sqrt{1 - \beta^2}}. \tag{B-8}$$

For a particle of zero rest mass (such as a photon or a neutrino), the following relations hold:

$$p = \frac{E}{c},$$ (B-9)

$$E = h\nu,$$ (B-10)

where ν is the frequency and h is Planck's constant.

The following table lists, for a number of values of β, the corresponding kinetic energies of electrons, π mesons, and protons. For each value of β the ratio of moving mass to rest mass is also given.

		Kinetic Energy in MeV		
β	m/m_0	Electron	π Meson	Proton
0.10	1.005	0.00257	0.710	4.72
0.20	1.021	0.0107	2.96	19.7
0.30	1.048	0.0247	6.81	45.3
0.40	1.091	0.0465	12.8	85.5
0.50	1.155	0.0791	21.8	145
0.60	1.250	0.128	35.2	235
0.70	1.400	0.205	56.4	375
0.75	1.512	0.262	72.2	480
0.80	1.667	0.341	94.0	625
0.85	1.898	0.459	127	843
0.90	2.294	0.661	182	1.21×10^3
0.95	3.203	1.13	311	2.07×10^3
0.96	3.571	1.31	363	2.41×10^3
0.97	4.114	1.59	439	2.92×10^3
0.98	5.025	2.06	568	3.78×10^3
0.99	7.089	3.11	859	5.71×10^3
0.995	10.01	4.61	1.27×10^3	8.45×10^3
0.999	22.37	10.9	3.01×10^3	2.00×10^4
0.9999	70.71	35.6	9.83×10^3	6.54×10^4
0.99999	223.6	114	3.14×10^4	2.09×10^5

Appendix C. Thermal Neutron Cross Sections

The data for this table are taken from the compilation "Neutron Cross Sections" by D. J. Hughes and R. B. Schwartz (BNL 325, 2nd ed., 1958, and Supplement No. 1, 1960, U. S. Government Printing Office, Washington, D. C.).

To use the table for the calculation of the rate of a nuclear reaction in a sample placed in a nuclear reactor, it must be realized that the magnitude of the flux of neutrons at any point in a reactor is given by the expression

$$n \int_0^\infty v P(v) \, dv = n\bar{v}, \tag{C-1}$$

where n is the density of neutrons at that point, $P(v) \, dv$ is the probability that a neutron will have a velocity between v and $v + dv$, and $\bar{v}$ is the average velocity. The rate $\mathbf{R}$ of a neutron reaction in a sample within which the attenuation of the neutron flux may be neglected is

$$\mathbf{R} = Nn \int_0^\infty P(v) \, \sigma(v) v \, dv, \tag{C-2}$$

where N is the number of target nuclei in the sample and $\sigma(v)$ is the cross section (usually velocity dependent) for the neutron reaction in question. In particular, if the neutron-reaction cross section is in the $1/v$ region (see section G of chapter 10),

$$\sigma(v) = \sigma_0 \frac{v_0}{v}, \tag{C-3}$$

then

$$\mathbf{R} = Nn \int_0^\infty P(v) \frac{\sigma_0 v_0}{v} v \, dv \tag{C-4}$$

$$= Nn v_0 \sigma_0. \tag{C-5}$$

Thus if the effective neutron flux in a reactor, regardless of its actual value, is given as the product of the *density* of neutrons times the velocity v_0, then the rate of any neutron reaction that is characterized by an inverse dependence of cross section on velocity may be computed from a knowledge of the cross section σ_0 at the particular velocity v_0. It is convenient that most neutron-

524

capture cross sections have a $1/v$ dependence in the thermal region; integration of (C-2) is required if they do not.

The velocity v_0 is taken as 2.2×10^5 cm sec^{-1}, the most probable velocity of a Maxwellian distribution at 20°C. Accordingly, the cross sections in the table are given in barns for this velocity, except in a few instances in which the values given refer to an average over a pile neutron spectrum; the latter values are shown in brackets. Cross sections that were determined by measurement of the transmission of neutrons through a sample (*absorption* cross section) are shown in boldface; the other values are *activation* cross sections, determined by measurements of the products of the neutron reactions.

Both activation and absorption cross sections for an isotope are given only when there is a substantial difference between the two values. Nearly all the activation cross sections given are for n, γ reactions; cross sections for other reactions are labeled (n, p), (n, α), or (n, f). When a cross section is given for a reaction leading to one of several isomers, the half-life of the product is listed in parentheses. An element cross section always refers to the natural isotopic mixture; an isotopic cross section is for the pure isotope.

The uncertainty in each last digit is ≤ 10.

H	**0.332**	Na^{23}	0.536
H^2	$5.7 \cdot 10^{-4}$	Mg	**0.069**
He^3	$\mathbf{5.5 \cdot 10^3}$ (n, p)	Mg^{24}	**0.033**
He^4	**0**	Mg^{25}	**0.28**
Li	**71**	Mg^{26}	0.027
Li^6	**945** (n, α)	Al^{27}	**0.241**
Li^7	0.035	Si	**0.16**
Be^7	$5.4 \cdot 10^4$ (n, p)	Si^{28}	**0.08**
Be^9	0.009	Si^{29}	**0.28**
B	**755**	Si^{30}	0.110
B^{10}	**4017** (n, α)	P^{31}	0.19
	<0.2 (n, p)	S	**0.52**
B^{11}	0.005	S^{32}	0.0018 (n, α)
C	**0.0037**	S^{33}	0.0015 (n, p)
C^{13}	0.0010	S^{34}	0.26
C^{14}	$<2 \cdot 10^2$	S^{36}	0.14
N	**1.88**	Cl	**33.8**
N^{14}	1.75 (n, p)	Cl^{35}	$0.3 \cdot 10^2$
	0.08 (n, γ)		0.19 (n, p)
N^{15}	$2.4 \cdot 10^{-5}$	Cl^{37}	0.6; $<5 \cdot 10^{-5}$ (n, α)
O	$<2 \cdot 10^{-4}$	Ar	**0.66**
O^{17}	0.4 (n, α)	Ar^{36}	6
O^{18}	$2.1 \cdot 10^{-4}$	Ar^{38}	0.8
F^{19}	0.009	Ar^{40}	0.53
Ne	<2.8	Ar^{41}	>0.06
Ne^{22}	0.04	K	**2.07**

K^{39}	**1.9**	Zn	**1.10**
K^{40}	**$0.7 \cdot 10^2$**	Zn^{64}	0.47
	3.8 (n, p)		$1.5 \cdot 10^{-6}$ (n, α)
K^{41}	**1.24**	Zn^{67}	**$6 \cdot 10^{-6}$ (n, α)**
Ca	**0.44**	Zn^{68}	0.099 (14 h)
Ca^{40}	**0.22**		1.0 (55 m)
Ca^{42}	**42**	Zn^{70}	0.08
Ca^{44}	0.72	Ga	**2.8**
Ca^{46}	0.25	Ga^{69}	1.4; **2.1**
Ca^{48}	1.1	Ga^{71}	**5.1**
Sc^{45}	22 (84 d)	Ge	**2.4**
Ti	**5.8**	Ge^{70}	**3.4**
Ti^{46}	**0.6**	Ge^{72}	**0.98**
Ti^{47}	**1.7**	Ge^{73}	**14**
Ti^{48}	**8.3**	Ge^{74}	**0.62**
Ti^{49}	**1.9**		0.040 (49 s)
Ti^{50}	0.14		0.21 (82 m)
V	**5.00**	Ge^{76}	**0.36**
V^{50}	**$3 \cdot 10^2$**		0.08 (54 s)
V^{51}	4.5		0.08 (11 h)
Cr	**3.1**	As^{75}	**4.3**
Cr^{50}	**16**	Se	**12.3**
Cr^{52}	**0.76**	Se^{74}	26; **50**
Cr^{53}	**18**	Se^{76}	**85**
Cr^{54}	0.38		7 (19 s)
Mn^{55}	13.3	Se^{77}	**42**
Fe	**2.62**	Se^{78}	**0.4**
Fe^{54}	**2.3**	Se^{80}	0.030 (57 m)
Fe^{56}	**2.7**		0.50 (18 m)
Fe^{57}	**2.5**	Se^{82}	**2**
Fe^{58}	1.01		0.05 (69 s)
Co	16 (10.5 m)		0.004 (25 m)
	20 (5.2 y)	Br	**6.7**
Co^{60m}	$1.0 \cdot 10^2$	Br^{79}	2.9 (4.5 h)
Co^{60}	6		8.5 (18 m)
Ni	**4.6**	Br^{81}	3.3
Ni^{58}	**4.4**	Kr	**31**
Ni^{60}	**2.6**	Kr^{78}	2.0
Ni^{61}	**2.0**	Kr^{80}	$0.9 \cdot 10^2$
Ni^{62}	**15**	Kr^{82}	$0.5 \cdot 10^2$
Ni^{64}	1.5	Kr^{83}	$2.0 \cdot 10^2$
Ni^{65}	20	Kr^{84}	0.1 (4.5 h)
Cu	**3.85**		0.06 (10 y)
Cu^{63}	**4.5**	Kr^{85}	<15
Cu^{65}	**2.2**	Kr^{86}	0.06
Cu^{66}	$[1.3 \cdot 10^2]$	Kr^{87}	$<6 \cdot 10^2$

Rb	**0.73**	Ag	**63**
Rb^{85}	0.91 (18.7 d)	Ag^{107}	**31, 45**
Rb^{87}	0.12	Ag^{109}	3.2 (253 d)
Rb^{88}	1.0		1.1 · 10^2 (24 s)
Sr	**1.21**	Cd	**2.45 · 10^3**
Sr^{84}	1.4 (65 d)	Cd^{106}	1.0
Sr^{86}	1.6 (2.8 h)	Cd^{110}	0.2 (49 m)
Sr^{88}	0.005	Cd^{112}	0.03 (14 y)
Sr^{89}	[0.5]	Cd^{113}	**2.00 · 10^4**
Sr^{90}	1.0	Cd^{114}	0.14 (43 d)
Y^{89}	**1.31**		1.1 (2.3 d)
Y^{90}	<7	Cd^{116}	1.5 (2.9 h)
Y^{91}	1.07	In	**191**
Zr	**0.185**	In^{113}	56 (50 d)
Zr^{90}	**0.10**		2.0 (72 s)
Zr^{91}	**1.6**	In^{115}	1.55 · 10^2 (54 m)
Zr^{92}	**0.3**		52 (14 s)
Zr^{93}	**<4**	Sn	**0.62**
Zr^{94}	0.078	Sn^{112}	1.3
Zr^{96}	0.053	Sn^{116}	0.006 (14 d)
Nb^{93}	1.0 (6.6 m)	Sn^{118}	0.010 (250 d)
	1.16	Sn^{120}	0.001 (>5 y)
Nb^{94}	15 (35 d)		0.14 (27 h)
Mo	**2.70**	Sn^{122}	0.16 (41 m)
Mo^{92}	<0.006		0.0010 (125 d)
Mo^{95}	**14**	Sn^{124}	0.004 (9.5 d)
Mo^{96}	**1.2**		0.2 (9.7 m)
Mo^{97}	**2.2**	Sb	**5.7**
Mo^{98}	0.51	Sb^{121}	**5.9**
Mo^{100}	0.20		7 (2.7 d)
Tc^{99}	**22**		0.19 (3.5 m)
Ru	**2.6**	Sb^{123}	**4.1**
Ru^{96}	0.21		0.03 (21 m)
Ru^{102}	1.4		0.03 (1.5 m)
Ru^{104}	0.7		2.5 (60 d)
Ru^{105}	0.2	Te	**4.7**
Rh^{103}	12 (4.4 m)	Te^{120}	**0.7 · 10^2**
	1.4 · 10^2 (42 s)	Te^{122}	**2.8**
Rh^{104} (4.4 m)	8 · 10^2		1.1 (104 d)
Rh^{104} (42 s)	0.4 · 10^2	Te^{123}	**4.1 · 10^2**
Pd	**8**	Te^{124}	**6.8**
Pd^{102}	5	Te^{125}	**1.6**
Pd^{108}	10.4 (13.6 h)	Te^{126}	0.09 (105 d)
	0.26 (4.7 m)		0.8 (9.3 h)
Pd^{110}	0.21 (22 m)	Te^{128}	0.015 (33 d)
	<0.05 (5.5 h)		0.13 (72 m)

Te^{130}	<0.008 (30 h)	Nd^{143}	**324**
	0.22 (25 m)	Nd^{144}	**5.0**
I^{127}	5.6; **7.0**	Nd^{145}	**60**
I^{129}	**32;** 24	Nd^{146}	**10;** 1.8 (11.1 d)
I^{130}	18	Nd^{148}	**3.4**
I^{131}	0.6 · 10^2	Nd^{150}	1.5
Xe^{124}	**74**	Pm^{147}	0.6 · 10^2
Xe^{128}	<5	Sm	**5.6 · 10^3**
Xe^{129}	**0.5 · 10^2**	Sm^{144}	<2
Xe^{130}	<5	Sm^{147}	**0.9 · 10^2**
Xe^{131}	**1.2 · 10^2**	Sm^{149}	**4.08 · 10^4**
Xe^{132}	0.2 (5.3 d)	Sm^{151}	[**1.24 · 10^4**]
Xe^{133}	1.9 · 10^2	Sm^{152}	**216**
Xe^{134}	0.2 (9.2 h)	Sm^{154}	5.5
Xe^{135} (9.2 h)	**2.72 · 10^6**	Eu	**4.3 · 10^3**
Xe^{136}	0.15	Eu^{151}	[**1.4 · 10^3** (9.3 h)]
Cs^{133}	3.0 (3 h)		**7.8 · 10^3**
	30 (2.15 y)	Eu^{152} (12.5 y)	[6 · 10^3]
Cs^{134}	134	Eu^{153}	**4.4 · 10^2**
Cs^{135}	8.7	Eu^{154}	[1.5 · 10^3]
Cs^{137}	<2	Eu^{155}	[1.4 · 10^4]
Ba	**1.2**	Gd	**4.60 · 10^4**
Ba^{130}	10	Gd^{152}	<125
Ba^{132}	7 (7.2 y)	Gd^{155}	**5.62 · 10^4**
Ba^{134}	**2**	Gd^{157}	**2.42 · 10^5**
Ba^{135}	**5.8**	Gd^{158}	3.9
Ba^{136}	**0.4**	Gd^{160}	0.8
Ba^{137}	**5.1**	Tb^{159}	**46**
Ba^{138}	0.7	Tb^{160}	[5.3 · 10^2]
Ba^{139}	4	Dy	**9.5 · 10^2**
La	**8.9**	Dy^{158}	1.0 · 10^2
La^{139}	8.2	Dy^{164}	2.0 · 10^3 (1.3 m)
La^{140}	3.1		8.0 · 10^2 (2.3 h)
Ce	**0.73**	Dy^{165}	[5 · 10^3]
Ce^{136}	0.6 (34 h)	Ho^{165}	**65**
	6 (8.7 h)		60 (27 h)
Ce^{138}	**9**	Er	**1.7 · 10^2**
	0.6 (140 d)	Er^{162}	2.0
Ce^{140}	0.31, **0.66**	Er^{164}	1.7
Ce^{142}	0.94	Er^{168}	2.0
Ce^{143}	[6.0]	Er^{170}	9
Pr^{141}	**11.3**	Tm^{169}	**127**
Pr^{142}	[18]	Tm^{170}	1.5 · 10^2
Pr^{143}	89	Yb	**37**
Nd	**46**	Yb^{168}	[1.1 · 10^4]
Nd^{142}	**18**	Yb^{174}	0.6 · 10^2

Yb^{176}	5.5	Au^{198}	$2.6 \cdot 10^4$
Lu	**112**	Au^{199}	$[0.3 \cdot 10^2]$
Lu^{175}	$0.4 \cdot 10^2$ (3.7 h)	Hg	**$3.8 \cdot 10^2$**
Lu^{176}	$4.0 \cdot 10^3$ (6.8 d)	Hg^{196}	$4.2 \cdot 10^2$ (23 h)
Hf	**105**		$9 \cdot 10^2$ (65 h)
Hf^{174}	**$1.5 \cdot 10^3$**	Hg^{198}	0.018 (44 m)
Hf^{176}	**$0.2 \cdot 10^2$**	Hg^{199}	**$[2.5 \cdot 10^3]$**
Hf^{177}	**$3.8 \cdot 10^2$**	Hg^{200}	[<**60**]
Hf^{178}	**75**	Hg^{201}	[<**60**]
Hf^{179}	**$0.6 \cdot 10^2$**	Hg^{202}	3.8
Hf^{180}	10	Hg^{204}	0.43
Ta^{181}	0.030 (16 m)	Tl	**3.4**
	19 (115 d)	Tl^{203}	**11.4; 8**
Ta^{182}	$[1.7 \cdot 10^4]$	Tl^{205}	**0.80**
W	**19.2**		0.10 (4.3 m)
W^{180}	10	Pb	**0.170**
W^{182}	**20**	Pb^{204}	**0.8**
W^{183}	**11**		0.7 ($3 \cdot 10^7$ y)
W^{184}	2.2 (74 d)	Pb^{206}	**0.025**
	2.0	Pb^{207}	**0.70**
W^{186}	**34**	Pb^{208}	0.0006
W^{187}	$0.9 \cdot 10^2$	Bi^{209}	**0.034**
Re	**86**		0.019 (5 d)
Re^{185}	**104**	Rn^{220}	[<0.2]
Re^{187}	**66**	Rn^{222}	[0.72]
	69 (17 h)	Ra^{223}	$[1.3 \cdot 10^2]$
Re^{188}	<2		$<1 \cdot 10^2$ (n, f)
Os	**15.3**	Ra^{224}	[12.0]
Os^{184}	$<2 \cdot 10^2$	Ra^{226}	[20]
Os^{190}	8 (15 d)		$<1 \cdot 10^{-4}$ (n, f)
Os^{192}	1.6	Ra^{228}	[36]
Os^{193}	$6 \cdot 10^2$		<2 (n, f)
Ir	**$4.4 \cdot 10^2$**	Ac^{227}	$8.0 \cdot 10^2$
Ir^{191}	$2.6 \cdot 10^2$ (1.4 m)		<2 (n, f)
	$7 \cdot 10^2$ (74 d)	Th^{227}	$1.5 \cdot 10^3$ (n, f)
Ir^{192} (74 d)	**$7 \cdot 10^2$**	Th^{228}	$[1.2 \cdot 10^2]$
Ir^{193}	$1.3 \cdot 10^2$ (19 h)		≤ 0.3 (n, f)
Pt	**8.8**	Th^{229}	45 (n, f)
Pt^{190}	**$1.5 \cdot 10^2$**	Th^{230}	21.4
Pt^{192}	$0.9 \cdot 10^2$ (4.4 d)		$\leq 1 \cdot 10^{-3}$ (n, f)
Pt^{194}	**1.2**	Th^{232}	**7.3; 7.6**
Pt^{195}	**27**		$<2 \cdot 10^{-4}$ (n, f)
Pt^{196}	0.87 (20 h)	Th^{233}	$[1.4 \cdot 10^3]$
Pt^{198}	3.9		[15] (n, f)
Pt^{199}	15	Th^{234}	[1.8]
Au^{197}	**98.8**		$<1 \cdot 10^{-2}$ (n, f)

Pa^{230}	$1.5 \cdot 10^3$ (n,f)	Pu^{239}	**$1.03 \cdot 10^3$**; 286
Pa^{231}	$2.0 \cdot 10^2$		742 (n,f)
	10 (n,f)	Pu^{240}	**$2.86 \cdot 10^2$**; $2.5 \cdot 10^2$
Pa^{232}	$[7.6 \cdot 10^2]$		0.03 (n,f)
	$7.0 \cdot 10^2$ (n,f)	Pu^{241}	**$1.40 \cdot 10^3$**; $[3.9 \cdot 10^2]$
Pa^{233}	29 (1.18 m)		$1.01 \cdot 10^3$ (n,f)
	22 (6.7 h)	Pu^{242}	**30**; 19
	<0.1 (n,f)		<0.2 (n,f)
Pa^{234} (1.18 m)	$\leq 5 \cdot 10^2$ (n,f)	Pu^{243}	$[1.7 \cdot 10^2]$
Pa^{234} (6.7 h)	$\leq 5 \cdot 10^3$ (n,f)	Pu^{244}	$[1.8]$
U	**7.68**	Pu^{245}	$[2.6 \cdot 10^2]$
	4.18 (n,f)	Am^{241}	**$6.3 \cdot 10^2$**
U^{230}	25 (n,f)		$[1.5 \cdot 10^2]$ (16.0 h)
U^{231}	$4 \cdot 10^2$ (n,f)		$[0.5 \cdot 10^2]$ (150 y)
U^{232}	$[3 \cdot 10^2]$		3.1 (n,f)
	$0.8 \cdot 10^2$ (n,f)	Am^{242m}	$2.50 \cdot 10^3$ (n,f)
U^{233}	**578**; 53	Am^{242}	**$[8.0 \cdot 10^3]$**
	525 (n,f)		$[6.4 \cdot 10^3]$ (n,f)
U^{234}	**105**; $0.9 \cdot 10^2$	Am^{243}	74
	≤ 0.65 (n,f)		$<7.5 \cdot 10^{-2}$ (n,f)
U^{235}	**678**, 101	Cm^{242}	$[20]$
	577 (n,f)		$[<5]$ (n,f)
U^{236}	**7**; 6	Cm^{243}	$[2.5 \cdot 10^2]$
U^{238}	**2.71**		$7.0 \cdot 10^2$ (n,f)
	$<5 \cdot 10^{-4}$ (n,f)	Cm^{244}	$[15]$
U^{239}	$[22]$	Cm^{245}	$[2.0 \cdot 10^2]$
	$[14]$ (n,f)		$1.9 \cdot 10^3$ (n,f)
Np^{234}	$9 \cdot 10^2$ (n,f)	Cm^{246}	$[15]$
Np^{236}	$2.8 \cdot 10^3$ (n,f)	Cm^{248}	$[6]$
Np^{237}	**170**; $1.9 \cdot 10^{-2}$ (n,f)	Bk^{249}	$[5.0 \cdot 10^2]$
Np^{238}	$1.60 \cdot 10^3$ (n,f)	Cf^{249}	**$[9.0 \cdot 10^2]$**; $[2.7 \cdot 10^2]$
Np^{239}	$[35]$ (7.3 m)		$[6.0 \cdot 10^2]$ (n,f)
	$[0.25 \cdot 10^2]$ (63 m)	Cf^{250}	$[1.5 \cdot 10^3]$
	<1 (n,f)	Cf^{251}	$[3 \cdot 10^3]$
Pu^{236}	$1.7 \cdot 10^2$ (n,f)	Cf^{252}	28
Pu^{237}	$2.5 \cdot 10^3$ (n,f)	Cf^{254}	$[<2]$
Pu^{238}	403	Es^{253}	$[3 \cdot 10^2]$
	16.8 (n,f)	Es^{254}	**$[2.7 \cdot 10^3]$**; $[<40]$

Appendix D.   References to Tabulations of Cross Sections
for Nuclear Reactions

The following list of references, although not exhaustive, will yield most of the data that are presently available on cross sections for various nuclear reactions.

A. Reactions induced by charged particles with energies below 50 MeV.

1. "Charged Particle Cross Sections; Hydrogen to Fluorine," N. Jarmie, J. Seagrave, and H. V. Argo, AEC Publication LA-2014 (1956), Office of Technical Services, Department of Commerce, Washington 25, D. C.
Experimental values of cross sections for reactions between charged particles with energies up to 30 MeV and targets between hydrogen and fluorine are presented in graphical form.

2. "Charged Particle Cross Sections; Neon to Chromium," D. B. Smith, N. Jarmie, J. D. Seagrave; AEC Publication LA-2424 (1960), Office of Technical Services, Department of Commerce, Washington 25, D. C.
Measured cross sections for all nuclear processes involving charged particles and targets from neon through chromium are presented in graphical form.

3. "Activation Cross-Section Survey of Deuteron-Induced Reactions," N. Baron and B. L. Cohen, *Phys. Rev.* **129,** 2636 (1963).
Tables of cross sections for various reactions between 18-20 MeV deuterons and various targets.

4. "Excitation Functions and Cross Sections," O. U. Anders and W. W. Meinke, Document No. ADI-4999, ADI Auxiliary Publications Project, Photoduplication Service, Library of Congress, Washington 25, D. C.
This material concerns excitation functions and, when available, cross sections for charged-particle reactions.

B. Reactions induced by neutrons with energies below 50 MeV.

1. "Neutron Cross Sections," D. J. Hughes and R. B. Schwartz, BNL-325, U. S. Government Printing Office, Washington, D. C., 1958.
Cross sections for (n, p), $(n, 2n)$, (n, α), and (n, γ) reactions with neutrons of various energies are given, mostly in graphical form.   New editions and supplements of this very complete and critical compilation are published periodically.

2. "Neutron Cross Sections," R. J. Howerton, 1959 Nuclear Data Tables (K. Way, Editor), Washington, D. C., National Academy of Sciences-National Research Council, 1959.
Cross sections for a wide variety of nuclear reactions induced by neutrons with a variety of energies.

C. Reactions induced by π mesons of various energies and by neutrons, protons, deuterons, and α particles with energies above 50 MeV.

1. "High-Energy Nuclear Reaction Cross Sections," Volumes I and II.   E. Bruninx, CERN Publications 61-1 and 62-9, Nuclear Physics Division; Geneva, Switzerland.

Tables of cross sections for reactions induced by particles with energies above 100 MeV.

2. "High-Energy Nuclear Reactions," J. M. Miller and J. Hudis, *Ann. Rev. Nuclear Sci.,* **9,** 159–202 (1959).

This review article contains in tabular form a bibliography of investigations of nuclear reactions induced by neutrons, protons, deuterons, and α particles at energies above 50 MeV, and by π mesons.

D. Thick-target yields

The following references contain compilations of yields in units of microcuries per microampere hour for the production of various radioactive isotopes in bombardments of appropriate thick targets with deuterons and alpha particles.   The targets are thick enough to stop the incident particles.

1. "Methods of Producing Radioiron," R. W. Dunn, *Nucleonics,* **10,** 8 (1952).

2. "Production and Isolation of Carrier-Free Radioisotopes," W. M. Garrison and J. G. Hamilton, *Chem. Revs.* **49,** 237 (1951).

3. "Production of Radionuclides," J. W. Irvine, Jr., *Nucleonics,* **3,** 5 (1948).

4. "Experimental Yields with 14-MeV Deuterons," E. T. Clarke and J. W. Irvine, *Phys. Rev.* **70,** 893 (1946).

Appendix E.   Table of Nuclides

In this table are listed all the stable and radioactive nuclides whose existence has been fairly reliably established, together with the best available published information on some of their characteristics.   Most of the data are taken from the "Nuclear Data Sheets," edited by the Nuclear Data Group of the National Academy of Sciences—National Research Council, Washington, D. C., through Vol. 5, No. 3.   Additional published data that have appeared in the literature up to summer 1963 are included.

Column 1 gives the chemical symbol and, as a superscript, the mass number of each nuclide.   For the first listed isotope of each element the atomic number is shown as a subscript.   Uncertainty in the mass assignment is indicated by a question mark in the superscript after the mass number.   A question mark after the symbol *and* mass number indicates uncertainty in the chemical identification of the activity.   An upper isomeric state of half-life greater than 10^{-3} sec is indicated by the symbol m following the mass number.   Metastable states of shorter half-life are not listed as separate entries.   If at a given Z and A there are two or more excited isomeric states, they are labeled m_1, m_2, etc., in order of increasing excitation energy.

Column 2 lists per cent isotopic abundances for the naturally occurring nuclides.   Column 3 gives isotopic masses, that is, atomic masses for the individual nuclides in atomic mass units (C^{12} scale).   The uncertainty in the last digit of any mass value listed is ≤ 5.   The mass values are largely taken from L. A. König, J. H. E. Mattauch, and A. H. Wapstra, "1961 Nuclidic Mass Table," *Nuclear Phys.*, **31**, 18(1962).

Directly measured nuclear spins are listed in column 4.   Spins obtained by paramagnetic resonance, nuclear magnetic resonance, and microwave spectroscopy methods are included; those inferred from radioactive decay processes (transition probabilities, spectrum shapes, angular correlations) are not.

Half-lives are given in column 5.   An attempt has been made to select a best value for each nuclide.   The symbols used in this column are y = years, d = days, h = hours, m = minutes, s = seconds.

Column 6 lists information on modes of decay, radiations emitted and their energies.   The symbols used are

α	alpha decay,
β^-	negatron (negative beta) decay,
β^+	positron (positive beta) decay,
EC	electron capture,
IT	isomeric transition,
SF	spontaneous fission
γ	gamma rays,
e^-	internal-conversion electrons,
n	neutrons.

For a given nuclide, different decay modes or types of radiation are listed in order of decreasing abundance when the relative abundances are known. Those modes known to occur in less than 10 per cent abundance are shown in parentheses.   The numbers following the symbols are the measured energies of the radiations in millions of electron volts; for β^- and β^+ decays the energies listed are the maximum β-particle kinetic energies for each transition.   The energies for each decay mode are given in order of decreasing abundance, and only those that occur in more than 10 per cent of the disintegrations are listed (except for an occasional, less abundant transition shown in parentheses).   In general, no more than three or four energies are shown for a given type of radiation.   When a greater variety of radiations is known, only the most abundant ones are listed, with the energy range sometimes stated for the others.   Sometimes only the notation "(others)" appears, indicating additional rays of the specified type, but each of abundance <10 per cent.

For gamma transitions the energy given is always the transition energy even if conversion electrons only are observed.   The approximate magnitude of internal conversion coefficients is sometimes indicated, especially for isomeric transitions, by the order and manner in which the symbols γ and e^- are written:

$\gamma(e^-)$	means internal conversion coefficient $\alpha = e^-/\gamma < 0.1$,
γe^-	means $0.1 < \alpha < 1$,
$e^-\gamma$	means $1 < \alpha < 10$,
$e^-(\gamma)$	means $\alpha > 10$.

For example, the notation IT $\gamma(e^-)$ 0.44 for Zn^{69m} indicates that the isomeric transition takes place largely by the emission of unconverted 0.44-MeV γ rays, the conversion coefficient being <0.1.

The letter (m) following a symbol or energy in column 6 indicates that the particular radiation is associated with the decay of a short-lived (10^{-9} to 10^{-3}s) isomeric state in radioactive equilibrium with the parent activity.

If a nuclide decays to one of a pair of isomers but not to the other, this is indicated.   For example, for Fe^{52} the notation "(to Mn^{52m})" indicates that the decay (by EC and β^+) goes to Mn^{52m}, but not to Mn^{52}.

Nuclide	Per Cent Abundance	Nuclidic Mass	Spin	Half-life	Decay Modes, Radiations, Energies
$_0n^1$		1.0086654	1/2	12.8m	β^- 0.782
$_1H^1$	99.9844–99.9867	1.0078252	1/2		
H^2	0.0133–0.0156	2.0141022	1		
H^3		3.0160494	1/2	12.26y	β^- 0.0186
$_2He^3$	1.3×10^{-4} (atm)	3.0160299	1/2		
	1.7×10^{-5} (well)				
He^4	~100	4.0026036	0		
He^6		6.01890	0	0.82s	β^- 3.51
$_3Li^6$	7.42	6.015126	1		
Li^7	92.58	7.016005	3/2		
Li^8		8.022488		0.84s	β^- 13; 2α
Li^9		9.028		0.17s	β^- ~8; n; 2α
$_4Be^7$		7.016931		53.6d	EC; γ 0.477
Be^8		8.005308		10^{-16}s	2α 0.047
Be^9	100	9.012186	3/2		
Be^{10}		10.013535		2.5×10^6y	β^- 0.555
Be^{11}		11.02166		13.6s	β^- 11.5, 9.3
$_5B^8$		8.024612		0.78s	β^+ 13.7; 2α
B^{10}	19.61	10.012939	3		
B^{11}	80.39	11.0093051	3/2		
B^{12}		12.014353		0.020s	β^- 13.37; (α, γ)
B^{13}		13.017779		~0.035s	β^-
$_6C^{10}$		10.01683		19.1s	β^+ 2.1; γ 0.72
C^{11}		11.011433	3/2	20.4m	β^+ 0.99
C^{12}	98.893	12.000000	0		
C^{13}	1.107	13.003354	1/2		
C^{14}		14.0032419	0	5720y	β^- 0.155
C^{15}		15.010600		2.4s	β^- 4.51, 9.82; γ 5.30
C^{16}		16.01470		0.74s	β^-; n
$_7N^{12}$		12.01871		0.011s	β^+ 16.6; (3α)
N^{13}		13.005739	1/2	10.0m	β^+ 1.19
N^{14}	99.634	14.0030744	1		
N^{15}	0.366	15.000108	1/2		
N^{16}		16.00609		7.38s	β^- 4.26, 10.4, 3.3; γ 6.13 (others)
N^{17}		17.00845		4.14s	β^- 3.7; n 0.9
$_8O^{14}$		14.008597		72s	β^+ 1.83; $^+$ 2.3
O^{15}		15.003072	1/2	2.0m	β^+ 1.72
O^{16}	99.759	15.9949149	0		
O^{17}	0.0374	16.999133	5/2		
O^{18}	0.2039	17.9991598	0		
O^{19}		19.003577		29.4s	β^- 3.25, 4.60; γ 0.20(m), 1.37
O^{20}		20.00407		13.6s	β^- 2.69; γ 1.06
$_9F^{17}$		17.002098		66s	β^+ 1.75
F^{18}		18.000950		110m	β^+ 0.649
F^{19}	100	18.998405	1/2		
F^{20}		19.999985		11.5s	β^- 5.41; γ 1.63
F^{21}		20.99997		4.32s	β^- 6.1, 4.4, 5.8; γ 0.35, 1.39
$_{10}Ne^{18}$		18.00572		1.5s	β^+ 3.2; (γ)
Ne^{19}		19.001892		17s	β^+ 2.24
Ne^{20}	90.92	19.9924404	0		
Ne^{21}	0.257	20.993849	3/2		
Ne^{22}	8.82	21.991385	0		
Ne^{23}		22.994475		38s	β^- 4.39, 3.95; γ 0.438 (others)
Ne^{24}		23.99360		3.4m	β^- 1.98; γ 0.472 (m), (0.878)

Nuclide	Per Cent Abundance	Nuclidic Mass	Spin	Half-life	Decay Modes, Radiations, Energies
$_{11}Na^{20}$		20.0089		0.38s	$\beta^+; \alpha > 2$
Na^{21}		20.99764		22s	β^+ 2.51; (γ)
Na^{22}		21.994435	3	2.58y	β^+ 0.544; EC; γ 1.274
Na^{23}	100	22.989773	3/2		
Na^{24}		23.990967	4	15.0h	β^- 1.39; γ 1.368, 2.753
Na^{24m}				0.02s	IT 0.472; $\beta^- \sim 6$
Na^{25}		24.9899		60s	β^- 3.8, 2.8; γ 0.98, 0.58, 0.40
Na^{26}		25.9917		1.0s	β^- 6.7; γ 1.82
$_{12}Mg^{22}$				3.9s	$\beta^+; \gamma$ 0.074, 0.59
Mg^{23}		22.99414		12s	β^+ 3.09, (2.64); $(\gamma$ 0.44)
Mg^{24}	78.70	23.985045	0		
Mg^{25}	10.13	24.985840	5/2		
Mg^{26}	11.17	25.982591	0		
Mg^{27}		26.984345		9.5m	β^- 1.75, 1.59; γ 0.834, 1.015 (others)
Mg^{28}		27.98388		21.3h	β^- 0.46; γ 0.032, 1.35, 0.40, 0.95
$_{13}Al^{24}$		24.0001		2.1s	β^+ 8.7, 4.5; γ 1.39, 2.73, 4.22 (others)
Al^{25}		24.99041		7.3s	β^+ 3.24; (γ)
Al^{26}		25.986900	5	7.4×10^5y	β^+ 1.16; (EC); γ 1.83 (others)
Al^{26m}			0	6.4s	β^+ 3.21
Al^{27}	100	26.981535	5/2		
Al^{28}		27.981908		2.3m	β^- 2.87; γ 1.78
Al^{29}		28.98044		6.6m	β^- 2.5, 1.3; γ 1.28, 2.43
Al^{30}		29.9812		3.3s	β^- 5.05; γ 2.26, 3.52
$_{14}Si^{26}$		25.9923		2.1s	β^+ 3.76, 2.94; γ 0.82
Si^{27}		26.98670		4.2s	β^+ 3.85; (γ)
Si^{28}	92.21	27.976927	0		
Si^{29}	4.70	28.976491	1/2		
Si^{30}	3.09	29.973761	0		
Si^{31}		30.975349		2.62h	β^- 1.48; (γ)
Si^{32}		31.97402		710y	$\beta^- \sim 0.1$
$_{15}P^{28}$		27.9917		0.28s	β^+ 10.6, others; γ 1.8–7.6
P^{29}		28.98182		4.3s	β^+ 3.95; (γ)
P^{30}		29.97832		2.50m	β^+ 3.24; (γ)
P^{31}	100	30.973763	1/2		
P^{32}		31.973908	1	14.3d	β^- 1.71
P^{33}				25d	β^- 0.24
P^{34}		33.9733		12.4s	β^- 5.1, 3.0; γ 2.1
$_{16}S^{30}$		29.9847		1.4s	β^+ 4.30, 4.98; γ 0.68
S^{31}		30.97960		2.6s	β^+ 4.39; (γ)
S^{32}	95.0	31.972074	0		
S^{33}	0.760	32.971460	3/2		
S^{34}	4.22	33.967864	0		
S^{35}		34.969034	3/2	87d	β^- 0.167
S^{36}	0.014	35.96709	0		
S^{37}		36.9710		5.1m	β^- 1.6, 4.8; γ 3.1
S^{38}		37.9712		2.87h	β^- 1.1; γ 1.88
$_{17}Cl^{32}$		31.9860		0.31s	β^+ 9.5, 5, 8.2; γ 2.2, 4.8 (others)
Cl^{33}		32.97745		2.6s	β^+ 4.5; (γ)
Cl^{34}		33.97376		1.6s	β^+ 4.4
Cl^{34m}				32.4m	β^+ 2.41, 1.24; IT γe^- 0.145; γ 2.14, 3.32, 1.16
Cl^{35}	75.53	34.968854	3/2		
Cl^{36}		35.96831	2	3.0×10^5y	β^- 0.714; (EC); (β^+)
Cl^{37}	24.47	36.965896	3/2		
Cl^{38}		37.96800		37.3m	β^- 4.81, 1.11, 2.77; γ 2.15, 1.60
Cl^{38m}				1s	IT γ 0.66

Nuclide	Per Cent Abundance	Nuclidic Mass	Spin	Half-life	Decay Modes, Radiations, Energies
Cl³⁹		38.96800		56m	β^- 1.91 (others); γ 1.27, 0.25, 1.52
Cl⁴⁰		39.9704		1.4m	β^- ~3.2, ~7.5; γ 1.46, 2.75, 6.0
₁₈Ar³⁵		34.97528		1.84s	β^+ 4.95; (γ)
Ar³⁶	0.337	35.967548	0		
Ar³⁷		36.966772	3/2	35.0d	EC
Ar³⁸	0.063	37.962725	0		
Ar³⁹		38.96432		270y	β^- 0.565
Ar⁴⁰	99.600	39.962384	0		
Ar⁴¹		40.96451		1.83h	β^- 1.20; γ 1.29
Ar⁴²		41.96304		34y	β^-
₁₉K³⁷		36.97336		1.2s	β^+ 5.1
K³⁸		37.96909		7.7m	β^+ 2.68; γ 2.16
K³⁸ᵐ				0.95s	β^+ 5.0
K³⁹	93.10	38.963714	3/2		
K⁴⁰	0.0118	39.964008	4	1.27 × 10⁹y	β^- 1.32; EC; γ 1.46; (β^+)
K⁴¹	6.88	40.961835	3/2		
K⁴²		41.96242	2	12.36h	β^- 3.55, 1.98; γ 1.52
K⁴³		42.96073	3/2	22.4h	β^- 0.83 (others); γ 0.619, 0.374 (others)
K⁴⁴		43.9620		22m	β^- 2.63, 4.9; γ 1.16, others
K⁴⁵?				34m	β^-
₂₀Ca³⁸		37.9758		0.66s	β^+; γ 3.5
Ca³⁹		38.97071		0.88s	β^+ 5.5
Ca⁴⁰	96.97	39.962589	0		
Ca⁴¹		40.06228	7/2	1.1 × 10⁵y	EC
Ca⁴²	0.64	41.958628			
Ca⁴³	0.145	42.958780	7/2		
Ca⁴⁴	2.06	43.955490			
Ca⁴⁵		44.956189		165d	β^- 0.255
Ca⁴⁶	0.0033	45.95369			
Ca⁴⁷		46.95451		4.5d	β^- 1.94, 0.66; γ 1.31 (others)
Ca⁴⁸	0.185	47.95236			
Ca⁴⁹		48.95566		8.8m	β^- 1.95, 0.89; γ 3.10, 4.05
₂₁Sc⁴⁰		39.9775		0.2s	β^+ 9.2; γ 3.75
Sc⁴¹		40.96925		0.55s	β^+ 5.6
Sc⁴²		41.96551		0.68s	β^+ 5.4
Sc⁴²ᵐ		41.96623		62s	β^+ 2.87; γ 0.44, 1.23, 1.52
Sc⁴³		42.96116	7/2	3.9h	β^+ 1.18, 0.82; EC; γ 0.37
Sc⁴⁴		43.95941	2	3.9h	β^+ 1.47; γ 1.16 (others); (EC)
Sc⁴⁴ᵐ		43.95970	6	2.44d	IT γe^- 0.27; (EC; γ)
Sc⁴⁵	100	44.955919	7/2		
Sc⁴⁶		45.95517	4	84d	β^- 0.357; γ 1.119, 0.887
Sc⁴⁶ᵐ		45.95532		20s	IT γe^- 0.142
Sc⁴⁷		46.95240		3.44d	β^- 0.44, 0.60; γ 0.160
Sc⁴⁸		47.95223		1.83d	β^- 0.65; γ 1.04, 1.31, 0.99, (0.18)
Sc⁴⁹		48.95002		57m	β^- 2.0; (γ)
Sc⁵⁰		49.9516		1.8m	β^- ~3.5; γ 1.17, 1.59, 0.51
Sc⁵⁰ᵐ				0.35s	IT γ 0.26
₂₂Ti⁴³		42.96850		0.6s	β^+ 5.8
Ti⁴⁴		43.95957		~50y	EC (to 3.9h Sc⁴⁴); γ 0.068, 0.078
Ti⁴⁵		44.95813	7/2	3.08h	β^+ 1.02; EC; (γ)
Ti⁴⁵ᵐ				0.006s	IT 0.28
Ti⁴⁶	7.93	45.952633			
Ti⁴⁷	7.28	46.95176	5/2		
Ti⁴⁸	73.94	47.947948			
Ti⁴⁹	5.51	48.947867	7/2		
Ti⁵⁰	5.34	49.944789			
Ti⁵¹		50.94662		5.9m	β^- 2.14; γ 0.32 (others)
Ti⁵²				12m	β^-

Nuclide	Per Cent Abundance	Nuclidic Mass	Spin	Half-life	Decay Modes, Radiations, Energies
$_{23}V^{46}?$		45.96023		0.4s	β^+ 6.05
V^{47}		46.95489		31m	β^+ 1.89; (γ)
V^{48}		47.95226		16.1d	β^+ 0.69; EC; γ 1.31, 0.99 0.95, 2.23
V^{49}		48.94852	7/2	330d	EC
V^{50}	0.24	49.947165	6	6×10^{15}y	β^- 0.4; EC; γ 0.78, 1.59
V^{51}	99.76	50.943978	7/2		
V^{52}		51.94480		3.77m	β^- 2.6; γ 1.44
V^{53}		52.94337		2m	β^- 2.5; γ 1.00, (1.29)
V^{54}				55s	β^- 3.3; γ 0.84, 0.99, 2.2
$_{24}Cr^{46}?$				1.1s	β^+
Cr^{48}		47.9538		23h	EC; γ 0.116, 0.305 (m)
Cr^{49}		48.95127		41.7m	β^+ 1.54, 1.39, 1.46; γ 0.089, 0.063, 0.15; (EC)
Cr^{50}	4.31	49.946051			
Cr^{51}		50.944786	7/2	27.8d	EC; $(\gamma$ 0.32$)$
Cr^{52}	83.76	51.940514			
Cr^{53}	9.55	52.940651	3/2		
Cr^{54}	2.38	53.938879			
Cr^{55}		54.9411		3.5m	β^- 2.85
Cr^{56}		55.9406		5.9m	β^- 1.50; γ 0.026, 0.083
$_{25}Mn^{50}$		49.9540		0.28s	β^+ 6.58
Mn^{50}				2.0m	β^+; γ 0.66–1.45
Mn^{51}		50.94820		45m	β^+ 2.2; (γ)
Mn^{52}		51.94556	6	5.7d	EC; β^+ 0.57; γ 0.747, 0.938, 1.434 (others)
Mn^{52m}		51.94597		21m	β^+ 2.63; γ 1.43 (others?); (IT 0.38)
Mn^{53}		52.94129	7/2	$\sim 2 \times 10^6$y	EC
Mn^{54}		53.94036	3	280d	EC; γ 0.835
Mn^{55}	100	54.938054	5/2		
Mn^{56}		55.93891	3	2.58h	β^- 2.84; 1.03, 0.72; γ 0.85, 1.81, 2.12 (others)
Mn^{57}		56.9383		1.7m	β^- 2.6; γ 0.12, 0.13
Mn^{58}				1.1m	β^-; γ 0.36–2.8
$_{26}Fe^{52}$		51.94812		8.3h	β^+ 0.80, EC (both to Mn^{52m}); γ 0.17
Fe^{53}		52.94558		8.9m	β^+ 2.8, 2.4, 1.6; γ 0.38 (others)
Fe^{54}	5.82	53.93962			
Fe^{55}		54.938302		2.60y	EC
Fe^{56}	91.66	55.93493			
Fe^{57}	2.19	56.93539	1/2		
Fe^{58}	0.33	57.93327			
Fe^{59}		58.93487		45d	β^- 0.46, 0.27; γ 1.10, 1.29 (others)
Fe^{60}				$\sim 1 \times 10^5$y	β^- 0.14 (to Co^{60m}); γ 0.027
Fe^{61}				6.0m	β^- 2.8; γ 0.29
$_{27}Co^{54}$		53.9484		0.19s	β^+ 7.34
Co^{54m}				1.5m	β^+ 4.5; γ 0.41, 1.1, 1.4
Co^{55}		54.94202	7/2	18h	β^+ 1.50, 1.03; EC; γ 0.94, 1.41, 0.48, (0.25)
Co^{56}		55.93987	4	77d	EC; β^+ 1.46; γ 0.845, 1.24, others 0.98–3.5
Co^{57}		56.93629	7/2	270d	EC; $\gamma(e^-)$ 0.122, $e^-(\gamma)$ 0.0144(m), $(\gamma$ 0.136$)$
Co^{58}		57.93575		71d	EC; β^+ 0.48; γ 0.81 (others)
Co^{58m}		57.93578		9.0h	IT $e^-(\gamma)$ 0.025
Co^{59}	100	58.933189	7/2		
Co^{60}		59.93381	5	5.26y	β^- 0.32; γ 1.173, 1.333
Co^{60m}		59.93387		10.5m	IT $e^-(\gamma)$ 0.059; (β^-)
Co^{61}		60.93243		1.65h	β^- 1.22; γ 0.07
Co^{62}		61.93395		13.9m	β^- 2.88, 0.88; γ 1.17, others
Co^{62}				1.7m	β^-; γ
Co^{63}				1.4h	β^-; γ

Nuclide	Per Cent Abundance	Nuclidic Mass	Spin	Half-life	Decay Modes, Radiations, Energies
Co^{63}				52s	β^- 3.6
Co^{64}				7.8m	β^-, γ 0.66, 0.97, 1.34 (others)
Co^{64m}				2m	γ
$_{28}$Ni^{56}				6.1d	EC; γ 0.164, 0.820, 0.755, 0.27, 0.48, 1.57 (others)
Ni^{57}		56.93976		37h	EC; β^+ 0.85 (others); γ 1.37, 0.13. 1.90 (others)
Ni^{58}	67.88	57.93534			
Ni^{59}		58.934344		8×10^4y	EC
Ni^{60}	26.23	59.93078			
Ni^{61}	1.19	60.93105	3/2		
Ni^{62}	3.66	61.92835			
Ni^{63}		62.92967		92y	β^- 0.067
Ni^{64}	1.08	63.92796			
Ni^{65}		64.93004		2.56h	β^- 2.10, 0.60, 1.01; γ 1.46, 1.11, 0.37 (others)
Ni^{66}		65.92909		55h	β^- 0.20
$_{29}$Cu^{58}		57.94447		3.2s	β^+ 7.48, 4.6; γ 1.45, 2.9
Cu^{59}		58.93950		81s	β^+ 3.7 (others); γ 1.31, 0.87, others
Cu^{60}		59.93738	2	24m	β^+ 2.00, 3.00, 3.9; (EC); γ 1.33, 1.76, 0.85 (others)
Cu^{61}		60.93344	3/2	3.32h	β^+ 1.22 (others); EC; γ 0.28, 0.66 (others)
Cu^{62}		61.93256	1	9.9m	β^+ 2.91; (EC); (γ)
Cu^{63}	69.09	62.92959	3/2		
Cu^{64}		63.92976	1	12.8h	EC; β^- 0.573; β^+ 0.656; (γ 1.34)
Cu^{65}	30.91	64.92779	3/2		
Cu^{66}		65.92887		5.1m	β^- 2.63, (1.59); (γ 1.04)
Cu^{67}		66.92776		61h	β^- 0.40, 0.48, 0.58; γ 0.18, 0.092 (others)
Cu^{68}				32s	β^- 3.0; γ 1.08–2.32
$_{30}$Zn^{60}				2.1m	β^+
Zn^{61}		60.9392		1.48m	β^+ 4.38, 3.9 (others); γ 0.48 (others)
Zn^{62}		61.93438		9.3h	EC; β^+ 0.67; γ 0.59, 0.042, 0.51 (others)
Zn^{63}		62.93321		38m	β^+ 2.34 (others); (EC); (γ 0.67, others)
Zn^{64}	48.89	63.929145	0		
Zn^{65}		64.92923	5/2	245d	EC; γ 1.11; (β^+ 0.33)
Zn^{66}	27.81	65.92605	0		
Zn^{67}	4.11	66.92715	5/2		
Zn^{68}	18.57	67.92486	0		
Zn^{69}		68.92665		55m	β^- 0.90
Zn^{69m}		68.92712		13.8h	IT $\gamma(e^-)$ 0.44
Zn^{70}	0.62	69.92535			
Zn^{71}		70.9280		2.4m	β^- 2.6, 2.1 (others); γ 0.51 (others)
Zn^{71m}				4.1h	β^- 1.5; γ 0.38, 0.49, 0.61
Zn^{72}				49h	β^- 0.30; γ 0.14, (0.19)
$_{31}$Ga^{64}		63.93674		2.6m	β^+ 6.1, 2.8; γ 0.98, 3.3, 0.80, 1.3, 2.2 (others)
Ga^{65}		64.93273		15m	β^+ 2.11, 1.39, 2.24, 0.82; EC; γ 0.054(m), 0.115 (others 0.09–2.33)
Ga^{66}		65.93160	0	9.5h	β^+ 4.16 (others); EC; γ 1.04, 2.75 (others)
Ga^{67}		66.92822	3/2	78h	EC; γe^- 0.093(m), γ 0.184, 0.091, 0.296 (others)
Ga^{68}		67.92800	1	68m	β^+ 1.89; EC; (γ)
Ga^{69}	60.4	68.92568	3/2		
Ga^{70}		69.92605	1	21m	β^- 1.6 (others); (γ)
Ga^{70m}				0.02s	IT 0.19

Nuclide	Per Cent Abundance	Nuclidic Mass	Spin	Half-life	Decay Modes, Radiations, Energies
Ga^{71}	39.6	70.92484	3/2		
Ga^{72}		71.92603	3	14.1h	β^- 0.64, 0.96, 0.56, 1.51; γ 0.835, 2.20 (others)
Ga^{72m}				0.04s	IT 0.10
Ga^{73}		72.9250		4.8h	β^- 1.19 (to 0.53s Ge^{73m}); γ 0.30
Ga^{74}		73.9272		8m	β^- 2.7, 1.1 (others); γ 0.60, 2.35 (others)
Ga^{75}				2m	β^- 3.3; γ 0.58 (others)
Ga^{76}				32s	β^- ~6; γ 0.57 (others)
$_{32}$Ge^{65}		64.9378		1.5m	β^+ 3.7; (γ 0.67, 1.7)
Ge^{66}		65.9348		2.7h	β^+ 1.3; EC; γ 0.046–0.71
Ge^{67}		66.9329		19m	β^+ 3.2, 2.3, 1.6; γ 0.170 (others)
Ge^{68}		67.929		280d	EC
Ge^{69}		68.92808		40h	EC; β^+ 1.22 (others); γ 1.12, 0.58, 0.88 (others)
Ge^{70}	20.52	69.92428	0		
Ge^{71}		70.92509	1/2	11d	EC
Ge^{71m}				0.020s	IT 0.023; γ 0.17 (m)
Ge^{72}	27.43	71.92174	0		
Ge^{73}	7.76	72.9234	9/2		
Ge^{73m_2}				0.53s	IT $e^-\gamma$ 0.054; $e^-(\gamma)$ 0.013 (m_1)
Ge^{74}	36.54	73.92115	0		
Ge^{75}		74.9228		82m	β^- 1.19, 0.92; γ 0.265, 0.20 (others)
Ge^{75m}				49s	IT $e^-\gamma$ 0.138
Ge^{76}	7.76	75.9214	0		
Ge^{77}		76.9236		11.3h	β^- 2.20, 1.38, 0.71; γ 0.042–2.32
Ge^{77m}				54s	β^- 2.90, 2.7; IT γe^- 0.159; γ 0.215
Ge^{78}		77.9227		2.1h	β^- 0.9; γ
$_{33}$As$^{68?}$				~7m	β^+
As^{69}		68.9323		15m	β^+ 2.9; γ 0.23
As^{70}		69.9313		50m	β^+ 1.35, 2.45; EC; γ 1.04, 2.0 (others)
As^{71}		70.92725		65h	EC; β^+ 0.81; γ 0.17 (m)
As^{72}		71.9264		26h	EC; β^+ 2.50, 3.34, 1.84; γ 0.835 (others 0.63–3.7)
As^{73}		72.9238		76d	EC (to 0.53s Ge^{73m_2})
As^{74}		73.92391		18d	β^+ 0.91, (1.51); β^- 1.36, 0.72; EC; γ 0.59, 1.36, 0.72, 0.63 (others)
As^{74m}				8s	IT 0.28
As^{75}	100	74.92158	3/2		
As^{75m}				0.017s	IT 0.305, 0.025; γ 0.280
As^{76}		75.92242	2	26.8h	β^- 2.97, 2.41 (others); γ 0.559 (others)
As^{77}		76.92067		38.7h	β^- 0.68; (γ)
As^{78}		77.9218		91m	β^- 4.1, 1.4; γ 0.615, 0.70, 1.31 (others)
As^{78m}				5.5m	IT γ 0.50
As^{79}		78.9210		9m	β^- 2.15 (others); γ 0.097 (others)
As^{80}		79.9230		15s	β^- 6.0, 5.4; γ 0.66, 1.64 (others)
As^{81}				33s	β^- 3.8
As^{85}				0.43s	β^-; n
$_{34}$Se^{71}		70.9320		5m	β^+ 3.4; γ 0.16
Se^{72}				8.4d	EC; γe^- 0.046
Se^{73}		72.9267		7.1h	β^+ 1.32; EC; γ 0.359 (m); γe^- 0.066 (m)
Se^{73m}				44m	β^+ 1.72; γ 0.25, 0.09, 0.58
Se^{74}	0.87	73.9224	0		

Nuclide	Per Cent Abundance	Nuclidic Mass	Spin	Half-life	Decay Modes, Radiations, Energies
Se^{75}		74.92251	5/2	120d	EC; γ 0.265, 0.136, 0.28. 0.40, 0.122 (others)
Se^{76}	9.02	75.91923	0		
Se^{77}	7.58	76.91993	1/2		
Se^{77m}				18.8s	IT γe^- 0.162
Se^{78}	23.52	77.91735	0		
Se^{79}		78.91852	7/2	6.5×10^4y	β^- 0.16
Se^{79m}				3.9m	IT $e^-(\gamma)$ 0.096
Se^{80}	49.82	79.91651	0		
Se^{81}		80.91786		18m	β^- 1.56 (others); (γ)
Se^{81m}				57m	IT $e^-(\gamma)$ 0.103
Se^{82}	9.19	81.9167	0		
Se^{83}		82.9189		25m	β^- 1.0, 1.8; γ 0.23, 0.35, 1.85, 2.29
Se^{83m}				69s	β^- 3.4, 1.5; γ 1.01, 2.02, 0.65, 0.35
Se^{84}				3.3m	β^- (to 32m Br^{84})
Se^{85}				39s	β^-
Se^{86}				16s	β^-
$_{35}Br^{74}$		73.9289		26m	β^+ 4.7; γ 0.64
Br^{75}		74.9254		1.6h	EC; β^+ 1.7, others; γ 0.29, (0.62)
Br^{76}		75.9242	1	16h	β^+ 3.1, 3.6 (others); EC; γ 0.56, 0.65, 1.22, 1.86 (others)
Br^{77}		76.92140	3/2	58h	EC; (β^+); γ 0.25 (m), 0.52, 0.58, 0.30 (others)
Br^{77m}				4.2m	IT $e^-\gamma$ 0.108
Br^{78}		77.9211		6.4m	β^+ 2.5, 1.9; (EC); γ 0.62
Br^{79}	50.54	78.91835	3/2		
Br^{79m}				4.8s	IT γe^- 0.21
Br^{80}		79.91854	1	17.6m	β^- 2.02, 1.38; γ 0.62 (others); (EC, β^+)
Br^{80m}			5	4.5h	IT $e^-(\gamma)$ 0.048; $e^-\gamma$ 0.036
Br^{81}	49.46	80.91634	3/2		
Br^{82}		81.91680	5	35.3h	β^- 0.44; γ 0.25–1.48
Br^{83}		82.91520		2.4h	β^- 0.94 (to Kr^{83m}); γ 0.05
Br^{84}		83.91655		32m	β^- 4.7, others; γ 0.88, 1.90, 3.9 (others)
Br^{84m}				6.0m	β^- 1.9, 0.8; γ 0.88, 1.46, 0.44, 1.89
Br^{85}		84.9154		3.0m	β^- 2.5
Br^{86}				54s	β^- 7.1, others; γ 1.57, 2.76
Br^{87}		86.9220		55s	β^- 2.6, 8.0; γ 5.4, others; (n)
Br^{88}				16s	β^-; (n)
Br^{89}				4.5s	β^-; (n)
Br^{90}				1.6s	β^-; (n)
$_{36}Kr^{74}$		73.9333		20m	β^+ 3.1
Kr^{75}				5.5m	
Kr^{76}				14.8h	EC; γ 0.039, 0.267, 0.316 (others)
Kr^{77}		76.92449		1.2h	β^+ 1.67, 1.86; EC; γ 0.131, 0.149, 0.281, 0.665
Kr^{78}	0.354	77.920368			
Kr^{79}		78.92009		34.5h	EC; β^+ 0.60; (γ 0.045–0.83)
Kr^{79m}				55s	IT 0.127
Kr^{80}	2.27	79.91639			
Kr^{81}		80.9166		2.1×10^5y	EC
Kr^{81m}				13s	IT γe^- 0.19
Kr^{82}	11.56	81.91348	0		
Kr^{83}	11.55	82.91413	9/2		
Kr^{83m}				1.9h	IT $e^-(\gamma)$ 0.033; $e^-(\gamma)$ 0.009 (m)
Kr^{84}	56.90	83.911504	0		
Kr^{85}		84.91243	9/2	10.6y	β^- 0.67; (γ)

Nuclide	Per Cent Abundance	Nuclidic Mass	Spin	Half-life	Decay Modes, Radiations, Energies
Kr⁸⁵ᵐ				4.5h	β^- 0.82; γ 0.150; IT $\gamma(e^-)$ 0.305
Kr⁸⁶	17.37	85.91062	0		
Kr⁸⁷		86.91337		78m	β^- 3.8, 1.3; γ 0.40, 2.57, 0.85 (others)
Kr⁸⁸		87.9142		2.8h	β^- 0.52, 2.7; γ 2.40, 0.19, 0.85, others
Kr⁸⁹				3.2m	β^- 3.9, 2.0; γ 0.21, 0.45, 0.60, others
Kr⁹⁰				33s	β^- 3.2, others; γ 0.12, 0.55, 1.13, 1.53 (others)
Kr⁹¹				10s	β^- 3.6, others; γ
Kr⁹²				3s	β^-
Kr⁹³				2.0s	β^-
Kr⁹⁴				1.4s	β^-
Kr⁹⁵				1–2s	β^-
Kr⁹⁷				~1s	β^-
₃₇Rb⁷⁹				24m	β^+; γ 0.15, 0.19; (EC)
Rb⁸⁰		79.9219		34s	β^+ 4.1, 3.5; γ 0.62; (EC)
Rb⁸¹		80.9190	3/2	4.7h	EC; β^+ 1.0, 0.58, 0.33; γ 1.1, 0.25, 0.45
Rb⁸¹ᵐ			9/2	32m	β^+ 1.4; IT 0.085
Rb⁸²		81.91796		1.3m	β^+ 3.15, 3.5, 4.2; (γ 0.77, 1.40); (EC)
Rb⁸²ᵐ			5	6.3h	EC; β^+ 0.80, 0.78, 0.67; γ 0.78, 0.62, 0.55, others
Rb⁸³			5/2	83d	EC; γ 0.53, 0.48, 0.046; [to Kr⁸³ᵐ]
Rb⁸⁴		83.91435	2	33d	EC; β^+ 0.80, 1.65; (β^-); γ 0.88 (others)
Rb⁸⁴ᵐ				20m	IT $\gamma(e^-)$ 0.216; γ 0.25, 0.47
Rb⁸⁵	72.15	84.91171	5/2		
Rb⁸⁶		85.91116	2	18.7d	β^- 1.78, (0.70); (γ 1.08)
Rb⁸⁶ᵐ				1.0m	IT γ 0.56
Rb⁸⁷	27.85	86.90918	3/2	5.7×10^{10}y	β^- 0.27
Rb⁸⁸		87.9112	2	18m	β^- 5.3, 3.3; γ 1.85, 0.91 (others)
Rb⁸⁹		88.9112		15m	β^- 1.6, 3.9 (others); γ 1.05, 1.26, 0.66, 2.2, 2.6 (others)
Rb⁹⁰		89.9144		2.8m	β^- 1.2–6.6; γ 0.5–5.2
Rb⁹¹				1.7m	β^- 4.6; γ 0.1
Rb⁹¹ᵐ				14m	β^- 3.0; γ
Rb⁹²				4.2s	β^-
Rb⁹³				5.2s	β^-
Rb⁹⁴				2.9s	β^-
₃₈Sr⁸⁰				1.8h	EC; γ 0.58
Sr⁸¹				29m	β^+
Sr⁸²				25d	EC
Sr⁸³				33h	EC; β^+ 1.15; γ 0.39, 0.76, others
Sr⁸⁴	0.56	83.91338			
Sr⁸⁵		84.9129		65d	EC; γ 0.514 (others)
Sr⁸⁵ᵐ		84.9132		70m	IT $e^-(\gamma)$ 0.0075; γ 0.225; EC; γ 0.15
Sr⁸⁶	9.86	85.9093	0		
Sr⁸⁷	7.02	86.9089	9/2		
Sr⁸⁷ᵐ				2.8h	IT γe^- 0.388; (EC)
Sr⁸⁸	82.56	87.9056	0		
Sr⁸⁹		88.9070		51d	β^- 1.46; (γ)
Sr⁹⁰		89.9073		29y	β^- 0.54
Sr⁹¹		90.9098		9.7h	β^- 1.09, 1.36, 2.67; γ 1.03, 0.75, 0.65, 1.41
Sr⁹²		91.9105		2.7h	β^- 0.55, 1.5; γ 1.37 (others)
Sr⁹³				8m	β^- 3.0–4.8; γ 0.60, 0.88, others 0.18–3.0
Sr⁹⁴				1.3m	β^-
Sr⁹⁵				0.7m	β^-

Nuclide	Per Cent Abundance	Nuclidic Mass	Spin	Half-life	Decay Modes, Radiations, Energies
Sr^{97}				short	β^-
$_{39}$Y^{82}				10m	β^+ 2
Y^{83}				7.4m	β^+
Y^{84}				40m	β^+ 2.5, 3.5; EC; γ 0.80, 0.98, 1.04 (others)
Y^{85}		84.9164		5.0h	β^+ 2.24, 2.01; EC; γ 0.23 (others); [to Sr^{85}]
Y^{85}		84.9164		2.7h	β^+ 1.54; γ 0.503 (others); [to Sr^{85m}]
Y^{86}		85.9148		14.6h	EC; β^+ 1.32 (others); γ 0.18–3.3
Y^{86m}				49m	IT $e^-(\gamma)$ 0.010; γ 0.210
Y^{87}		86.9107		80h	EC; γ 0.483; (β^+)
Y^{87m}				14h	IT γe^- 0.381
Y^{88}		87.9095		108d	EC; γ 1.84, 0.90; (β^+)
Y^{89}	100	88.9054	1/2		
Y^{89m}				16s	IT γ (e^-) 0.92
Y^{90}		89.9067	2	64h	β^- 2.27; (γ)
Y^{90m}				3.2h	IT $\gamma(e^-)$ 0.48; γ 0.20
Y^{91}		90.9069	1/2	58d	β^- 1.54; (γ)
Y^{91m}				50m	IT $\gamma(e^-)$ 0.551
Y^{92}		91.9085		3.6h	β^- 3.6 (others); γ 0.932 (others 0.07–2.4)
Y^{93}		92.9092		10h	β^- 2.89; $(\gamma$ 0.27–2.4)
Y^{94}		93.9115		20m	β^- 5.0; γ 0.56–3.5
Y^{95}				11m	β^-
Y^{96}				2.3m	β^- 3.5; γ 1.0, 0.7 (others)
Y^{97}				short	β^-
$_{40}$Zr^{86}				17h	EC; γ 0.24
Zr^{87}		86.9145		1.6h	β^+ 2.1; (γ); [to 14h Y^{87m}]
Zr^{88}				85d	EC; γ 0.394 (m)
Zr^{89}		88.9085		79h	EC; β^+ 0.90 [to Y^{89m}]
Zr^{89m}				4.2m	IT $\gamma(e^-)$ 0.59; (EC; β^+) $(\gamma$ 1.53)
Zr^{90}	51.46	89.9043			
Zr^{90m}				0.83s	IT 2.30
Zr^{91}	11.23	90.9052	5/2		
Zr^{92}	17.11	91.9046			
Zr^{93}		92.9061		1×10^6y	β^- 0.063, 0.034
Zr^{94}	17.40	93.9061			
Zr^{95}		94.9079		65d	β^- 0.36, 0.40 (others); γ 0.72, 0.76
Zr^{96}	2.80	95.908			
Zr^{97}		96.9107		17h	β^- 1.91, (0.45) [to Nb^{97m}]; (γ)
Zr^{98}				1m	β^-
$_{41}$Nb^{89}		88.9126		1.9h	β^+ 2.9
Nb^{89m}				0.8h	β^+
Nb^{90}		89.9109		14.6h	EC, β^+ 1.50, (0.65); γ 0.133–2.3
Nb^{90m_1}				24s	IT γe^- 0.12
Nb^{90m_2}				0.01s	IT γ 0.25 [to Nb^{90m_1}]
Nb^{91}		90.9070		long	EC
Nb^{91m}				62d	IT $e^-(\gamma)$ 0.104; (EC; γ)
Nb^{92}		91.9068		10.1d	EC; γ 0.93 (others); (β^+)
Nb^{92m}				$\sim10^7$y	
Nb^{93}	100	92.9060	9/2		
Nb^{93m}				3.7y	IT $e^-(\gamma)$ 0.029
Nb^{94}		93.9070		2.0×10^4y	β^- 0.50; γ 0.70, 0.87
Nb^{94m}				6.6m	IT $e^-(\gamma)$ 0.042; $(\beta^-; \gamma)$
Nb^{95}		94.9067		35.1d	β^- 0.16; γ 0.77
Nb^{95m}				90h	IT $e^-(\gamma)$ 0.235
Nb^{96}		95.9079		24h	β^- 0.69; γ 0.77, 0.56, 1.18, others
Nb^{97}		96.9078		72m	β^- 1.27; γ 0.67, (1.02)
Nb^{97m}				60s	IT $\gamma(e^-)$ 0.75

Nuclide	Per Cent Abundance	Nuclidic Mass	Spin	Half-life	Decay Modes, Radiations, Energies
Nb^{98}				51m	β^- 2.6, 3.5; γ 0.78, 0.72 (others 0.3–2.7)
Nb^{99}				2.5m	β^- 3.2; γ 0.10, 0.26
Nb^{99}				10s	
Nb^{100}				12m	β^- 3.1, 3.5; γ 0.53, 0.62
Nb^{101}				1.0m	β^-
$_{42}$Mo^{90}		89.9136		6h	EC; β^+ 1.2 [to 0.01s Nb^{90m_2}]
Mo^{91}		90.9117		15.5m	β^+ 3.3 [to Nb^{91}]
Mo^{91m}				65s	IT $\gamma(e^-)$ 0.65; β^+ 2.5, 2.8; γ 1.2, 1.5; (EC) [to Nb^{91m}]
Mo^{92}	15.84	91.9063			
Mo^{93}		92.9065		>2y	EC
Mo^{93m}		92.9091		6.9h	IT γe^- 0.264; γ 0.684, 1.48
Mo^{94}	9.04	93.9047	0		
Mo^{95}	15.72	94.9057	5/2		
Mo^{96}	16.53	95.9045	0		
Mo^{97}	9.46	96.9058	5/2		
Mo^{98}	23.78	97.9055			
Mo^{99}		98.9079		67h	β^- 1.23, 0.45 [to Tc^{99m}]; γ 0.740, 0.181 (*m*) (others)
Mo^{100}	9.63	99.9076			
Mo^{101}		100.9089		14.6m	β^- 2.2, others; γ 0.19, 1.02, 2.08, others
Mo^{102}				11.5m	β^- 1.2
Mo^{103}				70s	β^-
Mo^{104}				1.1m	β^-
Mo^{105}				40s	β^-
$_{43}$Tc^{92}				4.0m	β^+ 4.1; γ 1.54, 0.79, 0.33, 0.135, others
Tc^{93}		92.9099		2.7h	EC; β^+ 0.82; γ 1.35, 1.48 (others); [to Mo^{93}]
Tc^{93m}				44m	IT γe^- 0.39; EC [to Mo^{93}]; γ 2.7
Tc^{94}				4.9h	EC; (β^+); γ 0.87, 0.70, 0.85
Tc^{94m}		93.9094		52m	β^+ 2.5; EC; γ 0.87, 1.85 (others); IT
Tc^{95}		94.9075		20h	EC; γ 0.77 (others)
Tc^{95m}				60d	EC; γ 0.204, 0.584, 0.84 (others); $(\beta^+$; IT 0.04)
Tc^{96}		95.9077		4.3d	EC; γ 0.77, 0.84, 0.81, 1.12 (others)
Tc^{96m}				52m	IT $e^-(\gamma)$ 0.034; $(\beta^+$; $\gamma)$
Tc^{97}		96.9059		2.6 × 10^{6}y	EC
Tc^{97m}				90d	IT $e^-\gamma$ 0.097
Tc^{98}		97.907		1.5 × 10^{6}y	β^- 0.30; γ 0.77, 0.67
Tc^{99}		98.9064	9/2	2.1 × 10^{5}y	β^- 0.29
Tc^{99m}				6.0h	IT $e^-(\gamma)$ 0.0021; γ 0.140; $(\gamma$ 0.142)
Tc^{100}		99.9066		16s	β^- 3.4 (others); (γ)
Tc^{101}		100.9059		14.0m	β^- 1.3, (1.1); γ 0.307 (others 0.13–0.94)
Tc^{102}		101.9081		5s	β^- 4.1
Tc^{102}				4.5m	β^- 2; γ 0.47, 0.63, 1.07, 1.77, 1.98
Tc^{103}				50s	β^- 2.0, 2.2; γ 0.135, 0.215, 0.35
Tc^{104}				18m	β^- 1.8–5.3; γ 0.36–4.7
Tc^{105}				7.7m	β^- 3.5; γ 0.11
$_{44}$Ru$^{93?}$				52s	β^+
Ru^{94}				~57m	
Ru^{95}		94.9099		1.65h	EC; β^+ 1.2; γ 0.34, 1.4, 0.64, 0.15; [to 20h Tc^{95}]
Ru^{96}	5.51	95.90759			
Ru^{97}				2.9d	EC; γ 0.216 (others); [to Tc^{97}]
Ru^{98}	1.87	97.90528			

Nuclide	Per Cent Abundance	Nuclidic Mass	Spin	Half-life	Decay Modes, Radiations, Energies
Ru^{99}	12.72	98.90593	5/2		
Ru^{100}	12.62	99.90421			
Ru^{101}	17.07	100.90557	5/2		
Ru^{102}	31.61	101.90434			
Ru^{103}		102.9063		40d	β^- 0.21 (others); γ 0.498 (others 0.05–0.61); [to Rh^{103m}]
Ru^{104}	18.58	103.90543			
Ru^{105}		104.90768		4.44h	β^- 1.15, 1.08 (others); γ 0.73, 0.48, 0.67, 0.32 (others)
Ru^{106}		105.90733		1.0y	β^- 0.040; [to 30s Rh^{106}]
Ru^{107}				4.2m	β^- 4.6; γ 0.195, 0.37, 0.48, 0.86 (others)
Ru^{108}				4.6m	β^- 1.3; γ 0.165
$_{45}$Rh^{96}				~11m	
Rh^{97}				35m	β^+ 2.1; γ 0.19, 0.26, 0.42
Rh^{98}		97.910		8.7m	β^+ 2.5; γ 0.65
Rh^{99}		98.90818		16d	EC; γ 0.35, 0.090, 0.18, 0.53 (others); (β^+)
Rh^{99}				4.7h	EC; γ 0.33, 0.61 (others); (β^+)
Rh^{100}		99.90812		21h	EC; γ 0.54, 2.38, 0.82, 1.58, others; (β^+)
Rh^{101}				~7y	EC; γ 0.195, 0.125
Rh^{101m}				4.5d	EC; γ 0.31 (others); (IT 0.15)
Rh^{102}		101.9068		210d	EC; β^- 1.15; β^+ 1.28; γ 0.475 (others)
Rh^{102m}				~2.5y	EC; γ
Rh^{103}	100	102.90551	1/2		
Rh^{103m}				57m	IT $e^-(\gamma)$ 0.040
Rh^{104}		103.9066		42s	β^- 2.4; (γ)
Rh^{104m}				4.4m	IT $e^-(\gamma)$ 0.077; γe^- 0.051; (γ; β^-)
Rh^{105}		104.90567		36h	β^- 0.56, 0.25; γ 0.32 (others)
Rh^{105m}				30s	IT $e^-\gamma$ 0.130
Rh^{106}		105.90728		30s	β^- 3.54 (others); γ 0.513, 0.624 (others)
Rh^{106m}				2.2h	β^- 0.79, 0.95, 1.18, 1.62; γ 0.513, others 0.22–1.22
Rh^{107}		106.9067		22m	β^- 1.2; γ 0.307, 0.365 (others)
Rh^{108}				17s	β^- 4.0; γ 0.43, 0.62 (others)
Rh^{109}				~30s	β^-; γ 0.32, 0.49
Rh^{109m}				50s	IT 0.11
Rh^{110}				~3s	β^-
$_{46}$Pd^{98}				17m	EC
Pd^{99}		98.9124		22m	β^+ 2.0; EC; γ 0.14, 0.42, 0.67, 0.28
Pd^{100}				4.1d	EC; γ 0.081
Pd^{101}				8.5h	EC; γ 0.29, 0.59 (others); (β^+)
Pd^{102}	0.96	101.90562			
Pd^{103}		102.90611		17d	EC; γ 0.053 (others)
Pd^{104}	10.97	103.90398			
Pd^{105}	22.23	104.90507	5/2		
Pd^{106}	27.33	105.90348			
Pd^{107}		106.90512		7×10^6y	β^- 0.035 [to stable Ag^{107}]
Pd^{107m}				21s	IT γe^- 0.22
Pd^{108}	26.71	107.90388			
Pd^{109}		108.90595		13.6h	β^- 1.03 [to Ag^{109m}]; (γ)
Pd^{109m}				4.7m	IT γe^- 0.18
Pd^{110}	11.81	109.90516			
Pd^{111}		110.90766		22m	β^- 2.13 [to Ag^{111m}]; (γ)
Pd^{111m}				5.5h	IT $e^-\gamma$ 0.17; β^-; (γ 1.69)
Pd^{112}		111.9075		21h	β^- 0.28; $e^-\gamma$ 0.018

Nuclide	Per Cent Abundance	Nuclidic Mass	Spin	Half-life	Decay Modes, Radiations, Energies
Pd^{113}				1.4m	β^-
Pd^{114}				2.4m	β^-
Pd^{115}				45s	β^-
$_{47}$Ag^{102}				13m	β^+ 2.2; γ 0.55, 0.72 (others)
Ag^{103}		102.9078	7/2	1.1h	EC; β^+ 1.2; γ 0.15, 0.11
Ag^{103m}				5.7s	IT $e^-\gamma$ 0.135
Ag^{104}		103.90857	5	67m	EC; β^+ 0.99; γ 0.56, 0.77, 0.94, others
Ag^{104m}			2	29m	β^+ 2.70; γ 0.56; IT 0.02; EC
Ag^{105}		104.9068	1/2	40d	EC; γ 0.35, 0.28, others 0.064–1.09
Ag^{106}		105.9067	1	24m	β^+ 1.95, (1.45); γ 0.51, others 0.21–1.8; EC; (β^-)
Ag^{106m}			6	8.3d	EC; γ 0.51, others 0.22–2.63
Ag^{107}	51.35	106.90508	1/2		
Ag^{107m}				44s	IT $e^-(\gamma)$ 0.093
Ag^{108}		107.90594		2.4m	β^- 1.65; (EC; γ; β^+)
Ag^{108m}				>5y	EC; γ 0.72, 0.62, 0.43; (IT 0.031; γ 0.081)
Ag^{109}	48.65	108.90475	1/2		
Ag^{109m}				41s	IT $e^-(\gamma)$ 0.088
Ag^{110}		109.90609		24s	β^- 2.87, (2.21); (γ)
Ag^{110m}			6	253d	β^- 0.085, 0.53; γ 0.44–2.46; (IT 0.116)
Ag^{111}		110.90531	1/2	7.5d	β^- 1.05; (γ)
Ag^{111m}				1.2m	IT 0.065
Ag^{112}		111.9071	2	3.2h	β^- 4.0, 3.4 (others); γ 0.62 (others)
Ag^{113}		112.9065	1/2	5.3h	β^- 2.0; (γ 0.12–1.18)
Ag^{113m}				1.2m	IT; β^-; γ 0.14–0.70
Ag^{114}		113.9085		5s	β^- 4.6; γ 0.57
Ag^{114m}				2m	β^-
Ag^{115}		114.9087		21m	β^- 2.9; (γ)
Ag^{115m}				20s	β^- [to 2.3d Cd^{115}]
Ag^{116}				2.5m	β^- 5.0; γ 0.52, 0.70
Ag^{117}				1.1m	β^-
$_{48}$Cd^{103}				10m	β^+; γ 0.22, 0.62, 0.85
Cd^{104}				57m	EC; γ 0.084 (others)
Cd^{105}				55m	β^+ 0.80, 1.69; γ 0.025–2.3
Cd^{106}	1.22	105.90646			
Cd^{107}		106.90661	5/2	6.7h	EC [to Ag^{107m}]; (γ 0.85; β^+)
Cd^{108}	0.87	107.90418			
Cd^{109}		108.90492	5/2	470d	EC [to Ag^{109m}]
Cd^{110}	12.39	109.90300			
Cd^{111}	12.75	110.90418	1/2		
Cd^{111m_2}				49m	IT $e^-\gamma$ 0.150; γ 0.247 (m)
Cd^{112}	24.07	111.90275			
Cd^{113}	12.26	112.90440	1/2		
Cd^{113m}		112.9049		14y	β^- 0.58; (IT)
Cd^{114}	28.86	113.90336			
Cd^{115}		114.90542		2.3d	β^- 1.11, 0.59; γ 0.523, 0.490 (others) [to In^{115m}]
Cd^{115m}			11/2	43d	β^- 1.63 (others); (γ); [to In^{115}]
Cd^{116}	7.58	115.90476			
Cd^{117}		116.9074		50m	β^- 1.8, 2.3; γ 0.425; [to 1.9h In^{117m}]
Cd^{117}		116.9074		2.9h	β^- 1.0; γ 0.27–2.2
Cd^{118}				50m	β^- 0.8; [to 5s In^{118}]
Cd^{119}				2.7m	β^-; [to 2.3m In^{119}]
Cd^{119}				11m	β^- 3.5; [to 18m In^{119m}]
$_{49}$In^{106}				5.3m	β^+ 4.9, 2.7; γ 0.63, 0.86, 1.66, 0.99 (others)
In^{107}				30m	β^+ ~2; γ 0.22

Nuclide	Per Cent Abundance	Nuclidic Mass	Spin	Half-life	Decay Modes, Radiations, Energies
In^{108}		107.9097		57m	EC; β^+ 1.3; γ 0.15–1.05
In^{108m}				40m	β^+ 3.50, 2.66; γ 0.63, 0.84; EC
In^{109}		108.90709	9/2	4.3h	EC; γ 0.21, 0.63 (others); (β^+)
In^{109m}				1.3m	IT $\gamma(e^-)$ 0.66
In^{110}		109.9072		66m	β^+ 2.25; γ 0.66 (others); EC
In^{110m}			7	4.9h	EC; γ 0.94, 0.88, 0.66 (others)
In^{111}		110.9055	9/2	2.82d	EC; γ 0.173, 0.246
In^{111m}				~10m	IT γ 0.53
In^{112}		111.90552		14m	β^- 0.66; EC; β^+ 1.62; (γ 0.62)
In^{112m_1}				21m	IT 0.155
In^{112m_2}				0.04s	IT 0.31
In^{113}	4.28	112.90411	9/2		
In^{113m}			1/2	1.73h	IT γe^- 0.393
In^{114}		113.90489		72s	β^- 1.98; (EC; β^+; γ)
In^{114m_1}			5	50d	IT $e^-\gamma$ 0.191; (EC; γ)
In^{114m_2}				2.5s	IT 0.15
In^{115}	95.72	114.90386	9/2	5×10^{14}y	β^- 0.48
In^{115m}			1/2	4.5h	IT γe^- 0.336; (β^- 0.84)
In^{116}		115.9053		14s	β^- 3.3; (γ)
In^{116m_1}			5	54m	β^- 1.00, 0.87, 0.60; γ 1.27, 1.09, 0.41, 0.82, 2.09 (others)
In^{116m_2}				2.2s	IT $e^-\gamma$ 0.16; [to In^{116m_1}]
In^{117}		116.90452	9/2	38m	β^- 0.74; γ 0.56, 0.16; [to stable Sn^{117}]
In^{117m}			1/2	1.9h	β^- 1.77, 1.62; IT $e^-\gamma$ 0.31; γ 0.16 (others)
In^{118}		117.9063		5s	β^- 4.2, 3.0; γ 1.22
In^{118m}				4.4m	β^- 1.3, 2.1; γ 1.22, 1.04, 0.69, 0.8, 0.45, 0.21
In^{119}		118.9059		2m	β^- 1.6; γ 0.82, (0.71)
In^{119m}				18m	β^- 2.7, 1.8; γ 0.91; (IT 0.3)
In^{120}				3s	β^- 5.6
In^{120m}				44s	β^- 2.0, 3.3, 4.0; γ 1.02, 1.18, 0.87, others
In^{121}				30s	β^-; γ 0.94
In^{121}				3.1m	β^- 3.7
In^{122}				7.5s	β^- 4.5; γ 1.14, 1.00
In^{123}				10s	β^-; γ 1.10
In^{123}				36s	β^- 4.6
In^{124}				3s	β^- 5.2; γ 1.35, 1.00
$_{50}$Sn^{108}				9m	EC
Sn^{109}				18m	EC; β^+ ~1.5, >2.5; γ 0.34, 1.12, 0.52, 0.89
Sn^{110}				4.0h	EC; γ 0.283; [to 66m In^{110}]
Sn^{111}		110.9082		35m	EC; β^+ 1.5
Sn^{112}	0.96	111.90481			
Sn^{113}		112.90484		118d	EC; (γ 0.255); [to 1.7h In^{113m}]
Sn^{113m}				27m	IT $e^-\gamma$ 0.079
Sn^{114}	0.66	113.90276			
Sn^{115}	0.35	114.90335	1/2		
Sn^{116}	14.30	115.90174	0		
Sn^{117}	7.61	116.90294	1/2		
Sn^{117m}				14d	IT $e^-(\gamma)$ 0.159; $\gamma(e^-)$ 0.161
Sn^{118}	24.03	117.90160	0		
Sn^{119}	8.58	118.90330	1/2		
Sn^{119m_2}				250d	IT $e^-(\gamma)$ 0.065; $e^-(\gamma)$ 0.024 (m_1)
Sn^{120}	32.85	119.90219	0		
Sn^{121}		120.90424		27h	β^- 0.38
Sn^{121m}				~25y	β^- 0.42
Sn^{122}	4.72	121.90343			
Sn^{123}		122.90574		41m	β^- 1.26; γ 0.15

Nuclide	Per Cent Abundance	Nuclidic Mass	Spin	Half-life	Decay Modes, Radiations, Energies
Sn^{123}				125d	β^- 1.42, (0.34); (γ 1.08)
Sn^{124}	5.94	123.90526			
Sn^{125}		124.90776		9.5d	β^- 2.33 (others); (γ 0.23–1.97)
Sn^{125m}				9.7m	β^- 2.04 (others); (γ 0.33–1.4)
Sn^{126}				2×10^5y	β^-; γ 0.06, 0.067, 0.092
Sn^{127}				2.1h	β^-; γ 1.10, 0.82
Sn^{127m}				4m	
Sn^{128}				59m	β^- 0.80, 0.73; γ 0.50, 0.57, 0.072, 0.04
Sn^{129}				1.0h	β^-
Sn^{129}				8.8m	
Sn^{130}				2.6m	β^-
Sn^{131}				3.4m	β^-
Sn^{131}				1.6h	
Sn^{132}				2.2m	β^-
$_{51}Sb^{112}$				0.9m	β^+; γ 1.27
Sb^{113}				7m	β^+ 1.85, 2.42
Sb^{114}		113.9097		3.4m	β^+ 2.7, 4.0; γ 0.90, 1.30
Sb^{115}		114.9068		30m	EC; β^+ 1.5; γ 0.50
Sb^{116}		115.9070		15m	EC; β^+ 1.5, 2.3; γ 1.30, 0.90, 2.22
Sb^{116m}				60m	EC; β^+ 1.45; γ 1.29, 0.90, 0.40, 0.14, 0.11, 2.23
Sb^{117}		116.9049		2.8h	EC; γ 0.161; (β^+); [to stable Sn^{117}]
Sb^{118}		117.9060		5h	EC; γ 1.03, 1.22, 0.26, 0.040 (m)
Sb^{118m_1}				3.5m	β^+ 2.60; EC; (γ)
Sb^{118m_2}				0.9s	γ 0.14, 0.30, 0.38
Sb^{119}		118.90392		38h	EC; $e^-\gamma$ 0.024 (m)
Sb^{120}		119.90511		16m	EC; β^+ 1.70; (γ)
Sb^{120}				5.8d	EC; γ 0.089 (m), 0.20, 1.04, 1.18
Sb^{121}	57.25	120.90381	5/2		
Sb^{122}		121.90517	2	2.74d	β^- 1.40, 1.97; γ 0.564 (others); (EC; β^+)
Sb^{122m_3}				4.2m	IT $e^-(\gamma)$ 0.026; γ 0.077 (m_2), 0.061 (m_1)
Sb^{123}	42.75	122.90421	7/2		
Sb^{124}		123.90595	3	60d	β^- 0.62, 2.31, 0.23; γ 0.603, 1.69 (others 0.63–2.3)
Sb^{124m_1}				1.5m	IT $e^-(\gamma)$ 0.010; β^- 1.19; γ 0.51, 0.65, 0.60
Sb^{124m_2}				21m	IT $e^-(\gamma)$ 0.025
Sb^{125}		124.90525		2.7y	β^- 0.30, 0.12, 0.62; γ 0.035 0.67
Sb^{126}				12.5d	β^- 1.9, others; γ 0.29–0.99
Sb^{126m}				19m	β^- 1.9; γ 0.415, 0.665, 0.696; IT $\leq$0.03
Sb^{127}		126.90690		3.9d	β^- 0.80, 1.5, 1.1, 0.86; γ 0.46, 0.77, 0.25 (others)
Sb^{128}				8.6h	β^- 1.0; γ 0.16–1.18
Sb^{128m}				11m	β^- 2.5, 2.8; γ 0.32, 0.75
Sb^{129}				4.6h	β^- 1.87, others; γ 0.53, 0.16, 0.31, 0.79
Sb^{130}				33m	β^-; γ 0.19, 0.33, 0.82, 0.94
Sb^{130m}				7m	γ 0.20, 0.82 (others)
Sb^{131}				23m	β^-
Sb^{132}				2m	β^-
Sb^{133}				2.4m	β^-
$Sb^{134?}$				0.8m	β^-
$_{52}Te^{114}$				16m	
Te^{115}				6m	
Te^{116}		115.9087	0	2.5h	EC; (β^+); γ 0.094; [to 15m Sb^{116}]
Te^{117}		116.9087	1/2	1.0h	EC; β^+ 1.74; γ 0.72 (others)

Nuclide	Per Cent Abundance	Nuclidic Mass	Spin	Half-life	Decay Modes, Radiations, Energies
Te^{118}				6.0d	EC; [to 3.5m Sb^{118m}]
Te^{119}		118.90638	1/2	16h	EC; γ 0.645, (1.76); (β^+)
Te^{119m}			11/2	4.6d	EC; γ 0.153, 1.22, 0.271, 0.93, 1.10 (others)
Te^{120}	0.089	119.90402			
Te^{121}				17d	EC; γ 0.575, 0.506, 0.070; (β^+)
Te^{121m}				154d	IT $e^-(\gamma)$ 0.082; γ 0.213; (EC; γ)
Te^{122}	2.46	121.90305			
Te^{123}	0.87	122.90426	1/2	1.2×10^{13}y	EC
Te^{123m}				104d	IT $e^-(\gamma)$ 0.089; γe^- 0.159
Te^{124}	4.61	123.90281			
Te^{125}	6.99	124.90444	1/2		
Te^{125m_2}				58d	IT $e^-(\gamma)$ 0.109; $e^-(\gamma)$ 0.035 (m_1)
Te^{126}	18.71	125.90333			
Te^{127}		126.90521		9.3h	β^- 0.70; (γ)
Te^{127m}				105d	IT $e^-(\gamma)$ 0.089; $(\beta^-; \gamma)$
Te^{128}	31.79	127.90449			
Te^{129}		128.90657		72m	β^- 1.45, 0.99 (others); γ 0.027 (m), 0.47 (others)
Te^{129m}				33d	IT $e^-(\gamma)$ 0.106; β^-
Te^{130}	34.48	129.90623			
Te^{131}		130.90857		25m	β^- 2.14, 1.68 (others); γ 0.148, 0.45 (others)
Te^{131m}				1.2d	β^- 0.42, 0.57 (others); IT $e^-\gamma$ 0.182; γ 0.78, 0.84, 1.14, others
Te^{132}		131.90854		77h	β^- 0.22; γ 0.23, 0.053
Te^{133}				~2m	β^- ~2.4; γ
Te^{133m}				53m	β^- 1.3, 2.4; γ 0.31–0.97; IT 0.334
Te^{134}				42m	β^- ~1.2; γ 0.20, 0.26, 0.17, 0.08
Te^{135}				1.4m	β^-
$_{53}I^{117}$				10m	
I^{118}				17m	
I^{119}				19m	β^+; EC; γ
I^{120}				1.4h	β^+ 4.0; EC
I^{121}				2.1h	EC; β^+ 1.13; γ 0.21, others
I^{122}		121.9074		3.5m	β^+ 3.1; EC
I^{123}			5/2	13h	EC; γ 0.159 (others)
I^{124}		123.90622	2	4.2d	EC; β^+ 1.55, 2.15; γ 0.603, 1.69, 0.65 (others)
I^{125}		124.90460	5/2	60d	EC; $e^-(\gamma)$ 0.035 (m) [to stable Te^{125}]
I^{126}		125.90563	2	13.1d	EC; β^- 0.87 (others); (β^+); γ 0.665, 0.386 (others)
I^{127}	100	126.90447	5/2		
I^{128}		127.90583	1	25.0m	β^- 2.12, 1.66; γ 0.45 (others); (EC)
I^{129}		128.90498	7/2	1.6×10^7y	β^- 0.15; $e^-(\gamma)$ 0.038
I^{130}		129.90667	5	12.5h	β^- 0.60, 1.02; γ 0.53, 0.67, 0.74, 0.42, 1.15
I^{131}		130.90612	7/2	8.06d	β^- 0.60 (others); γ 0.364 (others)
I^{132}		131.90800	4	2.29h	β^- 0.80, 1.04, 1.61, 2.14 (others); γ 0.673, 0.78, others 0.24–2.7
I^{133}		132.9075	7/2	21h	β^- 1.22 (others); γ 0.53 (others)
I^{134}		133.90984		53m	β^- 2.41, 1.25, others; γ 0.85, 0.89 (others 0.14–1.8)
I^{135}			7/2	6.7h	β^- 1.0, 1.4, 0.5; γ 0.14–2.0
I^{136}		135.9147		84s	β^- 4.2, 5.6, 2.7, 7.0; γ 1.32, others 0.20–3.2

Nuclide	Per Cent Abundance	Nuclidic Mass	Spin	Half-life	Decay Modes, Radiations, Energies
I^{137}				24s	β^-; γ 0.39; (n)
I^{138}				6s	β^-; (n)
I^{139}				2s	β^-; (n)
$_{54}$Xe^{121}				40m	β^+ 2.77; EC; γ 0.096, 0.08, 0.13, 0.44
Xe^{122}				19h	EC; γ 0.090, 0.148, 0.239
Xe^{123}				1.8h	EC; β^+ 1.51; γ 0.148, 0.178, 0.33
Xe^{124}	0.096	123.9061			
Xe^{125}				18h	EC; γ 0.055, 0.075, 0.113, |0.188, 0.242 (others)
Xe^{125m}				55s	IT; γ 0.111, 0.075
Xe^{126}	0.090	125.90417			
Xe^{127}		126.9051		36.4d	EC; γ 0.203, 0.173, 0.37 (others)
Xe^{127m}				75s	IT $e^-\gamma$ 0.175; γ 0.125
Xe^{128}	1.919	127.90353			
Xe^{129}	26.44	128.90478	1/2		
Xe^{129m}				8.0d	IT $e^-(\gamma)$ 0.196; $e^-\gamma$ 0.040
Xe^{130}	4.08	129.90350			
Xe^{131}	21.18	130.90508	3/2		
Xe^{131m}				12d	IT $e^-(\gamma)$ 0.163
Xe^{132}	26.89	131.90416			
Xe^{133}		132.9055		5.27d	β^- 0.34; $e^-\gamma$ 0.081 (m); (γ)
Xe^{133m}				2.3d	IT $e^-\gamma$ 0.233
Xe^{134}	10.44	133.90539			
Xe^{135}		134.9070		9.2h	β^- 0.91; γ 0.25 (others)
Xe^{135m}				15.7m	IT γe^- 0.53
Xe^{136}	8.87	135.90721			
Xe^{137}				3.9m	β^- 3.5; γ 0.26, 0.45
Xe^{138}				17m	β^- 2.4; γ 0.42, 0.51, 1.78, 2.01
Xe^{139}				41s	β^- ~3.5, ~4.6; γ 0.22, 0.30, 0.17, 0.40
Xe^{140}				16s	β^-
Xe^{141}				1.7s	β^-
Xe^{142}				~1.5s	β^-
Xe^{143}				1s	β^-
Xe^{144}				1s	β^-
$_{55}$Cs^{123}				8m	β^+
Cs^{125}				45m	β^+ 2.05; EC; γ 0.112
Cs^{126}		125.9093		1.6m	β^+ 3.8; γ 0.38, 0.48; EC
Cs^{127}		126.9073	1/2	6.2h	EC; γ 0.406, 0.125 (others); (β^+)
Cs^{128}		127.90773		3.8m	β^+ 2.89, 2.45; γ 0.44 (others); EC
Cs^{129}			1/2	32h	EC; γ 0.37, 0.41 (others), $e^-(\gamma)$ 0.040
Cs^{130}		129.90672	1	30m	EC; β^+ 1.97; (β^-)
Cs^{131}		130.90547	5/2	9.69d	EC
Cs^{132}		131.9061	2	6.48d	EC; γ 0.668 (others); $(\beta^-; \beta^+)$
Cs^{133}	100	132.9051	7/2		
Cs^{134}		133.9065	4	2.1y	β^- 0.66, 0.086 (others); γ 0.605, 0.80, 0.57 (others)
Cs^{134m}			8	2.90h	IT $e^-\gamma$ 0.128; $e^-(\gamma)$ 0.010; (β^-)
Cs^{135}		134.9058	7/2	2.0×10^6y	β^- 0.21
Cs^{135m}				53m	IT $\gamma(e^-)$ 0.84; γ 0.78
Cs^{136}		135.9071	5	12.9d	β^- 0.34, (0.66); γ 0.83, 1.07, others 0.067-1.26
Cs^{137}		136.9068	7/2	30y	β^- 0.51, (1.18)
Cs^{138}		137.9102		32m	β^- 1.5-3.4; γ 1.43, 1.01, 0.46, 2.21 (others)
Cs^{139}		138.9132		9.5m	β^- 4; γ 1.28, 0.63
Cs^{140}				1.1m	β^-; γ 0.61
Cs^{141}				25s	β^-
Cs^{142}				~1m	β^-

Nuclide	Per Cent Abundance	Nuclidic Mass	Spin	Half-life	Decay Modes, Radiations, Energies
$_{56}Ba^{123}$				2m	
Ba^{125}				6.5m	
Ba^{126}				96m	EC; γ 0.225, 0.70
Ba^{127}				11m	β^+
Ba^{128}				2.4d	EC
Ba^{129}				2.1h	EC; γ 0.05–1.62
Ba^{129}				2.6h	EC; β^+ 1.43; γ
Ba^{130}	0.101	129.90625			
Ba^{131}				11.6d	EC; γ 0.055–1.7
Ba^{131m}				14.6m	IT $e^-(\gamma)$ 0.078; $\gamma(e^-)$ 0.107
Ba^{132}	0.097	131.9051			
Ba^{133}		132.9056		7.2y	EC; γ 0.355, 0.081, 0.302 (others)
Ba^{133m}				39h	IT $e^-\gamma$ 0.276; $e^-(\gamma)$ 0.012
Ba^{134}	2.42	133.9043			
Ba^{135}	6.59	134.9056	3/2		
Ba^{135m_1}				29h	IT $e^-\gamma$ 0.268
Ba^{135m_2}				0.33s	IT?; γ 0.80, 0.70
Ba^{136}	7.81	135.9044			
Ba^{137}	11.32	136.9056	3/2		
Ba^{137m}				2.6m	IT γe^- 0.662
Ba^{138}	71.66	137.9050			
Ba^{139}		138.9086		82.9m	β^- 2.34, 2.17 (others); γ 0.165 (others)
Ba^{140}		139.9105		12.8d	β^- 1.01, 0.5 (others); γ 0.030, 0.54 (others)
Ba^{141}		140.9137		18m	β^- 2.8 (others); γ 0.19, 0.29, 0.35, 0.46, 0.64 (others)
Ba^{142}				11m	β^- ~4; γ 0.08–1.8
Ba^{143}				12s	β^-
$_{57}La^{125}$				<1m	
La^{126}				1.0m	β^+; γ 0.26
La^{127}				3.8m	
La^{128}				4.2m	β^+; γ 0.28
La^{129}				7m	
La^{130}				9m	β^+; γ 0.36
La^{131}				60m	EC; β^+ 1.43, 1.94, 0.70; γ 0.11–0.88
La^{132}		131.9103		4.5h	β^+ ~3.8; γ 1.0–3.3
La^{133}		132.9080		4.0h	EC; β^+ 1.2; γ 0.8
La^{134}		133.9083		6.5m	EC; β^+ 2.7; γ 0.60
La^{135}		134.9067		19.8h	EC; (γ)
La^{136}		135.9074		10m	EC; β^+ 1.8; $(\gamma$ 0.83$)$
La^{137}				6×10^4y	EC
La^{138}	0.089	137.9068	5	1.1×10^{11}y	EC; γ 1.43; β^- 0.20; γ 0.81
La^{139}	99.911	138.9061	7/2		
La^{140}		139.9093	3	40.2h	β^- 1.34, others 0.42–2.20; γ 1.60, 0.49, 0.82, 0.33 (others)
La^{141}		140.9106		3.8h	β^- 2.4; (γ)
La^{142}				92m	β^- 4.0, others; γ 0.63, 2.4, others 0.87–3.4
La^{143}		142.9157		14m	β^- 3.3; γ 0.20–2.85
$_{58}Ce^{131}$				10m	EC; $(\beta^+; \gamma)$
Ce^{132}				4.2h	β^+
Ce^{133}				6.3h	EC; β^+ 1.3; γ 1.8
Ce^{134}				72h	EC
Ce^{135}				22h	EC; (β^+); γ 0.28
Ce^{136}	0.193	135.9071			
Ce^{137}				8.7h	EC; $e^-(\gamma)$ 0.010 (m); (γ)
Ce^{137m}				34.5h	IT $e^-\gamma$ 0.255; (EC; γ)
Ce^{138}	0.250	137.9057			
Ce^{138m}				0.009s	IT 0.30; γ 1.04, 0.80
Ce^{139}		138.9063		140d	EC; γ 0.166 (m)
Ce^{139m}				55s	IT $\gamma(e^-)$ 0.74
Ce^{140}	88.48	139.90528			
Ce^{141}		140.90801	7/2	32.5d	β^- 0.44, 0.58; γ 0.145

Nuclide	Per Cent Abundance	Nuclidic Mass	Spin	Half-life	Decay Modes, Radiations, Energies
Ce^{142}	11.07	141.9090		$\sim 5 \times 10^{15}$y	α 1.5
Ce^{143}		142.91217		33h	β^- 1.09, 1.38, others; γ 0.29, 0.057 (m), others 0.23–1.10
Ce^{144}		143.91343		284d	β^- 0.32, 0.19; γ 0.133 (others)
Ce^{145}		144.9162		3.0m	β^- 2.0
Ce^{146}		145.9183		14m	β^- 0.70; γ 0.32, 0.22, 0.14, 0.11 (others)
Ce^{147}				1.1m	β^-
Ce^{148}				0.7m	β^-
$_{59}Pr^{134}$				40m	γ 0.72
Pr^{135}				22m	β^+ 2.5; γ 0.08, 0.22, 0.30
Pr^{136}				1.1h	EC; β^+ 2.0; γ 0.17 (others)
Pr^{137}				1.5h	EC; β^+ 1.7
Pr^{138}				2.1h	EC; β^+ 1.4; γ 0.30, 0.80, 1.04 (others)
Pr^{139}		138.9085		4.5h	EC; β^+ 1.0; γ 1.3, 1.6
Pr^{140}		139.90878		3.5m	EC; β^+ 2.4; γ 1.2
Pr^{141}	100	140.90739	5/2		
Pr^{142}		141.90979	2	19.2h	β^- 2.15; (γ 1.57)
Pr^{143}		142.91063	7/2	13.7d	β^- 0.93
Pr^{144}		143.91310		17.3m	β^- 2.98; (γ)
Pr^{145}		144.9141		6.0h	β^- 1.80; γ 0.07–1.15
Pr^{146}		145.9172		25m	β^- 3.8, 2.3; γ 0.45, 1.49, 0.75 (others)
Pr^{147}				12m	β^-; γ 0.32, 0.58, 0.64, 0.92, 1.25
Pr^{148}				2m	β^-; γ 0.30
$_{60}Nd^{138}$?				22m	β^+ 2.4; γ
Nd^{139}				5.5h	EC; β^+; γ 1.3
Nd^{140}				3.3d	EC; γ 0.11–0.50
Nd^{141}		140.90932	3/2	2.5h	EC; (β^+ 0.78); (γ)
Nd^{141m}				63s	IT $\gamma(e^-)$ 0.76
Nd^{142}	27.11	141.90748			
Nd^{143}	12.17	142.90962	7/2		
Nd^{144}	23.85	143.90990		2.4×10^{15}y	α 1.83
Nd^{145}	8.30	144.9122	7/2		
Nd^{146}	17.22	145.9127			
Nd^{147}		146.91583	5/2	11.1d	β^- 0.81, 0.37; γ 0.091 (m), 0.53 (others)
Nd^{148}	5.73	147.9165			
Nd^{149}		148.9198	5/2	1.8h	β^- 1.1, 1.5, 0.95; γ 0.114, 0.210, 0.240, 0.112 (others)
Nd^{150}	5.62	149.9207			
Nd^{151}		150.9242		12m	β^- 2.0, 1.2, 1.8; γ 0.085–2.17
$_{61}Pm^{141}$		140.9132		22m	β^+ 2.6; EC; γ 0.20
Pm^{141m}				0.0022s	IT γe^- 0.43; γ 0.19
Pm^{142}		141.9126		30s	β^+ 3.8; γ 1.6
Pm^{143}		142.9108		280d	EC; γ 0.742
Pm^{144}				~ 400d	EC; γ 0.61, 0.70, 0.48
Pm^{145}		144.9123		18y	EC; γ 0.072, (0.067)
Pm^{146}		145.9145		1.9y	EC; β^- 0.78; γ 0.75, 0.45
Pm^{147}		146.91486	7/2	2.65y	β^- 0.225; (γ)
Pm^{148}		147.9171	1	5.4d	β^- 2.45, 0.99, 1.9; γ 0.55, 1.46, 0.91
Pm^{148m}			6	43d	β^- 0.39, 0.49, 0.68; γ 0.55, 0.63, 0.73, others; (IT)
Pm^{149}		148.9181	7/2	53h	β^- 1.07 (others); (γ)
Pm^{150}		149.9203		2.7h	β^- 2.3, 3.2; γ 0.33, others 0.41–3.08
Pm^{151}		150.9216	5/2	28h	β^- 0.33–1.2; γ 0.10, 0.34, others 0.026–0.95
Pm^{152}				6m	β^- 2.2; γ 0.12 (m), 0.24, ~ 1
Pm^{153}				5.5m	β^- 1.65; γ 0.125, 0.18
Pm^{154}				2.5m	β^- 2.5

Nuclide	Per Cent Abundance	Nuclidic Mass	Spin	Half-life	Decay Modes, Radiations, Energies
$_{62}Sm^{141?}$				~20d	EC
Sm^{142}				72m	EC; β^+ 1.0; (γ)
Sm^{143}		142.9145		8.7m	EC; β^+ 2.5; (γ)
Sm^{143m}				1.1m	IT 0.75
Sm^{144}	3.09	143.9116			
Sm^{145}		144.9130		340d	EC; $e^-(\gamma)$ 0.061 (m)
Sm^{146}		145.9129		5×10^7y	α 2.55
Sm^{147}	14.97	146.91462	7/2	1.1×10^{11}y	α 2.15
Sm^{148}	11.24	147.9146			
Sm^{149}	13.83	148.9169	7/2		
Sm^{150}	7.44	149.9170			
Sm^{151}		150.9197		~93y	β^- 0.076; (γ)
Sm^{152}	26.72	151.9195			
Sm^{153}		152.9217	3/2	47h	β^- 0.70, 0.64, 0.80; γ 0.103, 0.070 (others)
Sm^{154}	22.71	153.9220			
Sm^{155}		154.9247		22m	β^- 1.53; γ 0.105 (others)
Sm^{156}		155.9257		9.4h	β^- 0.43, 0.71; γ 0.087, 0.20, 0.165, (0.25)
Sm^{157}				0.5m	β^-; γ 0.57
$_{63}Eu^{144?}$				18m	β^+ 2.4
Eu^{145}				5.8d	EC; γ 0.89, 0.65, 0.23 (others); (β^+)
Eu^{146}				4.4d	EC; (β^+); γ 0.75, 0.64, 0.71, 0.67 (others)
Eu^{147}		146.9166		22d	EC; γ 0.12, 0.077, 0.20 (others); $(\alpha$ 2.88)
Eu^{148}				54d	EC; γ 0.55, 0.63, others 0.24–2.19
Eu^{149}				106d	EC; γ 0.022 (m), others to 0.558
Eu^{150}		149.9196		12.5h	EC; γ 0.33–2.02
Eu^{150}				>5y	EC; γ 0.334, 0.439 (others)
Eu^{151}	47.82	150.9196	5/2		
Eu^{152}		151.9215	3	12.5y	EC; β^- 0.71 (others); (β^+); γ 0.122 (m), 0.344, 1.41, 0.96, 1.11, 1.08, 0.78 (others)
Eu^{152m}			0	9.3h	β^- 1.87 (others); EC; γ 0.122, 0.84 (others); (β^+)
Eu^{153}	52.18	152.9209	5/2		
Eu^{154}		153.9228	3	16y	β^- 0.25–1.85; γ 0.123 (m), others 0.25–1.60
Eu^{155}		154.9228		1.81y	β^- 0.15–0.25; γ 0.019–0.14
Eu^{156}		155.9247		15d	β^- 0.50, 2.45; γ 0.089–2.20
Eu^{157}		156.9253		15h	β^- ~1.7; γ 0.041–0.73
Eu^{158}				60m	β^- 2.65
Eu^{159}				18m	β^- 2.2; γ 0.07–0.22
Eu^{160}				2.5m	β^- ~3.6
$_{64}Gd^{145}$				24m	EC; β^+ 2.5; γ 0.78, 1.05
Gd^{146}				50d	EC; γ 0.11, 0.15, 0.07
Gd^{147}				22h	EC; γ 0.23, 0.38, 0.94, 0.64, 0.78, 0.28 (others)
Gd^{148}		147.9177		84y	α 3.18
Gd^{149}		148.9189		9.5d	EC; γ 0.150, 0.35, 0.30, 0.75 (others); (α)
Gd^{150}		149.9185		2×10^6y	α 2.73
Gd^{151}				120d	EC; γ 0.022–0.35
Gd^{152}	0.200	151.9195		1.1×10^{14}y	α 2.14
Gd^{153}		152.9211		242d	EC; γ 0.103 (m), 0.070, 0.10 (others)
Gd^{154}	2.15	153.9207			
Gd^{155}	14.73	154.9226	3/2		
Gd^{156}	20.47	155.9221			
Gd^{157}	15.68	156.9239	3/2		
Gd^{158}	24.87	157.9241			

Nuclide	Per Cent Abundance	Nuclidic Mass	Spin	Half-life	Decay Modes, Radiations, Energies
Gd^{159}		158.9260	3/2	18h	β^- 0.95, 0.89, 0.59; γ 0.058, 0.36 (others)
Gd^{160}	21.90	159.9271			
Gd^{161}		160.9293		3.7m	β^- 1.60, (1.54); γ 0.36, 0.057, 0.315, 0.102 (others)
Gd^{162}?				>1y	β^-; γ 0.04–1.39
$_{65}Tb^{147}$				24m	β^+; γ 0.31, 0.15
Tb^{148}				70m	β^+ 4.6; γ 0.78, 1.12
Tb^{149}				4.1h	EC; β^+; α 3.95; γ 0.35, 0.16 (others)
Tb^{149m}				4.3m	α 3.99; IT
Tb^{150}				3.1h	β^+; γ 0.64 (others)
Tb^{151}		150.9230		18h	EC; γ 0.11–1.31; (α 3.44)
Tb^{152}				18h	EC; γ 0.12–1.05; (β^+)
Tb^{152m}				4m	EC; β^+; γ 0.24, 0.14; (α)
Tb^{153}				2.6d	EC; γ 0.016–0.99
Tb^{154}				21h	EC; β^+ 2.8; γ 0.123 (m), 0.24–2.48
Tb^{154}				8h	EC; γ 0.123–1.29
Tb^{155}				5.6d	EC; γ 0.019–0.72
Tb^{156}				5.4d	EC; γ 0.089–2.31
Tb^{156m}				5.5h	IT 0.088; (β^-)
Tb^{157}				>30y	EC
Tb^{158}		157.9250		>3y	EC; γ 0.04
Tb^{158m}				11s	IT $e^-(\gamma)$ 0.11
Tb^{159}	100	158.9250	3/2		
Tb^{160}		159.9268	3	73d	β^- 0.27–1.71; γ 0.88, 0.30, 0.97, 0.087 (others)
Tb^{161}		160.9272	3/2	6.9d	β^- 0.51, 0.45, 0.58; γ 0.049, 0.026, 0.057, 0.075 (others)
Tb^{162}?				2h	β^-
Tb^{163}				7m	β^-; γ 0.18
Tb^{164}				23h	β^-
$_{66}Dy^{149}$				~15m	EC; γ 0.17
Dy^{150}				8m	β^+; γ 0.39; α 4.21
Dy^{151}				18m	EC; γ 0.145; (α 4.06; β^+)
Dy^{152}		151.9244		2.3h	EC; β^+; γ 0.26; (α 3.66)
Dy^{153}		152.9254		6h	EC; γ 0.08–0.54; (α 3.48)
Dy^{154}		153.9248		10^6y	α 2.85
Dy^{154m}				13h	α 3.35
Dy^{155}				10h	EC; γ 0.065–1.66; (β^+)
Dy^{156}	0.052	155.9238			
Dy^{157}				8.5h	EC; γ 0.33 (others)
Dy^{158}	0.090	157.9240			
Dy^{159}		158.9254		144d	EC; $e^-(\gamma)$ 0.058, (γ)
Dy^{160}	2.294	159.9248			
Dy^{161}	18.88	160.9266	5/2		
Dy^{162}	25.53	161.9265			
Dy^{163}	24.97	162.9284	5/2		
Dy^{164}	28.18	163.9288			
Dy^{165}		164.9317	7/2	2.3h	β^- 1.28, 1.19 (others); γ 0.095 (others 0.043–1.08)
Dy^{165m}				1.3m	IT $e^-(\gamma)$ 0.106; (β^-; γ)
Dy^{166}		165.9329	0	82h	β^- 0.40 (others); γ 0.084, 0.054, 0.030 (others)
Dy^{167}				4.4m	β^-
$_{67}Ho^{151}$				36s	EC; α 4.51
Ho^{151m}				42s	α 4.60
Ho^{152}				2.4m	α 4.38
Ho^{152m}				52s	α 4.45
Ho^{153}				9m	EC; (α 3.92)
Ho^{154}				5.6m	EC; α 4.12
Ho^{155}				46m	β^+ 2.1; γ 0.14
Ho^{156}				~1h	EC; γ 0.138

Nuclide	Per Cent Abundance	Nuclidic Mass	Spin	Half-life	Decay Modes, Radiations, Energies
Ho^{158}				1.9h	β^+ 1.90, 2.98; γ 0.85
Ho^{159}				33m	EC; γ 0.057–0.309
Ho^{160}				28m	EC; γ 0.73, 0.96, 0.88, 0.65 (others); (β^+)
Ho^{160m}				5.3h	IT $e^-(\gamma)$ 0.060
Ho^{161}			7/2	2.5h	EC; γ 0.026–0.18
Ho^{162}		161.9288		12m	β^+ 1.14; EC; $e^-\gamma$ 0.081
Ho^{162m}				68m	IT $e^-(\gamma)$ 0.010; $e^-\gamma$ 0.058, 0.038; EC; γ 0.185, 0.081, 1.21
Ho^{163}		162.9284		>500y	EC
Ho^{163m}				0.8s	IT γe^- 0.30
Ho^{164}		163.9303		37m	β^- 0.99, 0.90; EC; γ 0.037, 0.073, 0.091
Ho^{165}	100	164.9303	7/2		
Ho^{166}		165.9324	0	27h	β^- 1.85, 1.77; γ 0.08 (m) (others)
Ho^{166m}		165.9324		>30y	β^- <0.10; γ 0.08–1.42
Ho^{167}		166.9331		3.0h	β^- 0.28, 1.0; γ 0.057–0.53
Ho^{168}				3.3m	β^- ~2.2; γ 0.85
Ho^{170}				45s	β^- ~3.1; γ 0.43
$_{68}$Er^{152}				11s	α 4.93
Er^{153}				36s	α 4.68
Er^{154}				4.5m	α 4.26
Er^{158}				2.5h	β^+ 1.30; γ 0.098–0.85
Er^{159}				~1h	EC; γ 0.048–0.30
Er^{160}				29h	EC
Er^{161}				3.1h	EC; γ 0.83, 0.21 (others 0.084–1.97)
Er^{162}	0.136	161.9288			
Er^{163}				75m	EC; γ 0.43, 1.1
Er^{164}	1.56	163.9293			
Er^{165}			5/2	10h	EC
Er^{166}	33.41	165.9304			
Er^{167}	22.94	166.9320	7/2		
Er^{167m}				2.5s	IT γe^- 0.21
Er^{168}	27.07	167.9324			
Er^{169}		168.9347	1/2	9.5d	β^- 0.34; $e^-(\gamma)$ 0.0084
Er^{170}	14.88	169.9355			
Er^{171}		170.9382	5/2	7.5h	β^- 1.05 (others); γ 0.308, 0.296, 0.112 (others)
Er^{172}		171.9396		50h	β^- 0.50, 0.38, 0.29; γ 0.050, 0.41, 0.61 (others)
$_{69}$Tm^{161}				30m	EC; γ 0.084, 0.144, 0.147, 0.17
Tm^{162}				77m	EC; γ 0.10, 0.24
Tm^{163}				1.9h	EC; β^+ 1.05, 0.40; γ 0.022–0.66
Tm^{164}		163.9335		2.0m	EC; β^+ 2.9; γ 0.091 (others)
Tm^{165}				30h	EC; γ 0.25, 0.29, 0.80, 0.34 (others)
Tm^{166}		165.9330	2	7.7h	EC; (β^+ 2.1); γ 0.073–2.10
Tm^{167}			1/2	9.4d	EC; γ 0.057 (m)
Tm^{168}				93d	EC; γ 0.075–1.64; (β^-)
Tm^{169}	100		1/2		
Tm^{170}		169.9359	1	127d	β^- 0.97, 0.88; $e^-\gamma$ 0.084 (m)
Tm^{171}		170.9366	1/2	1.9y	β^- 0.10; (γ 0.067)
Tm^{172}		171.9386		63.6h	β^- 0.28–1.92; γ 0.079 (others 0.18–1.61)
Tm^{173}				7.3h	β^- 0.9; γ 0.40, 0.47
Tm^{174}				5.5m	β^- 2.5
Tm^{175}				20m	β^- 2.0; γ 0.51
Tm^{176}				1.5m	β^- 4.2
$_{70}$Yb^{155}				1.6s	α 5.21
Yb^{164}				75m	EC
Yb^{166}		165.9333		56h	EC; γ 0.082

Nuclide	Per Cent Abundance	Nuclidic Mass	Spin	Half-life	Decay Modes, Radiations, Energies
Yb^{167}				18m	EC; γ 0.026–0.18
Yb^{168}	0.135	167.9339			
Yb^{169}				31d	EC; γ 0.008–0.31
Yb^{169m}				46s	IT 0.024
Yb^{170}	3.03	169.9349			
Yb^{171}	14.31	170.9365	1/2		
Yb^{172}	21.82	171.9366			
Yb^{173}	16.13	172.9383	5/2		
Yb^{174}	31.84	173.9390			
Yb^{175}		174.9414		4.2d	β^- 0.47, 0.07; γ 0.396 (others)
Yb^{175m}				0.072s	IT 0.495
Yb^{176}	12.73	175.9427			
Yb^{177}		176.9455		1.9h	β^- 1.38 (others); (γ 0.12 1.24)
Yb^{177m}				6.5s	IT $e^-\gamma$ 0.23; γe^- 0.104
$_{71}$Lu^{167}				55m	EC; γ 0.03–0.40; (β^+)
Lu^{168}				7.1m	EC; γ 0.087, 0.99, 0.90 (others); (β^+)
Lu^{168}				2.15h	EC; γ 0.087
Lu^{169}				1.5d	EC; γ 0.024–1.39
Lu^{170}		169.9387		2.0d	EC; γ 0.084 (m), 2.04 (others 0.15–3.02)
Lu^{171}				8.3d	EC; γ 0.020–1.50
Lu^{172}				6.7d	EC; γ 0.079–2.08
Lu^{172m}				3.7m	IT $e^-(\gamma)$ 0.042
Lu^{173}		172.9390		500d	EC; γ 0.079,0.101, 0.273 (others)
Lu^{174}		173.9406		3.6y	EC; γ 0.077, 1.23
Lu^{174m}				160d	IT $e^-(\gamma)$ 0.059; γ 0.067, 0.045
Lu^{175}	97.41	174.9409	7/2		
Lu^{176}	2.59	175.9427	7	3×10^{10}y	β^- 0.43; γ 0.31, 0.20, 0.088 (m)
Lu^{176m}			1	3.7h	β^- 1.20, 1.10; γ 0.088 (m)
Lu^{177}		176.9440	7/2	6.8d	β^- 0.50 (others), γ 0.113 (others)
Lu^{177m}				160d	$\beta^- \sim 1$; γ 0.208, others 0.10–0.41
Lu^{178}				22m	β^-; γ 0.33, 0.43, 0.56, 0.67, 0.78
Lu^{179}				4.6h	β^- 1.35, 1.08; γ 0.215
Lu^{180}				2.5m	β^- 3.3
$_{72}$Hf^{168}				22m	EC; γ 0.13, 0.17; (β^+)
Hf^{169}				1.5h	EC; γ 0.049, 0.12; (β^+)
Hf^{170}				12h	EC
Hf^{171}				11h	EC; γ 0.12–1.07
Hf^{172}				$\sim$5y	EC; γ 0.024, 0.12, 0.08 (others)
Hf^{173}				24h	EC; γ 0.12, 0.30 (others)
Hf^{174}	0.18	173.9403		2×10^{15}y	α 2.50
Hf^{175}				70d	EC; γ 0.343 (others)
Hf^{176}	5.20	175.9416			
Hf^{177}	18.50	176.9435			
Hf^{178}	27.14	177.9439			
Hf^{178m}				4s	IT $e^-(\gamma)$ 0.089; γ 0.427, 0.326, 0.214, 0.093 (m)
Hf^{179}	13.75	178.9460	9/2		
Hf^{179m}				19s	IT $e^-(\gamma)$ 0.161; γ 0.217
Hf^{180}	35.24	179.9468			
Hf^{180m}				5.5h	IT $e^-\gamma$ 0.058; γ 0.44, 0.33, 0.22, 0.093, 0.50
Hf^{181}		180.94908		45d	β^- 0.41 (others), γ 0.133 (m$_2$), 0.48 (m$_1$), 0.35, 0.14 (others)
Hf^{182}		181.9507		9×10^6y	β^-; γ 0.27
Hf^{183}		182.9538		65m	$\beta^- \sim 1.4$; γ

Nuclide	Per Cent Abundance	Nuclidic Mass	Spin	Half-life	Decay Modes, Radiations, Energies
Hf^{184}				2.2h	β^-
$_{73}Ta^{172}$				24m	EC; β^+
Ta^{173}				3.7h	EC; γ 0.090, 0.17, (β^+)
Ta^{174}				1.2h	β^+; EC; γ 0.13, 0.21, 0.28, 0.35 (others)
Ta^{175}				11h	EC; γ 0.050–1.64
Ta^{176}				8.0h	EC; γ 0.088 (m), 0.20 (others 0.091–2.9)
Ta^{177}		176.9447		57h	EC; γ 0.113 (others 0.05–1.06)
Ta^{178}		177.9459		2.2h	EC; γ 0.089–0.43
Ta^{178}				9.3m	EC; γ 0.093 (others); (β^+)
Ta^{179}		178.9461		1.6y	EC
Ta^{180}	0.0123	179.9475			
Ta^{180m}				8.1h	EC; γ 0.093; β^- 0.60, 0.70; $(\gamma$ 0.102)
Ta^{181}	99.9877	180.94798	7/2		
Ta^{182}		181.95014		115d	β^- 0.18–0.51; γ 0.033–1.61
Ta^{182m}				16m	IT $e^-(\gamma)$ 0.184; γ 0.172, 0.147 (others)
Ta^{183}		182.95144		5.0d	β^- 0.62 (others); γ 0.041–0.406
Ta^{184}		183.9538		8.7h	β^- 0.15–1.36; γ 0.11–1.2
Ta^{185}		184.9555		48m	β^- 1.72 (others); γ 0.175, 0.075, 0.100 (others)
Ta^{186}		185.9583		10.5m	β^- 2.2; γ 0.125–1.1
$_{74}W^{176}$				80m	EC; γ 1.3, 0.1; (β^+)
W^{177}				2.2h	EC; γ 0.45, 1.2
W^{178}				22d	EC; $(\gamma?)$
W^{179}				40m	EC; γ 0.031
W^{179m}				7m	IT 0.22
W^{180}	0.135	179.9470			
W^{180m}				0.005s	IT 0.24; γ 0.37
W^{181}		180.9482		126d	EC; $e^-(\gamma)$ 0.0063 (m); (other γ's)
W^{182}	26.41	181.94827			
W^{183}	14.40	182.95029	1/2		
W^{183m}				5.3s	IT 0.103; γ 0.11, 0.053, 0.046
W^{184}	30.64	183.95099			
W^{185}		184.9535		74d	β^- 0.43, (0.30); $(\gamma$ 0.125)
W^{185m}				1.7m	IT $e^-\gamma$ 0.125; γ 0.175, 0.075, 0.100
W^{186}	28.41	185.9543			
W^{187}		186.9574	3/2	24h	β^- 0.63, 1.32, 0.34; γ 0.69, 0.48 (others)
W^{188}		187.9587		65d	β^- 0.43, 0.34; γ 0.057, 0.15, 0.22, 0.26, 0.29
$_{75}Re^{177}$				17m	β^+
Re^{178}				15m	EC; β^+ 3.1
Re^{179}				20m	EC
Re^{180}		179.9501		2.4m	EC; $(\beta^+$ 1.1); γ 0.88, 0.11
Re^{180m}				20h	EC; β^+ 1.9
Re^{181}				20h	EC; γ 0.020–1.54
Re^{182}				13h	EC; γ 0.032–2.05
Re^{182}				64h	EC; γ 0.018–1.44
Re^{183}				68h	EC; γ 0.041–0.407
Re^{184}				35d	EC; γ 0.111, 0.90, 0.79, 0.89 (others)
Re^{184m}				165d	EC; (IT 0.217); γ 0.11–0.90
Re^{185}	37.07	184.9530	5/2		
Re^{186}		185.9551		89h	β^- 1.07, 0.93; γ 0.137 (others); (EC)
Re^{187}	62.93	186.9560	5/2	6×10^{10}y	β^- 0.001
Re^{188}		187.9582		17h	β^- 2.13, 2.0; γ 0.155 (others)
Re^{188m}				19m	IT $e^-(\gamma)$ 0.002, 0.016; γ 0.064, 0.106, 0.092

Nuclide	Per Cent Abundance	Nuclidic Mass	Spin	Half-life	Decay Modes, Radiations, Energies
Re^{189}				23h	β^- 0.98, others; γ 0.070, 0.15, 0.22, 0.25
Re^{190}		189.9622		2.8m	β^- 1.7; γ 0.19, 0.39, 0.57, 0.83
Re^{191}				9.8m	β^- 1.8
$_{76}$Os^{181}				23m	EC; γ 0.09, 0.10
Os^{181}				2.7h	EC; γ 0.23
Os^{182}				22h	EC; γ 0.510, 0.18, 0.26 (others); [to 13h Re^{182}]
Os^{183}				13.7h	EC; γ 0.114, 0.382 (m), 0.168 (others)
Os^{183m}				9.9h	EC; IT 0.171; γ 1.108, 1.102 (others)
Os^{184}	0.018	183.9526			
Os^{185}		184.9541		94d	EC; γ 0.65, 0.88 (others)
Os^{186}	1.59	185.9539			
Os^{187}	1.64	186.9560	1/2		
Os^{188}	13.3	187.9560			
Os$^{188m?}$				26d	IT
Os^{189}	16.1	188.9582	3/2		
Os^{189m}				5.7h	IT $e^-(\gamma)$ 0.031
Os^{190}	26.4	189.9586			
Os^{190m}		189.9604		10m	IT $e^-(\gamma)$ 0.038; γ 0.61, 0.50, 0.36, 0.19
Os^{191}		190.9612		14.6d	β^- 0.14; [to 4.8s Ir^{191m}]
Os^{191m}				14h	IT $e^-(\gamma)$ 0.074
Os^{192}	41.0	191.9614			
Os^{193}		192.9645		31h	β^- 1.13 (others); γ 0.139 (others 0.07–0.56)
Os^{194}				1.9y	β^-
Os^{195}				6.5m	β^- 2
$_{77}$Ir^{182}				15m	EC; γ 0.13–4.0; (β^+)
Ir^{183}				55m	EC; γ 0.24, others
Ir^{184}				3.2h	EC; γ 0.125–4.3; (β^+)
Ir^{185}				14h	EC; γ 0.037–1.10
Ir^{186}		185.9580		15.8h	EC; γ 0.071–2.89; (β^+ 1.94)
Ir^{187}				10.5h	EC; γ 0.010–0.99
Ir^{188}		187.9590		41h	EC; γ 0.155, others 0.32–2.22; (β^+ 1.66)
Ir^{189}				12d	EC; γ 0.070, 0.25, others 0.031–0.28
Ir^{190}		189.9608		12.3d	EC; γ 0.187–1.43
Ir^{190m}		189.9637		3.2h	EC; β^+ 2.04; [to 10m Os^{190m}]
Ir^{191}	37.3	190.9608	3/2		
Ir^{191m}				4.8s	IT $e^-(\gamma)$ 0.042; $e^-\gamma$ 0.129, 0.047, 0.082
Ir^{192}		191.9630		74.0d	β^- 0.67, 0.54, 0.24; γ 0.316, 0.47, 0.30 (others); (EC)
Ir^{192m_1}				1.45m	IT $e^-(\gamma)$ 0.057; (β^-)
Ir^{192m_2}				6 × 10^{2}y	IT 0.16 [to 74d Ir^{192}]
Ir^{193}	62.7	192.9633	3/2		
Ir^{193m}				12d	IT $e^-(\gamma)$ 0.080
Ir^{194}		193.9652		19h	β^- 2.24, 1.91 (others); γ 0.328 (others)
Ir^{194m}				0.032s	IT 0.115
Ir^{195}				2.3h	β^- ~1; γ 0.10–0.66
Ir^{197}				7m	β^- 1.5, 2.0; γ 0.50
Ir^{198}				50s	β^- 3.6; γ 0.78
$_{78}$Pt^{185}				1.2h	
Pt^{186}				2.9h	EC
Pt^{187}				2.2h	
Pt^{188}		187.9596		10d	EC; γ 0.195, 0.187, 0.055 (others); (α 3.9)
Pt^{189}				11h	EC; γ 0.072–0.80
Pt^{190}	0.0127	189.9599		7 × 10^{11}y	α 3.11

Nuclide	Per Cent Abundance	Nuclidic Mass	Spin	Half-life	Decay Modes, Radiations, Energies
Pt^{191}				3.0d	EC; γ 0.042–0.62
Pt^{192}	0.78	191.9614			
Pt^{193}		192.9633		0.3–500y	EC
Pt^{193m}				4.4d	IT e^-(γ) 0.135; e^-(γ) 0.013
Pt^{194}	32.9	193.9628	0		
Pt^{195}	33.8	194.96482	1/2		
Pt^{195m}				4.1d	IT e^-(γ) 0.130; e^-γ 0.031, 0.099
Pt^{196}	25.3	195.96498	0		
Pt^{197}		196.96736		20h	β$^-$ 0.67, 0.48; γ 0.077 (m), 0.19
Pt^{197m}				1.3h	IT 0.35; (β$^-$; γ 0.28)
Pt^{198}	7.21	197.9675			
Pt^{199}		198.9707		30m	β$^-$ 0.8–1.7; γ 0.074–0.96
Pt^{199m}				14s	IT γ(e^-) 0.39; e^-(γ) 0.032
Pt^{200}				11.5h	β$^-$
Pt^{201}				2.3m	β$^-$
$_{79}$Au^{185}				7m	EC
Au^{186}				12m	EC; γ 0.16, 0.22, 0.30, 0.40
Au^{187}				8m	EC
Au^{188}				8m	EC; γ 0.25, 0.33, 0.63; (α 5.1)
Au^{189}				30m	EC; γ
Au^{190}				40m	EC; γ 0.29 (others 0.30–3.46)
Au^{191}			3/2	3.4h	EC; γ 0.030–2.17
Au^{192}		191.9649	1	4.1h	EC; β$^+$ 2.2; γ 0.045–1.16
Au^{193}			3/2	17h	EC; γ 0.186, 0.112 (others 0.013–0.49)
Au^{193m}				3.9s	IT e^-(γ) 0.032; γ 0.26; (EC)
Au^{194}		193.9655	1	39h	EC; γ 0.328, 0.29, 0.62 (others 0.095–2.41); (β$^+$)
Au^{195}		194.96511	3/2	183d	EC; γ 0.099, 0.031, 0.13
Au^{195m}				31s	IT e^-(γ) 0.057; γ 0.20, 0.061, 0.26
Au^{196}		195.96655	2	6.2d	EC; γ 0.356, 0.333 (others); (β$^-$ 0.26)
Au^{196m}			12	9.7h	IT e^-(γ) 0.175; γ 0.148, 0.188, 0.085 (m) (others)
Au^{197}	100	196.96655	3/2		
Au^{197m}				7.2s	IT e^-(γ) 0.130; γ 0.28
Au^{198}		197.96824	2	2.70d	β$^-$ 0.96 (others); γ 0.412 (others)
Au^{199}		198.96865	3/2	3.15d	β$^-$ 0.30, 0.25; γ 0.158, 0.208
Au^{200}		199.9708		48m	β$^-$ 2.2, 0.7; γ 0.37, 1.23
Au^{201}		200.9719		25m	β$^-$ 1.5; (γ 0.53)
Au^{202}				25s	β$^-$
Au^{203}				55s	β$^-$ 1.9; γ 0.69
$_{80}$Hg^{185}				53s	EC; α 5.64
Hg^{186}				1.5m	EC; γ 0.13, 0.27, 0.35, 0.44
Hg^{187}				3m	EC; α 5.14; γ 0.18, 0.25, 0.40
Hg^{188}				3.7m	EC; γ 0.14
Hg^{189}				9m	EC; γ 0.17, 0.24, 0.32, 0.50
Hg^{190}				20m	EC; γ 0.14, 0.22
Hg^{191}				57m	EC; γ 0.25, 0.27
Hg^{192}				5h	EC; γ 0.031–0.31
Hg^{193}				6h	EC; γ 0.038–1.08
Hg^{193m}				11h	EC; γ 0.032–1.65; IT e^-(γ) 0.101
Hg^{194}		193.9657		146d	EC
Hg^{194m}				0.4s	IT; γ 0.134, 0.048
Hg^{195}			1/2	9.5h	EC; γ 0.061–1.17
Hg^{195m}			13/2	40h	EC; IT e^-(γ) 0.123; γ 0.016–1.24
Hg^{196}	0.146	195.96582			
Hg^{197}			1/2	65h	EC; γ 0.077 (m), (0.191)
Hg^{197m}			13/2	23h	IT e^-(γ) 0.165; γ 0.134; (EC; γ)
Hg^{198}	10.02	197.96677			

Nuclide	Per Cent Abundance	Nuclidic Mass	Spin	Half-life	Decay Modes, Radiations, Energies
Hg^{199}	16.84	198.96826	1/2		
Hg^{199m}		198.96883		44m	IT $e^-\gamma$ 0.370; γ 0.158
Hg^{200}	23.13	199.96834			
Hg^{201}	13.22	200.97031	3/2		
Hg^{202}	29.80	201.97063			
Hg^{203}		202.97285		47d	β^- 0.21; γ 0.279
Hg^{204}	6.85	203.97348			
Hg^{205}		204.9762		5.5m	β^- 1.65; (γ 0.20)
Hg^{206}		205.97747		8.5m	β^- 1.29
$_{81}Tl^{191}$				10m	EC
Tl^{192}				short	
Tl^{192m}				11m	IT 0.11; EC; γ 0.42
Tl^{193}				23m	EC; γ 0.24, 0.25, 0.26, 0.31 (others)
Tl^{194}				33.0m	EC; γ 0.43
Tl^{194m}				32.8m	EC; γ 0.097; IT
Tl^{195}			1/2	1.2h	EC; γ 0.037 (others); $\beta^+ \sim 1.8$
Tl^{195m}				3.5s	IT $e^-(\gamma)$ 0.099; γ 0.383, 0.393
Tl^{196}		195.9708		1.8h	EC; γ 0.426; β^1
Tl^{196m}		195.9750		1.4h	EC; γ 0.084 (others); (IT)
Tl^{197}			1/2	2.8h	EC; γ 0.152 (others); (β^+)
Tl^{197m}				0.54s	IT γe^- 0.222; γ 0.385, 0.387
Tl^{198}		197.9705	2	5.3h	EC; γ 0.412, others 0.19 2.78; (β^+)
Tl^{198m}			7	1.9h	IT 0.261; γ 0.28 (others); EC; γ 0.049–0.64
Tl^{199}		198.9694	1/2	7.4h	EC; γ 0.158 (m), others 0.037–0.49
Tl^{199m}				0.042s	IT 0.37
Tl^{200}		199.97097	2	26h	EC; γ 0.065–2.28; (β^+)
Tl^{201}		200.9708	1/2	73h	EC; γ 0.167 (others)
Tl^{202}		201.9721	2	12d	EC; γ 0.44 (others)
Tl^{203}	29.50	202.97233	1/2		
Tl^{204}		203.97389	2	3.80y	β^- 0.76; (EC)
Tl^{205}	70.50	204.97446	1/2		
Tl^{206}		205.97608		4.3m	β^- 1.57
Tl^{207}		206.97745		4.8m	β^- 1.44; (γ)
Tl^{208}		207.98201		3.1m	β^- 1.79, 1.28, 1.52; γ 2.61, 0.58, 0.51, 0.86 (others)
Tl^{209}		208.98530		2.2m	β^- 2.0; γ 0.12, 0.45, 1.56
Tl^{210}		209.99000		1.3m	β^- 1.97; γ 0.09–2.45
$_{82}Pb^{194}$				11m	EC; γ 0.20
Pb^{195}				17m	EC; [to 3.5s Tl^{195m}]
Pb^{196}				37m	EC; γ 0.19–0.50
Pb^{197m}				42m	EC [to 0.54s Tl^{197m}]; IT 0.234; γ 0.085
Pb^{198}				2.4h	EC; γ 0.031–0.87
Pb^{199}				1.5h	EC; γ 0.367, 0.353, 0.72
Pb^{199m}				12m	IT $e^-\gamma$ 0.42
Pb^{200}				21.5h	EC; γ 0.033–0.61
Pb^{201}				9.5h	EC; γ 0.33 (others 0.13–1.40)
Pb^{201m}				1.0m	IT γe^- 0.63
Pb^{202}		201.9722		$\sim 3 \times 10^5$y	EC
Pb^{202m}		201.9745		3.6h	IT 0.79, 0.13; γ 0.42, 0.96, 0.66 (others); (EC)
Pb^{203}		202.97321		52h	EC; γ 0.279 (others)
Pb^{203m}		202.97410		6.1s	IT γe^- 0.825
Pb^{204}	1.48	203.97307			
Pb^{204m_2}		203.97542		67m	IT $\gamma(e^-)$ 0.912; γ 0.375 (m_1), 0.899
Pb^{205}		204.97452		3×10^7y	EC
Pb^{205m}		204.97561		0.004s	IT $e^-(\gamma)$ 0.026; γ 0.99 (others)
Pb^{206}	23.6	205.97446	0		
Pb^{207}	22.6	206.97590	1/2		
Pb^{207m}		206.97765		0.8s	IT γe^- 1.06; γ 0.57

Nuclide	Per Cent Abundance	Nuclidic Mass	Spin	Half-life	Decay Modes, Radiations, Energies
Pb^{208}	52.3	207.97664	0		
Pb^{209}		208.98111		3.3h	β^- 0.64
Pb^{210}		209.98418		22y	β^- 0.015, 0.061; $e^-(\gamma)$ 0.046; (α)
Pb^{211}		210.98880		36.1m	β^- 1.36 (others); (γ 0.07–1.10)
Pb^{212}		211.99190		10.6h	β^- 0.34, 0.58; γ 0.239 (others)
Pb^{214}		213.9998		26.8m	β^- 0.59, 0.65; γ 0.053–0.35
$_{83}$Bi$^{196?}$				7m	EC; (α 5.83)
Bi$^{197?}$				2m	α 6.2
Bi^{199}			9/2	26m	EC; (α 5.47)
Bi^{200}			7	35m	EC; γ 0.46, 1.03
Bi^{201}			9/2	1.85h	EC
Bi^{201}				62m	EC; (α 5.15)
Bi^{202}			5	1.6h	EC; γ 0.42, 0.96
Bi^{203}		202.9768	9/2	11.8h	EC; γ 0.060–1.90; (β^+; α 4.85)
Bi^{204}		203.9777	6	11.2h	EC; γ 0.079–2.10
Bi^{205}		204.97742	9/2	15.3d	EC; γ 0.026–2.61; (β^+)
Bi^{206}		205.9783	6	6.24d	EC; γ 0.11–1.90
Bi^{207}		206.97847		30y	EC; γ 0.57, 1.06 (Pb^{207m}) (others)
Bi^{208}		207.97973		$\sim 8 \times 10^5$y	EC; γ 2.61
Bi^{208m}				0.0026s	IT 0.92; γ 0.51
Bi^{209}	100	208.98042	9/2		
Bi^{210}		209.98411	1	5.01d	β^- 1.16; (α)
Bi^{210m}				3×10^6y	α 4.95, 4.92 (others); γ 0.26 (m), 0.30 (m) (others)
Bi^{211}		210.98729		2.15m	α 6.62, 6.28; γ 0.35; (β^-)
Bi^{212}		211.99127		60.6m	β^- 2.25 (others); α 6.05 (others); $e^-(\gamma)$ 0.040 (other γ's)
Bi^{213}		212.99433		47m	β^- 1.39, 0.96; γ 0.44; (α 5.86)
Bi^{214}		213.99863		19.7m	β^- 1.51, 1.0, 3.18; γ 0.61–2.42; (α)
Bi^{215}		215.0019		8m	β^-
$_{84}$Po^{192}				0.5s	α 6.58
Po^{193}				4s	α 6.47
Po^{194}				13s	α 6.38
Po^{195}				30s	α 6.25
Po^{196}				1.8m	α 6.14
Po^{197}				4m	α 6.04
Po^{198}				7m	α 5.93
Po^{199}				12m	α 5.87
Po^{200}				11m	EC; (α 5.86, 5.75)
Po^{201}			3/2	18m	EC; (α 5.57, 5.67, 5.77)
Po^{202}			0	44m	EC; (α 5.57)
Po^{203}			5/2	42m	EC; (α 5.48)
Po^{204}			0	3.5h	EC; (α 5.37)
Po^{205}			5/2	1.8h	EC; (α 5.23)
Po^{206}		205.9805	0	8.8d	EC; γ 0.060–1.32; (α 5.22)
Po^{207}		206.98159	5/2	6.0h	EC; γ 0.100–2.06; (β^+; α 5.1)
Po^{208}		207.98126		2.9y	α 5.11; (EC; γ)
Po^{209}		208.98246	1/2	103y	α 4.88; (EC; γ)
Po^{210}		209.98287	0	138.4d	α 5.305
Po^{211}		210.98665		0.52s	α 7.448 (others); (γ)
Po^{211m}		210.98804		25s	α 7.14 (others); γ 1.06, 0.57 (Pb^{207m})
Po^{212}		211.98886		3×10^{-7}s	α 8.78
Po^{212m}		211.99201		46s	α 11.7 (others); (γ)
Po^{213}		212.99284		4.2×10^{-6}s	α 8.34
Po^{214}		213.99519		1.6×10^{-4}s	α 7.69
Po^{215}		214.99947		0.0018s	α 7.37; (β^-)
Po^{216}		216.00192		0.16s	α 6.78

Nuclide	Per Cent Abundance	Nuclidic Mass	Spin	Half-life	Decay Modes, Radiations, Energies
Po^{217}				<10s	α 6.54
Po^{218}		218.0089		3.05m	α 6.00; (β^-)
$_{85}At^{200?}$				0.9m	α 6.41, 6.46
At^{201}				1.5m	α 6.35
At^{202}				3.0m	EC; α 6.13, 6.23
At^{203}				7.4m	EC; α 6.09
At^{204}				9.3m	EC; (α 5.95)
$At^{204?}$				25m	EC
At^{205}				26m	EC; α 5.90
At^{206}				30m	EC; (α 5.70; γ)
$At^{206?}$				2.9h	EC
At^{207}		206.9857		1.8h	EC; α 5.75
At^{208}		207.9865		1.6h	EC; γ 0.66, 0.17, 0.25; (α 5.65)
$At^{208?}$				6.2h	EC
At^{209}		208.98614		5.5h	EC; γ 0.78, 0.54, 0.20, 0.091; (α 5.64)
At^{210}		209.9870		8.3h	EC; γ 0.047–1.60; (α)
At^{211}		210.98750	9/2	7.21h	EC; α 5.86; (γ)
At^{212}				0.30s	α 7.66, 7.60; γ 0.063
At^{212m}				0.12s	α 7.82, 7.88; γ 0.063
At^{213}		212.9931		<2s	α 9.02
At^{214}		213.9963		2×10^{-6}s	α 8.78
At^{215}		214.99866		$\sim10^{-4}$s	α 8.00
At^{216}		216.00240		3×10^{-4}s	α 7.79
At^{217}		217.00465		0.018s	α 7.05
At^{218}		218.00855		1.3s	α 6.69 (others)
At^{219}		219.0114		0.9m	α 6.27; (β^-)
$_{86}Rn^{204}$				3m	α 6.28
Rn^{206}				6.5m	α 6.25; EC
Rn^{207}				11m	EC; (α 6.12)
Rn^{208}				22m	EC; α 6.14
Rn^{209}				30m	EC; α 6.04
Rn^{210}		209.9897		2.7h	α 6.04; EC
Rn^{211}		210.99060		16h	EC; γ 0.032–1.82; α 5.78 (others)
Rn^{212}		211.99073		25m	α 6.26
Rn^{213}				0.019s	α 8.13
Rn^{215}		214.9987		$\sim10^{-6}$s	α 8.6
Rn^{216}		216.00023		4.5×10^{-5}s	α 8.01
Rn^{217}		217.00392		5.4×10^{-4}s	α 7.68
Rn^{218}		218.00559		0.019s	α 7.12 (others); (γ)
Rn^{219}		219.00952		3.92s	α 6.81, 6.55 (others); γ 0.27 (others)
Rn^{220}		220.01140		54s	α 6.28; (γ)
Rn^{221}				25m	β^-; α 6.00
Rn^{222}		222.0175		3.82d	α 5.49 (others); (γ)
$_{87}Fr^{205?}$				$\sim$4s	α 6.83
Fr^{206}				16s	α 6.74
Fr^{207}				19s	α 6.74
Fr^{208}				37s	α 6.59
Fr^{209}				54s	α 6.62
Fr^{210}				2.6m	α 6.50
Fr^{211}				3.1m	α 6.52
Fr^{212}		211.9961		19m	EC; α 6.39, 6.41, 6.34; γ
Fr^{213}				34s	α 6.77
Fr^{214}				0.004s	α 8.55
Fr^{215}				<0.001s	α 9.4
Fr^{217}		217.0048		<2s	α 8.3
Fr^{218}		218.0075		0.005s	α 7.85
Fr^{219}		219.00925		0.02s	α 7.30
Fr^{220}		220.01233		28s	α 6.69
Fr^{221}		221.01418		4.8m	α 6.33, 6.11; γ 0.22
Fr^{222}				15m	β^-; (α)
Fr^{223}		223.01980		22m	β^- 1.15; γ 0.049, 0.080 (others) (α 5.34)

Nuclide	Per Cent Abundance	Nuclidic Mass	Spin	Half-life	Decay Modes, Radiations, Energies
$_{88}$Ra^{212}				18s	α 6.90
Ra^{213}				2.7m	α 6.74, 6.61
Ra^{214}				2.6s	α 7.17
Ra^{215}				0.0016s	α 8.7
Ra^{219}		219.0100		~0.001s	α 8.00
Ra^{220}		220.01097		0.023s	α 7.45 (others); (γ)
Ra^{221}		221.01386		28s	α 6.61, 6.75, 6.66, 6.57; γ 0.15, 0.18 (others)
Ra^{222}		222.01536		38s	α 6.56, 6.23; γ 0.33
Ra^{223}		223.01856		11.7d	α 5.71, 5.60 (others); γ 0.031–0.58
Ra^{224}		224.02022		3.64d	α 5.68 (others); (γ)
Ra^{225}		225.02352		14.8d	β^- 0.32; γe^- 0.040
Ra^{226}		226.0254		1622y	α 4.78 (others;) (γ)
Ra^{227}		227.02922		41m	β^- 1.30 (others); (γ)
Ra^{228}		228.03123		6.7y	β^- 0.055; (γ?)
Ra^{229}				~1m	β^-
Ra^{230}				1h	β^- 1.2
$_{89}$Ac^{213}				~1s	α 7.42
Ac^{214}				12s	α 7.12, 7.18, 7.24
Ac^{221}		221.0157		<2s	α 7.54
Ac^{222}		222.0178		4.2s	α 6.96
Ac^{223}		223.01912		2.2m	α 6.64; EC
Ac^{224}		224.0217		2.9h	EC; γ 0.22, 0.13; α 6.17
Ac^{225}		225.02314		10.0d	α 5.82, 5.78 (others); γ 0.037 (others)
Ac^{226}		226.0262		29h	β^- 1.17; γ 0.23, 0.16, 0.07; EC; γ 0.25, 0.18
Ac^{227}		227.02781	3/2	22y	β^- 0.046; (α; γ)
Ac^{228}		228.03117		6.13h	β^- 1.11, 0.45, 2.18; γ 0.057–1.64
Ac^{229}				66m	β^-
Ac^{230}				<1m	β^- 2.2
Ac^{231}		231.0386		15m	β^- 2.1; γ 0.085–0.71
$_{90}$Th^{223}		223.0209		0.9s	α 7.55
Th^{224}		224.02138		1.1s	α 7.17, 6.9 (others); γ 0.18 (others)
Th^{225}		225.0237		8.0m	α 6.47 (others); EC; γ
Th^{226}		226.02489		31m	α 6.33, 6.22; γ 0.11 (others)
Th^{227}		227.02777		18.2d	α 5.98, 6.04 (others); γ 0.030–0.33
Th^{228}		228.02675		1.9y	α 5.42, 5.34; $e^-(\gamma)$ 0.084; (γ)
Th^{229}		229.03163		7.3×10^3y	α 4.84, others; γ 0.20, 0.15
Th^{230}		230.0331		7.5×10^4y	α 4.68, 4.61; $e^-(\gamma)$ 0.068, (γ)
Th^{231}		231.03635		25.6h	β^- 0.30, 0.22, 0.14; γ 0.017–0.23
Th^{232}	100	232.03821		1.39×10^{10}y	α 4.01, 3.95; $e^-(\gamma)$ 0.059
Th^{233}		233.04143		22.4m	β^- 1.23; (γ)
Th^{234}		234.0436		24.1d	β^- 0.19, 0.10; γ 0.029, 0.063, 0.091
Th^{235}				<5m	β^-
$_{91}$Pa^{224}				~0.6s	α 7.75
Pa^{225}				0.8s	α 7.24
Pa^{226}		226.0278		1.8m	α 6.81
Pa^{227}		227.02885		38m	α 6.46; EC
Pa^{228}		228.03100		22h	EC; (α); γ 0.057–1.89
Pa^{229}		229.03195		1.5d	EC; (α 5.69)
Pa^{230}		230.03437		17d	EC; β^- 0.41; γ 0.053–0.95; (α; β^+)
Pa^{231}		231.03594	3/2	3.48×10^4y	α 5.00, 4.94, 5.02, 4.72 (others); γ 0.027–0.38
Pa^{232}		232.03861		1.31d	β^- 0.26, 0.37 (others); γ 0.047–1.15
Pa^{233}		233.04011	3/2	27.0d	β^- 0.25, 0.15; γ 0.016–0.42
Pa^{234}		234.0434		6.7h	β^- 0.14, 0.28 (others); γ 0.043–1.68

Nuclide	Per Cent Abundance	Nuclidic Mass	Spin	Half-life	Decay Modes, Radiations, Energies
Pa^{234m}				1.18m	β^- 2.33 (others); (IT 0.021; γ)
Pa^{235}		235.0454		24m	β^- 1.4
Pa^{236}				12m	β^- 3.3
Pa^{237}		237.0510		39m	β^- 2.30, others; γ 0.88, 0.46, 0.92 (others)
$_{92}$U^{227}		227.0309		1.3m	α 6.8
U^{228}		228.03128		9.3m	α 6.68; EC; γ
U^{229}		229.0332		58m	EC; 6.36, 6.33 (others); γ 0.029 (others)
U^{230}		230.03393		21d	α 5.88, 5.81; $e^-(\gamma)$ 0.072; (γ)
U^{231}		231.0363		4.3d	EC; γ 0.018–0.22; (α 5.45)
U^{232}		232.03717		74y	α 5.32, 5.26; $e^-(\gamma)$ 0.058; (γ)
U^{233}		233.03950	5/2	1.62×10^5y	α 4.82, 4.77 (others); $e^-(\gamma)$ 0.043; (γ)
U^{234}	0.0056	234.0409		2.48×10^5y	α 4.77, 4.72 (others); $e^-(\gamma)$ 0.053; (γ); (SF)
U^{235}	0.7205	235.04393	7/2	7.13×10^8y	α 4.39 (others); γ 0.18, 0.14, 0.10 (others); (SF)
U^{235m}				26m	IT $e^-(\gamma)$ <0.0001
U^{236}		236.04573		2.4×10^7y	α 4.50, 4.45; γ 0.05; (SF)
U^{237}		237.04858		6.75d	β^- 0.25 (others); γ 0.06, 0.21, others 0.027–0.43
U^{238}	99.274	238.0508		4.51×10^9y	α 4.19 (others); (γ 0.045); (SF)
U^{239}		239.0543		23.5m	β^- 1.21; γ 0.074
U^{240}		240.05670		14h	β^- 0.36 (others); γ 0.044
$_{93}$Np^{231}		231.0383		50m	α 6.28
Np^{232}				13m	EC; γ
Np^{233}		233.0406		35m	EC; (α 5.53)
Np^{234}		234.0428		4.4d	EC; γ 0.043–1.61; (β^+)
Np^{235}		235.04407		410d	EC; (α 5.02; γ)
Np^{236}				$>5 \times 10^3$y	
Np^{236m}		236.04662		22h	EC; β^- 0.52; $e^-\gamma$ 0.045
Np^{237}		237.04803	5/2	2.20×10^6y	α 4.78, 4.76 (others); γ 0.087, 0.019, 0.030 (others)
Np^{238}		238.0509	2	2.10d	β^- 1.24, 0.26 (others), 0.044–1.03
Np^{239}		239.05294	5/2	2.35d	β^- 0.33, 0.44 (others); γ 0.013–0.49
Np^{240}		240.0562		7.3m	β^- 2.18, 1.60, 1.30; γ 0.56, 0.04, 0.60 (others)
Np^{240}				63m	β^- 0.89; γ 0.085–1.16
Np^{241}		241.0582		16m	β^- 1.36
Np^{241}				3.4h	
$_{94}$Pu^{232}		232.0411		36m	EC; α 6.58
Pu^{233}		233.0427		20m	EC; (α 6.30)
Pu^{234}		234.0433		9.0h	EC; (α 6.19)
Pu^{235}		235.0453		26m	EC; (α 5.85)
Pu^{236}		236.04607		2.85y	α 5.76, 5.72 (others); γ 0.047 (others); (SF)
Pu^{237}		237.04828		45d	EC; γ 0.033, 0.060 (others); (α)
Pu^{237m}				0.18s	IT $e^-(\gamma)$ 0.145
Pu^{238}		238.0495		86.4y	α 5.49, 5.45 (others); γ 0.044 (others); (SF)
Pu^{239}		239.05216	1/2	2.44×10^4y	α 5.15, 5.13, 5.10 (others); γ 0.053, 0.013 (others); (SF)
Pu^{240}		240.05397		6580y	α 5.16, 5.12 (others); γ 0.045 (others); (SF)
Pu^{241}		241.05671	5/2	13.0y	β^- 0.021; (α; γ)
Pu^{242}		242.0587		3.8×10^5y	α 4.90, 4.86; γ 0.045; (SF)
Pu^{243}		243.0620		5.0h	β^- 0.58, 0.49; γ 0.084, 0.054, 0.012 (others)
Pu^{244}				8×10^7y	α; (SF)
Pu^{245}				10.1h	β^-

Nuclide	Per Cent Abundance	Nuclidic Mass	Spin	Half-life	Decay Modes, Radiations, Energies
Pu^{246}		246.0702		11d	β^- 0.15, 0.33; γ 0.047, 0.027 (others)
$_{95}$Am^{237}		237.0498		1.3h	EC; (α 6.01)
Am^{238}				1.86h	EC; γ 0.98, 0.58, 1.35, 0.37 (others)
Am^{239}		239.0530		12.1h	EC; γ 0.225, 0.275; (α 5.77)
Am^{240}				51h	EC; γ 1.00, 0.90 (others)
Am^{241}		241.05669	5/2	458y	α 5.48, 5.43 (others); γ 0.060 (m), 0.017, 0.013
Am^{242}		242.0595	1	16h	β^- 0.63, 0.67; EC; γ 0.042, 0.045
Am^{242m}				150y	IT e^-(γ) 0.049; (α; γ)
Am^{243}		243.06138	5/2	8.0×10^3y	α 5.27, 5.22 (others); γ 0.075 (others)
Am^{244}		244.0645		10.1h	β^- 0.38; γ 0.043–0.74
Am^{244m}				26m	β^- 1.5; (EC)
Am^{245}		245.06631		2.0h	β^- 0.90; γ 0.036–0.26
Am^{246}		246.0698		25m	β^- 1.31, 1.60 (others); γ 0.035–1.06
$_{96}$Cm^{238}		238.0530		2.5h	EC; α 6.50
Cm^{239}				2.9h	EC; γ 0.19
Cm^{240}		240.05550		27d	α 6.26; (SF)
Cm^{241}		241.0575		35d	EC; γ 0.48 (others); (α 5.95)
Cm^{242}		242.0588	0	162d	α 6.11, 6.07 (others); γ 0.044 (others); (SF)
Cm^{243}		243.06138		32y	α 5.78, 5.74 (others); γ 0.28, 0.23 0.21 (others); (EC)
Cm^{244}		244.06291		18y	α 5.80, 5.76 (others); γ 0.043 (others); (SF)
Cm^{245}		245.06534		9.3×10^3y	α 5.36, 5.45 (others); γ 0.17, 0.13
Cm^{246}		246.0674		5.5×10^3y	α 5.37; (SF)
Cm^{247}				1.6×10^7y	α
Cm^{248}				4.7×10^5y	α 5.05; SF
Cm^{249}		249.0758		64m	β^- 0.86
Cm^{250}				$\sim 2 \times 10^4$y	SF
$_{97}$Bk^{243}		243.0629		4.5h	EC; γ 0.84, 0.96, 0.74 (others); (α)
Bk^{244}				4.4h	EC; γ 0.90, 0.20 (others); (α)
Bk^{245}		245.0662		5.0d	EC; γ 0.25 (others); (α)
Bk^{246}				1.8d	EC; γ 0.82 (others)
Bk^{247}		247.0702		7×10^3y	α 5.51, 5.67 (others); γ 0.084, 0.27
Bk^{248}		248.0730		16h	β^- 0.65; EC
Bk^{249}		249.07484		314d	β^- 0.13; (α; γ)
Bk^{250}		250.0785		3.2h	β^- 0.73, 1.76; γ 0.99 1.03 (others)
$_{98}$Cf^{244}		244.06593		25m	α 7.17
Cf^{245}		245.0679		44m	EC; α 7.11
Cf^{246}		246.0688		36h	α 6.75, 6.71 (others); γ 0.042 (others); (SF)
Cf^{247}				2.5h	EC; γ 0.32, 0.42, 0.46
Cf^{248}		248.07235		350d	α 6.26, 6.22; γ 0.04; (SF)
Cf^{249}		249.07470		360y	α 5.81 (others); γ 0.40, 0.34 (others); (SF)
Cf^{250}		250.0766		13y	α 6.02, 5.98; γ 0.043; (SF)
Cf^{251}				$\sim$800y	α; γ 0.18
Cf^{252}				2.55y	α 6.11, 6.07; α 0.042, 0.10; (SF)
Cf^{253}		253.0850		17d	β^- 0.27
Cf^{254}				60d	SF
$_{99}$Es^{245}				1.2m	α 7.7
Es^{246}				7.3m	α 7.35; EC
Es^{248}				25m	EC; (α 6.87)

Nuclide	Per Cent Abundance	Nuclidic Mass	Spin	Half-life	Decay Modes, Radiations, Energies
Es^{249}		249.0762		2h	EC; (α 6.76)
Es^{250}				8h	EC
Es^{251}		251.0799		1.5d	EC; (α 6.48)
Es^{252}		252.0829		~140d	α 6.64
Es^{253}		253.08469		20d	α 6.63 (others); γ 0.09, 0.38, 0.39, 0.43; (SF)
Es^{254}		254.0881		39h	β^- 0.48, 1.13 (others); γ 0.044, 0.69, 0.65 (others); (EC)
Es^{254}				250d	α 6.42; γ 0.062 (m)
Es^{255}				40d	β^-
Es^{256}				<1h	β^-
$_{100}$Fm^{248}		248.0772		0.6m	α 7.8
Fm^{249}				2.5m	α 7.9
Fm^{250}		250.0795		30m	α 7.43; (EC)
Fm^{251}				7h	EC; (α 6.89)
Fm^{252}		252.0827		23h	α 7.04 (others); (γ)
Fm^{253}				5d	EC; α 6.94
Fm^{254}		254.0870		3.24h	α 7.20, 7.16 (others); γ 0.041 (others); (SF)
Fm^{255}				19.9h	α 7.03 (others); (γ; SF)
Fm^{256}				3.1h	SF
Fm$^{257?}$				11d	SF
$_{101}$Md^{255}		255.0906		~30m	EC; α 7.34
Md^{256}				~1.5h	EC
$_{102}$^{254}				3s	α 8.3; SF
$_{102}$^{256}				8s	α
$_{103}$Lw^{257}				8s	α 8.6

Name Index

Page numbers in *italics* refer to the chapter bibliographies.

Subject Index

Absolute counting, *see* Disintegration rate
Absorption edges for X rays, 116
Accelerating tubes, 357–358
Accelerators, as neutron sources, 379–381
  as X-ray sources, 377–379
  linear, 359–362
  measurement of beam current in, 397–400
  measurement of beam energies in, 395–397
  targeting in, 391–395
Actinide series, 217–218
Actinium, discovery of, 4
Actinium series, 8, 11
Actinometers, 122
Activation analysis, 208–209
Active deposit, 12
Age of earth, 500–501
Ages of meteorites, 499–501, 507–508
Ages of minerals, from K^{40}-Ar^{40} content, 499
  from Pb^{206}-Pb^{207} content, 498
  from ratio of radiogenic lead to Pb^{204}, 500
  from Rb^{87}-Sr^{87} content, 500
  from thorium-lead content, 498
  from uranium-helium content, 497–498
  from uranium-lead content, 498
Alpha counting, absolute, 417–418
Alpha decay, 49–50, 222–233
  Coulomb barrier penetration in, 225–226
  decay constant for, 223–228
  hindered, 228–232
  spectra in, 222, 228–230
  theory of, 223–228
Alpha-decay energies, systematics of, 232–233
Alpha particle, binding energy of, 28
Alpha-particle energies, determination of, 94–98, 136, 140
Alpha particles, back-scattering of, 418
  counting of, 136, 140, 417–418
  long-range, 222–223
  magnetic deflection of, 4
  ranges of, 3, 94–98
  Rutherford scattering of, 18–20

Alpha particles, straggling of, 102
  tracks of, in cloud chamber, 150
Alternating-gradient focusing, 374
Alternating-gradient synchrotron, 374–376
  beam intensity of, 375
  multiple traversals in, 375
  targeting in, 375
Amplitude, of scattered rays, 332
Angular correlations, 261–263
  chemical effects on, 466–469
  in gamma decay, 261–263, 466–469
Angular momentum, effect on level density, 347
  effect on nuclear reactions, 348
  in stripping reactions, 350–351
Anion exchange, 406–408
Annihilation of positrons, *see* Positrons
Anticoincidence shielding, for background reduction, 148, 503
Antineutrinos, 252–253, *see* Neutrinos
Antiprotons, from synchrotron, 373
Artificial elements, 215–216
Artificial radioactivity, discovery of, 13, 14
Astatine, 215
  in natural radioactive families, 9–11
Atomic number, 20
Atomic structure, models of, 18–20
Atomic weight, semiempirical formula for, 43
Atomic weight scale, 26
Atomic weights, from mass spectroscopy, 26–27
  of nuclides, 535–566
Auger effect, 52
  chemical consequences of, 214
Average life, 69
Average value, 166
  estimate of, 167
  of disintegration rate, 173
  of observed counting rate, 181
  precision of estimate, 168

573